9th International Symposium on Steel Bridges 2018

IOP Conference Series: Materials Science and Engineering
Volume 419

Prague, Czech Republic
10 – 11 September 2018

ISBN: 978-1-5108-7198-4
ISSN: 1757-8981

Printed from e-media with permission by:

Curran Associates, Inc.
57 Morehouse Lane
Red Hook, NY 12571

CURRAN ASSOCIATES INC.
proceedings
.com

Some format issues inherent in the e-media version may also appear in this print version.

Printed by Curran Associates, Inc. (2018)

For permission requests, please contact the Institute of Physics
at the address below.

Institute of Physics
Dirac House, Temple Back
Bristol BS1 6BE UK

Phone: 44 1 17 929 7481
Fax: 44 1 17 920 0979

techtracking@iop.org

Additional copies of this publication are available from:

Curran Associates, Inc.
57 Morehouse Lane
Red Hook, NY 12571 USA
Phone: 845-758-0400
Fax: 845-758-2633
Email: curran@proceedings.com
Web: www.proceedings.com

Table of contents

Volume 419

9th International Symposium on Steel Bridges
10–11 September 2018, Prague, Czech Republic

Accepted papers received: 20 August 2018
Published online: 20 September 2018

Preface

Papers

Bridge maintanance

Highway bridges

Historical bridges

Pedestrian bridges

Railway bridges

Refurbishment

Research and inovations

Research and materials

Structural connections

IOP Conf. Series: Materials Science and Engineering **419** (2018) 011001 doi:10.1088/1757-899X/419/1/011001

Foreword

We all realize that bridges are symbols of art and science, architecture and but mainly engineering skills. They symbolize links between people, communities, and nations. Bridge building is a high-ranked profession.

Although the competition from other building materials has become stronger, steel bridges offer a lot of advantages. Steel offers efficient solutions in terms of initial costs, sustainability and life cycle costs. This is shown by the success of composite bridges in particular for short and medium spans. The coming high-strength steels bring reduction of self-weight and decrease of construction depth. Steel structures fulfil the highest degree of aesthetic quality. And steel is the futuristic material to continue connecting our lives under growing human population challenges.

Steel bridges are innovative. In order to exchange the newest trends in steel bridge construction, international experts have come together for 9th International Symposium on Steel Bridges in Prague 2018 organised by Czech Constructional Steelwork Association (CAOK) in coordination with European Convention for Constructional Steelwork (ECCS). It will promote the exchange knowledge, findings, innovations, and experiences in design, fabrication and erection.

I would like thank to contributors to this publication of 9th International Symposium on Steel Bridges in Prague 2018 providing us with the latest deep knowledge on steel bridges. I am sure that the symposium contributes to the technological development of steel bridges. I wish us great lectures full of ideas and visions and a lot of fruitful discussions.

František Wald,
Chairman of Scientific Committee

IOP Conf. Series: Materials Science and Engineering **419** (2018) 011002 doi:10.1088/1757-899X/419/1/011002

Peer review statement

All papers published in this volume of *IOP Conference Series: Materials Science and Engineering* have been peer reviewed through processes administered by the proceedings Editors. Reviews were conducted by expert referees to the professional and scientific standards expected of a proceedings journal published by IOP Publishing.

Problems and innovative solutions considering "open type" bridge deck structures in old steel riveted railway bridges

Martina Nikolova [1], Lazar Georgiev [2]

[1] PhD student Eng., Eng. PhD Student in Roads and Transport Structural Facilities Department, FTE, UACEG, Hr. Smirnenski blvd 1, Sofia 1046, Republic of Bulgaria, e-mail: nikolova23@abv.bg;
[2] Assoc. Prof. D-r Eng., Roads and Transport Structural Facilities Department, FTE, University of Architecture Civil Engineering and Geodesy, Hristo Smirnenski Blvd, Sofia 1046, Republic of Bulgaria, ldg_77@abv.bg, ; lazar_fte@uacg.bg;

E-mail: ldg_77@abv.bg

Abstract There are 331 steel bridges with total length 11201,70m in exploitation in the railway infrastructure of the Republic of Bulgaria. Significant part of those bridges is with "open type" deck including longitudinal and cross beams. The stress state of a standard bridge deck of an old steel riveted truss bridge along Sofia - Varna railway line at km 17+536 (representing wide class of bridges in exploitation) is analysed in this paper in accordance with the prescriptions of the Eurocode design standard. Results, obtained from different FEM (with Frame and Shell elements) indicate that the load bearing capacity of longitudinal and cross beams is significantly insufficient, which is due to increased loads and some specificity of the construction details. On the other hand there is an overage of load bearing capacity for the elements of the main truss and the effect of an innovative concept for replacement of the existing "open type" deck with new reinforced concrete plate is investigated and analyzed.

1. Introduction.

There are 331 steel bridges with total length 11201,70m in exploitation in the railway infrastructure of the Republic of Bulgaria. Significant part of those bridges are riveted with "open type" deck. Old steel riveted bridges were designed in accordance with the Admissible Stresses Method which is much conservative in comparison with the Limit States Method in the present design standards. Thereby there is a reserve in the load bearing capacity of the elements which can compensate the increased live loads [3], [4]. Series of investigations shows that the remaining fatigue life of the main girders of part of these bridges can be prolonged by adequate rehabilitation and strengthening which leads to the possibility of rational time distribution of the restricted funds for maintenance by the responsible institutions – NRIC (National Railway Infrastructure Company) [1], [2], [4]. At the same time because of the details specificity and the increased values of the horizontal live loads serious problems with the load bearing capacity of the elements of the "open type" decks of the old riveted railway bridges are registered [3]. .

At the Figure 1 are presented cross sections of a typical old riveted truss railway bridge. The bracing system is connected only with the 2L profiles, forming the upper chord of the longitudinal beam, the out of plain inertia moments are very low. This leads to high stresses and insufficient load bearing capacity for the noising force effects. As it is described in [3] in spite of the conservative design standards and material reserves a serious deficiency of load bearing capacity of the deck members, dangerous for the exploitation of the bridge, is registered.

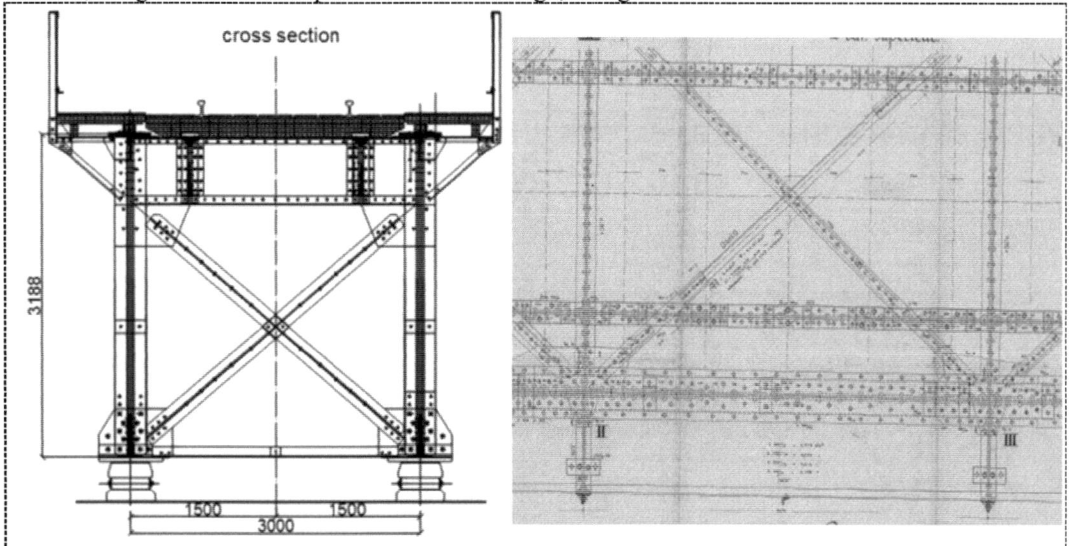

Figure 1. Bridge deck cross section (left) and part of deck plan (right – original drawing) of an old steel riveted bridge with truss main girder along Sofia-Varna railway line at km 17+536 and 31m span length.

2. Numerical example.

In order to analyze the quantitative aspect of the problems with "open type" bridge decks described in point 1 a two different FEM of the old riveted bridge from Figure 1 in SAP2000 - M1 and M2 are developed – Figure 2. The elements of the main truss girder in both variants are modeled as beam (Frame) finite elements with rigid connection in joints. In model M1 longitudinal girders are modeled as beam (Frame) finite elements (FE), simply supported in vertical direction on the cross girders because the structural connection is realized only between webs. In model M2 the elements of the bridge deck are modeled as Shell FE considering the correct geometry and connection between webs which leads to semi-rigid connection between longitudinal girders and cross girders. In model M2 the connection between the cross girders and the verticals of the main riveted truss considers the real zone of connection by adjusting the mesh of finite elements of the Shell web of cross beams and Frame elements of the vertical of main truss girder – Figure 3.

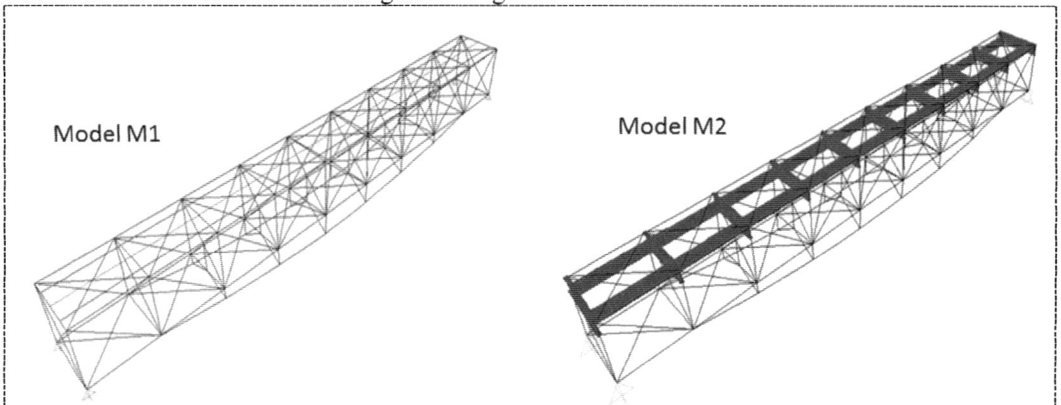

Figure 2. FEM models - M1(Frame) and M2(Shell).

IOP Conf. Series: Materials Science and Engineering **419** (2018) 012001 doi:10.1088/1757-899X/419/1/012001

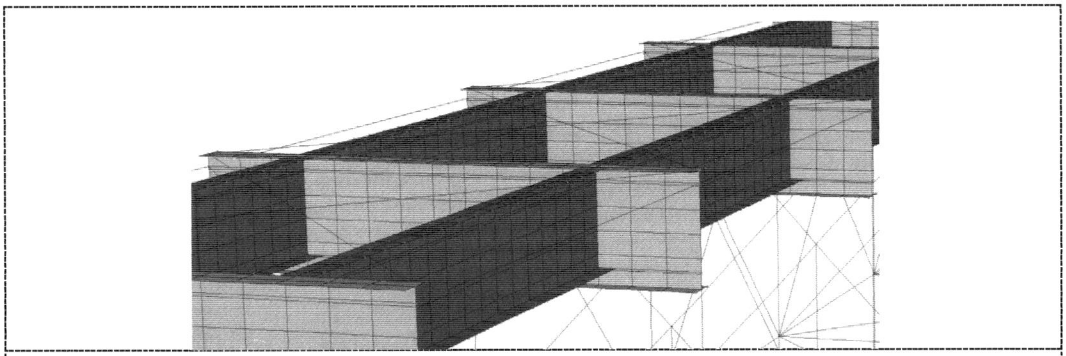

Figure 3. Connection between cross beams and the vertical of the main truss in the model M2.

Most unfavorable load combinations of loads and effects in accordance with [5] and reduced in accordance with [6] for bridges in exploitation, used in Germany – self weight G_k of the steel bridge superstructure; railway track, sidewalks and parapets; classified vertical (LM71 and SW/2, in accordance with BDS EN 1991-2) and horizontal (noising force L, braking inertia force Q_{lbk} and acceleration inertia force Q_{lak}) live loads. The characteristic maximum values of edge stresses in midsection of longitudinal girder and cross girders obtained from the analyzed loads and effects in both models M1 and M2 are presented on Figure 4 and Figure 5. It must be mentioned that higher values for edge stresses in longitudinal girder are obtained from noising force. It is expected because the bracing system for the noising force is connected only with the upper 2L profiles, forming the upper chord of the longitudinal girder and considering the negligible torsion stiffness the noising force is transferred by the out of plane bending of those 2L profiles.

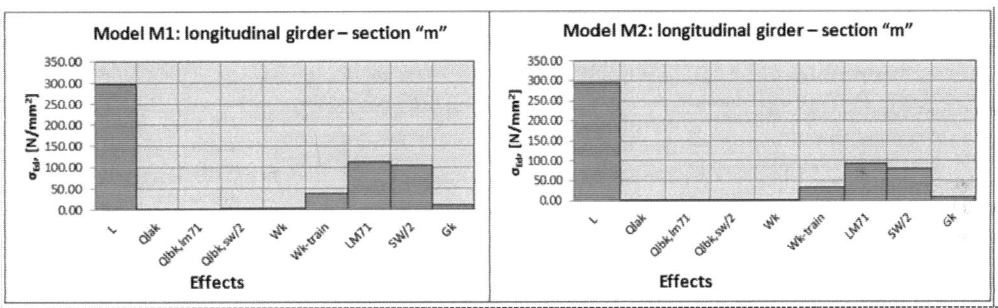

Figure 4. Stresses from characteristic values of the analysed loads and effects for longitudinal girder from models M1 and M2.

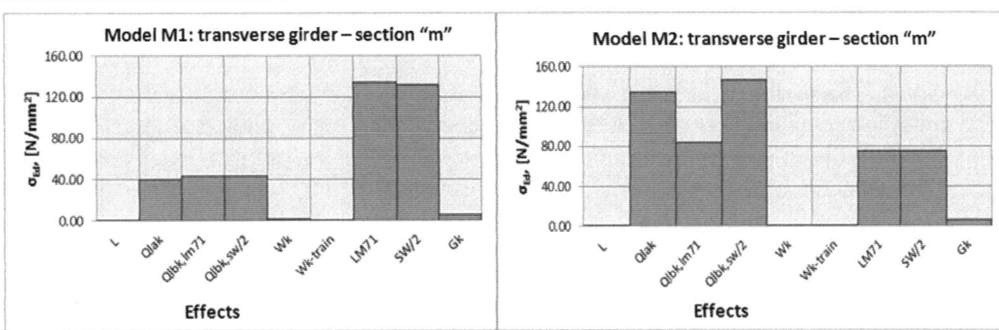

Figure 5. Normal stresses from characteristic values of the considered loads and effects presented for cross beams for model M1 and M2

The percent difference between the characteristic normal edge stresses in longitudinal and crossbeam are presented obtained from the considered loads for model M1 (Frame) and model M2 (Shell) are presented in Table 1. The differences for noising force are negligible, but from the classified vertical load LM71 are significant. The reason probably is that the connection between crossbeam and the vertical of the main truss in model M1 is realized in a particular node and in model M2 is realized in a zone with height equal of the height of the crossbeam – Figure 3. In the longitudinal beam as it is expected for the midsection m the normal stresses are higher in comparison with these obtained from M2, because the elastic stiffness of the connection between the webs of longitudinal beam and crossbeam is not considered in model M1. The normal stresses in the support section of the longitudinal beam are caused by the interaction with the main truss girder. It is seen from Table 1 that because of the specific connection detail between the upper bracing system and the longitudinal beam the normal stresses caused by noising force are much higher than those obtained from the classified vertical loads UIC71 as it was mentioned above.

Table 1. Maximal edge normal stresses for various effects

LONGITUDINAL BEAM				TRANSVERSE BEAM			
[N/mm^2]	FRAME	SHELL	Δ [%]	[N/mm^2]	FRAME	SHELL	Δ [%]
$\sigma_{m,LM71}$	113	93	22.01	$\sigma_{m,LM71}$	135	75	79
$\sigma_{s,LM71}$	50.60	51.51	1.80	$\sigma_{m,Qlbk,sw/2}$	44.01	146.85	234
$\sigma_{m,L}$	296.85	297.00	0.05				

In Table 2 are presented the most unfavorable values of the utilization factor $\eta_1 = \sigma_{Ed} / f_{yd}$ (σ_{Ed} - design value of the edge normal stress from the most unfavorable load combination, calculated in models M1 and M2, f_{yd} - design value of the yielding stress) for the edges of the cross sections of the longitudinal beam midsection (m) and support (s) and cross beam midsection (m) in accordance with loads defined in [5] and the decreased loads in [6] (by means of decreasing γ_Q and α).

Table 2. Utilisation factor η_1 for members of the bridge deck

	Model M1: FRAME - η_1		Model M2: SHELL - η_1	
	γ=1.45;α=1.21	γ=1.3;α=1.00	γ=1.45;α=1.21	γ=1.3;α=1.00
LONGITUDINAL BEAM - m	4.03	3.38	3.73	3.12
LONGITUDINAL BEAM - s	1.67	1.40	2.84	2.38
TRANSVERSE BEAM - m	2.48	2.06	2.37	1.97

It is seen from Table 2 that in accordance with [5] the load bearing capacity of the longitudinal beam is ~3,0÷4,0 times lower than the standardized design minimum and for the cross beam it is ~2,0÷2,5 times lower. In the case of rehabilitation and strengthening of old riveted railway bridges in relatively good condition of the main girder and considering such a significant load bearing capacity deficiency for elements of the deck solutions for replacement of the "open type" old deck must be analyzed.

3. Variant for replacement of the old steel deck of "open type" with reinforced concrete, connected for composite action with elements of the bridge.

The developed detailed design [3] shows that significant part of the members of the old steel riveted bridge described in point 2 has sufficient load bearing capacity in respect to defined in [5] loads and effects even in the case of the replacement of the existing steel riveted deck with reinforced concrete plate (with thickness 200÷230mm and concrete C45/55), connected for a composite action with the upper chord of the old riveted truss bridge – Figure 6. It must be mentioned that the increasing of the self-weight is significant. The self-weight of the old bridge in present condition is 24,24kN/m (including elements of the structure and railway track with timber sleepers) and after replacement of the deck with reinforce concrete plate and the self-weight becomes 64,4 kN/m. If UHPFRC (Ultra High Performance Fiber Reinforced Concrete) is used for the new deck plate the unfavorable effect of self-weight increasing can be reduced [3]. Variant for 150mm thick prestressed in transverse direction UHPFRC (with characteristic compression strength $150N/mm^2$ and modulus of elasticity $50000N/mm^2$ [7]) deck plate is developed. The weight of this 150mm UHPFRC deck plate is about 50% lower than a 200-230mm thick RC plate from C45/55 (29,7kN/m for the standard RC and 19,7kN/m for the UHPFRC plate), which is favorable in respect to the elements of the main truss girder, bearings and the substructure of the bridge. The self-weight of the bridge structure after replacement the deck with 200-230mm thick RC - C45/55 is 64,4kN/m and for transversely prestressed UHPFRC plate is 54,3kN/m (about 18% lower self-weight of the superstructure).

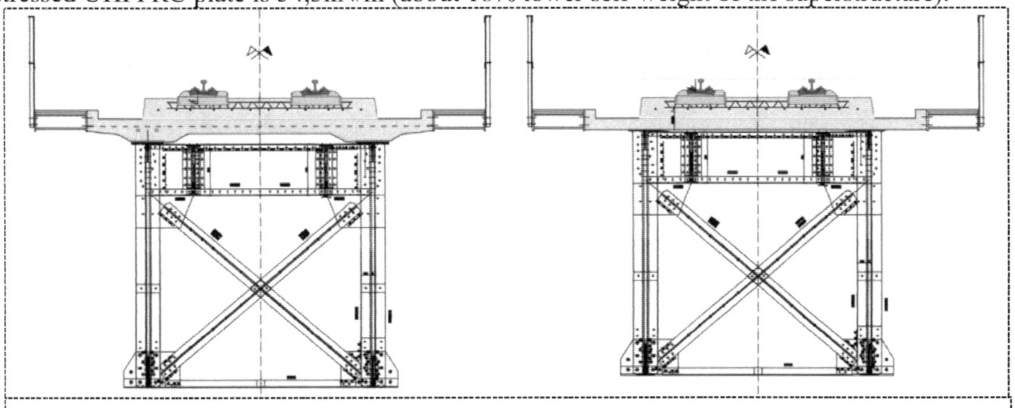

Figure 6. Cross section of the bridge in variant with reinforced concrete plate C45/55(left) and prestressed in transverse direction plate with 150mm height of UHPFRC with fck=150N/mm2 [7] (right) connected for composite action with the upper chord of the truss.

4. Conclusion

In series of cases the old steel railway bridges can be rehabilitated to correspond the present exploitation and design standards and this is a solution more rational than replacement. In this case the mentioned before serious problems with the deck structure must be solved.

The simplified static models, used for old riveted bridge analysis, are in most cases with sufficient accuracy for the main truss girder elements can lead to significant inaccuracy for the elements of the deck – longitudinal beams and crossbeams. The bridge deck of "open type" of an old riveted truss bridge along the railway line Sofia – Varna at km 17+536 is analyzed. The connection between the riveted members of the deck is complex and cannot be correctly modeled in 3D FEM with beam (frame) elements (in both variants – fixed and free connection) – M1. For comparison a 3D FEM in which the members of the deck are modeled with Shell elements and members of the main truss are modeled with Frame elements is developed – M2. Serious differences between the results for models M1 and M2 is registered for the elements of the deck. The reason is that in the model with Shell elements M2 the connection between cross beam and vertical of the main truss and the connection of longitudinal beam and cross beam are more correctly approximated in comparison with model M1. In spite of this both models shows that the load bearing capacity of the old riveted deck elements is insufficient in accordance with the present European design standard [5] and even in comparison with the reduced demands of [6], used for assessment of old bridges in Germany.

In series of cases (represented by the analyzed in this paper old riveted railway bridge) considering the specificity of the details and the increased values of the defined in Eurocodes horizontal longitudinal and transverse live loads a serious deficiency of the load bearing capacity of the deck elements is evaluated (in the particular case 3,0÷4,0 times for longitudinal beams and 2,0÷2,5 times for cross beams). In spite of the conservativeness of the defined in Eurocodes horizontal live loads in the relatively rare case of heavy trains passing with high speed local deck destruction or plastic deformations can occur with all related unfavorable consequences.

In [3] and in the present paper the variant of an old riveted deck of "open type" replacement with reinforced concrete plate connected for composite action with elements of the old bridge is analyzed. In spite of the serious self weight increase concerning the composite action the load bearing capacity of the main girder can be satisfied with relatively low strengthening needs. Using UHPFRC (with characteristics in accordance with [7]) for the new transversely prestressed plate the increase in the self weight is 18% lower. This solution is optimal with respect to the bearings and the substructure.

The problem with the load bearing capacity of riveted "open type" bridge decks is serious and must be analyzed for wider range of old riveted bridges in order to minimize the risk of potential failures by means of in time adequate rehabilitation and strengthening.

References

[1] Karsten Geissler, "Assessment of Old Steel Bridges, Germany", Structural Engineering International 4/2002
[2] Dulevski E., Ivanov S., Georgiev L., "Possibilities for reconstruction of the old bridges in accordance with present exploitation conditions", Railway Transport, 5/2004 in Bg
[3] Georgiev L. and others, "Investigation of the behavior of steel bridges with "open type" deck in the case of replacement the deck with reinforced concrete plate connected for composite action with members of the old structure, Scientific Report, Contract БН-175/2015, CNIP in UACEG, 2015, in Bg
[4] Němec,I., Kolař,Vl., Ševčik, I., Vlk, Z., Blauwendraat, J., Buček, J., Teplý, B., Novák,D., Štembera, V., Finite Element Analysis of Structures, Shaker Verlag, 2009
[5] БДС EN 1991-2. Eurocode 1: Actions on structures - Part 2: Traffic loads on bridges.
[6] Richtlinie 805 -Tragsicherheit bestehender Brückenbauwerke. Bekanntgabe 3 zur Richtlinie 805, gültig ab 01.09.2002, DB Netz AG, 2002
[7] "Ultra High Performance Reinforced Concrete. Recommendations", AFGC/SETRA http://www.ductal.com, Revised Edition 2013.

IOP Conf. Series: Materials Science and Engineering **419** (2018) 012002 doi:10.1088/1757-899X/419/1/012002

Reinforced structures type "Filler Beam" – advantages in reconstruction of bridge superstructures.

Alexander Jiponov [1], Lazar Georgiev[2]

[1] Alexander Konstantinov Jiponov, Assistant Professor and PhD Student in "Roads and Transport Structural Facilities Department", FTE, UACEG, 1 Hristo Smirnenski Blvd, Sofia 1046, Republic of Bulgaria

[2] Lazar Dimitrov Georgiev, D-r Eng., Associated Professor in "Roads and Transport Structural Facilities Department", FTE, UACEG, 1 Hristo Smirnenski Blvd, Sofia 1046, Republic of Bulgaria

E-mail: ajiponov@abv.bg; ldg_77@abv.bg

Abstract. Reinforced concrete superstructure type "filler beam" would be appropriate in future reconstruction of such structures with span length of 5 - 30m. They can be applied in road and railway bridges as their advantages are significant. They have a small construction height, simplified installation, steel elements are simplified and produced in factory conditions, less formwork, as well as easy maintenance in operation. This type of structure would be appropriate in urban environment as besides the above – mentioned advantages, it also has low noise emission during the passage of vehicles.

IOP Conf. Series: Materials Science and Engineering **419** (2018) 012002 doi:10.1088/1757-899X/419/1/012002

1. INTRODUCTION.

The main types of bridge superstructures used in our country are made from reinforced concrete (monolithic or prefabricated) with normal or prestressed reinforcement, from steel or from combined steel and reinforced concrete sections.
For a number of reasons, the most widespread system in Bulgaria is the one of prestressed GT beams for spans of 20, 40 and even 60m (Viaduct Bebresh on Hemus motorway). During the operation of bridges of this type due to the accumulation of several unfavorable factors (Some omissions or "optimizations" in the design, insufficient quality of the execution, poor maintenance) after only about 30 years of operation (according to BDS EN 1990 the regulated service life for bridges is 100 years) some serious damage is registered. In some cases, even partial or total replacement of the top structure is necessary (Viaduct Eleshnitza on Hemus motorway, bridge on km 67 on Hemus motorway).
On the secondary roads can be located also a significant number of operational bridges with shorter spans, up to 20m, made from ordinary concrete. Poor execution, combined with poor concrete quality and inadequate maintenance are the main reasons why a significant part of these bridges in Bulgaria are in unsatisfactory technical condition. In a number of cases, partial or complete replacement of the superstructures is necessary. In this case, the question of choosing a feasible new superstructure is important. The basic requirements in these cases are low cost, high reliability, fast execution, minimizing the negative effects on traffic in the area of the facility (especially in urban conditions).
For spans up to 20-25m (for road bridges), the standard monolithic reinforced concrete construction (with or without prestressing) has the lowest cost in terms of input material, but is a serious step back from the other listed criteria of innovative (for the practice in our country) solutions, such as the filler beam system (rolled steel 2T profiles, imbedded in the reinforced concrete).
In the case of a monolithic reinforced concrete solution, it is necessary to have a densely placed scaffolding to maintain the formwork and this adversely affects the traffic under the facility, which is extremely undesirable especially in urban conditions – Figure 1.

Figure 1. Densely placed scaffolding and limiting the traffic during the execution of a bridge of prestressed concrete at crossroad with heavy traffic in Sofia – 2014.

In the area of 20-22m span lengths, the combined steel-reinforced concrete superstructure is usually more expensive than monolithic reinforced concrete (by input material), but in European and world practice such a solution is often preferred for its speed of execution, lack of necessity of densely placed scaffolding (depending on the solution, the temporary supports may be completely absent), minimizing the formwork and supporting it on the self-supporting steel structure. With the filler beam system, a minimum design height can be achieved by providing the same bending stiffness as compared to other bridge superstructure designs. To speed up the execution time, it is possible parts of the superstructure to be pre-prepared and assemble them in situ – Figure 2. With the bridge static system being a continuous beam, the system with imbedded steel profiles can be successfully implemented for railway bridges with span lengths up to 30-35m and road bridges with span lengths up to 40-50m. The implementation of this system is possible also for road bridges in their transversal direction. Superstructures with imbedded asymmetrical steel beams are also possible. The limited application of the above-mentioned efficient construction solutions for bridge superstructures in Bulgaria is mainly due to the lack of experience of a significant number of the designers and builders in the field of combined steel-reinforced concrete bridges. For this reason, it is not uncommon for the old well-known reinforced concrete solutions to be applied outside their area of technical and economic efficiency.

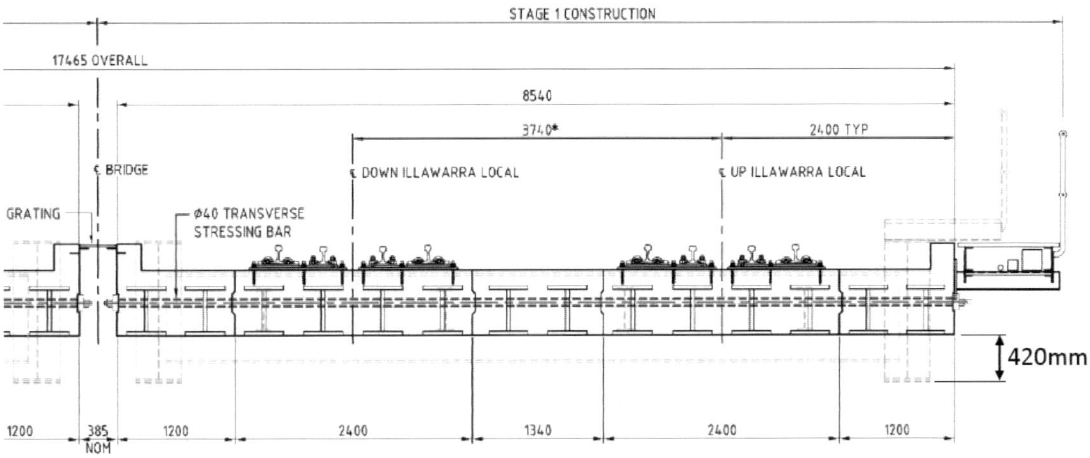

Figure 2. Substitution of old riveted railway bridge with 13,1m span (dashed) with prestressed in transversal direction superstructure of imbedded steel profiles (filler beam)

Figure 3. Railway bridge superstructure from imbedded steel profiles in transversal direction

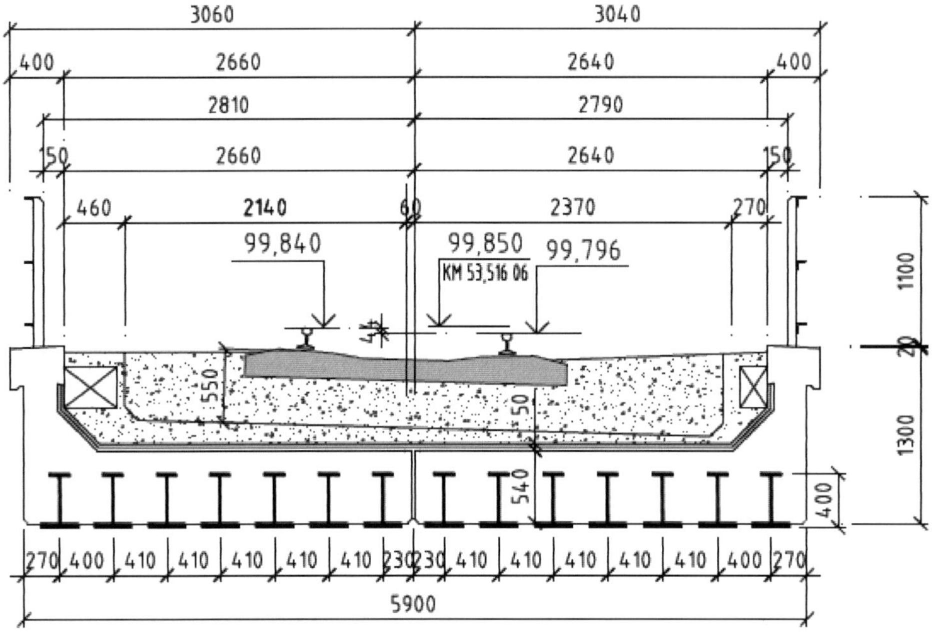

Figure 4. Railway bridge with 10m span from imbedded asymmetrical steel beams

2. OPTIONS FOR APPLICATION OF BRIDGE SUPERSTRUCTURES WITH IMBEDDED STEEL PROFILES IN BULGARIA

2.1. Proposal for repair of facilities, part of a rehabilitation project on Route III - 811 "II - Bogyovtsi - Slivnitsa - Rakita - Breznik - Pali Lula" with a total length of 42,520 km.

a) Bridge at km. 12 + 839

The superstructure of the facility has a total length of 7.00 meters and a span of 6.50 meters. The transversal gauge of 6,00 m includes a roadway with a width of 4,60 m and two pavements of 0,70 m each. The superstructure consists of 4 monolithic beams, combined with a slab, and in the middle of the span there is a cross beam. The structure is in a dangerous condition and in our opinion, the best solution is to be completely replaced with a new superstructure with imbedded profiles. The advantages are that the construction height is less, which would ease the hydraulic regime in the area of the facility. The difference is about 20 centimetres, which would allow the construction of a normal reinforced concrete bearing coffin as stated in the proposal.

Noticed disadvantages:

i) lack of concrete cover for the beams;
ii) extremely poor quality of the concrete even when it was poured;
iii) poor connection between the bridge and the embankments at both ends of the facility;
iv) very poor state of safety fences;
v) railings are missing.

Rehabilitation / reconstruction measures:

i) dismantling the existing structure;
ii) implementation of new reinforced concrete bearing coffin;
iii) implementation of facing for the invert and protecting the area under the bridge from undermining;
iv) installation of the new construction according to the chosen option;
v) making transition areas of gabion mattresses;
vi) shaping of longitudinal and transverse joints in the new road surface.

Figure 5. Existing situation

Figure 6. Existing situation

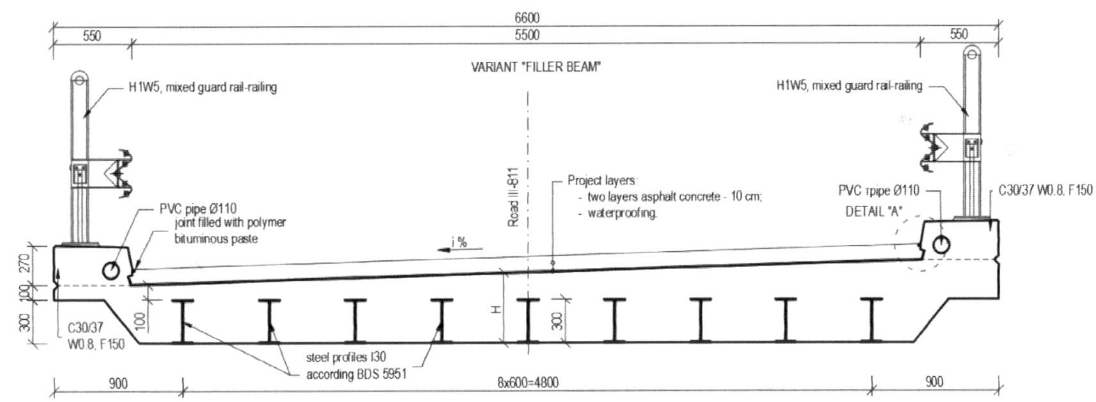

Figure 7. Proposal for a new superstructure.

b) Bridge at km. 19 + 760

The facility has a total length of 7.40 m and a span of 6.00 meters. The transversal gauge of 5.40 m includes a 4.50 m roadway and two pavements of 0.45 m each. The superstructure consists of 4 monolithic beams, combined with a slab, and in the middle of the span there is a cross beam. The structure is in a dangerous condition and in our opinion, the best solution is to be completely replaced with a new superstructure with imbedded profiles. The construction height is smaller, which would ease the hydraulic regime in the area of the facility. The difference is about 30 centimetres, which would allow the construction of a normal reinforced concrete bearing coffin as stated in the proposal.

Noticed disadvantages:

i) lack of concrete cover for the beams;
ii) extremely poor quality of the concrete even when it was poured;
iii) poor connection between the bridge and the embankments at both ends of the facility;
iv) very poor state of safety fences;

Rehabilitation / reconstruction measures:

i) dismantling the existing structure;
ii) implementation of new reinforced concrete bearing coffin;
iii) strengthening / widening of the existing abutments, incl. invert facing;
iv) installation of the new construction according to the chosen option;
v) making transition areas of gabion mattresses;
vi) shaping of longitudinal and transverse joints in the new road surface.

Figure 8. Existing situation

Figure 9. Existing situation

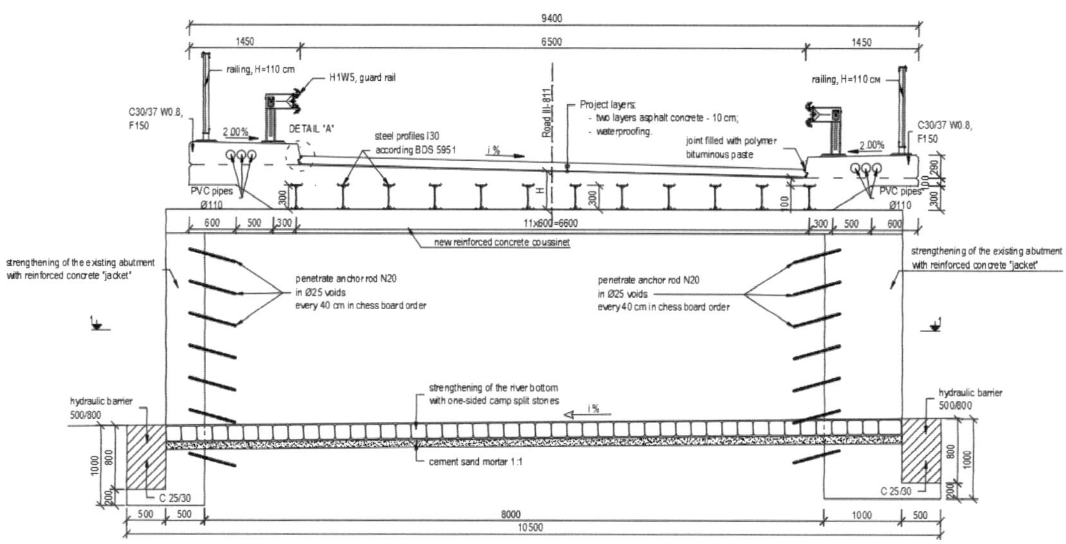

Figure 10. Proposal for a new superstructure

2.2. Proposal for replacing the existing superstructure in the conditions of a single-track railway in operation from the line Plovdiv - Burgas (near Chirpan).

a) **Bridge at km 67+363** – In this case, the old superstructure is without ballast bed. The main beams are solid steel beams in good condition and their incorporation in the new superstructure as "filler beams" is appropriate and reduces the cost of reconstruction. The beams are cut into two halves, which leads to reduction of the overall height and allows the implementation of gravel bed.

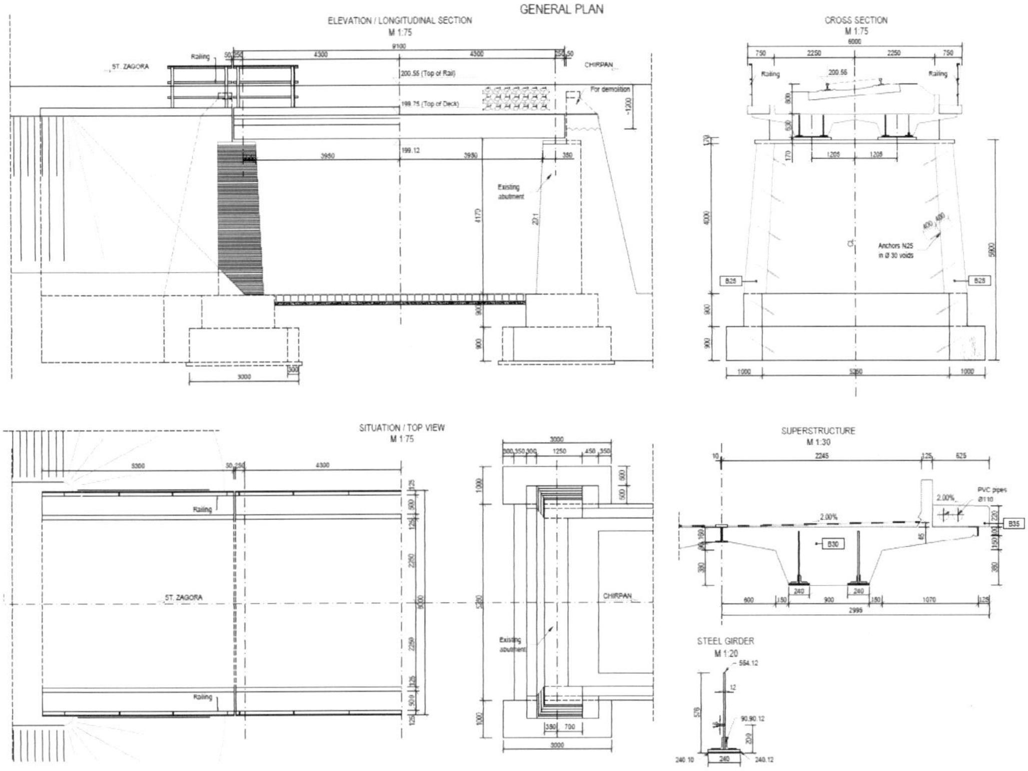

Figure 11. General plan

Technology of embodiment

Figure 12. Temporary bridging and dismantled construction

Figure 13. Cutting the existing structure and embedding it into a new one filled beams

Figure 14. Beams and "soft" reinforcement of semi - structure

Figure 15. Mounting of the second semi – structure

3. CONCLUSION.

The replacement of old bridge superstructures and implementation of new ones with the filler beams system has a certain number of advantages, compared with the old well-known reinforced concrete solutions for bridges with short and middle-length spans.

3.1. For superstructures with filler beams the needed bending stiffness is achievable at minimal construction height, which is a serious advantage at tight conditions. Their possible implementation is for spans up to 30-35m for road bridges and spans up to 40-50m for railway bridges.

3.2. The steel profiles are self-supporting and can be used for supporting the needed formwork. With this, besides the serious formwork reduction and the needed work for it, a significant reduction or even elimination of the necessary scaffolding can be achieved (for monolithic reinforced concrete superstructure a densely placed scaffolding is needed for supporting the formwork and taking the weight of the soft concrete). For replacement of superstructures or construction of new bridges in urban areas, the most significant factors are the time needed for implementation and minimizing or complete elimination of the temporary scaffolding, which prevents the normal traffic conditions under the facility.

3.3. An extra construction time reduction is possible, if parts of the filler beams superstructure are manufactured in a factory.

3.4. An advantage of the filler beam implemented bridges is the relatively low maintenance costs as well as lower sensitivity from hits on the superstructure from vehicles passing under it.

3.5. The application of the filler beams system for replacing an existing superstructure of bridges with short or middle-length spans has a certain number of advantages over the old well-known solutions in our country. The limited application of the above-mentioned efficient construction solutions for bridge superstructures is mainly due to the lack of experience of a significant number of the designers, builders and institutions in the field of combined steel-reinforced concrete bridges. In the future, this problem should be overwhelmed by making the necessary efforts from the aforementioned groups, in interest of the possibilities for implementation of rational and effective solutions for bridge superstructures, which solutions should correspond to the social needs and resources.

IOP Conf. Series: Materials Science and Engineering **419** (2018) 012002 doi:10.1088/1757-899X/419/1/012002

Literature

[1] Oliver Hechler, Andreas Girkes, "Efficient reconstruction of bridges with small and medium spans", http://sections.arcelormittal.com/uploads/tx_abdownloads/files

[2] Ken Maxwell, "Upgrading Of Railway Bridge At Wolli Creek, Sydney", 7th Australian Small Bridges Conference 2015

[3] Vincent Kvočák, Viktória Kožlejová, Katerina Chupayeva, "State of the Art in the Utilization of Deck Bridges with Encased Filler-Beams in the Standard Construction Practice", SSP - JOURNAL OF CIVIL ENGINEERING Vol. 8, Issue 1, 2013.

[4] M. Sc. C. Eng. Konstantin Jiponov, Assoc. Prof. Eng. Mario Galabov, "Reconstruction of steel railway bridges in the conditions of exploitation"

Failure analysis and repair assessment of a steel box girder bridge

Maciej Kulpa[1] and Tomasz Siwowski[1]

[1] Rzeszow University of Technology, Poznanska 2, Rzeszow 35-959, Poland

E-mail: kulpa@prz.edu.pl

Abstract. The paper focuses on the failure of a motorway steel box girder bridge. After almost entire bridge had been completed (excluding deck equipment), there was a few month break in its construction. During this break, in two support sections of the steel box girder, an unusual failure occurred comprising the significant deformation (buckling) and local cracking of the steel webs. In the paper, the analysis of failure causes and the assessment of how these failures impacted the stress distribution in box girders have been presented. After construction of the bridge had been resumed, the failures were repaired. The steel superstructure after repair was also subjected to detailed analysis regarding change of stress distribution in repaired girder elements. All analysis were carried out using advanced numerical FEM models, allowing to consider sequences of particular loads (including those not typical for bridges) and repair works. Conclusions drawn from these analysis concern both the proper shaping of the steel bridge structure as well as the methodology of assessing impact of this type of failure on stress distribution in the structure and effectiveness of the applied repair method.

1. Introduction

Studying past failures can be useful in mitigating the incidence and potential of future adverse events [1]. The causes of failure in bridges have been often divided into seven categories [2]. The first three are limited knowledge, design errors and other aspects of human error. The remaining four are natural hazards, accidents, overloading and deterioration. Design errors include cases where there was evidence of incorrect design assumptions, wrong estimation of loads, mistakes in calculations etc. Limited knowledge captures the cases where there was insufficient understanding of a failure mode at the relevant time, such as aerodynamic instability or a structural/material problem such as brittle fracture, fatigue or buckling. The remaining aspects of human errors are distinguished from design errors and cover those failures caused by negligence, ignorance, as well as poor workmanship, wrong assembly sequence, etc. Natural hazards encompass cases where failures have taken place due to extreme loading such as flooding, storms or very high winds. Accidents pertain to vehicle impacts, fire and explosions. Corrosion is the principal cause of deterioration in the case of steel bridge failures leading to collapse. However, in many cases, failure is induced by a combination of causes. In these cases, an attempt has been made to identify as far as possible the primary cause, and to classify the case accordingly. It has been also the case presented in the paper.

The paper focuses on the failure of a motorway steel box girder bridge. The analysis of failure causes and the assessment of how these failures impacted the stress distribution in box girders have been presented. The steel superstructure after repair was also subjected to detailed analysis regarding change of stress distribution in repaired girder. All analysis were carried out using advanced numerical FEM models. Conclusions drawn from these analysis concern both the proper shaping of the bridge superstructure as well as the methodology of assessing impact of this type of failure on stress distribution in the structure and effectiveness of the implemented repair method.

2. Bridge description

The motorway bridge has total length of 212.0 m and consists of two independent parallel three-span structures with theoretical spans of 65.0 m + 80.0 m + 65.0 m. Both, north and south structures are practically the same, except for the total width of the deck, which is 18.95 m and 17.55 m in two parallel structures respectively. The superstructure of both bridges has a steel box girder and accompanying concrete deck slab acting compositely with a girder. The supports of the bridge are massive, reinforced concrete wall abutments and pillars, located parallel to the river with the angle of 80° to the bridge longitudinal axis.

The depth of the box girder is variable and ranges from 2.20 m in the mid-span to 3.50 m over the intermediate supports (Fig.1). The concrete deck slab has variable thickness ranging from 0.25 m to 0.56 m and is transversely prestressed with cables spaced at 1.0 m. The bottom plate and both webs of the steel box girder are stiffened with closed ribs and the transverse inner bracing is mounted every 4.0 m to prevent girder's cross-section deplanation. Additionally, over the supports the double steel box cross-beams are located and the reinforced concrete bottom slab is cast along the negative moment's section, connected with the steel bottom plate. The steel girder is made of S355J2+N mild steel. The deck slab is made of C35/45 concrete reinforced with B500A steel rebars and is prestressed transversely with Y1860 steel cables. The same concrete and reinforcement were used for casting the bottom slab.

After almost entire bridge had been completed (excluding deck equipment), there was a few month break in its construction due to the contractor's bankruptcy and withdrawal. After that break had been terminated the detailed inspection of the existing structures was carried out by the succeeding contractor. In two support regions of the north steel box girder an unusual failures were detected comprising the significant deformation (buckling) and local cracking of steel webs just over the both pillars. Steel web cracking was placed around the rectangular section of cross-beams located on the other side of the web (Fig.2).

The analysis carried out by the authors aimed to re-check the limit states of the superstructure according to the codes used in the basic design to assess and confirm the bridge safety and to find failure causes, before the works were resumed. Additionally, the independent checking of the repair method proposed by the contractor was also provided.

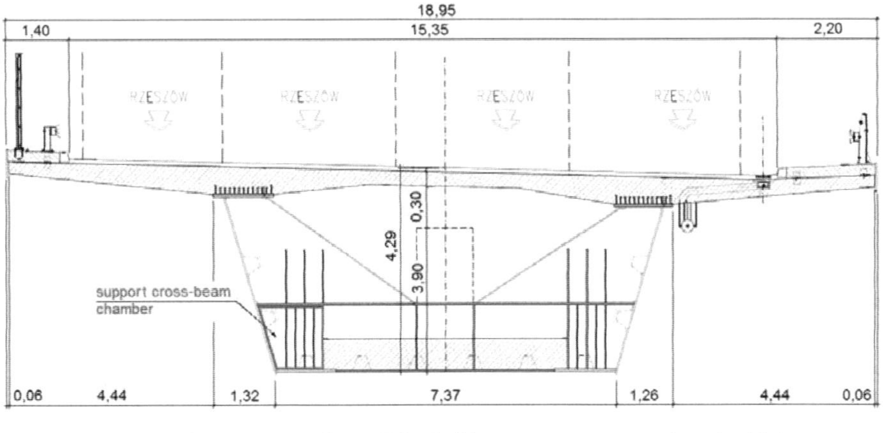

Figure 1. Cross-section of the bridge superstructure (north side)

Figure 2. Side view of the bridge (left) and close-up on web failure over pillar (right).

3. Re-checking of the superstructure

The first goal of the analysis was to check the safety of the superstructure according to the codes used in the basic design. The global FE models of the superstructure were developed in the *Sofistik* system (Fig.3). The four-node shell elements were applied to model the box girder and its internal support cross-beams as well as the RC top and bottom slabs. The concrete reinforcement and transverse prestressing cables were also modelled in the subsequent concrete elements. The beam finite elements were used to discretize the longitudinal stiffening ribs of the webs and bottom flange as well as the box internal bracing. All finite elements had an assigned thickness or cross-section corresponding to the actual geometry of steel plates / shapes and the material parameters were assumed according to material testing results, obtained in the quality assurance procedures during construction.

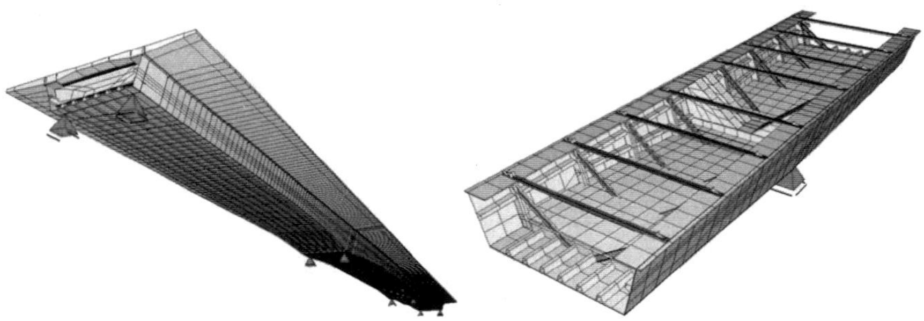

Figure 3. FE models of the superstructure.

The FE models were loaded with permanent loads (self-weight, equipment, prestressing, rheological effects etc.) and variable loads (live load, climatic effects) according to the Polish bridge standard used in the basic design. The actual phases of the bridge construction and their sequence implemented on-site were taken into account in FEM calculation, as follows: assembly of the steel members on temporary supports, removal of temporary supports after connecting members, casting the bottom slab, casting the deck slab and the final transverse prestressing of the deck slab. Concrete shrinkage and creep in the bottom and top slab were included in the particular construction phases. In the final phase of the evaluation the equipment weight and live loads as well as the environmental impacts were taken into account.

The global FE models were used for a linear analysis: the maximum values of internal forces were determined and then the basic ultimate limit states were checked according to the Polish bridge standards. Several load combinations were applied taking into account f.e. temperature effects. The equivalent stresses were determined based on the H-M-H theory and compared to the ultimate strength of the relevant material. Particularly, the entire steel superstructure was examined in order to check its appropriate strength and stability. As a result of the analysis based on FEM application, it was found that the superstructure was properly designed both during construction and in service condition. Thus any design error and limited knowledge case were excluded as the main causes of the detected failures. The remaining aspects of human errors as caused by negligence, ignorance, as well as poor workmanship, wrong assembly sequence, etc. had to be considered in searching the actual failure cause.

4. Failure analysis of the girder web

After excluding the design error and limited knowledge as the failure cause the supplemental inspection of the superstructure was undertaken to find another potential cause of web buckling and cracking. The unexpected water flooding inside the steel box in the support zones was discovered when the winter had finished (Fig.4). The authors hypothesized that the failure of steel webs might be caused by water trapped in cross-sections of the support crossbeam and frozen at low temperatures in winter. Therefore the next stage of numerical analysis was aimed at checking this hypothesis.

Figure 4. Water flooding inside the steel box girder in the support zones.

Basing on the on-site inspection and design drawings of the superstructure, a local FE model of the damaged part of the structure over the support was developed. The model consisted of one chamber constituting a part of the support cross-beam (Fig. 5). The total number of 6611 four-node shell elements with dimensions of 0.05 × 0.05 m were used to discretize the steel structure under consideration. The stress-strain curve described in Eurocode 3 was used to describe the non-linear behaviour of structural steel. The concrete bottom slab was discretized with 8450 eight-node solid elements with dimensions of 0.05 × 0.05 × 0.05 m. A linear isotropic material model according to Eurocode 2 was adopted for C35/45 concrete. The bottom slab reinforcement were also modelled in the subsequent concrete elements.

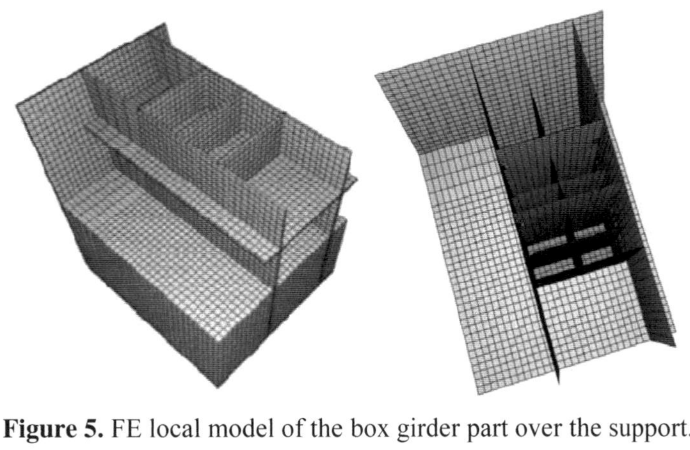

Figure 5. FE local model of the box girder part over the support.

The decisive issue in the failure analysis was to model the appropriate loads, suspected to be a main cause of failure. Due to water appearance inside the box girder an atmospheric ice loads due to freezing rain or snow shall be considered in the assessment of ice-sensitive parts of structure, as the closed steel chambers of support cross-beam. In the areas where records or experience indicate that freezing water produces large loads, the site-specific studies shall be used [3]. The load was modelled as the surface pressure applied to steel plates creating the chamber. The pressure was gradually increased up to the lack of the convergence of FE model. By means of the geometrically linear and materially non-linear FE analysis the maximum pressure value was determined at which the yielding of steel occurred and the iteration was not convergent (Fig. 6)

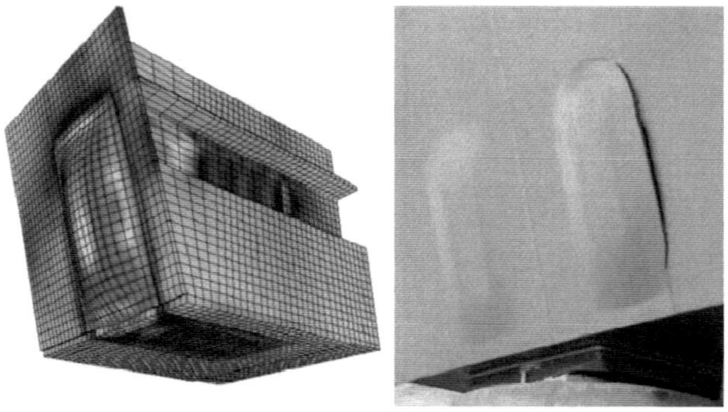

Figure 6. Numerically determined web deformation and stress distribution against actual web deformation and failure.

Based on the FE analysis results the displacement plot of the web centre point depending on the applied pressure of freezing water was established (Fig. 7). The yielding of web steel occurred at the approx. 6 MPa pressure and at approx. 12.5 MPa the plot approached the asymptote, which meant cracking of the web. The corresponding strain reached 48.5 ‰, which was a value many times higher than yielding point of 1.6 ‰ for S355 steel.

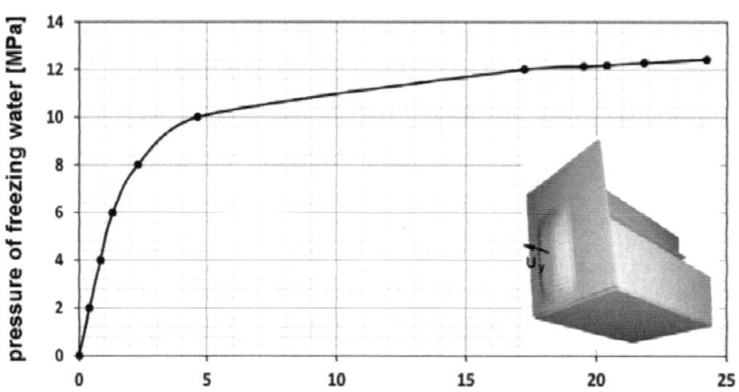

Figure 7. Displacement of the web centre point depending on the applied pressure of freezing water.

The numerical analysis clearly revealed that the hypothesis the freezing water had caused buckling and cracking of steel web was good justified. The detailed on-site inspection and actual deformation measurement confirmed also that hypothesis - observed damages resulted from freezing water filling the support cross-beam chamber. The analysis showed that the pressure applied on the internal surfaces of the chamber associated with the increase in the volume of freezing water could cause significant deformation of the web.

5. Repair method assessment

The web repair method prepared by the contractor comprised (Fig. 8):

- cutting damaged/deformed web plates and replaced them with the new ones, welded in the same place,
- welding of vertical stiffening ribs on the girder's web,
- welding of vertical stiffening ribs on the support cross-beam's webs,
- drilling drainage holes with diameter of 30 mm in the bottom plate of the cross-beam's chamber.

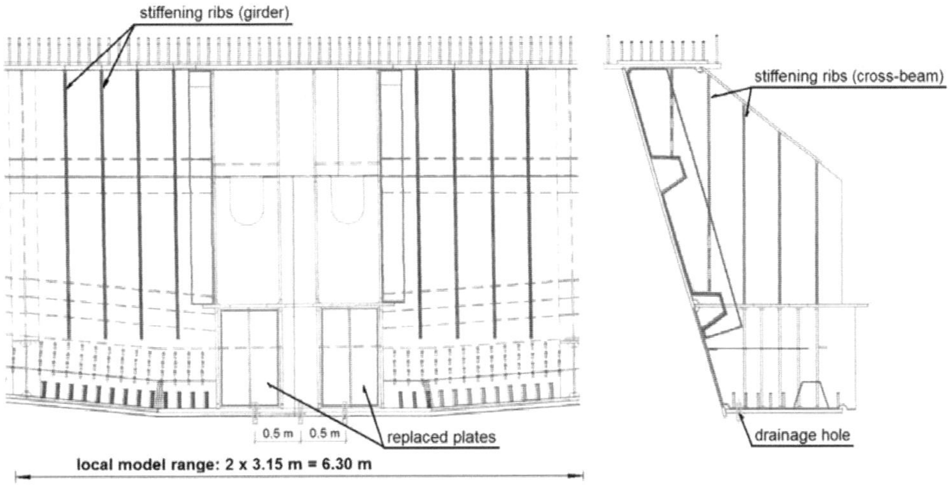

Figure 8. Repair method of the web: stiffening ribs (red, blue), replaced plates (green) and drainage holes (yellow).

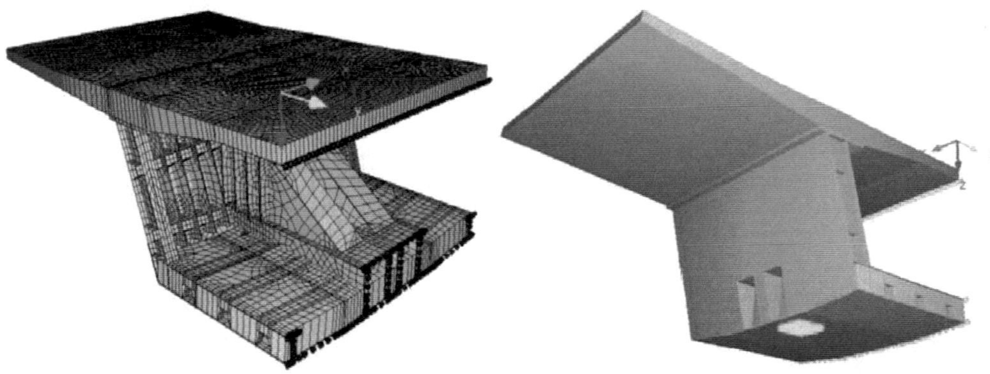

Figure 9. The local model of the support zone after repair: mesh (left) and the phase of damaged web plates removal (right)

To confirm that the proposed method of repair is correct, a numerical analysis was carried out using two models: the global one, used for safety analysis of the entire structure (Fig. 3) and the local one, which discretized the support zone of the superstructure (Fig. 5). The local model comprised a bigger part of the box girder than the local model used for failure analysis: the half of the box girder (one web) and the girder section of ± 3.15 m from the support axis. The subsequent concrete deck slab and the RC bottom slab were modelled as well (Fig. 9). In contrary to the first local model the possibility of transverse cracking of concrete deck slab was taken into account and the cracked concrete with the reduced stiffness in the longitudinal direction and tension stiffening effect of reinforcement were assumed in the calculation. The second local model was loaded with internal forces derived from the safety analysis carried out by means of the global model (Fig. 3). The girder support on bearings was modelled using elastic finite elements with infinitely high stiffness densely situated on the surface corresponding to the bearing plates. The entire local model consisted of 15 388 shell elements including 13 077 nodes.

The FE analysis was divided into two phases of repair:

- phase 1 - strengthening of the girder webs and the cross-beam with vertical ribs while cutting-off the damaged part of webs,
- phase 2 - repair of the webs with welded plates, i.e. fixing the holes.

Assuming two phases of repair works, loads acting on the girder were divided into two groups. In phase 1 the self-weight, early concrete shrinkage and transverse prestressing of the deck slab were considered. In phase 2 the equipment self-weight, final concrete shrinkage and live loads were additionally taken into account.

The linear FE analysis was carried out and the maximum values of equivalent stresses in superstructure were determined according to the H-M-H theory. On Fig. 10 the total equivalent stresses for both phases of repair works are presented. After repair and strengthening the maximum equivalent stresses in the girder web at the support zone reached approximately 180 MPa. It should be noted that the concentrated stresses in the corners of the model result from concentrated external forces and should not be treated as overstressing the web. The equivalent stresses in vertical stiffening ribs were also checked. In any case, the equivalent stresses did not exceed the yield strength of the S 355 steel. Therefore, it was confirmed that in any phase of the repair works as well as in service the ultimate limit state (safety) of the repaired zone would not be exceeded.

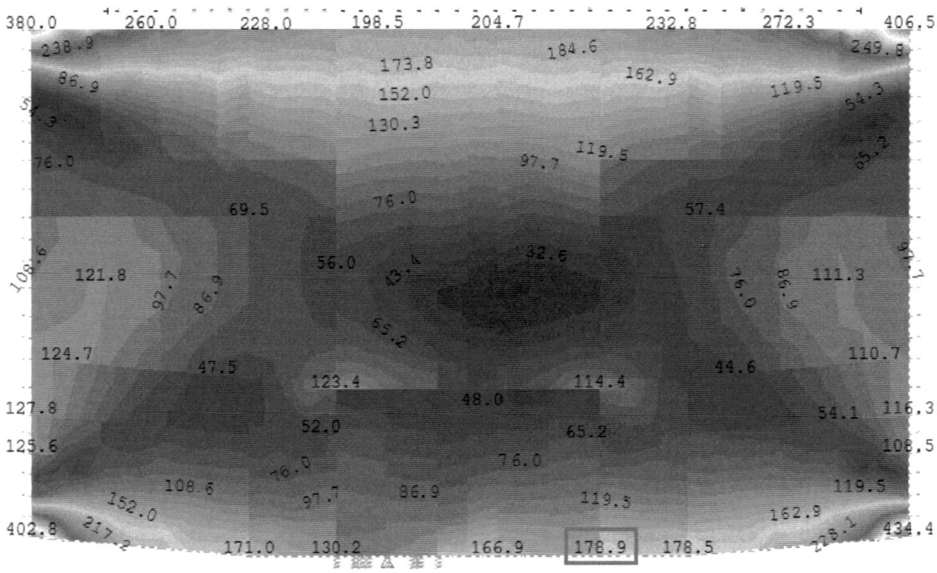

Figure 10. Summary of the equivalent stresses in the girder web in both phases (maximum stress in the support zone approx. 180 MPa).

6. Summary

The subject of this paper is an untypical failure of a motorway composite box girder bridge. After the unplanned break in bridge construction the contractor discovered the significant deformation and local cracking of the steel web in the support zones. The detailed FEM analysis aimed at determining the safety status of the superstructure and the damage causes as well as assessing the repair method was carried out. Analysis of the damage causes was preceded by a detailed analysis of structural performance and the hypothesis that the failure may result from freezing water collected in the chambers, theoretically empty, located at the lowest point of the box section. Such failures of steel structures with box section members occur often as a result of negligence, ignorance, as well as poor workmanship, wrong assembly sequence, etc. [2], [3].

As a result of the conducted analysis, it was confirmed that the assumed hypothesis was correct and it was found that the direct cause of deformation and cracking of steel web was the freezing water filling the closed cross-beams at low temperature. The following contributed to the failure:

- design errors in shaping the steel box girder including lack of drainage of the support zone and lack of drainage of the chamber of the box-shaped cross-beams located at the lowest point along the length of the superstructure;
- negligence during construction, enabling the inflow and long-term accumulation of large amounts of water in the supporting zone of the box section;
- lack of the appropriate superstructure protection during the break of construction related to the change of the contractor.

These design errors and former contractor's negligence resulted in the need to design and implement a complex repair project, which in turn resulted in increased bridge construction costs and extended time for completion.

References

[1] Scheer J 2010 *Failed Bridges: Case Studies, Causes and Consequences* (Berlin: Ernst & Sohn Verlag für Architektur und technische Wissenschaften GmbH & Co.)

[2] Imam B M and Chryssanthopoulos M K 2012 Causes and consequences of metallic bridge failures *Structural Engineering International* **22** pp 93-98

[3] Sobala D, Siwowski T and Żółtowski K 2005 Genesis of damages of a railway truss bridge - analisys of bridge carrying capacity *Proc. XXII Conf. on Structural Failures (Miedzyzdroje)* ed M Kaszynska (Szczecin: West Pomeranian University of Technology) pp 511-520 (in Polish)

IOP Conf. Series: Materials Science and Engineering **419** (2018) 012004 doi:10.1088/1757-899X/419/1/012004

Lifetime extension of existing steel bridges

Alessio Pipinato[1], Roberto Pavan[1]

[1]AP&P srl, Minadois Str. 20, Rovigo 45100, Italy

E-mail: info@pipinatoandpartners.it

Abstract. Prolonging the life of existing steel trusses all over the world is a incoming necessity for all managing authorities wherever these structures have reached a particular state of use. The unsatisfactory behavior of these structures under the current type and increasing number of loads make clearer that innovative solutions are needed. For this purpose, the realization of composite decks is needed, in order to redistribute the common actual load onto a structure conceived with historical codes, characterized by lower design live loads. The possibilities of retrofitting bridges with this solution studied in a recent research program are presented and compared in this study. Increasing traffic density and vehicle weight have been increasing in recent decades. Many existing truss bridges, all over the world, were not designed for the high service loads and the increased number of load cycles that they are exposed to today

1. Introduction

The increasing traffic density and vehicle weight have been increasing in recent decades. Many existing truss bridges, all over the world, were not designed for the high service loads and the increased number of load cycles that they are exposed to today. Some bridges will have to be either strengthened or replaced in the next decades. As late as the 1980s, steel truss bridges in Europe were often constructed with a non-composite concrete slab on top of steel girders. Different alternatives of interventions for truss bridges are presented in this paper, according to the different truss type to be reinforced. Sometimes the bridge should be retrofitted by deck replacing, other times a new structural configuration should be adopted, or in some cases only little interventions are needed in order to strengthen a truss. This paper presents a brief overview of different alternatives to strengthen existing trusses, ever considering the optimization of costs and material savings.

2. Strengthening old truss road bridges

The investigated structure is a two-lane roadway steel bridge truss. The overall bridge length is 120 m through three spans (40 m each one). The superstructure consists of riveted built-up truss members. Lower chords are inverse T-shaped sections, diagonals and upper chords are C-built-up elements with battens (stiffening brackets), while struts are I-section shaped built-up elements composed of four L-shaped elements and a plate. The deck is realized with longitudinal stringers and transverse floor beams. The floor beams have a fixed distance of 4 m, while the stringers are at 1.15 m one to the other. Top and bottom double-L bracings provide adequate stiffening of the structure. Built-up members, realized with plates, L- profiles or C profiles, are connected by hot riveting and connection joints are made of gusset riveted plates. The main structure and the member details have been completely revealed by an on-site geometrical survey. The bridge structure was modeled using the finite element method (FEM) software Midas (2005), using only beam elements. Rigid links (rigid body) were used to represent eccentricities of the elements. Overall, the entire bridge model consists of about 2000 beam elements. A Young's modulus of 210,000 MPa (N/mm2), Poisson's ratio of 0.3 and a material density value of 7850 kg/m3 (weight density of 76.98 kN/m3) were used for the analyses. All beam member sections were modeled as the as-built structure, as measured during the geometrical survey. The bridge is subjected to permanent loads, such as the self-weight of steel elements and of nonstructural elements and to variable loads such as temperature and traffic. At first, the bridge was checked referring to the Italian Ministerial Decree 09/06/1945, n. 6018 (DM 6018, 1945) and then to current Italian and Eurocode. Historical tests on the bridge have been used to calibrate the FEM model. In this study, retrofit alternatives chosen from the following solutions are considered:

a. making composite an existing non-composite deck;
b. building an orthotropic deck;
c. building a new concrete deck, directly connected to the main trusses.

All these strategies are normally combined with steel- only interventions (including cover-plating, element replacement). The described procedures are finalized to redistribute the live loads adequately onto the deck with a new or modified deck. As can be observed in the following, the beneficial use of a rigid deck is often the best solution able to extend the bridge life adequately. In the case of Italy, the bridge category shifts from the 1st to the 2nd class:

- AD1 Load, 1st category: Load Model 1, EN 1991- 2:2003 or
- AD2 Load, 2nd category: Load Model 1, EN 1991- 2:2003 taking a reduction of 20% for all loads of Lane number 1.

In the present study, the reference retrofit solutions considered are:

- BR00) The existing bridge is calculated with the historical live load, HS-LOAD, without any retrofit intervention;
- BR01) The existing bridge is calculated with the actual code live load, AD1-LOAD, without any retro- fit intervention;
- BR02) The existing bridge is calculated with the actual code live load, AD2-LOAD, without any retro- fit intervention;
- RROOR1-2) The bridge is retrofitted with the introduction of an orthotropic deck laying between stringers, with open ribs considering both the AD1- Load and AD2-Load;
- RROCR1-2) The bridge is retrofitted with the introduction of an orthotropic deck laying between stringers, with closed ribs considering both the AD1- Load and AD2-Load;
- RRA1-2) The bridge is retrofitted with the introduction of a concrete deck considering both the AD1- Load and AD2-Load; welded shear studs are introduced in the bridge to connect the new concrete deck with the stringers (three □20 mm studs/m along all stringers, except for stringers along sections 1, 9–13, 19–23, 29–31 where six □20 mm studs/m have been placed); a parametric analysis is developed varying both the concrete deck thickness (with a fixed strength of C40/50) among 10 cm (RRA1, 2–10), 15 cm (RRA1, 2–15), 20 cm (RRA1, 2–20), 25 cm (RRA1, 2–25), 30 cm (RRA1, 2–30) and the deck concrete strength (fixing the deck thickness at the lowest value of 10 cm) among C30/37 (RRA1, 2-R1), C35/45 (RRA1, 2-R2), C40/50 (RRA1, 2-R3), C45/55 (RRA1, 2-R4), C55/67 (RRA1, 2-R5);
- RRA1-2-I) The bridge is retrofitted with the introduction of a concrete deck considering both the AD1- Load and AD2-Load, fixing the concrete strength at C40/50 and the deck thickness at 100 mm; moreover, the steel-to-steel intervention -cover plating- are introduced adopting S355 new members;
- RRB1-2-I) The bridge is retrofitted with the introduction of a concrete deck considering both the AD1- Load and AD2-Load, introducing a UHPC concrete of C90/105 strength class and fixing the deck thickness at 50 mm; moreover, the steel-to-steel intervention -cover plating- are introduced adopting S355 new members;

IOP Conf. Series: Materials Science and Engineering **419** (2018) 012004 doi:10.1088/1757-899X/419/1/012004

- RRC1-2-I) The bridge is retrofitted with the introduction of a concrete deck considering both the AD1- Load and AD2-Load, introducing a UHPC concrete of C150/160 strength class and fixing the deck thickness at 30 mm; moreover, the steel-to-steel intervention are introduced adopting S355 new members. According to the analysis performed, the optimal solution is represented by the RRB1-I, which has been compared in terms of utilization rate to the main solution descripted above (maximum ratios Ed/Rd defined as the minimum safety factor of all ultimate limit state-ULS checks).

Figure 1. The road-bridge over the river Adige.

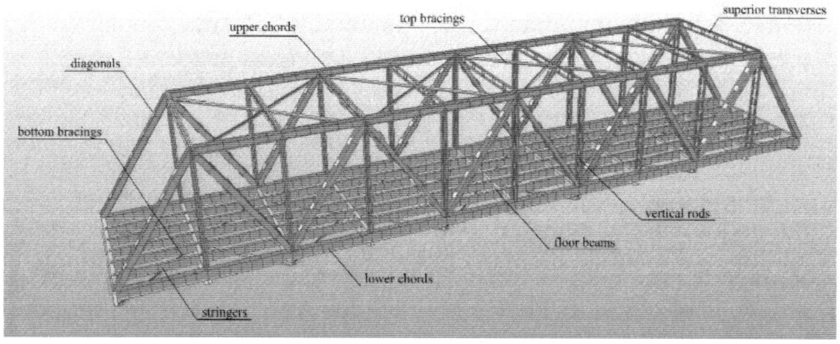

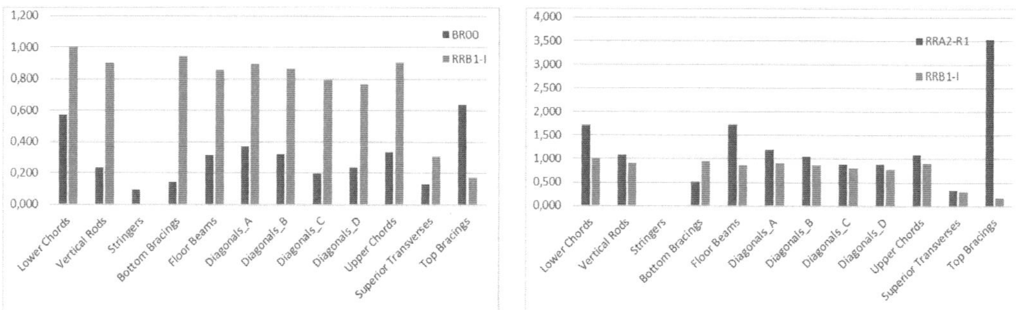

Figure 2: Comparison of B00 vs. RRB1-I model (on left) Comparison of RRA2-R1 vs. RRB1-I (on right)

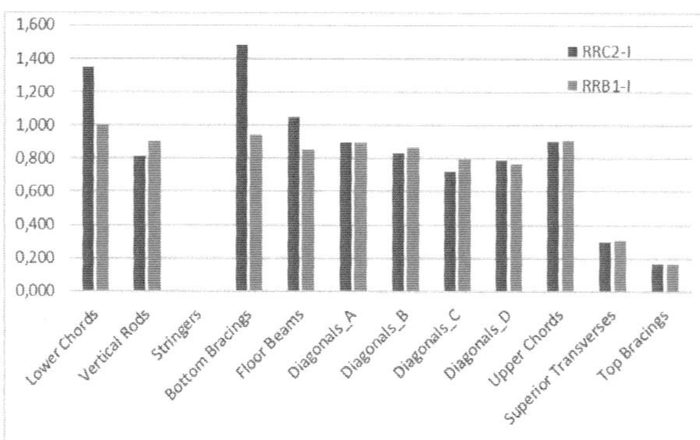

Figure 3: Comparison of RRB1-Ivs. RRB1-I model.

3. Strengthening for fatigue induced damages in trusses: a parametric analysis

Traffic running on bridges produces a stress spectrum that may cause fatigue damage. This stress spectrum depends on the geometry of the vehicles, the axle loads, the vehicle spacing, the composition of the traffic, and its dynamic effects. However, the increasing loads on truss bridges (magnitude and quantity), both railways and roadways, imply the increasing accumulation of damage in iron or steel member, that could lead to sudden failure. A parametric analysis has been performed onto two existing bridges in Italy, the case study of the par. 2.4 (Br1) and the railway Adige bridge (Br2), in order to check which are the more suitable and effective solution for bridge repair, in order to prolong their fatigue life. After the retrofit solution proposed, all bridges are checked according EN1993-1-9:2005. The following solutions have been considered:

- RRF1: the bridge is retrofitted with the introduction of a concrete deck, made of UHPC concrete; the cover-plating technique is adopted in hot spot fatigue stresses;
- RRF2: the bridge is retrofitted with the introduction of a concrete deck, made of UHPC concrete; hot spot stress truss-nodes are encased in new UHPC nodes;

For every RRFi (i=1, 2) retrofit solution, a limit of adding no more new material than the 25% of the original weight of the bridge is fixed.

For railway bridges, an unballasted solution is provided for RRF1-RRF2.

A wide number of other parameters are varied in order to analyze the sensitivity of the model to these variations. According to the analysis performed, the optimal solution depends on the bridge type, which has been compared in terms of utilization rate to the main solution descripted above (maximum ratios Ed/Rd defined as the minimum safety factor of all ultimate limit state-ULS checks).

Figure 4. The railway bridge over the river Adige.

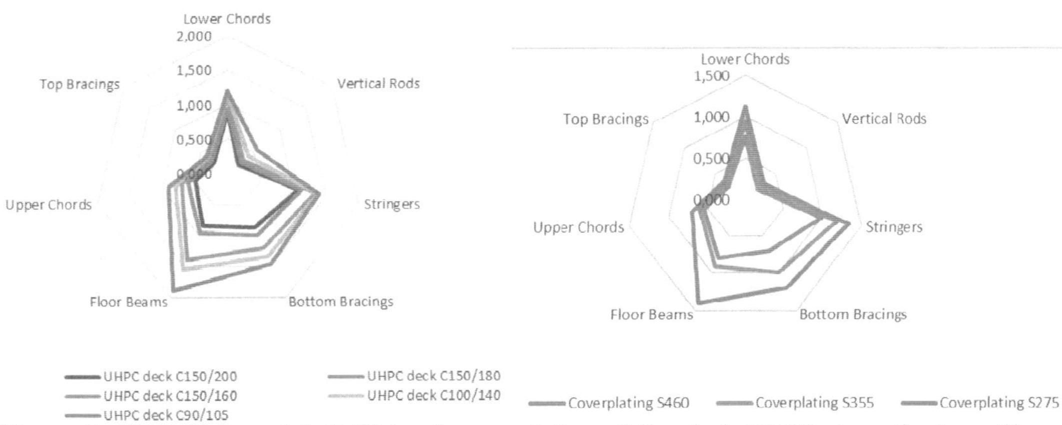

Figure 5: RRF1-Br1 model, Ed/Rd values: variation of the deck UHPC strength class, 50 mm thickness; cover-plating with S275 steel grade (on left), RRF1-Br1 model, Ed/Rd values: variation of the cover plating steel strength class; UHPC deck grade C150/160, 50 mm thickness (on right).

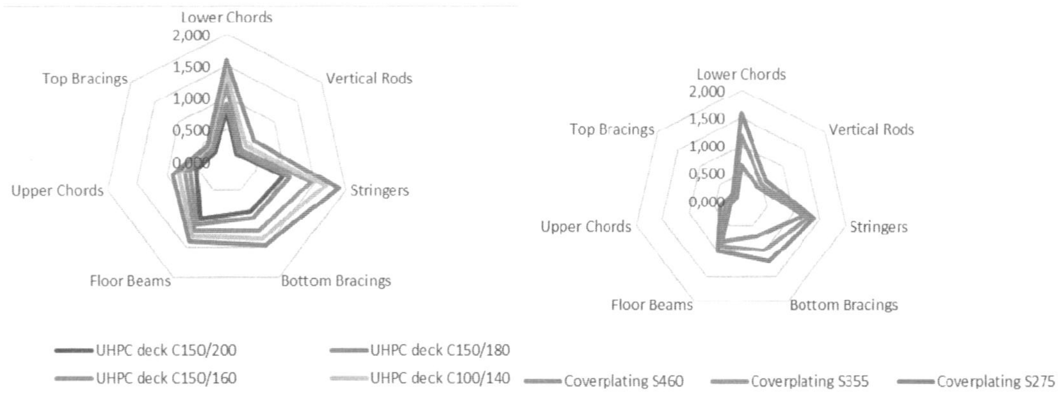

Figure 6: RRF1-Br2 model, Ed/Rd values: variation of the deck UHPC strength class, 80 mm thickness; cover-plating with S275 steel grade (on left), RRF1-Br2 model, Ed/Rd values: variation of the cover plating steel strength class; UHPC deck grade C150/160, 80 mm thickness (on right).

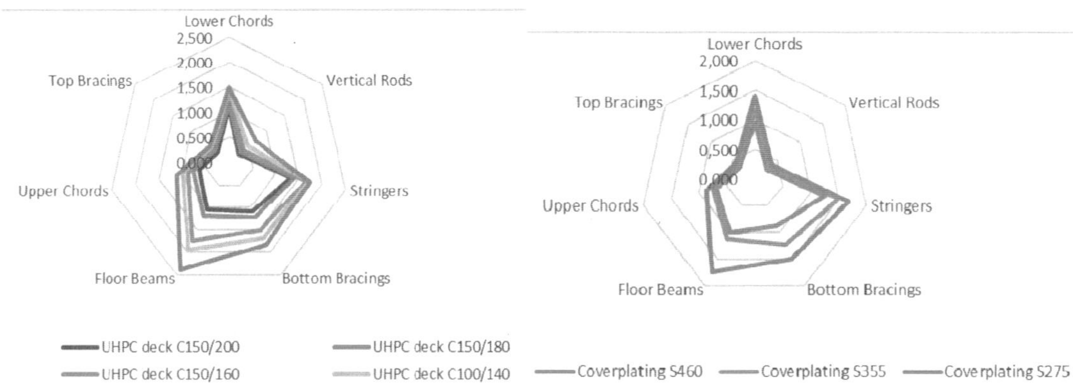

Figure 7: RRF2-Br1 model, Ed/Rd values: variation of the deck UHPC strength class, 50 mm thickness; cover-plating with S275 steel grade (on left), RRF2-Br1 model, Ed/Rd values: variation of the cover plating steel strength class; UHPC deck grade C150/160, 50 mm thickness (on right).

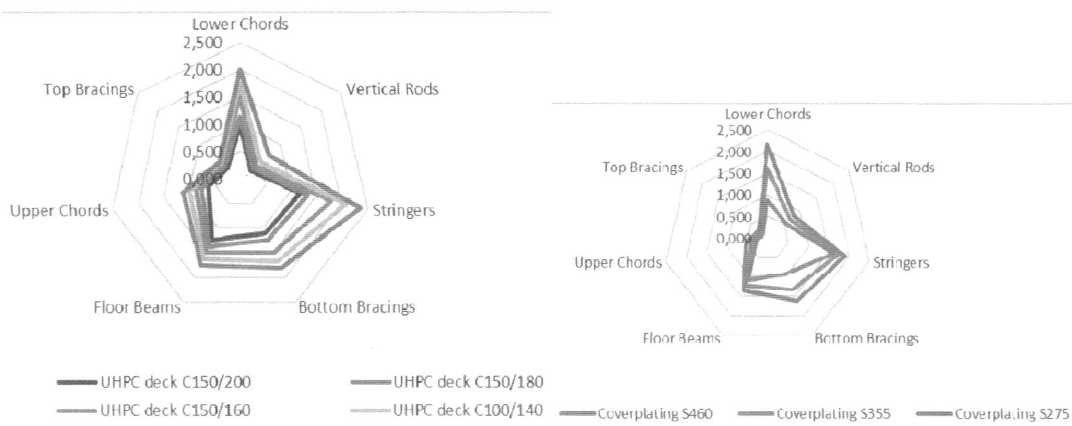

Figure 8: RRF2-Br2 model, Ed/Rd values: variation of the deck UHPC strength class, 80 mm thickness; cover-plating with S275 steel grade (on left), RRF2-Br2 model, Ed/Rd values: variation of the cover plating steel strength class; UHPC deck grade C150/160, 80 mm thickness (on right).

4. Conclusion

In this paper different strategies in order to prolong the lifetime of existing steel truss bridges have been presented. The study presented a parametric analysis to prolong the fatigue strength of existing bridges, to discover the most suitable retrofit solution, and to compare different retrofit techniques. As a matter of fact, the whole solutions presented are finalized to demonstrate that it is possible to prolong the life of existing trusses, with the innovative techniques illustrated herein.

Acknowledgments

This research was supported by the European Union, represented by the European Commission Directorate General for Research and Innovation, Research Fund for Coal & Steel (RFCS) program, under grant agreement no. RFSR-CT-2015-00025. This funding is gratefully acknowledged.

References

[1] DM 6018 (1945). Italian Ministerial Decree 09/06/1945 n. 6018, Ministero dei Lavori Pubblici, Roma

[2] EN 1993-1-9 (2006). Eurocode 3: Design of steel structures - Part 1-9: Fatigue. CEN European Committee for Standardization, Brussels, Belgium

[3] EN 1991-.2 (2003). Eurocode 1: Actions on structures - Part 2: Traffic loads on bridges. CEN-European Committee for Standardization, Brussels, Belgium

[4] MIDAS (2016). Midas User Manual. MIDAS Information Technology Co., Korea.

IOP Conf. Series: Materials Science and Engineering **419** (2018) 012005 doi:10.1088/1757-899X/419/1/012005

Lightweight solution for existing steel movable bridge retrofit and repair

Antonella Ruzzante[1], Roberto Pavan[1]

[1]AP&P srl, Minadois Str. 20, Rovigo 45100, Italy

E-mail: info@pipinatoandpartners.it

Abstract. Prolonging the life of existing steel movable bridges is a paramount problem for managing authorities, where these specific types of bridges have reached a prolonged state of use. The unsatisfactory behaviour of these structures under the current type and increasing number of loads make clearer that innovative solutions are needed. For this purpose, the realization of lightweight composite decks is needed, in order to redistribute the common actual load onto a structure conceived with historical codes, characterized by lower design live loads. The possibilities of retrofitting bridges with this solution studied in a recent research program are presented and compared in this study.

1. Introduction

The increasing traffic density and vehicle weight have been clearly noted worldwide in recent decades. Many existing movable bridges, all over the world, were not designed for the high service loads and the increased number of load cycles that they are exposed to today. Some bridges will have to be either strengthened or replaced in the next decades. Movable bridges have proved to be an economical solution to the problem of how to carry a rail line or highway across an active waterway. A "closed" movable bridge has closed the waterway to marine traffic, while an "open" bridge has opened the water-way to marine traffic. Highway bridges are typically de-signed to remain in the closed position and only to be opened when required by marine traffic. However, movable railroad bridges can be designed to remain in either the open or closed position, depending on how frequently they are used by train traffic. The difference is important as different wind and seismic load design conditions are used to design for a bridge that is usually open vs. one that is usually closed. The three major categories of movable bridges are swing, bascule, and vertical lift. This list is not exclusive and there are other types, such as jackknife, reticulated, retracting, and floating that are not common and will not be described here. Movable bridges are designed to be operated following a set protocol, and this protocol is incorporated into the control system as a series of permissive inter-locks. The normal sequence of operation is as follows. Vessel signals for an opening, usually through a marine radio but it can be through a horn. For a highway bridge the operator sounds a horn, activates the traffic signals, halting traffic, lowers the roadway gates, then lowers the barrier gates. For a rail bridge the operator needs to get a permissive signal from the train dispatcher. After the barrier gates are lowered, a permissive signal allows the operator to withdraw the locks and/or wedges and lifts and, once that is completed, to open the span. The vessel then can proceed through the opening. To close the bridge, the steps are reversed. The controls are operated from a control desk and Note that the control desk includes a position indicator to demonstrate the movable span(s) position as well as an array of push buttons to control the operation. A general objective in designing such a desk is to have the position of the buttons mimic the sequence of operations. Typically, the but-tons are lit to indicate their status. or every type of movable bridge there are a wide amount of interventions needed both every year, both during long time scheduling maintenance. The most frequent interventions lie on the moving equipment maintenance and checks, while the structural maintenance and eventual retrofit refers to longer period.

Figure 1: In service movable bridge.

2. Retrofit solutions: literature survey

A special category of steel structures where the fatigue induced damage is relevant for the life cycle of the structure, deals with movable bridges. These are daily subjected to high level stresses induced by load cycles of passing vehicles: the incrementing speeds and concurring dynamic amplifications, and the incremental number of loads and number of vehicles, represented a critical structural situation especially for those bridges in principle lines and with some decades of exercise. The literature for these detailed analysis and damage assessment is not so extensive, and relegated to a small number of specialists: guidelines for the assessment of historical existing steel structures are deepen in Kuhn et al. (2008) and Brühwiler et al. (2012), in a sort of pre normative design document, respectively in the European and Swiss context; while well-established authorities in this framework have studied riveted bridges Fisher et al. (1984), Brühwiler et al. (1990), other studies are focused on large scale experimentations and damage assessment procedures (Pipinato et al 2008b; Pipinato et al. 2009; Pipinato et al. 2010a; Pipinato et al 2011c; Pipinato 2011; Pipinato 2012a); a detailed procedures for accounting for remaining fatigue life is provided in Imam et al. (2005), while other studies are more focused on the structural analysis of single bridges (Pipinato et al. 2012a; Pipinato et al. 2012a; Pipinato 2010b; Pipinato et al. 2012e; Pipinato et al. 2010a; Pipinato and Modena 2010 Pipinato et al. 2008) even considering special combination of external loadings, and damage accumulation including fatigue and seismic loads (Pipinato 2012b); or FRP interventions for build up new structures or repairing solutions Pipinato (2012d); particular cases of damages are included also in some hint studies (Pipinato et al. 2012e; Pipinato et al. 2012c) re-porting particular analysis solutions that can help to solve bridge damaged systems; retrofit procedures and load analysis also for high speed bridges are reported in (Pipinato 2013a,b,c), also coupling damage as in the aforementioned studies (Pipinato 2012c; Pipinato et al 2011a; Pipinato et al 2010b) a particular subset of these studies includes damage analysis and control of existing and historical bridges (Pipinato 2012d; Pipinato et al.

2013); reliability analysis also considering dynamic effects are described in (Imam et al. 2012; Imam et al. 2012; Kaliyaperumal et al. 2011; Imam and Righiniotis 2010), while a more focused study on the probabilistic fatigue evaluation of riveted bridges is included in Imam et al (2008); the fatigue analysis of riveted bridge connections using the theory of critical distances and relative numerical modelling are described in (Righiniotis et al. 2008; Imam et al. 2007); while a general introduction on the specific argument could be found in (Imam et al. 2006; Imam et al. 2005); more focused and specialist deepens deals with the fracture reliability of a typical Northridge steel moment resisting connection (Righiniotis and Imam 2004), the fatigue reliability of riveted connections in bridges (Imam et al. 2006, 2008), the ef-fect of climate change on the deterioration of steel bridges (Kallias and Imam 2013), the risk assessment of steel bridges under the influence of changing environ-mental conditions (Kallias and Imam 2012), the fatigue assessment of a bridge detail using dynamic analysis and probabilistic fracture mechanics (Imam and Kaliyaperumal 2012), the modelling of failure consequences for robustness evaluation (Chryssanthopoulos et al. 2011), the review of metallic bridge failure statistics (Imam and Chryssanthopoulos 2010), the effects on the fatigue reliability of deteriorating riveted bridges (Imam et al. 2009a), the global-local finite element analysis of riveted bridge connections for fatigue evaluation (Imam et al. 2009b), the probabilistic fatigue load spectra for riveted bridges (Imam et al. 2007), the probabilistic fatigue life estimates for riveted bridges (Imam et al. 2006), the effect of fracture on the reliability of a moment resisting frame under earthquake loading (Righiniotis and Imam 2006), the probabilistic fatigue life estimates for riveted bridges (Imam et al. 2006), the remaining fatigue life estimates for riveted bridges (Imam et al. 2005), and the connection fixity effects on stress histories in riveted rail bridges (Imam et al. 2004).

3. Movable bridge deck intervention

A special As a fact, the most damaged structure of movable bridges is represented by the deck. A crucial part of the design of movable bridges is to limit the moving dead load which affects the size of the counterweight, the overall size of the main structural members, and, to a lesser extent, the machinery depending upon the type of movable bridge. For movable railroad bridges this is typically not a problem, as movable span decks are de-signed with open decks (timber ties on stringers) and the design live load is such a large part of the overall de-sign load that the type of deck is not an issue. For high-way bridges, however, the type of deck needs to be carefully selected to provide a minimum weight while providing an acceptable riding surface. Early movable spans used timber decks, but they are relatively heavy and have poor traction and wear. Timber was replaced by open steel grid, a good solution that is both light-weight and long wearing. In addition, the open grid reduced the exposed wind area, particularly for bascule bridges in the open position. However, with higher driving speeds, changes in tires and greater congestion, steel grid deck has become the source of accidents, particularly when wet or icy. Now most new movable bridge decks are designed with some type of solid sur-face. Depending on the bridge, this can be a steel grid partially filled with concrete or epoxy, an orthotopic deck, lightweight concrete, or the “exodermic” system. Aluminum and composite decks are also now being developed and may prove to be a good solution. While orthotropic decks would seem to be a good solution, as the deck can be used as part of the overall structural system, they have not yet seen widespread use in new designs.

Consequently, a number of factors that should be considered in the selection of a lightweight solid deck system to replace steel open grid deck on typical movable bridges. These factors include (URS 2012):

- Costs including:
  o Deck fabrication and installation,
  o Modifications to the bascule leaf steel framing, bascule pier, flanking spans, counterweight, and trunnion assemblies required to implement the new deck system,
  o Design and construction inspection for the new deck system,
  o Future maintenance and inspection.
- Functionality and Safety including:

o Load capacity (i.e. support of legal loads and permit-ted overloads),
o Improved riding surface and skid resistance for vehicular and bicycle traffic,
o Reduced traffic generated noise.
- Maintenance including:
o Ease of repair and/or replacement of all or portions of the deck,
o Need for periodic maintenance (e.g. reapplication of coatings and/or replacement of wearing surfaces).
- Service-life and Durability including:
o Deck and/or deck component (e.g. wearing service, fasteners, etc.) service-life,
o Resistance to corrosion, fatigue, wear, impact, fire, ultraviolet light, and chemicals,
o Accommodation of thermal movements.
- Constructability including:
o Disruption to traffic (e.g. overall construction duration, maintenance of traffic),
Ability to accommodate fabrication and installation tolerances,
o Sensitivity to environmental conditions during construction,
o Shipping, storage and handling,
o Specialized inspection requirements during fabrication and installation.
- Other Risks including:
o Familiarity of the product and technology (e.g. years' experience, previous bridge installations, quantity and quality of applicable research, endorsement by bridge design community),
o Availability of design tools and construction techniques,
o Financial and technical support from supplier(s),
o Product availability (e.g. opportunities for competitive bidding, sole source and/or patents).

These new lightweight solid decks to replace steel open grid decks on typical movable bridges include:
a) Sandwich Plate System
The Sandwich Plate System is a proprietary and patented composite laminate plate material consisting of thin metal faceplates continuously bonded to a polyurethane elastomer core. The elastomer core transfers forces between the metal faceplates by way of shear, analogous to an I-beam when subjected to flexure with the steel plates acting as the flanges and the elastomer core as the web. The elastomer core also prevents local buckling of the metal faceplates in compression, which permits development of full yield of the metal face-plates. Panels are very thin (1 to 2 inches thick) com-pared to traditional bridge decks. The use of these thin panels is possible through a combination of panel bending and membrane action (i.e. in-plane tension). Separation of the thin metal plates with a lightweight elastomer core more efficiently increases strength and stiffness of the plate, while minimizing the increase in weight of the plate (URS 2012).
b) Aluminum Orthotropic Deck
Aluminum alloys have much to offer for bridge deck applications, and continue to be used, primarily over-seas, where their lightweight, high strength-to-weight ratio and excellent corrosion resistance satisfy service requirements. Aluminum is manufactured by an extrusion process, which permits highly customized sections that can be optimized for a given application. The lack of widespread use of aluminum, despite the structural and maintenance advantages, is primarily due to the high initial cost of the material (URS 2012).
c) Fiber Reinforced Polymer (FRP) Composite Deck
Fiberglass reinforced plastic (FRP) is a composite material that has a number of characteristics that makes it a good alternative to replace steel open grid deck on bascule bridges. FRP composites can be designed to pro-vide a wide range of mechanical properties including tensile, flexural, impact and compressive strengths. FRP composites are generally lightweight (i.e. approximately 20% of steel) with high strength-to-weight ratio and are generally resistant to creep and fatigue. FRP composites are versatile and can be fabricated in many different configurations, customized specifically for a given application (e.g. bridge decks), designed and fabricated such that the strength is oriented to meet specific configuration and loading demands, made into larger more complex structural components and systems that simplifies connections and minimizes the number of components that

must be handled. The relatively low modulus of elasticity of FRP composites make them relatively flexible compared to other bridge deck materials and thus deflections generally control the design of FRP bridge decks. FRP composite materials are susceptible to degradation from ultraviolet light and chemicals; however, the materials can be formulated with additives to improve the resistance to these effects (URS 2012).

d) Ultra-high Performance Concrete (UHPC) thin deck

Ultra-High Performance Concrete (UHPC) is a material that has recently gained attention in the bridge industry. UHPC provides a combination of superior properties including strength, ductility, durability that permits design and construction of thinner sections and longer spans that are lighter and more efficient, while providing improved durability and impermeability against corrosion, abrasion and impact. UHPC has potential to be used in an alternative deck design to replace steel open grid deck. The high concrete compressive and tensile strength allows design and construction of a lightweight waffle slab bridge deck with thin elements (slab and ribs) (URS 2012).

e) Making composite existing non-composite decks

Many older steel bridges are constructed with a non-composite concrete slab over steel girders. A potentially economical means of strengthening these floor systems is to connect the existing concrete slab and steel girders to permit the development of composite action. Composite action means that the steel girder and concrete can help each other out carrying the loads more efficiently than in the original nanocomposite condition. To achieve the benefits of composite action, the existing steel girder must be connected to the existing concrete slab to permit the transfer of shear forces at the steel-concrete interface.

4. Conclusion

Many existing movable bridges, all over the world, were not designed for the high service loads and the in-creased number of load cycles that they are exposed to today. Some bridges will have to be either strengthened or replaced in the next decades. For this reason, in this paper specific intervention are presented, including the following: bridge deck traditional intervention; light deck intervention (including SPS systems, aluminium orthotropic deck, FRP composite deck UHPC thin deck, making composite existing non composite solutions). According to the specific situation, span, loads, traffic, service life to be maintained, the proper solution or a combination thereof should be designed.

Acknowledgments

This research was supported by the European Union, represented by the European Commission Directorate General for Research and Innovation, Research Fund for Coal & Steel (RFCS) program, under grant agreement no. RFSR-CT-2015-00025. This funding is gratefully acknowledged.

References

[1] Brühwiler E., Smith I. F. C., Hirt M. A., 1990. Fatigue and Fracture of Riveted Bridge Members. Journal of Structural ngineering (ASCE), 116 (1), 198-214.

[2] Brühwiler E., Vogel T., Lang T., Lüch P., 2012. Swiss Standards for Existing Structures. Structural Engineering International 2/2012 Technical Report, pp. 275-280.

[3] Chryssanthopoulos MK, Janssens V, Imam B., 2011. Modelling of Failure Consequences for Robustness Evaluation'. IABSE/IASS London, UK: IABSE–IASS Symposium: Taller, Longer, Lighter

[4] Fisher J. W., Yen B. T., Wang D., 1984. NCHRP Report No 302 : Fatigue and Fracture Evaluation for Rating Riveted Bridges. Transportation Research Board, National Research Council, Washington, D.C.

[5] Imam B, Chryssanthopoulos MK, Frangopol DM., 2009b. System effects on the fatigue reliability of deteriorating riveted railway bridges'. CRC Press Safety, Reliability and Risk of Structures, Infrastructures and Engineering Systems: Proceedings of the Tenth International Conference on Structural Safety and Reliability, Osaka, Japan: ICOSSAR2009

[6] Imam B, Chryssanthopoulos MK., 2010. A review of metallic bridge failure statistics'. CRC Press Bridge Maintenance, Safety and Management: Proceedings of the Fifth International IABMAS Conference, Philadelphia, USA: IABMAS 2010, pp. 3275-3282.
[7] Imam B, Righiniotis TD, Chryssanthopoułos MK, Bell B., 2006. Analytical fatigue assessment of riveted rail bridges'. Proceedings of the Institution of Civil Engineers: Bridge Engineering, 159 (3), pp. 105-116.
[8] Imam B, Righiniotis TD, Chryssanthopoulos MK, Bell B., 2006. Probabilistic fatigue life estimates for riveted railway bridges'. Taylor & Francis Advances in Bridge Maintenance, Safety Management, and Life-Cycle Performance: Proceedings of the Third International Conference, Porto, Portugal: IABMAS '06
[9] Imam B, Righiniotis TD, Chryssanthopoulos MK., 2004. Connection fixity effects on stress histories in riveted rail bridges'. Taylor & Francis Bridge Maintenance, Safety, Management and Cost: Proceedings of the 2nd International Conference on Bridge Maintenance, Safety and Management, Kyoto, Japan: Iabmas'04.
[10] Imam B, Righiniotis TD, Chryssanthopoulos MK., 2005. Fatigue assessment of riveted railway bridges'. Korean Society Of Steel Construction Steel Structures, 5 (5), pp. 485-494.
[11] Imam B, Righiniotis TD, Chryssanthopoulos MK., 2005. Remaining fatigue life estimates for riveted railway bridges'. Thomas Telford (ICE Publishing) Bridge Management Five: Proceedings of the 5th International Conference on Bridge Management, Guildford, UK: 5th International Conference on Bridge Management, pp. 417-425.
[12] Imam B, Righiniotis TD, Chryssanthopoulos MK., 2009b. Global-local finite element analysis of riveted railway bridge connections for fatigue evaluation'. Taylor and Francis Bridge Maintenance, Safety Management, Health Monitoring and Informatics - IABMAS '08, Seoul, Korea: Proceedings of the 4th International Conference on Bridge Maintenance
[13] Imam B, Righiniotis TD, Chryssanthopoulos MK., 2014. Fatigue crack initiation in riveted railway bridge connections. Philadelphia, USA: Proceedings of the 1st International Conference on Fatigue and Fracture in the Infrastructure, Bridges and Structures of the 21st Century
[14] Imam B, Righiniotis TD, Chryssanthopoulos MK., 2006. Fatigue reliability of riveted connections in railway bridges. American Society of Mechanical Engineers Glasgow, UK: 3rd ASRANet International Colloquium
[15] Imam B. M., Chryssanthopoulos MK, Frangopol DM., 2012. Fatigue system reliability analysis of riveted railway bridge connections. Structure and Infrastructure Engineering, 8 (10), pp. 967-984.
[16] Imam BM, Chryssanthopoulos MK., 2012. Causes and consequences of metallic bridge failures. Structural Engineering International: Journal of the International Association for Bridge and Structural Engineering (IABSE), 22 (1), pp. 93-98.
[17] Imam BM, Kaliyaperumal G., 2012. Fatigue assessment of a railway bridge detail using dynamic analysis and probabilistic fracture mechanics. Taylor & Francis Group Bridge Maintenance, Safety, Management, Resilience and Sustainability - Proceedings of the Sixth International Conference on Bridge Maintenance, Safety and Management, Lake Como, Italy: The 6th International Conference on Bridge Maintenance, Safety and Management (IABMAS 2012), pp. 3206-3213.
[18] Imam BM, Righiniotis TD, Chryssanthopoulos MK, Bell B. 2006. Probabilistic fatigue life estimates for riveted railway bridges'. Proceedings of the 3rd International Conference on Bridge Maintenance, Safety and Management - Bridge Maintenance, Safety, Management, Life-Cycle Performance and Cost, , pp. 499-500.
[19] Imam BM, Righiniotis TD, Chryssanthopoulos MK., 2007. Numerical modelling of riveted railway bridge connections for fatigue evaluation'. ELSEVIER Engineering Structures, 29 (11), pp. 3071-3081.
[20] Imam BM, Righiniotis TD, Chryssanthopoulos MK., 2007. Probabilistic fatigue load spectra for riveted railway bridges'. Taylor & Francis Proceedings of the 10th International Conference on Applications of Statistics and Probability in Civil Engineering, Tokyo, Japan: 10th International Conference on Application of Statistics and Probability in Civil Engineering, pp. 73-74.
[21] Imam BM, Righiniotis TD, Chryssanthopoulos MK., 2008. Probabilistic fatigue evaluation of riveted railway bridges'. Journal of Bridge Engineering, 13 (3), pp. 237-244.
[22] Imam BM, Righiniotis TD., 2010. Fatigue evaluation of riveted railway bridges through global and local analysis'. Elsevier Journal of Constructional Steel Research, 66 (11), pp. 1411-1421.
[23] Imam, B., Righiniotis, T.D., Chryssanthopoulos, M.K., 2005. Remaining fatigue life estimates for

riveted railway bridges. Proceedings of the 5th International Conference on Bridge Management, p 417-425, 2005, Bridge Management 5: Inspection, Maintenance, Assessment and Repair - Proceedings of the 5th International Conference on Bridge Management
[24] Kaliyaperumal G, Imam B, Righiniotis T., 2011. Advanced dynamic finite element analysis of a skew steel railway bridge'. Elsevier Sci Ltd Engineering Structures, 33 (1), pp. 181-190.
[25] Kallias AN, Imam B., 2012. Risk assessment of steel bridges under the influence of changing environmental conditions'. Faculty of Construction and Environment & Research Institute for Sustainable Urban Development, Hong Kong Polytechnic University Proceedings of the First International Conference on Performance-based and Life-cycle Structural Engineering (PLSE 2012): 5-7 December 2012, Hong Kong, China, Hong Kong, China: First International Conference on Performanced-based and Life-cycle Structural Engineering
[26] Kallias AN, Imam B., 2013. Effect of climate change on the deterioration of steel bridges. International Association for Bridge and Structural Engineering IABSE Symposium Report, Rotterdam, The Netherlands: IABSE Rotterdam Spring Conference 2013: Assessment, Upgrading and Refurbishment of Infrastructures, pp. 1731-1738.
[27] Kuhn, B., Lukic´ , M., Nussbaumer, A., Gunther, H.-P., Helmerich, R., Herion, S., Kolstein, M.H., Walbridge, S., Androic, B., Dijkstra, O. and Bucak, O., 2008. Assessment of existing steel structures: Recommendations for estimation of remaining fatigue life. Joint report prepared under the JRC-ECCS cooperation agreement for the evolution of Eurocode 3 (programme of CEN/TC 250). Editors: G. Sedlacek, F. Bijlaard, M. Ge´radin, A. Pinto and S. Dimora. Background documents in support to the implementation, harmonization and further development of the Eurocodes First Edition, February 2008 EUR 23252 EN-2008.
[28] Pipinato A., 2010b. Structural analysis and fatigue life assessment of the Paderno Arch Bridge. IABMAS 2010 - Bridge Maintenance, Safety, Management, Resilience and Sustainability - Proceedings of the Sixth International Conference on Bridge Maintenance, Safety and Management , pp. 1003-1007. Philadelphia, USA, 11-15 July 2010. ISBN: 9780415877862.
[29] Pipinato A., 2012a. Life cycle assessment of existing steel bridges considering corrosion and fatigue coupled problems. IABMAS 2012 - Bridge Maintenance, Safety, Management, Resilience and Sustainability - Proceedings of the Sixth International Conference on Bridge Maintenance, Safety and Management , pp. 1003-1007. Stresa, Lake Maggiore, Italy, 8-12 July 2012. ISBN: 9780415621243.
[30] Pipinato A., 2012b. Coupled damage in assessing the lifetime of bridge and viaducts. Life-Cycle and Sustainability of Civil Infrastructure Systems - Proceedings of the 3rd International Symposium on Life-Cycle Civil Engineering, IALCCE 2012 , pp. 2292-2297. Vienna, Austria, October 3-6, 2012. ISBN 9780415621267.
[31] Pipinato A., 2012d. Assessment procedures and strengthening of an existing metal bridge. IABMAS 2012 - Bridge Maintenance, Safety, Management, Resilience and Sustainability - Proceedings of the Sixth International Conference on Bridge Maintenance, Safety and Management , pp. 3651-3655. Stresa, Lake Maggiore, Italy, 8-12 July 2012.
[32] Pipinato A., Pellegrino C. and Modena C., 2010b. Residual service life of existing railway bridges. IABMAS 2010 - Bridge Maintenance, Safety, Management, Resilience and Sustainability - Proceedings of the Sixth International Conference on Bridge Maintenance, Safety and Management , pp. 1003-1007. Philadelphia, USA, 11-15 July 2010.
[33] Pipinato A., 2010a. Step level procedure for remaining fatigue life evaluation of one railway bridge. Baltic Journal of Road and Bridge Engineering, 5(1), pp. 28-37. DOI: 10.3846/bjrbe.2010.04
[34] Pipinato, A., 2011. Assessment of existing bridges: Safety and security issues [Problemi di sicurezza nelle valutazioni strutturali di ponti esistenti]. Ingegneria Ferroviaria, 66(4), 355-371.
[35] Pipinato, A., 2012c. Coupled safety assessment of cable stay bridges. Modern Applied Science, 6(7), pp. 64-78. DOI: 10.5539/mas.v6n7p64
[36] Pipinato, A. 2013a. Retrofit procedures of bridge networks against seismic actions in Italy. Durability of Bridge Structures , pp. 215-228. Proceedings of the 7th New York City Bridge Conference, 26-27 August 2013. Edited by Khaled M . Mahmoud, CRC Press 2013 Print ISBN: 978-1-138-00112-1
[37] Pipinato, A. (2013b). Moving load and fatigue analysis of a long span high speed railway bridge. Advanced Materials Research 629 , pp. 403-408. 10.
[38] Pipinato, A. 2013c. Moving loads bridge response: Structural analysis and eurocode provisions.

Advanced Materials Research 629 , pp. 409-417. 10.
[39] Pipinato, A., & Modena, C., 2010. Structural analysis and fatigue reliability assessment of the Paderno bridge. Practice Periodical on Structural Design and Construction, 15(2), 109-124. DOI: 10.1061/(ASCE)SC.1943-5576.0000037
[40] Pipinato, A., Molinari, M., Pellegrino, C., Bursi, O. S., & Modena, C. 2011c. Fatigue tests on riveted steel elements taken from a railway bridge. Structure and Infrastructure Engineering, 7(12), 907-920. DOI: 10.1080/15732470903099776
[41] Pipinato, A., Pellegrino, C., & Modena, C. (2008). Scheduled maintenance actions and residual life evaluation of railway infrastructures for metallic structure bridges [Interventi di manutenzione programmata e valutazione della vita residua delle infrastrutture ferroviarie da ponte a struttura metallica]. Ingegneria Ferroviaria, 63(2), 125-134.
[42] Pipinato, A., Pellegrino, C., & Modena, C., 2008b. Evaluation of fatigue strength of one riveted historical railway bridge. Bridge Maintenance, Safety Management, Health Monitoring and Informatics IABMAS '08 Proceedings of the Fourth International IABMAS Conference, Seoul, Korea, July 13-17 2008. Edited by Hyun-Moo Koh and Dan M . Frangopol, Taylor & Francis 2008, Print ISBN: 978-0-415-46844-2.
[43] Pipinato, A., Pellegrino, C., & Modena, C., 2010a. Structural analysis of historical metal bridges in Italy. Advanced Materials Research, 133-134, 525-530. DOI: 10.4028/www.scientific.net/AMR.133-134.525
[44] Pipinato, A., Pellegrino, C., & Modena, C., 2011a. Fatigue assessment of highway steel bridges in presence of seismic loading. Engineering Structures, 33(1), 202-209. DOI: 10.1016/j.engstruct.2010.10.008
[45] Pipinato, A., Pellegrino, C., & Modena, C., 2012a. Assessment procedure and rehabilitation criteria for the riveted railway Adige Bridge. Structure and Infrastructure Engineering, 8(8), 747-764. DOI: 10.1080/15732479.2010.481674
[46] Pipinato, A., Pellegrino, C., & Modena, C., 2012c. Structural Analysis of the Cantilever Construction Process in Cable-Stayed Bridges. Periodica Polytechnica: Civil Engineering 56 (2) , pp. 141-166. DOI: 10.3311/pp.ci.2012-2.02
[47] Pipinato, A., Pellegrino, C., & Modena, C., 2012d. Fatigue behaviour of steel bridge joints strenghtened with FRP laminates. Modern Applied Science 6 (10) , pp. 1-15. ISSN: 1913-1844. DOI: 10.5539/mas.v6n10p1
[48] Pipinato, A., Pellegrino, C., & Modena, C., 2013. Residual life of historic riveted steel bridges: an analytical approach. ICE-Bridge Engineering, UK Institution of Bridge Engineers. DOI: 10.1680/bren.11.00014
[49] Pipinato, A., Pellegrino, C., and Modena, C., 2012e. Fatigue Damage Estimation in Existing Railway Steel Bridges by Detailed Loading History Analysis. ISRN Civil Engineering, 2012, 1-13. Volume 2012 (2012), Article ID 231674, 13 pages.
[50] Pipinato, A., Pellegrino, C., Bursi, O. S., & Modena, C., 2009. High-cycle fatigue behavior of riveted connections for railway metal bridges. Journal of Constructional Steel Research, 65(12), pp. 2167-2175. DOI: 10.1016/j.jcsr.2009.06.019
[51] Pipinato, A., Pellegrino, C., Fregno, G., & Modena, C., 2012b. Influence of fatigue on cable arrangement in cable-stayed bridges. International Journal of Steel Structures, 12(1), 107-123. DOI: 10.1007/s13296-012-1010-5
[52] Pritchard B., 1992. Bridge Design for Economy and Durability – Concepts for new, strengthened and replacement bridges. London: Thomas Telford Service Ltd.
[53] Righiniotis TD, Imam B., 2004. Fracture reliability of a typical Northridge steel moment resisting connection. Engineering Structures, 26 (3), pp. 381-390.
[54] Righiniotis TD, Imam B., 2006. Effect of fracture on the reliability of a moment resisting frame under earthquake loading. Millpress Proceedings of 9th International Conference on Structural Safety and Reliability, Rome, Italy: ICOSSAR '05
[55] Righiniotis TD, Imam BM, Chryssanthopoulos MK., 2008. Fatigue analysis of riveted railway bridge connections using the theory of critical distances'. Elsevier Engineering Structures, 30 (10), pp. 2707-2715.
URS, 2012. Bascule bridge lightweight solid deck retrofit research project. Deck Alternative Screening Report Final Fpid 419497-1-B2-01, Florida

IOP Conf. Series: Materials Science and Engineering **419** (2018) 012006 doi:10.1088/1757-899X/419/1/012006

Assessment of existing steel bridges: codes and standard

Erica Siviero[1], Roberto Pavan[1]
[1]AP&P srl, Minadois Str. 20, Rovigo 45100, Italy

E-mail: info@pipinatoandpartners.it

Abstract. The assessment of existing structures adopting a specific code or standard is a technical procedure that is provided only in some nation. The possibility of harmonizing these type of procedure is needed in Europe, together with a completely new plan of infrastructural renewal. In this study new insight on this theme are provided, together with some code proposal for the new Eurocode era. Many existing bridges all over the world were not designed for the high service loads and the increased number of load cycles that they are exposed to today. Some bridges will have to be either strengthened or replaced in the next decades. Advancing their age, bridges could be subjected to a large variety of not designed actions or could also be not accurately maintained, leading to a particular state in which they need particular care and attention, in order to remain in service. Moreover, existing structures were not exempted from serious failures and engineers had regularly to suffer setbacks. These setbacks contributed in many cases, however, to research and advancements in better understanding structural behaviour and developing new theories (Kuhn et al. 2010). In these cases, the assessment procedures are needed.

1. Introduction

Many existing bridges all over the world were not designed for the high service loads and the increased number of load cycles that they are exposed to to-day. Some bridges will have to be either strengthened or replaced in the next decades. Advancing their age, bridges could be subjected to a large variety of not designed actions or could also be not accurately maintained, leading to a particular state in which they need particular care and attention, in or-der to remain in service. Moreover, existing structures were not exempted from serious failures and engineers had regularly to suffer setbacks. These setbacks contributed in many cases, however, to research and advancements in better understanding structural behavior and developing new theories (Kuhn et al. 2010). In these cases, the assessment procedures are needed. The assessment of an existing structure aims at producing evidence that it will function safely over a specified residual service life. It is mainly based on the results of assessing hazards and load effects to be anticipated in the future, and of assessing material properties and geometry taking into account the pre-sent state of the structure (JCSS 2001). Guidelines for existing structures exist in a large number of countries. Thereby many countries have presented documents for particular categories of structures. In Canada, Germany, the Netherlands, Switzerland, UK and USA such guidelines have been prepared at a detailed level. In any assessment, the problem of fixing risk acceptance criteria is difficult since it must be compatible to codes for new structures (limit state analysis, safety factor format, etc.), and with national determined parameters (generally partial safety factor values) (Kuhn et al. 2010). As far as specific code and standards are not available, technical guide or pre-standard documents should be used. Moreover, procedures coming from evidenced based or field related experience, are quite extensively used also by bridge and infrastructures owner. Basing on past studies and research, a simplified approach has been developed by the author and applied in recent investigations of existing bridges. This approach is reported in the following paragraphs.

Figure 1: A multi span historical truss bridge.

2. The assessment procedure of existing bridges

Ist level: The preliminary evaluation is the first level of investigation aiming at removing existing doubts about the safety of existing structures, adopting fairly simple methods and identifying critical parts or members in the structure. In order to identify critical members, it is necessary to carry out an intensive study of the available original design documents, along with a visual inspection of the structure and a photographic survey. The inspection procedure is often coded by infrastructural agencies manuals and procedures; however, at least the following points must be checked:

- The bridge construction is conforming to the original drawings and/or differences between as-built and drawings.
- Bridge modification during service (rehabilitation, strengthening, changes in the static system, etc.).
- Presence of any visual evidence of degradation (damaged expansion joints, supports, corrosion, cracks, vibration or loose rivets, collision, lack of structural members etc.).

Moreover, if available, inspection and maintenance reports can be used, and reference should be made to the evaluation report. The preliminary evaluation should include codes and recommendations analysis procedure where available, and conservative assumptions where information is lacking or doubtful. In this way, critical construction details can be identified.

IInd level: The aim of the detailed investigation is to update the information obtained in other analysis by carrying out a refined assessment, especially for those members for which adequate safety was not confirmed by preliminary evaluation. At this stage a specialized consultant should assist. In this phase a FEM numeric model of the entire structure is developed. Based on the current code provisions, the structure should be recalculated, and verification tables should report whether the structural members are safe or not. Concerning specific issues, such as the fatigue and seismic behaviour of the bridge, detailed code provisions should be referred to. From this step-level investigation, non-destructive testing (NDT) could be used in order to characterize the basic material properties of the structure. The final report of the investigation should establish whether the structure is verified against specific issues and has sufficient static strength against actual loadings.

IIIrd level: In case of key structures that have major consequences in terms of risks or costs related to a decision, a team of experts is needed in order to carefully check the conclusions and proposals reached in the last phase. Discussions and further assessments using specific tools can also be carried out to help reach decisions. At this level, on-site testing could be adopted in order to provide the dynamic identification of the structure, as reported in the following example.
IVth level: This advanced level of investigation should be reserved for recurrent bridges along infrastructural nets, in which a rational procedure of analysis and intervention could help in determining if retrofitting interventions could be adopted, or if rational dismantling large scale operations are required. The procedure is based on a detailed survey of the existing bridge, a FEM analysis, a code verification procedure, NDT diffused sampling, and, based on these data on real scale testing of one case study structure, aims to determine the global static and cyclic behaviour of the bridge. In specific cases, on-site dynamic identification could be performed. Concerning the fatigue assessment, in this case a LEFM investigation is required. Concerning seismic analysis, non-linear analysis is required. Specific material testing analysis should be performed dealing with the case analysed. The advanced testing result should report on the various analysis performed and should clearly state verification results indicating the specific retrofit needed for recurrent interventions. An advanced testing operation is reported in the following section, dealing with a recurrent existing bridge type in service along the railway lines.

3. Assessment case studies

As reported in the aforementioned step level classification, various stages of analysis could be adopted by bridge owners or managing agencies. Some examples have been developed by the author and could be found in detail in Pipinato (2016) the assessment of a new bridge is presented in order to evaluate different typologies of construction alternatives to gain different lifetime; while in Pipinato (2014) the assessment and rehabilitation of steel railway bridges using fiber-reinforced polymer (FRP) composites in the particular situation of the rehabilitation of metallic civil infrastructure using fiber reinforced polymer (FRP) composites is evaluated. The residual life of historic riveted steel bridges could be also assessed with an analytical approach or introducing an orthotropic steel deck design to extend the lifetime of plate and box girder bridge and viaducts, as explained in Pipinato et al. 2014 and in Pipinato (2014a). And however steel bridge exhibited in the past a wide amount of problems not evidenced in other bridges (Pipinato 2014b), for this particular type of structure a variety of technologies to repair and retrofit are available: this is the case reported for the fracture propagation and life cycle design of girders (Pipinato 2014c), or for advanced solutions of the corrosion protection (Pipinato 2014d). For network of bridges, however, an integrated solution has been presented in order to understand what is the possibility of prioritizing the bridge retrofit (Pipinato 2013a). A particular asset of bridges in which the lifecycle assessment should be performed from the design stage is the case of long span bridges, where the traffic is represented also from exceptional moving loads and the consequent fatigue expected lifetime could fail in less than ten years from the construction stage if a correct detailing is not perceived (Pipinato 2013b-2013c-2010d); similar issues comes also from the original construction alternative choses, as demonstrated in Pipinato et al. 2012a. The particular case of fatigue sensitive materials of course is more recurrent in steel bridges where retrofit and assessment alternatives are in part applied (Pipinato 2012b), in part not well investigated, as is the case of coupled damages (seismic and fatigue, Pipinato 2012a,b, Pipinato et al. 2012c,d,e,f, Pipinato et al. 2011c,d). Wherever there is the possibility to investigate with full scale tests existing bridges, e.g. where managing agencies are interested to know the structural behaviour of recurrent structures along their network, this could be an interesting occasion to understand how materials decay could influence the static and fatigue behaviour of bridges: this is the case both of railway (Pipinato et al. 2011a,b, Pipinato 2011, Pipinato 2010b, Pipinato et al. 2010a,b,c,e Pipinato 2010, Pipinato et al. 2009, 2008a,b, Pipinato et al. 2007) and roadways (Pipinato 2010a). A particular and further subset is represented by monumental and large bridge structures (Pipinato and Modena 2010): in this case, a long and time-consuming but useful procedure should be performed, by evaluating al the assessment steps described before, in order to understand in detail the aforementioned problems that could influence the final retrofit design decision.

4. Estimation of knowledge level and confidence factor

The application of the aforementioned assessment procedures, lead to consequences in the structural analysis of the structure. Generally, as much information and tests have been performed onto the bridge, as much knowledge has been accumulated onto the structure, leading to more accurate data on the structural properties of materials, and on the structural behaviour of the bridge itself. On the basis of this, a knowledge level listing could be drafted out, and consequently the so-called confidence factors could be applied, in a very similar procedure to the Italian existing standard for the assessment of existing buildings (Ministerial Decree, 2008).

KL1-Knowledge level 1:

Geometry: the geometry of the structure is known or from a survey or from original drawings. In the latter case a visual survey is required to confirm the documentation. Collected data should be sufficient for developing a linear analysis.

Structural details: details are not available from drawings and are designed according the practice of the construction age. A limited in situ verification is requested. Local resistance verification are required.

Materials property: material properties are not available, neither from the original documents, nor from original tests. Usual values of the practice are adopted, validated from limited in-situ tests.

Structural analysis type: The safety assessment in the case of limited knowledge is usually performed by static and dynamic linear analysis methods.

KL2-Knowledge level 2:

Geometry: the geometry of the structure is known or from a survey or from original construction drawings. In the latter case a visual survey is required to confirm the documentation. Collected data should be sufficient for developing a linear or a non-linear analysis.

Structural details: details are available from original drawings or from not original construction documents. In this last situation, a limited verification of common details is performed. Local resistance verification are performed.

Materials property: material properties are available, from the original documents, or from original tests. Extensive in situ tests are performed to validate the original data whereas they are available from design and not from tests.

Structural analysis type: The safety assessment in the case of limited knowledge is usually performed by static and dynamic linear or non-linear analysis methods.

KL3-Knowledge level 3:

Geometry: the geometry of the structure is known or from a survey or from original drawings. In the latter case a visual survey is required to confirm the documentation. Collected data should be sufficient for developing a linear or a non-linear analysis.

Structural details: details are available from original drawings or from original construction documents. Available data enable the implementation of a complete linear FEM model, performing local resistance verification, or the implementation of a non-linear model.

Materials property: material properties are available, from the original documents, or from original tests, or from an extensive in situ testing program. In the first case, limited in situ tests are developed to confirm the historical values; if the values obtained are minor than the original, an extensive in situ program is settled and performed. Available data enable the implementation of a complete linear FEM model, performing local resistance verification, or the implementation of a non-linear model.

Structural analysis type: The safety assessment in the case of limited knowledge is usually performed by static and dynamic linear or non-linear analysis methods.

The confidence factor range should be calibrated according to the engineering knowledge level of the structure. Suggested values are in the following range:

- Confidence factor α: 1.30-1.35
- Confidence factor β: 1.20-1.25
- Confidence factor γ: 1.00-1.10

Although this procedure is suggested, it remains on the engineering choice of the structural engineer to establish the use of the confidence factor of the corresponding knowledge level, or in alternative, to use a lower confidence factor, when doubts arise from calculation. The confidence factor (CF) is considered a partial safety factor to be applied to the design strength in order to take into account for the lack of knowledge when analysing existing structures:

$Fd = fm / (CF*\gamma m)$.

5. Bridge rating worldwide

The AASHTO LRFR Guide Manual is the first bridge load rating method in the United States to be based on modern principles of structural reliability and limit states design. The essential ingredients of a reliability-based design and evaluation include probabilistic models of the structural resistance and loads and a method for analysing the reliabilities (or, conversely, the limit state probabilities) that are relevant to each bridge limit state. Such methods and tools have been applied to developing the AASHTO LRFD Specifications and the LRFR Guide Manual and are expected to be relevant also to improve rating methods for other countries. Details are available in the archival literature (Nowak, 1999; Moses, 2001).

Section 7 of the Austrian Bridge Design Standard provides rating guidelines with a commentary. The concept of rating is based on the limit state design philosophy and both serviceability and ultimate limit states are considered. The ultimate action is defined as an action that has a 5% probability of being exceeded during the design life, which represents an average return interval of 2000 year; while the survivability action is defined as one having 5% probability being exceeded per year, corresponding to a return interval of 20 years.

Document BD 21/01, Assessment of Highway Bridges and Structures, adopts a limit state format with appropriate partial safety factors for condition evaluation of most highway bridges except for cast iron bridges and masonry arch bridges. It is stipulated that bridges built after 1965 should normally be evaluated for serviceability as well as for the ultimate limit states; bridges constructed before 1965 do not need to be assessed for service limit states. Requirements for fatigue endurance however are not included in the standard and the reason stated is that the past stress history of each structure, which could profoundly influence fatigue limit checking, cannot generally be determined to the accuracy level required for assessment.

The GBSM (2012) is a reference document dedicated in particular for railway bridges. However, it shows a clear and easy procedure for the bridge analysis.

6. Conclusion

The assessment of existing structures adopting a specific code or standard is a technical procedure that is provided only in some nation. The possibility of harmonizing these type of procedure is needed in Europe, together with a completely new plan of infrastructural renewal. From the analysis of the documents presented, a review of existing assessment approaches and states rating practices have revealed a number of research issues that must be addressed for existing bridges evaluations (GDT 2009):

- The bridge may contain archaic structural materials. Design documentation may be missing.
- Material strengths in situ may be vastly different from the standardized or nominal values assumed in design. On the one hand, concrete strength can increase by as much as 150% beyond the 28-day standard basis due to continued hydration; on the other hand, the strength can deteriorate due to

aggressive environmental attack from physical or chemical mechanisms. Failure to consider best estimates of strength and the time-dependent nature of the structural strength and stiffness invariably will lead to an erroneous estimate of in situ strength.

- Analytical approaches to bridge evaluation usually (but not always) yield a conservative measure of actual load-carrying capacities [Bakht and Jaeger, 1990]. This conservatism is the result of assumptions made in the analysis regarding load sharing, composite action, support conditions and nonlinear behaviour, in addition to differences in material strengths noted above.
- Discrepancies among the different approved rating methods were noted. These discrepancies were confirmed by preliminary rating calculations performed on bridges. The reasons for these differences must be completely understood and addressed in developing new guidelines or codes.
- Satisfactory bridge performance over a period of years of service provides additional information not available at the design stage. This information should be taken into account in designing in-service inspection programs and in making decisions regarding upgrading and rehabilitation.
- Current condition assessment relies heavily on visual inspection. More quantitative models of structural deterioration have been developed but have yet to be incorporated in condition assessment procedures.
- A test load must be a significant fraction of the expected maximum live load for the proof test to be informative and to lower subsequent risk. If the test load is increased to an informative level, the probability of damaging the bridge during the load test increases as well. This trade-off between information gained and likelihood of damage must be part of the decision to load-test a bridge rather than relying on other rating methods.
- Uncertainties in loads and resistances at the design stage are reflected in the safety factors (or load and resistance factors). At the evaluation stage, uncertainties can be either greater (e.g., due to deterioration) or less (measured properties; successful load test). These uncertainties must be identified and analysed.

Acknowledgments

This research was supported by the European Union, represented by the European Commission Directorate General for Research and Innovation, Research Fund for Coal & Steel (RFCS) program, under grant agreement no. RFSR-CT-2015-00025. This funding is gratefully acknowledged.

References

[1] GDT (2009). Condition Assessment Of Existing Bridge Structures. GDOT Project No. RP05 – 01. Report Of Task 1 – Appraisal Of State-Of-The-Art Of Bridge Condition Assessment. By Naiyu Wang, Bruce R. Ellingwood, Abdul-Hamid Zureick, Curtis O'Malley, Georgia Institute Of Technology School Of Civil And Environmental Engineering.
[2] GBSM (2012). Guideline for Bridge Safety Management. Rail Safety, Canada Transport.
[3] Ministerial Decree (2008). Norme Tecniche per le Costruzioni DM 14/01/2008, e Circolare esplicativa 617/2009, Ministero delle Infrastrutture, Roma
[4] Nowak, A. S. (1999). "Calibration of LRFD Bridge Design Code." NCHRP Report 368, Transportation Research Board, National Research Council, Washington D.C.
[5] Moses, F., (2001). "Calibration of Load Factors for LRFR Bridge Evaluation." NCHRP Report 405, Transportation Research Board, National Research Council, Washington D.C.
[6] Kühn B., M. Lukić, A. Nussbaumer, H.-P. Günther, R. Helmerich, S. Herion, M.H. Kolstein, S. Walbridge, B. Androic, O. Dijkstra, Ö. Bucak (2010). Assessment of Existing Steel Structures: Recommendations for Estimation of Remaining Fatigue Life. Joint Report Prepared under the JRC – ECCS cooperation agreement for the evolution of Eurocode 3 (programme of CEN / TC 250).

[7] JCSS (2001). Probabilistic Assessment of Existing Structures, Joint Committee on Structural Safety, 2001.
[8] Pritchard B. (1992). Bridge Design for Economy and Durability – Concepts for new, strengthened and replacement bridges. London: Thomas Telford Service Ltd.
[9] Pipinato, A. (2016). Structural analysis and design of a multispan network arch bridge. Proceedings of the Institution of Civil Engineers: Bridge Engineering, 169 (1), pp. 54-66. DOI: 10.1680/jbren.14.00013
[10] Pipinato, A. (2014). Assessment and rehabilitation of steel railway bridges using fibre-reinforced polymer (FRP) composites. Rehabilitation of Metallic Civil Infrastructure Using Fiber Reinforced Polymer (FRP) Composites: Types Properties and Testing Methods, pp. 373-405. DOI: 10.1533/9780857096654.4.373
[11] Pipinato, A., Pellegrino, C., Modena, C. (2014). Residual life of historic riveted steel bridges: An analytical approach. Proceedings of the Institution of Civil Engineers: Bridge Engineering, 167 (1), pp. 17-32. Cited 9 times. DOI: 10.1680/bren.11.00014
[12] Pipinato, A. (2014a). Orthotropic steel deck design to extend the lifetime of plate and box girder bridge and viaducts. Bridge Maintenance, Safety, Management and Life Extension - Proceedings of the 7th International Conference of Bridge Maintenance, Safety and Management, IABMAS 2014, pp. 1329-1336.
[13] Pipinato, A. (2014b). Steel bridges: Codes, design specifications and construction issues. Bridge Maintenance, Safety, Management and Life Extension - Proceedings of the 7th International Conference of Bridge Maintenance, Safety and Management, IABMAS 2014, pp. 1891-1898.
[14] Pipinato, A. (2014c). Longitudinal web stiffening in steel girders: Fracture propagation and life cycle design
[15] (2014) Bridge Maintenance, Safety, Management and Life Extension - Proceedings of the 7th International Conference of Bridge Maintenance, Safety and Management, IABMAS 2014, pp. 1886-1890.
[16] Pipinato, A. (2014d). Steel bridge coatings and corrosion protection. Bridge Maintenance, Safety, anagement and Life Extension - Proceedings of the 7th International Conference of Bridge Maintenance, Safety and Management, IABMAS 2014, pp. 1899-1905.
[17] Pipinato, A. (2013a). Retrofit procedures of bridge networks against seismic actions in Italy. Durability of Bridge Structures, pp. 215-228.
[18] Pipinato, A. (2013b). Moving load and fatigue analysis of a long span high speed railway bridge.Advanced Materials Research, 629, pp. 403-408.
[19] Pipinato, A. (2013c). Moving loads bridge response: Structural analysis and eurocode provisions (2013) Advanced Materials Research, 629, pp. 409-417. 10.4028/www.scientific.net/AMR.629.409
[20] Pipinato, A., Pellegrino, C., Modena, C. (2012a). Structural Analysis of the Cantilever Construction Process in Cable-Stayed Bridges (2012) Periodica Polytechnica: Civil Engineering, 56 (2), pp. 141-166. DOI: 10.3311/pp.ci.2012-2.02
[21] Pipinato, A., Pellegrino, C., Modena, C. (2012b). Fatigue behaviour of steel bridge joints strenghtened with FRP laminates. Modern Applied Science, 6 (10), pp. 1-15. DOI: 10.5539/mas.v6n10p1
[22] Pipinato, A., Pellegrino, C., Modena, C. (2012c). Coupled damage in assessing the lifetime of bridge and viaducts. Life-Cycle and Sustainability of Civil Infrastructure Systems - Proceedings of the 3rd International Symposium on Life-Cycle Civil Engineering, IALCCE 2012, pp. 2292-2297.
[23] Pipinato, A. (2012a). Coupled safety assessment of cable stay bridges. Modern Applied Science, 6 (7). DOI: 10.5539/mas.v6n7p64
[24] Pipinato, A., Pellegrino, C., Modena, C. (2012d). Assessment procedure and rehabilitation criteria for the riveted railway Adige Bridge (2012) Structure and Infrastructure Engineering, 8 (8), pp. 747-764. Cited 21 times. DOI: 10.1080/15732479.2010.481674
[25] Pipinato, A. (2012b). Assessment procedures and strengthening of an existing metal bridge. (2012) Bridge Maintenance, Safety, Management, Resilience and Sustainability - Proceedings of the Sixth International Conference on Bridge Maintenance, Safety and Management, pp. 3651-3655.
[26] Pipinato, A., Pellegrino, C., Modena, C. (2012e). Life cycle assessment of existing steel bridges considering corrosion and fatigue coupled problems. Bridge Maintenance, Safety, Management, Resilience and Sustainability - Proceedings of the Sixth International Conference on Bridge Maintenance, Safety and Management, pp. 1003-1007.
[27] Pipinato, A., Pellegrino, C., Fregno, G., Modena, C. (2012f). Influence of fatigue on cable arrangement in cable-stayed bridges. International Journal of Steel Structures, 12 (1), pp. 107-123. DOI:

10.1007/s13296-012-1010-5
[28] Pipinato, A., Molinari, M., Pellegrino, C., Bursi, O.S., Modena, C. (2011a). Fatigue tests on riveted steel elements taken from a railway bridge (2011) Structure and Infrastructure Engineering, 7 (12), pp. 907-920. DOI: 10.1080/15732470903099776
[29] Pellegrino, C., Pipinato, A., Modena, C. (2011b). A simplified management procedure for bridge network maintenance. Structure and Infrastructure Engineering, 7 (5), pp. 341-351. DOI: 10.1080/15732470802659084
[30] Pipinato, A. (2011). Assessment of existing bridges: Safety and security issues [Problemi di sicurezza nelle valutazioni strutturali di ponti esistenti] (2011) Ingegneria Ferroviaria, 66 (4), pp. 355-371.
[31] Pipinato, A., Pellegrino, C., Modena, C. (2011c). Fatigue assessment of highway steel bridges in presence of seismic loading (2011) Engineering Structures, 33 (1), pp. 202-209. DOI: 10.1016/j.engstruct.2010.10.008
[32] Pipinato, A., Pellegrino, C., Modena, C. (2011d). Correlation between fatigue effects and seismic actions in highway steel bridges (2011) Proceedings of the 8th International Conference on Structural Dynamics, EURODYN 2011, pp. 1118-1123.
[33] Pipinato, A. (2010a). Structural analysis and fatigue life assessment of the Paderno Arch Bridge. Bridge Maintenance, Safety, Management and Life-Cycle Optimization - Proceedings of the 5th International Conference on Bridge Maintenance, Safety and Management, pp. 3480-3484.
[34] Pipinato, A. (2010b) Structural analysis and fatigue life assessment of the Paderno Arch Bridge Bridge Maintenance, Safety, Management and Life-Cycle Optimization - Proceedings of the 5th International Conference on Bridge Maintenance, Safety and Management, pp. 3486-3490.
[35] Pipinato, A., Pellegrino, C., Modena, C. (2010a) Residual service life of existing railway bridges Bridge Maintenance, Safety, Management and Life-Cycle Optimization - Proceedings of the 5th International Conference on Bridge Maintenance, Safety and Management, pp. 1444-1452.
[36] Pipinato, A., Pellegrino, C., Modena, C. (2010b) Structural analysis of historical metal bridges in Italy Advanced Materials Research, 133-134, pp. 525-530. DOI: 10.4028/www.scientific.net/AMR.133-134.525
[37] Pipinato, A., Pellegrino, C., Modena, C. (2010c). Residual service life of existing railway bridges Bridge Maintenance, Safety, Management and Life-Cycle Optimization - Proceedings of the 5th International Conference on Bridge Maintenance, Safety and Management, pp. 1455-1462.
[38] Pipinato, A., Modena, C. (2010). Structural analysis and fatigue reliability assessment of the paderno bridge (2010) Practice Periodical on Structural Design and Construction, 15 (2), pp. 109-124. DOI: 10.1061/(ASCE)SC.1943-5576.0000037
[39] Pipinato, A. (2010). Step level procedure for remaining fatigue life evaluation of one railway bridge. Baltic Journal of Road and Bridge Engineering, 5 (1), pp. 28-37. DOI: 10.3846/bjrbe.2010.04
[40] Pellegrino, C., Pipinato, A., Freno, G., Modena, C. (2010d). Arranging geometric configuration of cable-stayed bridges taking fatigue into account (2010) Large Structures and Infrastructures for Environmentally Constrained and Urbanised Areas, pp. 692-693.
[41] Pipinato, A., Gueli, M., Pellegrino, C., Modena, C. (2010e). Innovative methods for railway steel bridge remaining life estimation (2010) Large Structures and Infrastructures for Environmentally Constrained and Urbanised Areas, pp. 384-385.
[42] Pipinato, A., Pellegrino, C., Bursi, O.S., Modena, C. (2009). High-cycle fatigue behavior of riveted connections for railway metal bridges. Journal of Constructional Steel Research, 65 (12), pp. 2167-2175. DOI: 10.1016/j.jcsr.2009.06.019
[43] Pipinato, A., Pellegrino, C., Modena, C. (2008a). Evaluation of fatigue strength of one riveted historical railway bridge. Bridge Maintenance, Safety, Management, Health Monitoring and Informatics - Proceedings of the 4th International Conference on Bridge Maintenance, Safety and Management, pp. 283-290. DOI: 10.1201/9781439828434.ch148
[44] Pipinato, A., Pellegrino, C., Modena, C. (2008b). Scheduled maintenance actions and residual life evaluation of railway infrastructures for metallic structure bridges [Interventi di manutenzione programmata e valutazione della vita residua delle infrastrutture ferroviarie da ponte a struttura metallica]. Ingegneria Ferroviaria, 63 (2), pp. 125-134.
[45] Pipinato, A., Pellegrino, C., Modena, C. (2007). Static load tests on a historical steel railway bridge (2007)Proceedings of the 9th International Conference on Steel, Space and Composite Structures, pp. 217-224.

"FEVE based coating systems: Protecting Steel Bridges for over 30 years"

Floris B. Kooistra[1], Ben Runhaar[2], Charlène Danoux[1], Kiyoshi Kasahara[1] and Takasi Takayanagi.[3]

[1]AGCCE, Sciencepark 408, 1098 XH Amsterdam, the Netherlands
[2]AGCCE, Zuidplein 80, 1077 XV Amsterdam, the Netherlands
[3]AGC, 1-12-1, Yurakucho 1-chome Chiyoda-ku, 100-8405 Tokyo, Japan

E-mail: floris.kooistra@agcce.com

Abstract. Fluoro Ethylene Vinyl Ether (**FEVE**) resins were commercialized by AGC in 1982 under the brandname *Lumiflon* as the first soluble fluorinated resins for coating formulations. FEVE resins are renowned for their extremely high durability due to the high C-F bond energy. Therefore, they do not degrade under the influence of UV radiation from the sun. Coatings based on FEVE resins show outdoor durability performance of over 30 years. This paper discusses the basic chemistry behind the extreme durability of *Lumiflon* based coating systems. Two case studies of actual steel bridge applications will be shown. The Daiichi- Mukaiyama bridge, which was newly constructed and coated in 1987, and the Tokiwa bridge, which was recoated with a *Lumiflon* based system in 1986. Through these examples, this paper demonstrates that FEVE based coated systems provide the best possible protection of steel bridges.

1. Introduction

Fluoroethylene vinyl ether (FEVE) resins were developed in Japan in the late 1970's and were first commercialized in 1982. FEVE resins are amorphous A-B type copolymers with repeating units of fluoroethylene and substituted vinyl ether (see Figure 1). Unlike pure fluoropolymers, FEVE resins are soluble in solvents due to the vinyl ether groups. Solvent solubility transforms FEVE resins from high performance polymers into high performance resins for coatings. [1]

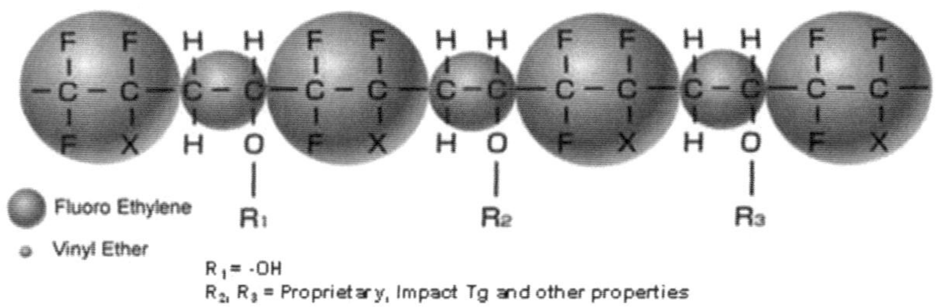

Figure 1. Alternating Structure of FEVE Resins

FEVE resins have characteristics of both fluoropolymers and hydrocarbons. The fluoroethylene groups are the strength of the FEVE resin. These groups are responsible for the polymer's high resistance to UV degradation. The C-F bond is very strong (see Table 1). The energy of this bond is higher than the energy of UV radiation at 290nm which is ~411KJ/mol. The alternating pattern, shown in figure 1, is critical for the extreme UV resistance properties. The chemically stable and UV resistant fluoroethylene units sterically and chemical protect the neighbouring vinyl ether units. [1]

Table 1. Bond energy of Fluoro-Chemicals and Commodity Chemicals [2]

Resin	C-C Chain	*KJ/mol*	C-F. C-H	*KJ/mol*
Fluoro Compound	CF3-CF3	414	**F**-CF2-CH3	523
Fluoro Compound	CF3-CH3	424	CF3CH2-**H**	447
Commodity Chemical	CH3-CH3	379	CH3CH2-**H**	411

The vinyl ether groups make FEVE polymers useable as resins for paint. Without the vinyl ether groups, FEVE resins would not be soluble in solvent. This solubility is what allows FEVE resins to be used in a wide array of coating formulations that can be applied in factory or on-site settings. A wide range of curing temperatures can be employed ranging from Room Temperature to 230 °C. The vinyl ether groups also contribute to high gloss and allow for functional groups, like hydroxyl groups, to be incorporated into the structure. Below is a table showing typical properties of FEVE resins. [1]

Table 2. Typical Properties of FEVE Resins

Property	Value
Fluorine Content	20-30 wt%
OH Value	47-170 mg KOH/g
COOH Value	0-15 mg KOH/g
Molecular Weight	M_n = 15000-100000
Specific Gravity	1.4 – 1.5
Morphology	Glassy (T_g = 20-50^{0}C
Solubility Parameter (cal'd)	8.8

2. Durabiltiy

As explained above FEVE resins are renowned for their extremely high durability due to the high C-F bond energy. Therefore, they do not degrade under the influence of UV radiation from the sun. Fluorourethane topcoats have been tested in both accelerated and natural weathering. The following figures show the weathering performance typical of fluorourethane coatings based on FEVE resins. Figure 2 shows a comparison of a FEVE coating to polysiloxane and acrylic urethane based coatings in QUV accelerated weathering. The FEVE coating clearly outperforms the other resins which are renowned on their own in the industry for their good weatherability. Furthermore Figure 3 shows South Florida exposure of a clear and a yellow FEVE coating. This data clearly shows the excellent performance of the FEVE based coating technology achieving a gloss retention of around 70% after 10 years of exposure.

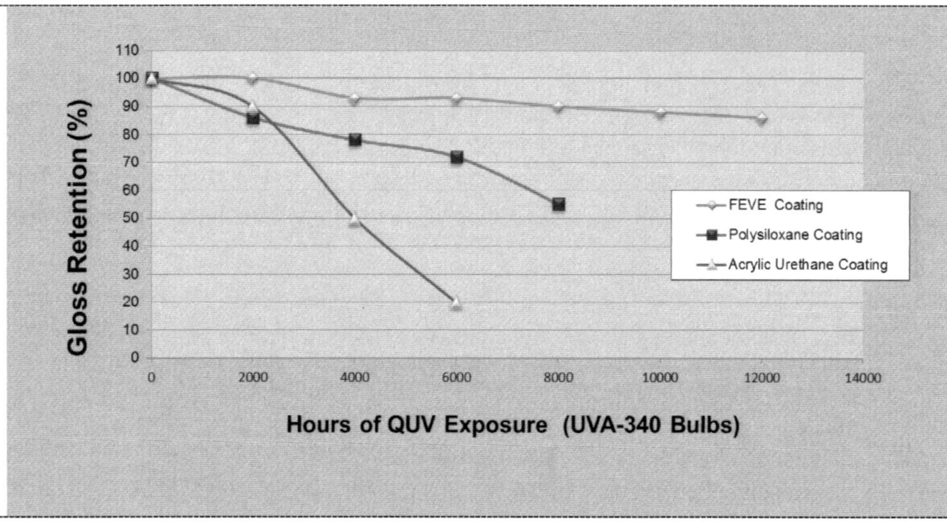

Figure 2. QUV Exposure of an FEVE Based Coating

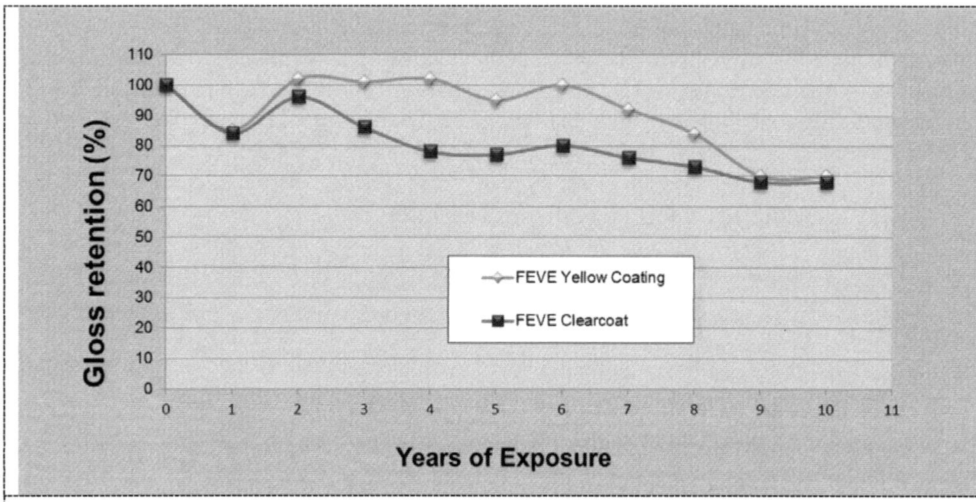

Figure 3. South Florida Exposure of an FEVE Based Coating

Besides excellent gloss and color retention FEVE topcoats also offer a great benefit in terms of general protection of underlying substrates. Due to the high UV resistance and lack of degradation of the resin the coating itself will stay intact much longer than for example an acrylic urethane based topcoat. Figures.4 and 5 show film consumption for the fluoro-polymer coating and polyurethane coating in cross section after 15 year of exposure. A portion of the coating was covered with tape and thus was not exposed to sunlight. Under the tape the film was not damaged. The film thickness of each topcoat after 15 years outdoor exposure could be compared to the original film thickness under the tape. After 15 years, the fluoro-polymer topcoat has lost about 1.1μm total (less than 0.1 μm/year), while the polyurethane topcoat has lost 22-28μm (about 2μm/year).

IOP Conf. Series: Materials Science and Engineering **419** (2018) 012007 doi:10.1088/1757-899X/419/1/012007

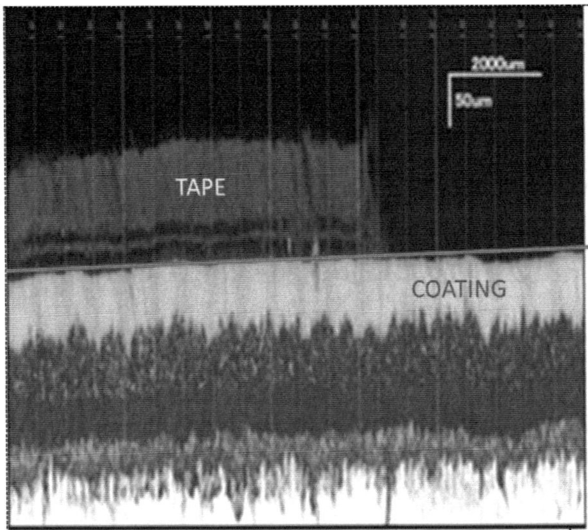

Figure 4. FEVE white coating. 1 micron of erosion after 15 years of exposure

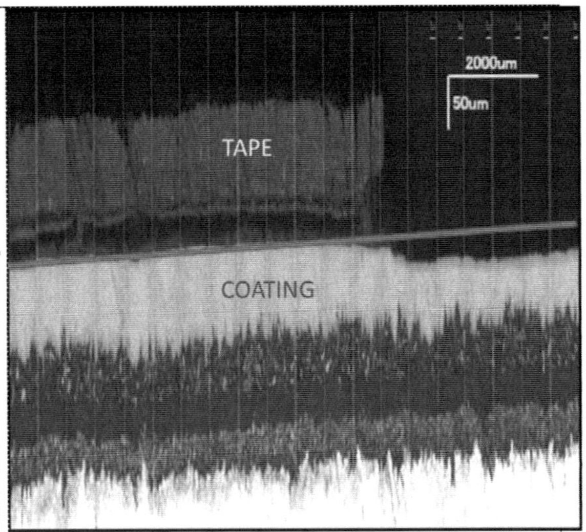

Figure 5. Polyurethane white coating. 22-28 micron of erosion after 15 years of exposure.

In Figure 6, two sections of the coating are analysed by Imaging IR (IRT7000[3]). That measurement can detect amide (II) absorbance and quantify it in comparison to theC-H band or C-F band. A chemical map is generated, which shows the distribution of the amide bond through the cross-section. The color gradient of this map directly relates to the concentration of isocyanates. In the case of the Fluoro-polymer, comparing the cross-section which has been light-sealed (map A) and the cross-section which has been exposed (map B) shows that the decomposition of isocyanates is very limited. Indeed, the same intensity in the gradient can be observed with the red color present through the depth of both cross-sections. In the case of the Polyurethane, the protected cross-section (maps C) and the exposed cross-section (map D) show large differences. In the map of the exposed cross-section, the yellow color indicates that the isocyanate concentration is much lower in comparison to that of the protected cross-section in red. UV light has induced the degradation of the isocyanate in the polyurethane coating, even at a 20μm depth from the surface

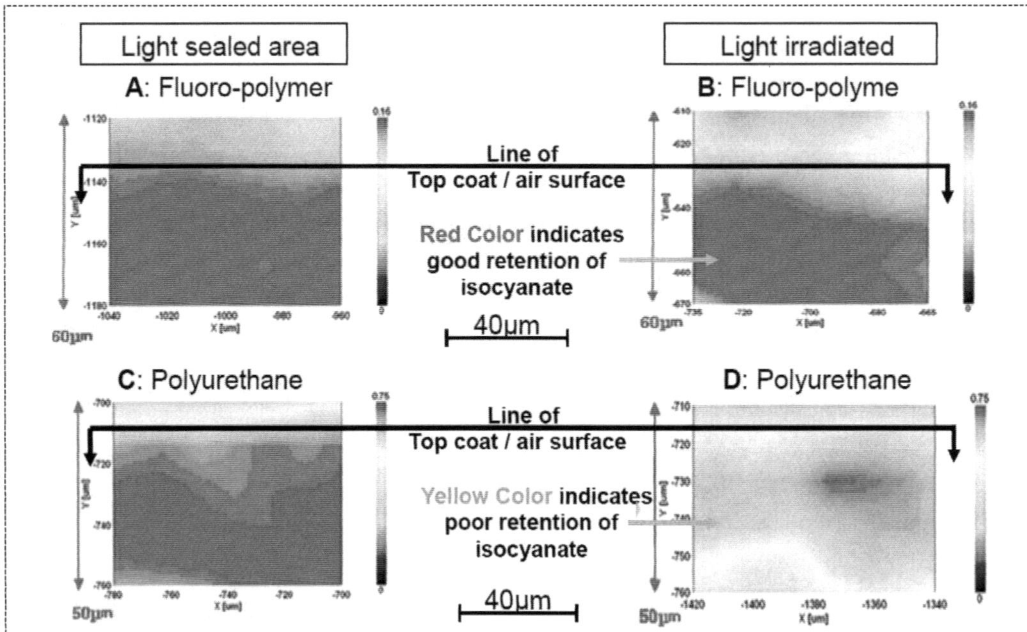

Figure 6. Isocyanates retention of each cross section of coats about mild light seal or irradiated surface area

Due to their high durability, FEVE based coatings are highly suitable for use in the protection of steel bridges. In Europe, only Italy has an optional fluorinated coating specification (ANAS IT.CDTG.05.18). Most other European countries specify coatings towards ISO 12944-5 which focusses on corrosion protection but does not mention gloss or color retention. The recently released ISO 12944-9 includes a category for protective systems offering a durability of 25 years or longer. FEVE resins are mentioned as a possible technology to achieve this. In contrast the Japanese Road Authorities allow bridges only to be coated with topcoats based on fluorinated resins due to their extremely long lifetimes. Maintenance cycles and resulting bridge closures, which generate an enormous economic cost, are therefore minimized. Nowadays over 80% of all bridges in Japan are coated with these coating systems. AGC has tracked the performance of *Lumiflon* based bridge coatings in Japan for over 30 years. These coatings still perform admirably after such long time without ever being re-coated. The expected lifetime of these coating systems is now estimated to be 60 years or more, far surpassing polyurethane and polysiloxane technologies that are commonly used throughout Europe.

3. Case Studies

In this section two case studies will be shown to demonstrate the excellent performance of *Lumiflon* based bridge coatings. The first example is the Daiichi- Mukaiyama bridge, which was newly constructed and coated in 1987. The second example will show the Tokiwa bridge which was repainted with a *Lumiflon* in 1986. Both bridges were never recoated and have been monitored from the start of the application until now. Both show excellent color and gloss retention and overall performance.

3.1. Daiichi-Mukaiyama bridge

The Daiichi-Mukaiyama bridge was newly build in 1987 (see figure 7). The primer and middle coats were shop applied and the topcoat system, based on a *Lumiflon* resin system, was applied on site. Color and gloss measurements were taken initially and after 22 and 30 years. Furthermore the coating was inspected for chalking and other degradation signs. As is demonstrated in Table 3 and Figure 7 the paint system performed admirably after 30 years of service life. A gloss retention of over 70% was measured after wiping the coating free of dust and dirt. Also, no rust peeling or cracking was observed. Color measurements could not be compared due to a change in colorimetric techniques throughout the years. However the red color still appears as bright as when it was applied 30 years ago.

Figure 7. Daiichi-Mukaiyama bridge

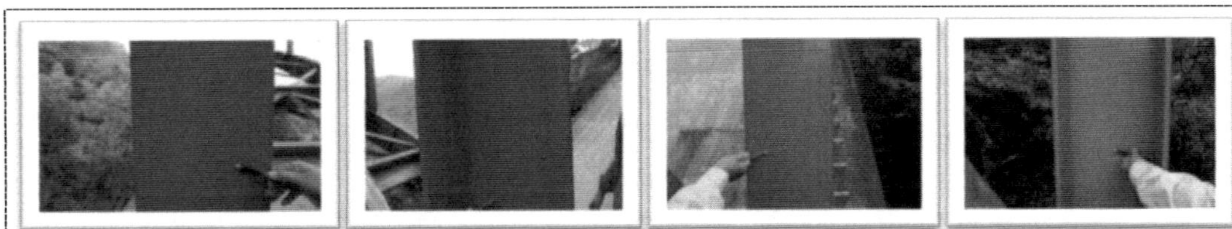

Figure 8. Daiichi-Mukaiyama bridge inspection after 30 years

Table 3. Gloss Retention.

		Initial	*22 years*	*30 years*
Non wiped	Measured value	52.4	46.5	28.3
	Retention	-	88.7%	54%
Wiped	Measured value	52.4	49.9	38.7
	Retention	-	92.5%	7.8%

Table 4. Appearance Observation after 30 years.

	Part Investigated	*Rust*	*Peeling*	*Crack*
Upstream Side	Brace	0	0	0
	Cross Girder	0	0	0
	Arch Rib	0	0	0
Downstream Side	Brace	0	0	0
	Cross Girder	0	0	0
	Arch Rib	0	0	0

3.2. Tokiwa bridge

The Tokiwa bridge was repainted in 1986 (see Figure 9). The original paint system was a chlorinated rubber coating applied in 1977. For the repainting the surface was prepared according to an ISO st3 level (3rd B grade). The performance of the coating was then followed for 30 years. As can be seen in table 5 and fig.10 the gloss and color retention are exemplary. At some locations some rust was found (see Table 6). However it should be noted that at the time of application no zinc rich primers were used. Also this particular bridge is located in the north of Japan and is therefore regularly exposed to road salt due to icy winter conditions. Overall the performance of the coating during its 30 year lifespan has been formidable.

Figure 9. Tokiwa Bridge

2years (1988) 22years (2008) 30years (2016)

Figure 10. Tokiwa bridge inspection

Table 5. Gloss Retention.

		Initial	*22 years*	*30 years*
Non wiped	Measured value	75.2	66.2	52.7
	Retention	-	88%	70.1
Wiped	Measured value	75.2	77.6	73.2
	Retention	-	103.2	97.3

Table 6. Appearance Observation after 30 years.

	Main Girder	*Rust*	*Peeling*	*Crack*
Upstream Side	Upper flange / lower side	0	0	0
	Web Plate	0	0	0
	Lower flange / upper side	1	1	0
	Lower flange / lower side	1	1	0
	Joint Part	0	0	0
Downstream Side	Upper flange / lower side	0	0	0
	Web Plate	0	0	0
	Lower flange / upper side	2	1	0
	Lower flange / lower side	2	1	0
	Joint Part	0	0	0

4. Conclusion

FEVE based resins are extremely durable in outdoor conditions due to the high strength of the C-F bond. These resins find applications in coating systems where long lifetimes, color and gloss retention and corrosion protection are required. Hence they are extremely suitable for the formulation of bridge coatings. Both the accelerated weathering data and the actual outdoor performance of coatings based on *Lumiflon* resins systems show unmatched performance with respect to color and gloss retention but also in regards to protection of the underlying substrate against the elements.

References

[1] Kristen Blankenship, "Formulation Techniques Using FEVE Resins in Waterborne and High Solids Coatings," *Proceedings of the Forty-Second Annual International Symposium of Waterborne, High Solids, and Powder Coatings Symposium*, p. 251 (2015).

[2] E. Bure, "Smart Fluorinated Organic Molecules", *Molecular structure and energetics vol.3*, Chap.4 pp141-191(1986)

[3] H. Tanabe, S. Nakayama, A. Ikeda, T. Shinohara "The durability of fluoro-polymer top coating system" *The Electrochemical Society Symposium volume 89-13*, (1989)

IOP Conf. Series: Materials Science and Engineering **419** (2018) 012008 doi:10.1088/1757-899X/419/1/012008

Retrofit against fatigue cracking at web penetration details with a slot in steel girder

Chihiro Sakamoto[1], Masahiro Sakano[2], Hideyuki Konishi[3] and Takashi Fujii[4]

[1]Japan civil engineering consultants association, Osaka, Japan
[2]Kansai University, Suita, Osaka, Japan
[3]Japan Bridge Association, Osaka, Japan
[4]Ministry of Land, Infrastructure and Transport, Himeji, Japan
E-mail: sakamoto_ssd@yahoo.co.jp

Abstract. Fatigue cracking in steel girder web penetration details is so dangerous that it can break steel girders. The purpose of this study is to clarify fatigue behavior of web penetration details with cross beam flange welded to upper or lower surface of a slot by fillet welding, and verify the effect of retrofitting methods against fatigue cracking in web penetration details. As a result, fatigue cracks were initiated at the web-side toe of the fillet weld along cross beam lower flange edges both in upper and lower welded specimen. In the upper welded specimen, fatigue cracks were propagated into the web plate diagonally upward almost perpendicularly to the maximum principal stress direction. Whereas in the lower welded specimen, fatigue cracks were propagated into the web plate vertically downward faster than in the case of the upper welded specimen, and broke the lower flange. The fatigue life of the lower welded specimen was shorter than that of the upper welded specimen. As a preventive measure, weld toe grinding can extend fatigue life a little, but cannot prevent crack initiation perfectly. As a post cracking measure, a stop hole can stop crack propagation of long cracks, whereas crack removal and shaping up method by grinding can prevent crack propagation of short cracks.

1. Introduction

Fatigue cracking in steel girder web penetration details is so dangerous that it can break steel girders [1]. One-meter-long crack was detected in Yamazoe Bridge in 2006 [2]. Since a number of highway bridges with such web penetration details may exist in Japan, it is of urgent importance to grasp these fatigue strength properties. However, few fatigue tests have been reported on steel girder web penetration details [3], [4]. The previous study reported fatigue behavior of steel girder web penetration details in which cross beam flange was welded to upper surface of a slot by fillet welding [5]. Then, the purpose of this study is to clarify fatigue behavior of the web penetration details with cross beam flange welded to upper or lower surface of a slot by fillet welding, and verify the effect of retrofitting methods against fatigue cracking in web penetration details.

2. Experimental procedure

2.1. Specimen

We designed and fabricated two type steel girder specimens with web penetration details where a cross beam lower flange was connected to each upper or lower surface of a slot by fillet welding. Figure 1 shows configurations and dimensions of each specimens; upper welded specimens (U1, U2), lower welded specimens (L1, L2), and Figure 2 shows the location of strain gauges. The number of specimens is two for each type. The specimen has a slot in the lower center of steel girder web. The material is JIS 490YA steel. Submerged arc welding was used to connect web and flanges, while CO_2 arc welding was used to connect the other members.

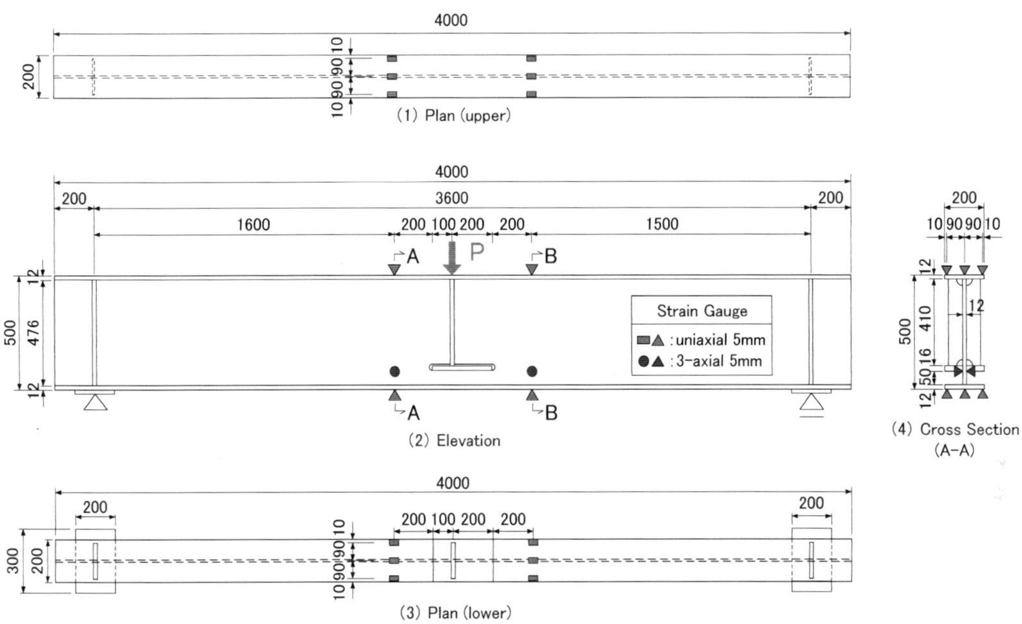

Figure 1. Configurations and dimensions of specimen

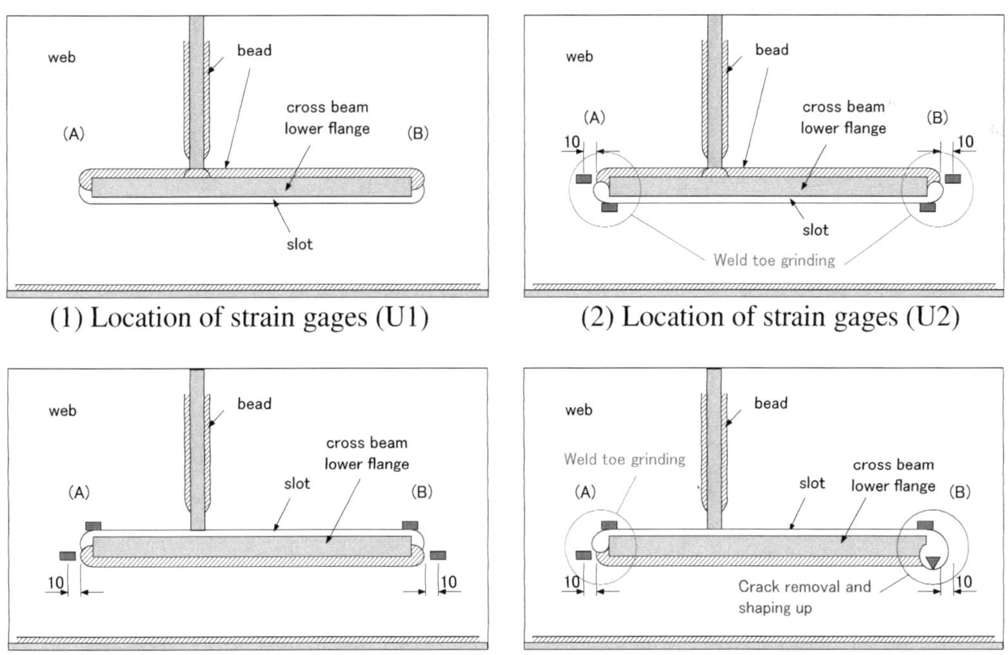

(1) Location of strain gages (U1) (2) Location of strain gages (U2)

(3) Location of strain gages (L1) (4) Location of strain gages (L2)

Figure 2. location of strain gauges

IOP Conf. Series: Materials Science and Engineering **419** (2018) 012008 doi:10.1088/1757-899X/419/1/012008

2.2. *Retrofitting method*

2.2.1. *Preventive measures*

(1) Weld toe grinding
Weld toes were ground using a bar grinder in the upper welded specimen U2 (A and B), and the lower welded specimen L2 (A). Photo 1 shows the condition of ground weld toe of U2 specimen (A).
(2) Crack removal and shaping up
Initial defects were removed and shaped up using a bar grinder in the lower welded specimen L2 (B). Photo 2 shows the condition after crack removal and shaping up of L2 specimen (B).

Before grinding (A) After grinding (A)

Photo 1. Condition of ground weld toe (U2 (A))

Photo 2. After crack removal and shaping up method (L2 (B))

2.2.2. *Post cracking measure*

(1) Stop hole and high tension bolt
A stop hole (SH) was drilled and a high tension bolt (HTB) was tightened at the crack tip of long cracks, in order to stop crack propagation in the upper welded specimen U1 (A).
(2) Crack removal and shaping up method
Short fatigue cracks were removed and shaped up using bar grinder, in order to prevent crack re-initiation in the lower welded specimen L1 (A).

2.3. *Static loading test*

First, we conducted static loading test in 3-point bending condition to grasp the stress distributions around web penetration details. 3-axis strain gauges were pasted on both surfaces of the web plate horizontally 100mm away from cross beam lower flange edges to avoid the influence of the stress concentration near the welded joint. Three 1-axis strain gauges were pasted on the upper surface of upper flange and the lower surface of lower flange in the cross section where 3-axis strain gauges were pasted. The test load was set to 100kN so that the maximum tensile stress of about 50MPa might be generated in the lower flange considering the live-load stress.

2.4. *Fatigue test*

Table 1 gives fatigue test steps. Fatigue test was conducted in 3-point bending condition as static loading test to clarify fatigue cracking behavior and fatigue life. The loading frequency was 6Hz. The load range was set to be 100kN, with the maximum load of 300kN and the minimum load of 200kN, considering the dead-load stress. In the case of a fatigue crack was propagated in one of two tested area, a stop hole was drilled and a high strength bolt was tightened to continue fatigue tests. Magnetic Particle Test (MT) was applied to detect fatigue cracks at weld toe along cross beam lower flange edges.

Table 1. Fatigue test steps

		Preventive Measure	Post Cracking Measure
U1	A	as weld	SH, or SH and HTB
	B	as weld	-
U2	A	weld toe grinding	-
	B	weld toe grinding	-
L1	A	as weld	-
	B	as weld	-
L2	A	weld toe grinding	crack removal and shaping up
	B	crack removal and shaping up	-

3. Test results

3.1. *Static loading test results*

Figure 3 shows principal stress distributions around web penetration details. Nominal stresses were obtained from bending moment and shearing force calculated according to the beam theory neglecting the cross beam lower flange and stiffeners. It was confirmed that the measured maximum principal stresses nearly equal to the calculated values, although the measured value was about 2～6% larger than the calculated one.

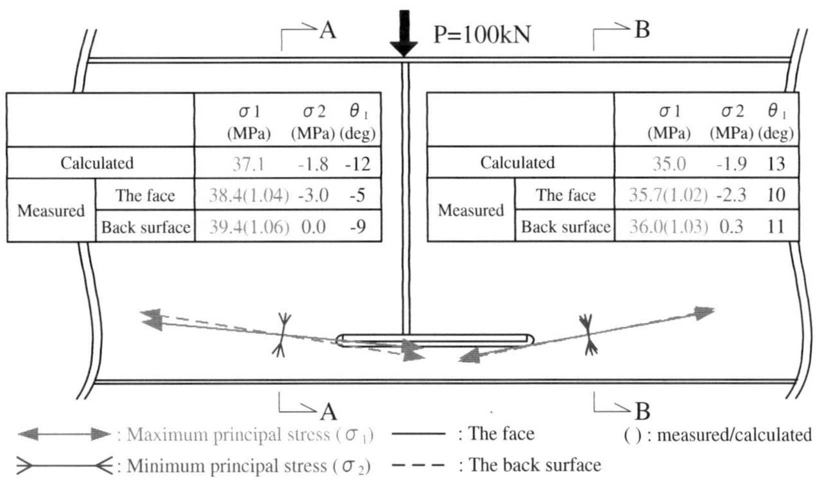

Figure 3. Principal stress distributions around web penetration details

3.2. Fatigue test results

3.2.1. Fatigue cracking behavior

(1) Fatigue crack initiation behavior

Figure 4 shows relationship between the fatigue crack length and the number of loading cycles for specimens U1 and L1. Fatigue cracks were initiated at the web-side toe of the fillet weld along cross beam lower flange edges.

(2) Fatigue crack propagation behavior

In the U1 specimen (A), fatigue cracks were propagated into the web plate diagonally upward almost perpendicularly to the maximum principal stress direction. In the L1 specimen (A), fatigue cracks were propagated into the web plate vertically downward faster than in the case of the upper welded specimen, and broke the lower flange. Because the bending stress increases as approaching the lower flange of the main girder.

Figure 5 shows relationship between the fatigue crack length and the number of loading cycles of specimens U2 and L1. In the U2 specimen, crack propagation rate ($\triangle a/\triangle N$) becomes smaller as approaching the neutral axis. Because the bending stress decreases as approaching the neutral axis.

(3) Fatigue life

Figure 6 shows the relationship between bending stress range and fatigue life of the specimens U1, U2, L1 and L2. In the U1 and L1 specimens (as weld), fatigue life when the crack was detected at web-side toes of the fillet welds along cross beam lower flange edges (Nd) is 1/5 of class H' (the lowest fatigue class in Japanese Specifications for Highway Bridges [6]), fatigue life when the crack was propagated into the web plate (Nw) was less than class H', and fatigue life when the crack length reaches 30mm (N30) satisfies class H'.

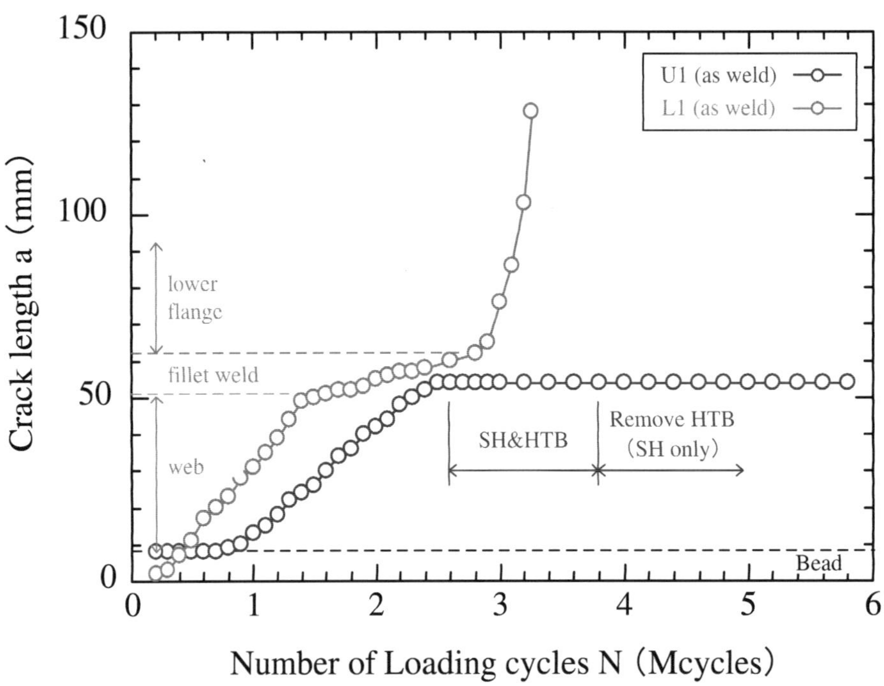

Figure 4. Relationship between the fatigue crack length and the number of loading cycles of U1 and L1

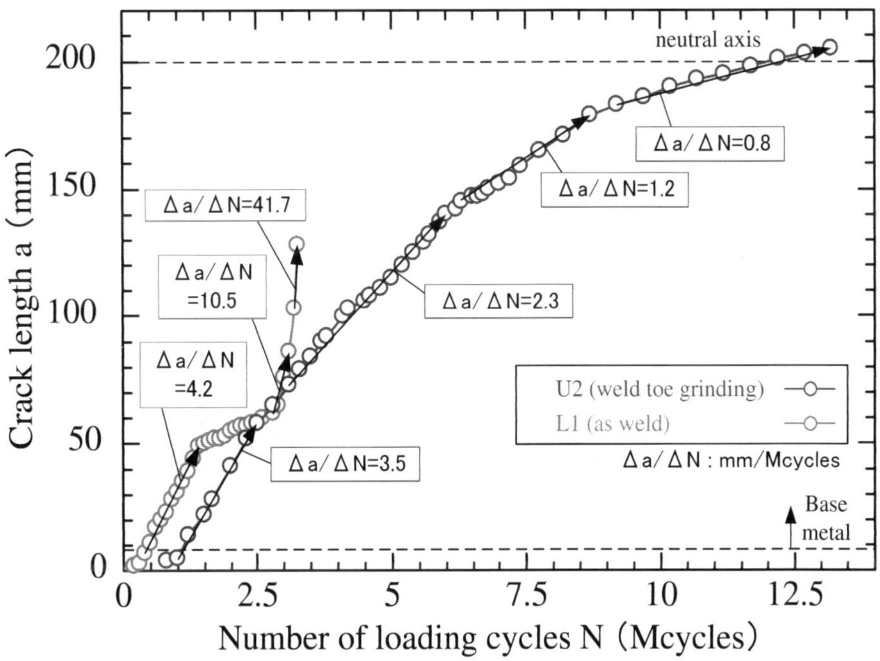

Figure 5. Relationship between the fatigue crack length and the number of loading cycles of U2 and L1

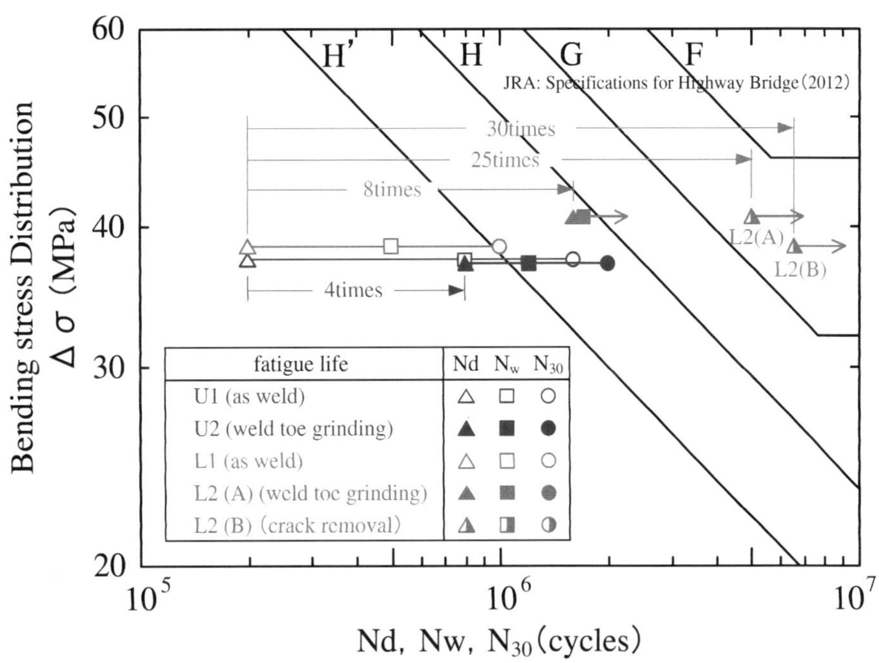

Figure 6. Fatigue life of specimens U1, U2, L1 and L2

3.2.2. Effectiveness of retrofitting method
(1) Preventive measure
a) Weld toe grinding
Crack initiation life of U2 specimen was four times of that of U1 specimen (as weld) as shown in Figure 6, but fatigue cracks were propagated, similarly as U1 specimen, into the web plate diagonally upward almost perpendicularly to the maximum principal stress direction.
Figure 7 shows relationship between the fatigue crack length and the number of loading cycles of specimens L1 and L2. Crack initiation life of L2 specimen (A) is eight times of that of L1 specimen (B), but fatigue cracks were propagated, similarly as L1 specimen, into the web plate vertically downward.
Therefore, this method can extend fatigue life a little, but cannot prevent fatigue crack initiation perfectly as shown Figure 6.
b) Crack removal and shaping up method
In the L2 specimen (B), no fatigue cracks were detected (see Figures 6 and 7). Therefore, crack removal and shaping up method can prevent fatigue crack initiation perfectly.
(2) Post cracking measure
a) Stop hole and high tension bolt, or stop hole only
In the U1 specimen (A), both of SH method and SH and HTB method can stop crack propagation of long cracks as shown in Figure 4.
b) Crack removal and shaping up method
In the L2 specimen (A), crack removal and shaping up method by grinding can stop crack propagation of short cracks as shown in Figures 6 and 7.

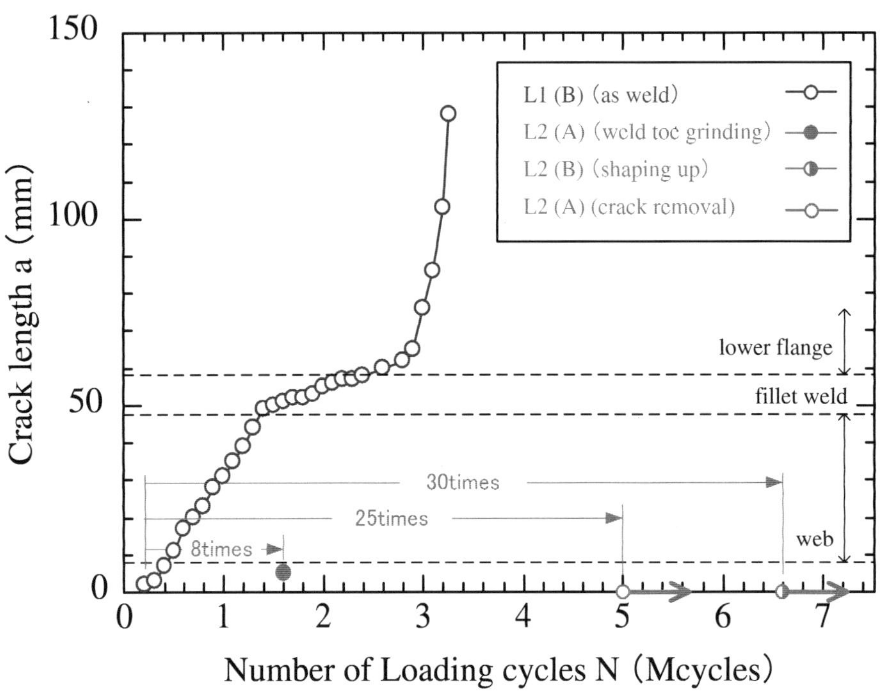

Figure 7. Relationship between the fatigue crack length and the number of loading cycles of L1 and L2

4. Conclusion

The principal results obtained through this study are as follows;

(1) Fatigue cracks were initiated at the web-side toe of the fillet weld along cross beam lower flange edges both in upper and lower welded specimen.

(2) In the upper welded specimen, fatigue cracks were propagated into the web plate diagonally upward almost perpendicularly to the maximum principal stress direction, whereas in the lower welded specimen, fatigue cracks were propagated into the web plate vertically downward faster than in the case of the upper welded specimen, and broke the lower flange.

(3) The fatigue life of the lower welded specimen was shorter than that of the upper welded specimen.

(4) As a preventive measure, weld toe grinding can extend fatigue life a little, but cannot prevent crack initiation perfectly.

(5) As a post cracking measure, a stop hole can stop crack propagation of long cracks, whereas crack removal and shaping up method by grinding can prevent crack propagation of short cracks.

For existing bridges, stop hole and crack removal are recommended to stop crack propagation. For new bridges, web penetration details with a slot should not be used. Further researches are necessary on countermeasures to prevent fatigue crack initiation perfectly.

References

[1] Fisher, J. W.: Fatigue and Fracture in Steel Bridges, Chapter 6, Wiley-Interscience, 1984.

[2] Nara National Highway (http://www.kkr.mlit.go.jp/nara/). (in Japanese)

[3] Sakano, M., Mikami, I., Murayama, H., Moriwaki, K.: Fatigue Strength of Railway Plate Girder with Web Penetration, Journal of Structural Engineering, Vol.41A, pp.965-973, 1995. (in Japanese)

[4] Sakano, M., Hozumi, M., Shimora, T., Mikami, I.: Long Term Fatigue Strength of Web Gusset Joint in Floor-Beam-to-Main-Girder Connection, Steel Construction Engineering, Vol.5, No.18, pp.31-40, 1998. (in Japanese)

[5] Yoshida, N., Sakano, M., Konishi, H., Fujii, T.: Fatigue Test of Steel Girder Web Penetration Details with a Slit, Sustainable Solutions in Structural Engineering and Construction, pp.95-100, ISEC Press, 2014.

[6] Japan Road Association: Specifications for Highway Bridges, Part I (General) and II (Steel bridges), 2012. (in Japanese)

Effectiveness of GPS technology in monitoring of traffic-induced response of highway steel bridge

Piotr Górski[1], Peter Breuer[2], Eduard Konopka[2] and Monika Napieraj[1]

[1]Opole University of Technology, Faculty of Civil Engineering and Architecture, Katowicka 48, Opole 45-061, Poland
[2]Stuttgart University of Applied Sciences, Department of Geomatics, Computer Science and Mathematics, Schellingstraβe 24, Stuttgart 70174, Germany

E-mail: p.gorski@po.opole.pl

Abstract. Despite some disadvantages, regarding the limited accuracy, the Global Positioning System (GPS) is widely used in monitoring of long-span bridges. This paper is concerned with the analysis of the capabilities and limitations of the use of the GPS technology in monitoring of semi-static and dynamic vertical responses of a highway steel cable-stayed bridge excited by traffic loads. The field test was conducted on the Maxau Rhine Bridge, located over the Rhine River near Karlsruhe, Germany. The so-called baseline method with the Post-Processing mode was applied using two units Leica Viva GS15. The basic characteristics of the GPS background noise before and after filtering were examined in the time and frequency domain. The natural frequencies of the bridge were extracted based on the vertical dynamic displacements recorded in the mid-span of the bridge using GPS and accelerometer. It was shown that the GPS receivers were able to measure only two primary modal frequencies of the bridge. The obtained experimental results proved that the GPS system is an efficient tool in measuring the vertical semi-static displacements of the bridge with satisfactory accuracy. Moreover, GPS is capable of tracking the vertical vibrations of long-period bridges to a few millimeters.

1. Introduction

Nowadays, the technology of Global Positioning System (GPS) became an alternative technique to monitor the semi-static and dynamic displacements of long-span cable-stayed or suspension bridges due to various ambient excitations, e.g. [1-3], and, based on it, to detect the potential structural damage of such structures, e.g. [4]. This is due to some important advantages of the GPS technology. Among other, the continuous long-term monitoring using GPS can be automated and performed regardless of the weather and visibility conditions. Moreover, the GPS system is able to measure structural displacements simultaneously in three directions, i.e. in two horizontal and one vertical directions. Due to the geometry of the satellite constellation the measurement accuracy in the vertical direction is about twice smaller than in the horizontal direction. Besides these advantages, the GPS measurement accuracy is susceptible to several kinds of error sources such as the noise of the GPS signals, limited data sampling rate, multipath effect, and GPS signals shielding effect due to shielding of a certain area of the sky by local obstacles. Recent advances in the sampling capability of the GPS technology allow us to monitor the total response of long-period engineering structures with 20 Hz, or even with 50÷100 Hz. Despite the constantly developing technology of GPS receivers the following question

IOP Conf. Series: Materials Science and Engineering **419** (2018) 012009 doi:10.1088/1757-899X/419/1/012009

arises: what is currently the effectiveness of the GPS technology in monitoring of bridge deformations?

This paper is concerned with the analysis of the capabilities and limitations of the use of the GPS technology in monitoring of semi-static and dynamic vertical responses of a highway steel cable-stayed bridge excited by daily traffic loads. The field test was conducted on the Maxau Rhine Bridge, located over the Rhine River near Karlsruhe, Germany. Two units Leica Viva GS15 as reference and rover units were applied, tracked the satellite signals at the data frequency equal to 20 Hz. The basic characteristics of the GPS background noise were examined in the time and frequency domain. The natural frequencies of the bridge were identified based on the recorded vertical dynamic displacements of the bridge. It was shown that the GPS receivers were able to measure only two primary modal frequencies of the bridge. To validate the GPS results, the dynamic response of the bridge was also recorded by an uniaxial accelerometer type PCB 3711E112G with the sampling rate of 200 Hz. In order to improve the measurement accuracy of the bridge deformations by GPS technology the filtering procedure using the eighth-order Type 1 Chebyshev band-pass digital filter with a selected pass-band frequency was adopted. The obtained experimental results proved that the GPS system is an efficient tool in measuring the vertical semi-static displacements of the bridge with satisfactory accuracy, and that GPS is capable of tracking the vertical vibrations of long-period bridges excited by traffic loads to a few millimeters.

2. GPS vertical background noise

The basic characteristics of GPS background noise was examined based on a static test using two GPS Leica Viva GS15 receivers. In the test the Kinematic On the Fly (KOF) method was applied in the Post-Processing Kinematic Mode (PPK-GPS) to determine the baseline (3D-vector) between two GPS units, i.e. reference and rover receivers. Both receivers were fixed in a stable position on concrete pillars at a distance of 15 m away from each other. One should be aware that increasing the distance between the rover and reference GPS units to several hundred meters does not affect the accuracy of measurements [5]. The static test was carried out with the sampling rate of 20 Hz in good GPS conditions. It means that a clear view to the sky on both GPS units was ensured and a good satellite configuration was prevailing defined by so-called the Geometric Dilution of Precision (GDOP) factor changing in the range from 1.7 to 3.7. The number of available satellites was 7÷10.

The representative time series of the background noise $B_V(t)$ of GPS system in the vertical direction recorded during a time interval of 33 min and 20 s (2000 seconds) are given in Fig. 1. Since the measured baseline between two GPS receivers was unchanged in time, it was assumed that any measured deviations from the expected baseline were considered as the signal background noise. For $B_V(t)$ the statistical characteristics were calculated and are also given in Fig. 1.

$B_V(t)$ was analyzed by the Fast Fourier Transform (FFT) and its Power Spectral Density (PSD) function is presented in Fig. 2. It is shown that $B_V(t)$ contains two components, i.e. (1) the low-frequency noise component $L_V(t)$, visible as a non-stationary trend with the dominant noise energy corresponding to the frequency range limited to 0.05 Hz, and (2) the high-frequency noise component $R_V(t)$, visible as a random noise with the frequency over 0.05 Hz and the spectrum close to the white noise type spectrum. In this case the values of the spectrum are in the range between 10^{-6} and 10^{-5} m^2s.

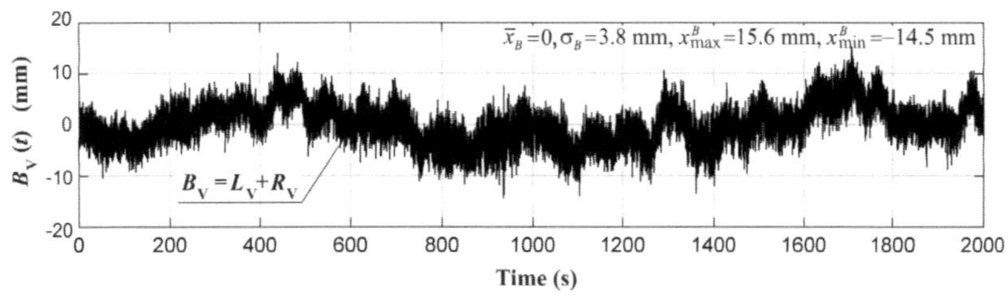

Figure 1. Time series of vertical background noise $B_V(t)$ of GPS system recorded during measurements of the position of a stationary point over 33 min and 20 s (2000 seconds).

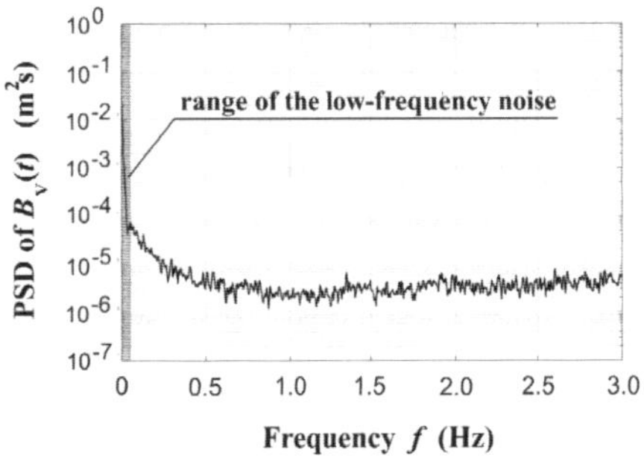

Figure 2. PSD of vertical background noise $B_V(t)$ of GPS system.

Hence, the GPS background noise $B_V(t)$ can be expressed as follows:

$$B_V(t) = L_V(t) + R_V(t) \tag{1}$$

where $L_V(t)$ is the low-frequency noise of GPS system, $R_V(t)$ is the high-frequency noise of GPS system, and t is the time tracked according to the inverse of the sampling rate of GPS system.

In the baseline method the $L_V(t)$ component is mainly caused by the changing number of available satellites, their geometrical configurations and multipath effect appearing during the measurements. The $R_V(t)$ component is mainly caused by the noise of the GPS signals. In order to study the effect of two mentioned components, i.e. $L_V(t)$ and $R_V(t)$ on the GPS measurement accuracy, $B_V(t)$ was filtered using the eighth-order Type 1 Chebyshev high-pass digital filter with pass frequency range below or over 0.05 Hz, respectively. The obtained both $L_V(t)$ and $R_V(t)$ components and their statistical characteristics are shown in Fig. 3. $L_V(t)$ is changeable between -5.8 and 8.7 mm, while $R_V(t)$ between -11.8 and 8.9 mm with the standard deviation of 1.9 mm. Such the accuracy level determines the sensitivity threshold of the used GPS receivers in monitoring of the quasi-static and dynamic structural displacements.

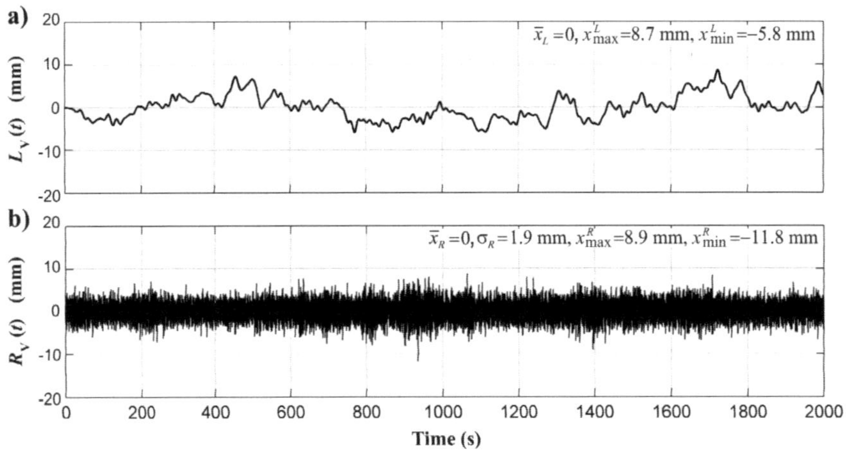

Figure 3. Time series of vertical (a) low-frequency $L_V(t)$ and (b) high-frequency $R_V(t)$ components of background noise of GPS system recorded during measurements of the position of a stationary point over 33 min and 20 s.

3. Maxau Rhine Bridge and experimental set-up

The considered cable-stayed Maxau Rhine Bridge is located over the Rhine River near Karlsruhe in Germany. The bridge carries highway traffic along the Federal highway B10 with two three-lane carriageways, footpath and bicycle lanes on both sides. The design traffic velocity is 80 km/h.

The streamlined box girder of the bridge is made of steel with the total length of 292 m, width of 35.3 m, and deck depth of 3.0 m. The bridge comprises two continuous spans with lengths of 175.2

and 116.8 m, supported by a single rectangular 48,0 m tall (above a concrete pillar) pylon made of steel. The cross-section of the pylon is a box-type and has a rectangular shape with outer dimensions of 2.6x2.0 m. The pylon is founded on a reinforced concrete 24.7 m tall central pillar. The longitudinal layout of six stay cables is made in a semi-fan type system. All suspension cables are arranged in two layers in a flat rectangle and consist of 18 steel strands, each with a diameter of 82 mm. The general view and longitudinal section of the Maxau Rhine Bridge with the most important dimensions are shown in Fig. 4(a-b).

The purpose of the test was to measure the vertical response of the bridge, i.e. the semi-static and dynamic displacements caused by daily traffic loads using both GPS technology and accelerometer and, based on it, to investigate the natural frequencies of the bridge. The second goal of the field test was to investigate the capabilities and limitations of the use of the GPS technology in monitoring of the semi-static and dynamic vertical responses of the bridge.

The field test was carried out on June 16, 2016. The KOF method was applied using GPS Leica Viva GS15 equipment with two dual-frequency receivers temporarily installed as a reference and rover units. The reference receiver was mounted on a tripod about 250 m northeastern from the bridge. The rover receiver was located in the mid-span of the main span of the bridge and was installed on the handrail of the existing railing at the footpath lane on the south side. The position and installation way of the GPS rover receiver on the railing of the bridge are shown in Fig. 4(b-c). The measurements were carried out using the baseline method in which both GPS receivers tracked the satellite signals simultaneously with the same data frequency equal to 20 Hz. This frequency rate is appropriate to analyze dynamic displacements of the Maxau Rhine Bridge with an expected dominating frequency of about 0.5 Hz. The bridge displacements were processed using the PPK-GPS mode. During GPS monitoring the quality of satellite constellation geometry was favorable and was very similar to the conditions of the static test, described in Section 2. The GDOP factor was changed in the range from 1.7 to 2.9, while the number of available satellites was 7÷10. During the field test the air temperature was constant with the value of 11°C, while full cloudiness and windless period were observed. In such conditions the wind and solar radiation influence were negligible.

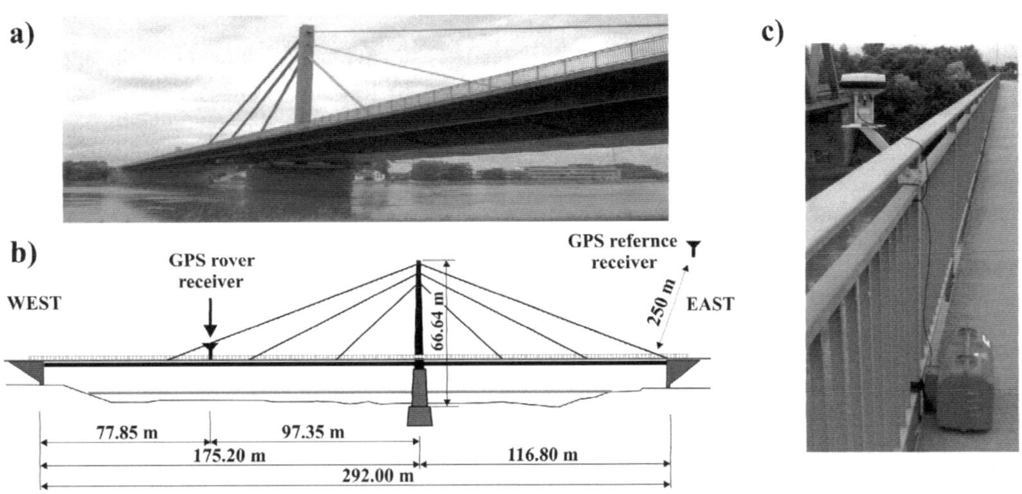

Figure 4. Maxau Rhine Bridge over the Rhine River near Karlsruhe, Germany: (a) general view, (b) longitudinal section with main dimensions and position of the GPS rover receiver, and (c) installation way of the GPS rover receiver on the barrier rail of the bridge.

To validate the GPS results, the dynamic response of the bridge was also recorded by an uniaxial accelerometer type PCB 3711E112G with the sampling rate of 200 Hz. The sensor was located inside the box girder of the bridge below the outside position of the GPS rover receiver. The high sensitivity accelerometer was based on the Micro-Electro-Mechanical System technology. The measurement range of the sensor was ±2 g, the frequency range 0–400 Hz, the broadband resolution 0.1 mg rms in 0.5 to 100 Hz bandwidth, and the spectral noise 15 μg/√Hz in 1 to 100 Hz bandwidth. Such sensitivity is adequate for studying dynamic behaviour of the considered bridge.

4. Measurement results and their analysis

Fig. 5(a) depicts the total vertical displacements of the bridge under daily traffic conditions recorded using GPS system in the mid-span of the bridge. The field test was performed over 33 min and 20 s with the sampling rate of 20 Hz. Fig. 5(b-c) shows two components of the measured vertical displacements, i.e. the low-frequency component, namely the semi-static component, and the high-frequency component, namely the dynamic component of the bridge displacements. Both components were extracted from the total displacements by applying adequate post-processing filtering procedure using the eighth-order Type 1 Chebyshev high-pass digital filter with pass frequency range below or over 0.20 Hz, respectively. The semi-static component was caused mainly by the weight of moving vehicles, especially heavy vehicles, while the dynamic component was caused by the dynamic effects due to moving vehicles, random characteristics of the pavement roughness and bridge-vehicle interaction impact. The recorded bridge displacements, depicted in Fig. 5, were related to the position of the measurement point in a static equilibrium of the bridge (dotted line). This position was determined based on the geometrical data of the bridge without taking into account any external loads.

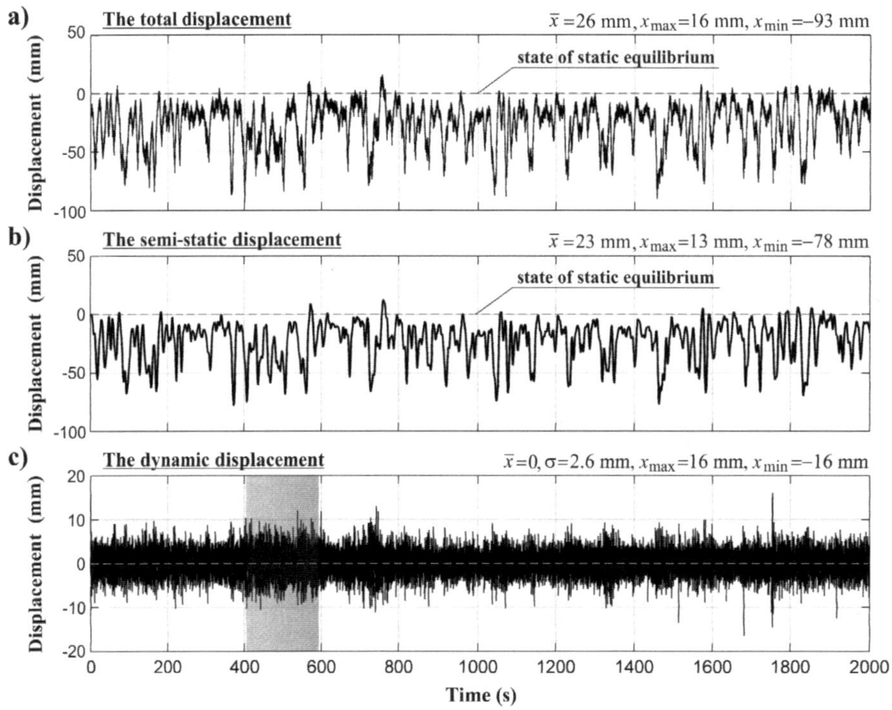

Figure 5. Vertical displacements of the bridge due to daily traffic loads, recorded using GPS system in the mid-span of the bridge on June 16, 2016 over 33 min and 20 s with the sampling rate of 20 Hz: (a) total displacements, (b) semi-static and (c) dynamic components of the displacements.

Regarding the static equilibrium the measured maximum semi-static displacement of the bridge was 78 mm with the mean value of 23 mm, while the amplitude of the dynamic displacement was ±16 mm with the standard deviation of 2.6 mm. It should be noticed that both measured components of the bridge displacements were significantly greater than the sensitivity threshold of GPS system recognized in Section 2 (see Fig. 3). However, one should be aware that the measured bridge responses are always overlaid by the background noise of GPS system.

Fig. 6 shows the vertical vibrations of the bridge due to daily traffic loads, recorded using GPS system in the mid-span in an enlarged time scale, during 180 s-interval, corresponding to the fragments indicated in Fig. 5(c). In this figure it is obviously visible that the measured dynamic response is disturbed by the GPS background noise. The displacements shown in Fig. 5(c) were analyzed by the FFT. The PSD function was estimated for the total length of interval time series of recorded displacements. Fig. 7 shows the PSDs of the recorded vertical displacements of the bridge and the GPS vertical background noise $B_V(t)$, as depicted in Fig. 2. The PSD of the measured bridge displacement is dominated by two well-separated peaks with f_1 = 0.520 Hz (T_1 = 1.923 s) and f_2 = 1.012 Hz (T_2 = 0.988 s), representing two primary natural frequencies of the considered bridge in the vertical direction. Because of the limited nominal measurement accuracy of GPS technology higher natural frequencies were not detected. Based on the analysis of PSD, the limit frequency of the semi-static components of the bridge displacements was identified as 0.2 Hz.

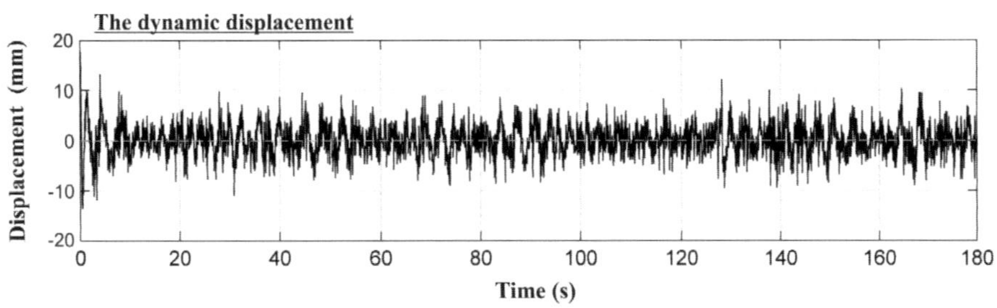

Figure 6. Vertical dynamic displacements of the bridge due to daily traffic loads, recorded using GPS system in the mid-span in an enlarged time scale corresponding to the fragments indicated in Fig. 5(c).

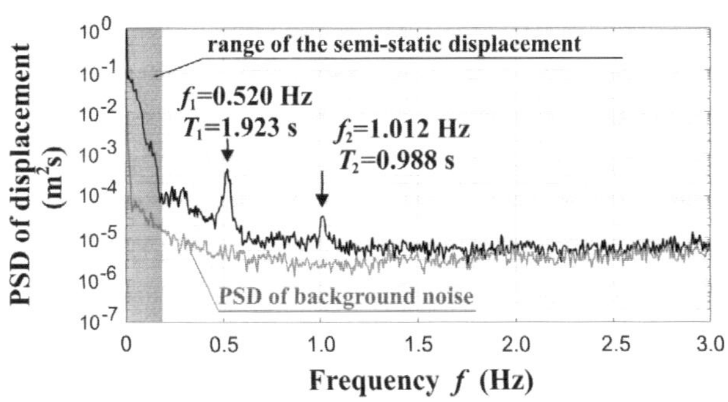

Figure 7. PSDs of recorded vertical dynamic displacements of the bridge by the GPS system and vertical background noise $B_V(t)$ of GPS system.

To validate the GPS results, the PSD of vertical accelerations of the bridge excited by traffic loads was determined and depicted in Fig. 8. The accelerations were recorded in the mid-span of the bridge on June 16, 2016 over 33 min and 20 s with the sampling rate of 200 Hz. A total of eight modal frequencies of the bridge were identified within the frequency range of 0÷4 Hz. The identified two primary modal frequencies of the bridge using the GPS system match well with those identified using the accelerometer. The maximum frequency difference was about 1.3%. Based on the analysis of the PSD of the bridge accelerations the frequency components of traffic excitations were also recognized within the frequency range of 8÷16 Hz. Thereby, it could be concluded that the modal frequencies of bridges can be satisfactorily identified when the clear separation between natural frequencies of structure and vehicle excitations is observed.

5. Improvement of GPS measurement accuracy by filtering process

In order to increase the measurement accuracy of the vertical dynamic displacements of the bridge using GPS technology the filtering procedure using the eighth-order Type 1 Chebyshev band-pass

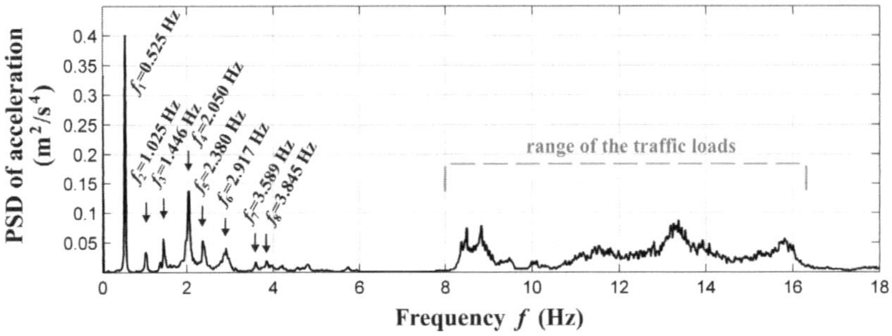

Figure 8. PSD of vertical accelerations of the bridge due to daily traffic loads, recorded using the accelerometer in the mid-span of the bridge on June 16, 2016 over 33 min and 20 s with the sampling rate of 200 Hz.

digital filter with a selected pass-band frequency was adopted. Since two primary mode frequencies of the bridge were extracted on the PSD (see Fig. 7), a band-pass filtering process may reduce significantly the GPS background noise. For the considered dynamic displacements the filter pass-band range of 0.4÷1.1 Hz was selected to cover two identified modal frequencies of the bridge. The efficiency of the proposed filtering process to de-noise the GPS data was validated in the papers [6,7].

Fig. 9 depicts the high-frequency component $R_V(t)$ of GPS background noise (see Fig. 3b) after band-pass filtering process, denoted as $F_V(t)$. The pass-band filtering process leads to a fourfold reduction of high-frequency GPS noise. In this case the accuracy of de-noised vertical dynamic displacements is about ±3 mm with the standard deviation of 0.5 mm. The dynamic displacements of the bridge (see Fig. 5c) after filtering process using the eighth-order Type 1 Chebyshev band-pass filter with pass-band 0.4÷1.1 Hz and pass-band ripple 1 dB are shown in Fig. 10. Fig. 11 depicts the PSDs of filtered bridge displacements and the GPS background noise $F_V(t)$ for the time series shown in Figs 10(a) and 9, respectively.

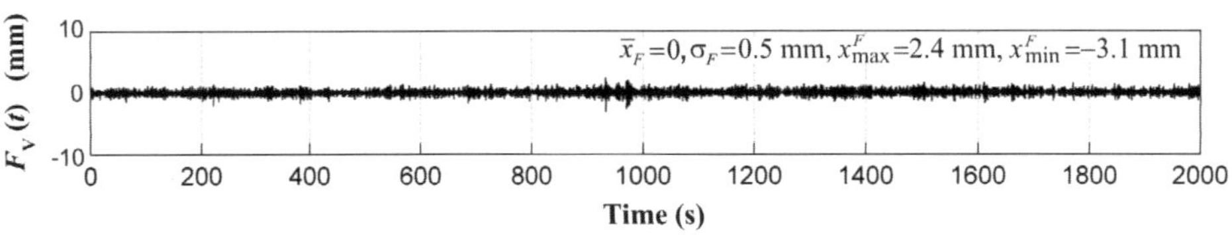

Figure 9. Filtered high-frequency component $R_V(t)$ of GPS background noise.

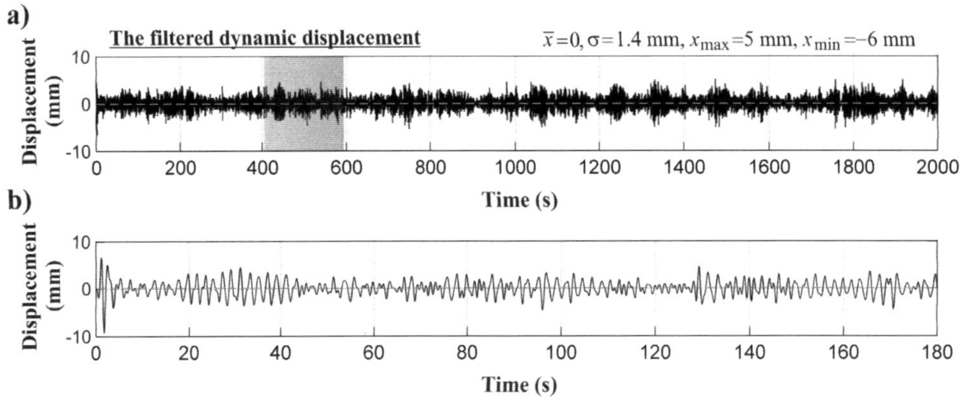

Figure 10. (a) Filtered dynamic component of the vertical displacements of the bridge and (b) the filtered dynamic displacements in an enlarged time scale corresponding to the fragments indicated in Fig. 10(a).

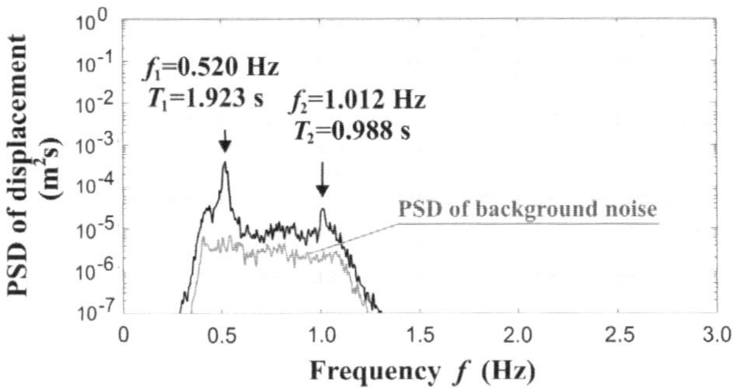

Figure 11. PSDs of filtered vertical dynamic displacements of the bridge and filtered vertical background noise $F_V(t)$ of GPS system.

6. Concluding remarks

The major outcome of the present study clearly proved that the GPS technology could detect the semi-static and dynamic components of the bridge response in the vertical direction under daily traffic excitations. The obtained results confirmed that the maximum values of the measured bridge displacements are significantly greater than the accuracy limit of the GPS system in the vertical direction. Based on the results of a static test using two GPS Leica Viva GS15 units it was found that the GPS measurement accuracy in the vertical direction was about ±9 mm for the low-frequency component (the noise frequency below 0.05 Hz), and ±12 mm for the high-frequency component (the noise frequency over 0.05 Hz). The filtering process using the eighth-order Type 1 Chebyshev band-pass digital filter with pass-band 0.4÷1.1 Hz and passband ripple 1 dB improved the measurement accuracy of the vertical dynamic displacements to ±3 mm.

The measured maximum semi-static component of the bridge displacements in the vertical direction due to the daily traffic loads was about 80 mm, while the amplitude of the dynamic displacements was about 16 mm. Based on the recorded bridge displacements using the GPS system two primary natural frequencies of the bridge could be extracted from FFT analysis, i.e. f_1 = 0.520 Hz and f_2 = 1.012 Hz. These frequencies match well with those detected using the accelerometer.

References

[1] Vazquez G E, Gaxiola-Camacho J R, Bennett R, Guzman-Acevedo G M and Gaxiola-Camacho I E 2017 Structural evaluation of dynamic and semi-static displacements of the Juarez Bridge using GPS technology *Measurement* **110** pp 146-153

[2] Han H, Wang J, Meng X and Liu H 2016 Analysis of the dynamic response of a long span bridge using GPS/accelerometer/anemometer under typhoon loading *Engineering Structures* **122** pp 238-250

[3] Yi T-H, Li H-N and Gu M 2013 Experimental assessment of high-rate GPS receivers for deformation monitoring of bridge *Measurement* **46** pp 420-432

[4] Miao C, Wang M, Tian H, Feng Z and Chen C 2015 Damage alarming of long-span suspension bridge based on GPS-RTK monitoring *Journal of Central South University* **22** pp 2800-2808

[5] Breuer P, Chmielewski T, Górski P and Konopka E 2002 Application of GPS technology to measurements of displacements of high-rise structures due to weak winds *Journal of Wind Engineering and Industrial Aerodynamic* **90** pp 223-230

[6] Górski P 2015 Investigation of dynamic characteristics of tall industrial chimney based on GPS measurements using Random Decrement Method *Engineering Structures* **83** pp 30-49

[7] Górski P 2017 Dynamic characteristic of tall industrial chimney estimated from GPS measurement and frequency domain decomposition *Engineering Structures* **148** pp 277-292

Estimate of the steel bridges fatigue life by application of the fracture mechanics

J M Djoković[1], R R Nikolić[2], J Bujnák[3] and B Hadzima[4]

[1] University of Belgrade, Technical Faculty of Bor, V. Jugoslavije 12, 19210 Bor, Serbia
[2] University of Žilina, Research Center, Univerzitna 8215/1 Žilina, Slovakia
[3] University of Žilina, Faculty of Civil Engineering, Univerzitna 8215/1 Žilina, Slovakia
[4] University of Žilina, Research Center, Univerzitna 8215/1 Žilina, Slovakia

E-mail: ruzicarnikolic@yahoo.com

Abstract. The bridge elements are subjected to the low-amplitude cyclic loading, causing the fatigue as the main problem that limits the carrying capacity and remaining working life of a structure. Identifying the parts prone to fatigue, inspection control, preventive maintenance, successful reinforcing and repair plans, secure the satisfying characteristics of a bridge during its working life.
For a new structure is important that elements subjected to fatigue are so designed to avoid the danger of the fatigue fracture. The fracture mechanics represents a useful tool for analyzing the fatigue crack growth in steel elements. The relevant parameters for the remaining working life estimate are the load level and frequency and the stress concentration.
A method, based on the Paris' law, for predicting the remaining fatigue working life of steel bridges, is proposed and verified on an actual bridge. The considered bridge was found to be safe against the fatigue fracture, both for the present and the future increased load, since it was verified that it has the reserve with respect to fatigue life. Based on obtained results, one can conclude that the fatigue working life of a bridge and its elements can be estimated based on application of the LEFM (linear elastic fracture mechanics) concept.

1. Introduction

Bridges are the very important part of the modern society infrastructure. In Europe, as well as in the rest of the developed world, the fast development of industry in the first half of the XIX century has caused construction of numerous roads and railways and bridges on them. Many of bridges constructed at those times are still in operation, despite the fact that their technical life has expired. Based on analysis of the age profile of the steel railway bridges in [1], it was established that about 70 % of those bridges are more than 50 years old, while about 30 % are more than 100 years old. Besides that, those bridges are now exposed to much higher requests related to traffic intensity and axle loading than in the times of their construction.

The main problem in steel bridges is fatigue, which limits the carrying capacity and the remaining working life of the existing structure. Recognizing the bridge parts prone to fatigue, inspection controls, preventive maintenance, as well as successful reinforcing and repair plan, provide for the good characteristics of bridges during their working life. It is also very important for the new structures, that elements of structures, which have tendency to fatigue, to be properly designed to avoid the danger of appearance of fatigue cracks.

Haghani et al. [2] have presented more than 100 cases of damages, which were analyzed and grouped according to the mechanisms of appearance or the type of the structural element, where the fatigue crack was registered. In about 90 % of cases the cause of damage was the crack due to deformation. This type of the fatigue damage is the most frequently the consequence of accidental and unpredictable restraints that appear due to mutual interactions of the various bridge elements. In addition, the weaker, not-reinforced elements of the structure, as well as the decrease in stiffness at points of individual bridge elements connections/extensions, also contribute to appearance of fatigue cracks. The calculation methods and rules of design usually do not provide sufficient data how to check and prevent this type of fatigue.

During their working life, the road and railway traffic produces large number of repeated load cycles to the bridge elements, which are, due to that, becoming sensitive to fatigue. The fatigue fracture appears as a consequence of a component's long-term exposure to the cyclic loading. There are three phases of the fatigue fracture: crack initiation, crack growth and fracture. The crack can appear due to surface defects of a component, caused by manual or machine processing, threading or due to appearance of the slip bands or dislocations, which are the result of the previous load cycles. In the crack growth phase, it continues to propagate due to loads that the component is subjected to. The fracture (failure) phase appears abruptly and it occurs when the component's undamaged portion without a crack cannot sustain the applied loads anymore. In Figure 1 is presented the schematic appearance of the cross-section of a component that has suffered the fatigue fracture, [3].

On the fatigue fracture surface in Figure 1, one can notice the initial surface defect, then mainly smooth surface, which is formed during the fatigue crack growth and the rough surface, which corresponds to the final fracture. Based on the surface appearance in the fatigue fracture, it is possible to discover the initial crack, since the lines in the fatigue fracture zone are bending curving backwards towards the fracture initiation point.

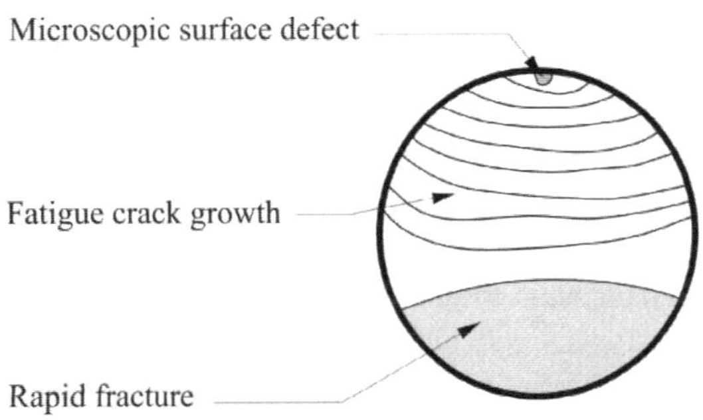

Figure 1. Appearance of the fatigue fracture surface.

2. Application of the fracture mechanics to estimate the fatigue working life

The fatigue working life of a bridge structural element can be predicted by application of the fracture mechanics, based on the fact that the stress field at the crack tip can be described by a single parameter, the stress intensity factor, *K*, or in the case of the cyclic loading, the change of the stress intensity factor ΔK, i.e. the difference between values of the stress intensity factor at maximum and minimum load. The latter parameter is defined as:

$$\Delta K = \Delta\sigma \cdot Y \cdot \sqrt{\pi a} \tag{1}$$

where $\Delta\sigma$ – is the range of the applied load, a – is the crack length and Y – is the dimensionless parameter, which depends on the sample geometry and applied load.

Change of the stress intensity factor ΔK, based on the Paris' law, determines the number of cycles of the applied load until the appearance of fracture. The Paris' law is, for the case of the stable crack growth, defined by the following expression, [4]:

$$\frac{da}{dN} = C(\Delta K)^m, \qquad (2)$$

where: da/dN – is the crack growth rate, da – is the crack length change that is within the range from the initial length to the critical length, which causes the fracture, N – is the number of the load cycles, C and m – are the material constants. The remaining working life is obtained by integration of equation (2):

$$N = \int_{a_i}^{a_{cr}} \frac{da}{C(\Delta K)^m}, \qquad (3)$$

where: a_i – is the initial crack length and a_{cr} is the critical crack length.

The unstable crack growth appears when the stress intensity factor K_I becomes greater than the experimentally determined material property – the fracture toughness K_{Ic}, namely when the crack reaches the critical length $a = a_{cr}$. Based on equation (1), the critical crack length can be determined as:

$$a_{cr} = \frac{1}{\pi}\left(\frac{K_{Ic}}{\sigma_c \cdot Y_{cr}}\right), \qquad (4)$$

where: Y_{cr} – is the geometrical factor for a_{cr} and σ_c – is the highest load level. The fracture toughness for the structural steels does not vary strongly during the fatigue working life and can be adopted as about $K_{Ic} = 3000$ Nmm$^{-3/2}$, [5]. The geometrical factor for the plate with the edge crack, subjected to the bending moment is, [6]:

$$Y = \frac{\sqrt{\frac{2 \cdot w}{\pi \cdot a}\tan\left(\frac{\pi \cdot a}{2 \cdot w}\right)}}{\cos\left(\frac{\pi \cdot a}{2 \cdot w}\right)} \cdot \left(0.923 + 0.199 \cdot \left(1 - \sin\left(\frac{\pi \cdot a}{2 \cdot w}\right)\right)^3\right), \qquad (5)$$

where w – is the connecting plate thickness.

The stress intensity factor K, governs the crack growth rate and the size of the plastic zone around the crack tip. If the plastic zone size or the applied load are of the order of magnitude of the crack length or the yield stress, respectively, the assumptions of the linear elastic fracture mechanics (LEFM) would not be valid, due to the extreme plasticity at the crack tip. However, the plastic zone size is in the cyclic loading usually smaller than in monotonous loading, so the application of the LEFM concept for analysis of the fatigue crack growth is justified, [7].

The fatigue working life up to fracture, of a structural element, can be divided into three phases, as shown in Figure 2, [8]. As can be seen from Figure 2, the dependence da/dN on ΔK in the log-log coordinate system is presented by the S-shaped curve (the sigmoidal curve). The curve is asymptotically approaching the limits of the crack growth ΔK_{th} and ΔK_C. The stress intensity factor range threshold value ΔK_{th} is the limit value below which the fatigue crack does not grow. At the ΔK_C value the fracture occurs.

The S curve spreads over three regions. In the first region (I), the dominant influence on the crack growth rate is exhibited by the microstructure, average stress and environment. The plastic zone size at the crack tip is of the order of the grain size and the deformation is crystallographic. In the second region (II), the plastic zone size is larger than the grain size. Influences of the microstructure and the sample thickness are small in this phase, while the influences of stress, environment and cyclic loading frequency are strong. This is the phase of the stable crack growth and equation (2) of the fatigue crack growth is valid.

In the third region (III) – close to the phase of the unstable crack growth, the strong influence is exhibited by the microstructure, average stress and the sample thickness, while the influence of environment is small.

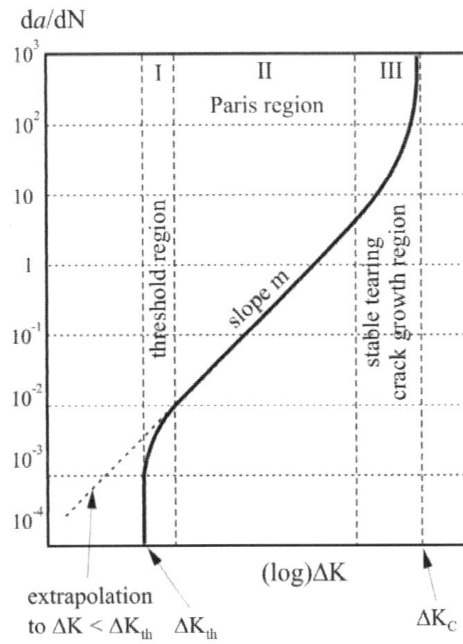

Figure 2. Phases of the fatigue crack growth, [8].

Determination of the stress intensity range threshold, needed for the beginning of the fatigue crack growth ΔK_{th} is a time-consuming and expensive process, since it is considered that the threshold is reached when the crack at sufficient number of cycles still does not grow. The ΔK_{th} values are usually smaller than 10 MPa·m$^{1/2}$, [9].

3. An example of the method application and discussion of results

The method defined in the previous section was checked on an example of a bridge on the railroad between Žilina and Košice in Slovakia, Figure 3. The bridge carries one railroad track and both passenger and freight trains are passing over it.

The bridge was commissioned in 1941 and it consists of three spans: the two side spans of 29.4 m by the left and the right bank and the middle span of 57.4 m, Figure 4. The bridge is built of steel of the tensile strength σ_M = 360 N/mm^2, with the elasticity modulus of E = 210000 N/mm^2 and the yield stress of σ_T = 235 N/mm^2.

The trains with axial load of 250 kN represent the worst possible case - trains of type 11, Table D.2: – Heavy traffic mix with 25 t (250 kN) axles), [10]. It is assumed that 5824 trains cross the bridge annually (16 trains per day × 7 working days per week × 52 weeks a year). The critical cross-section for analysis is the cross-section 3' (Figure 4 (b)). The upper flange is made of the 13 mm thick plate. It is connected by the rivets to the wall of dimensions 2800×15 mm and the pair of angles of dimensions L160×160×15 mm. The lower flange is made of angles only. The parameters used for analysis of the fatigue crack growth are $C = 9.14\cdot10^{-12}$, $m = 3$, K_{th} = 8 MPa·m$^{1/2}$ and K_{Ic} = 120 MPa·m$^{1/2}$, [11].

Figure 3. The bridge on the Žilina - Košice railway in Slovakia over river Váh in Strečno.

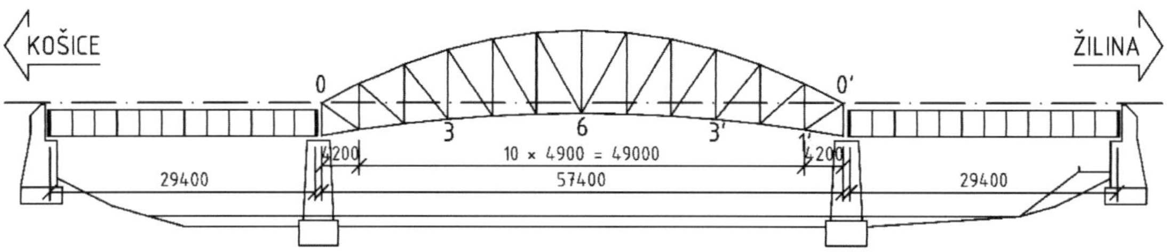

(a)

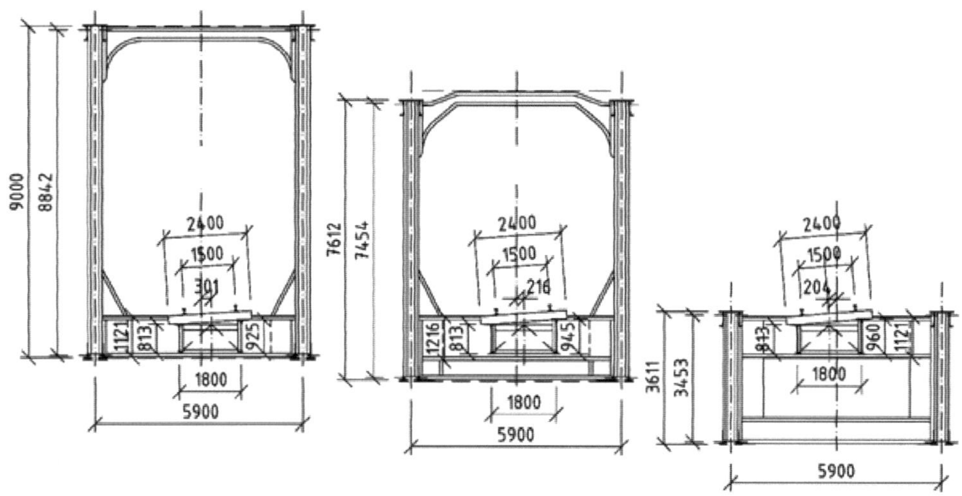

Cross-section 6 Cross-section 3' Cross-section 0'

(b)

Figure 4. Geometry of the bridge [mm]: (a) Elevation and (b) Cross section.

The diagram of the crack growth rate *da/dN* [m/cycle] dependence on the relative crack length, *a/W*, presented in Figure 5, is obtained by application of the programming routine *Mathematica*®.

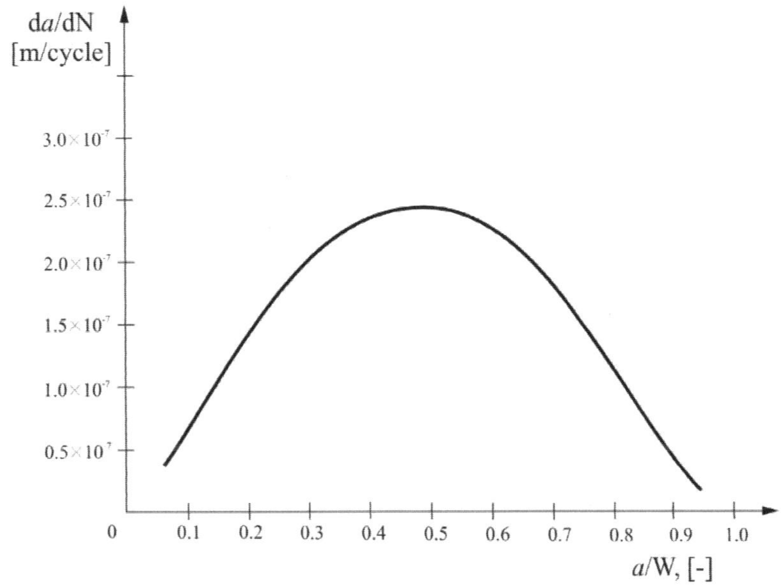

Figure 5. Crack growth rate in terms of the relative crack length.

From diagram in Figure 5 can be seen that the crack growth rate increases with the relative crack length increase up to 50 % and then it decreases, what corresponds to the beginning of fracture. It can also be noticed that the fatigue crack starts to grow when the initial crack reaches 7 % of the plate thickness.

In Figure 6 is shown the number of cycles needed for the crack growth in terms of the relative crack length. The diagram is obtained according to equation (3) and by application of the programming package *Mathematica*®.

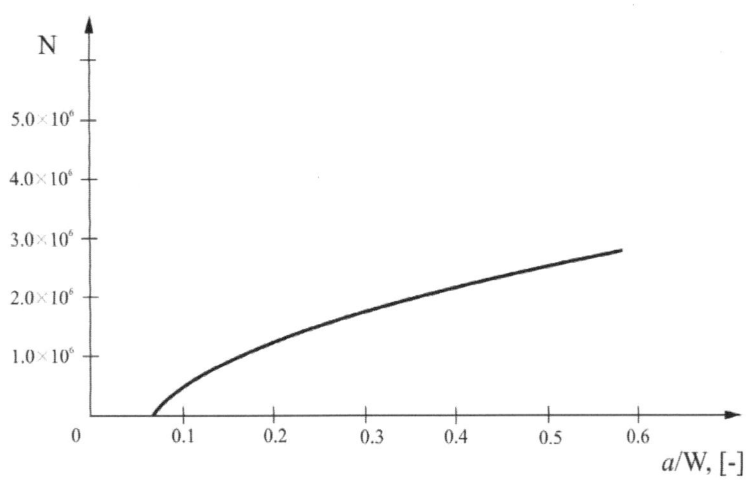

Figure 6. Number of load cycles needed for the fatigue crack growth vs. the relative crack length.

From Figure 6 can be seen again that the fatigue crack starts to grow when the initial crack length reaches 7 % of the plate thickness. In addition, one can read-off the value of the fatigue life of the plate, from the moment of the fatigue crack appearance. That value for the considered bridge is about 3 million cycles. For the assumed traffic level that amounts to 515 years.

4. Conclusion

The paper presented a method for calculation of the remaining working life of bridge components, based on application of the linear elastic fracture mechanics. The conditions for the LEFM application are that the cyclic plastic zone around the fatigue crack is small and that the crack would propagate only if the difference between values of the stress intensity factor at the maximum and the minimum load is greater than the stress intensity factor value needed for the fatigue crack growth. The number of load cycles that the component can still sustain was calculated based on the Paris' law.

The method was applied for calculating the remaining working life of a component of a real bridge on the railway between Žilina and Košice in Slovakia. It was shown that the component of a bridge, commissioned in 1941, with the present cyclic load due to traffic, can still be operational for over 3 million cycles, i.e. 515 years. This, of course, is assumed if the adequate maintenance, repair plans and operations of the bridge structure were regularly executed.

Acknowledgements

This research was partially financially supported by European regional development fund and Slovak state budget by the project "Research Centre of the University of Žilina" and by the Ministry of Education, Science and Technological Development of Republic of Serbia through grant: ON 174004.

References

[1] Bien J, Elfgren L and Olofsson J 2007 *Sustainable Bridges. Assessment for Future Traffic Demands and Longer Lives* (Wroclav: DWE)

[2] Haghani R, Al-Emrani M and Heshmati M 2012 Fatigue-Prone Details in Steel Bridges *Buildings* **2** 456-476

[3] Sachs N W 2005 Understanding the surface features of fatigue fracture: how they describe the failure cause and the failure history *Journal of Failure Analysis and Prevention* **5** 11-15

[4] Paris P C and Erdogan F 1963 A Critical Analysis of Crack Propagation Laws *Journal of Basic Engineering* **D85** 528-534

[5] Kühn B, Lukić M, Nussbaumer A, Günther H-P, Helmerich R, Herion S, Kolstein M H, Walbridge S, Androic B, Dijkstra O and Bucak Ö 2008 Assessment of Existing Steel Structures: Recommendations for Estimation of Remaining Fatigue Life *JRC Scientific and Technical reports (EUR 23252 EN)*

[6] British Standards Institution (BS) 1999 Guide on methods for assessing the acceptability of flaws in metallic structures *British Standards (BS 7910 1999)*

[7] ASM Handbook 1996 **19**: *Fatigue and Fracture* (Novelty, Ohio: ASM International)

[8] Schijve J 2003 Fatigue of Structures and Materials in the 20th Century and the State of the Art *International Journal of Fatigue* **25** 679-702

[9] Stephens R I, Fatemi A, Stephens R R and Fuchs H O 2001 *Metal Fatigue in engineering* (New York: Wiley Interscience)

[10] European Committee for Standardization (CEN) 2003 Actions on Structures; Traffic loads on bridges *Eurocode 1, Part 2 (EN 1991-2)*

[11] Ramsamooj D N and Shugar T A 2001 Model prediction of fatigue crack propagation in metal alloys in laboratory air *International journal of fatigue* **23** S287-S300

Potential modelling and verification of bridge superstructures behaviour

J Bujnák[1], J. Gocál[1] and J Odrobiňák[1]

[1] University of Žilina, Faculty of Civil Engineering, Univerzitna 8215/1, 01026 Žilina, Slovakia

E-mail: jan.bujnak@fstav.uniza.sk

Abstract. During design and analysis process, many simplified technical approaches are routinely used in the application of theories to practice. With the use of modern structural analysis computer programs, the reliable design alternative, providing the most probable response of a bridge structure due to a range of designed loads can be identified. Field measurements should be performed to determine the actual load effect and to verify the applied analytical models. Modelling principles and suggest some guidelines as well as consideration in finite element based structural analysis are illustrated in the paper on selected bridge structures. Measurements for monitoring structural response, collected data are shown using tests carried out on composite bridge superstructures.

1. Introduction

Design of bridges is a complex task. Initially, the selection of modelling methodology, nowadays most often based on finite elements, is essential and followed by a description of the structural geometry, definition of the material and section properties of the components making up the structure, next by description of load situations acting on the structure. Conditions of the piers or abutments at the support points must be examined and implemented into the structural analysis model properly. With the increased power of computers, coupled with higher demand placed on the accuracy and efficiency of structural designs, more-detailed understanding of the basic principles and assumptions associated with the use of modern computer programs for structural analysis is required.

Obviously, when designing, there is not usually enough time to verify all details. In many cases, design of certain parts of the bridge structure is executed on the basis of experience. As a result numerous simplifying assumptions are used in a multitude of analytical methods available to structural analysts. Therefore loading tests are habitually necessary, especially in the case of bridge superstructures of unusual conception. The main purpose of testing is to verify calculation model and assumptions of designer as well as suitability of transformation model. To satisfy all required code criteria, still more sophisticated analyses are needed. In the paper, three rather uncommon bridge superstructures are presented. Firstly a composite overpass, next a superstructure consisting of steel grid with concrete deck and finally quite large double beams highway bridge structural modelling as well as experimental assessments are given.

2. Steel-concrete highway overpass

2.1. Structural concept

Composite bridge, whose concrete deck is supported by two steel girders, which carries road to slim piers, was especially preferable to common prestressed concrete bridge [1]. After consideration of the technical aspects, it was decided that the bridge structural system would consist of a total of four spans with lengths 21.0+31.189+23.811+21.0 m. The roadway has a width of 7.2 m between barriers and the total deck width is 8.7 meters. Thus, this composite superstructure defines the two-lane overcrossing over its whole 103.0 m length. The steel built-up girders of I-section, axially 4.0 m spaced were selected for the bridge having the variable slender web depth from 1.30 to 1.80 m with parabolic haunches over piers (Figure 1). Transverse stiffeners welded to inner sides of the web were placed for meeting slenderness requirements and preventing local buckling. The inconstant area of flanges was used to save material where the bending moment would be smaller or larger in a span. Especially, the top flanges that act with the concrete slab are of the constant 350 mm width and proportioned by varying thickness from 20 to 50 mm. For increasing the flexural strength of cross-sections, the bottom flanges are 650 mm width, and vary in the thickness from 30 to 40 mm. Low-alloy structural carbon steel S355 grade has been used for steel bridge structural parts. Bracings consisting of rectangular hollow diagonals and horizontal channels and acting as a truss provide lateral stability of the girder bridge and distribute vertical loads. Reinforced concrete with 28-days compressive strength 35 N.mm^{-2} was used in 330 mm thick slab with haunches near girders. Shear stud connectors Ø 22/150 from steel grade S235 at the interface between the concrete slab and structural steel should ensure a full composite action.

Figure 1. Overpass structure after erection.

Figure 2. Structural finite element modelling

Three pairs of reinforced concrete two-column piers of square-shape section provide vertical supports for spans at intermediate points. In addition to transferring vertical loads to the foundations, the piers should resist higher lateral actions caused by potential earthquake events in this low seismic area. The seat-type abutments were constructed separately from the bridge superstructure as the reinforced earth-retaining structure. The bridge superstructure seats on the support stems through pot bearings with teflon sliding surfaces accommodating translational movement. The fixed bearing allowing only rotations but restricting movements is located at the central pier.

2.2. Transformation models and testing results

Curvature and real direction of bearing displacement had to be appropriately implemented into the model. The concrete deck was approximated with shell finite elements, as it can be seen in Figure 2. Seven different thicknesses of elements were adopted for better account of variable deck. Steel girders and cross truss beams as well as support diaphragms were meshed by beam elements, including real eccentricities from bridge slab.

Deflections in mid-spans of steel beams and bearing settlement were measured during loading test. Applied test load should simulate a traffic situation, which produce deformation and internal forces large enough to satisfy given criteria of relevant codes, e.g. STN 73 6209 [2]. Generally, an efficiency relation η is specified, as a ratio between elastic experimental values S_e of any observed parameter activated by testing loading and calculated values S_{cal} produced by code ideal load scheme $\eta = S_e/S_{cal}$. In the case of this bridge structure, the eight trucks weighty 28 tons, as testing load were imposed in the second largest span for achieving a deflection S_e = 17.6 mm and corresponding η = 0.97. A level of residual part of deflection S_r produced by irreversible deformation of members and connections is the supplementary important criterion. This remaining deformation has to be fewer than 10 % of total measured values S_r/S_{tot} < 0.10. The total measured values are given by the summation $S_{tot} = S_e + S_r$. This additional condition was also satisfied.

Figure 3. Strain gauges location at the support cross-section of left girder.

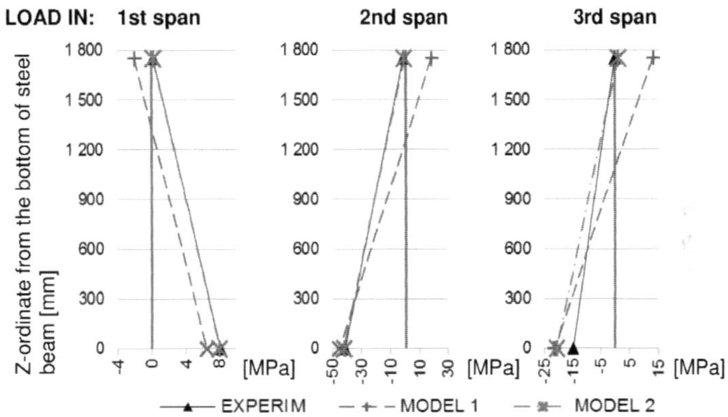

Figure 4. Stresses distribution through height of left girder above middle pier under three chosen load position.

In addition, the extra field strain measurements have been executed in centre of the second largest span and at the pier bridge cross-section. For this purpose, nineteen strain gauges were used as illustrated in Figure 3. In vicinity of these segments, the superstructure was modelled with a higher degree of mesh refinement to achieve the well-defined force distribution. Firstly, our numerical finite element model 1 shown in Figure 2 takes into account at the 15 % of the hogging spans adjacent to the support only the steel girder and longitudinal deck reinforcement, as given in EN 1994-2 [3]. Over the internal pier, alternatively model 2 has considered unchanging effective width in hogging region, despite the fact that concrete is subject to cracking for a number of reasons, including direct loading and shrinkage. Stress distributions in Figure 4 indicate that numerical model 2, considering constant concrete effective width of the deck could provide results closer to the measured values. The investigation confirmed evidence that, when transverse reinforcement appropriate to the shear connector spacing is provided, even the cracked slab is able to transfer shear to longitudinal reinforcement at a distance of several slab thicknesses on either side of the steel girders. As conclusion, it can be confirmed that the theoretical model could reproduce suitably real flexural bridge behaviour and the criteria imposed for loading test had been satisfied.

3. Steel-concrete composite expressway bridge

3.1. Structural concept

Two similar parallel continuous composite expressway bridges have two end spans of 21.0 m and 20.0 m as well as intermediate ones of 26.0 or 25.5 m extending over fourteen spans. Thus, the superstructure of every one bridge is 339.0 m long with connected concrete deck 13.55 m wide [4]. The roadways have a width of 11.25 m, beginning per transit curves and in the second spans proceeding via circular bents of 930 m road radius. Figure 5 shows a typical bridge cross-section consisting of the deck 300 mm thick of concrete class C 35/45 and four built-up girders of I-sections axially 3.0 m spaced. The steel plate girders selected for each bridge have a slender web 1.2 m depth increasing at 1.8 m by haunches next to intermediate supports. Transverse intermediate stiffeners welded to both sides of the web were placed in the third parts of spans allowing webs to develop buckling capacity. The top flange acting with the concrete deck is of the constant 360 mm width and proportioned by varying thickness from 20 to 40 mm to respect bending moment's differences. The bottom flanges of even 650 mm width remain of constant 40 mm thick. Structural carbon steel S355 grade have been used for steel bridge structural parts. Cross bracings above piers consists of angel's diagonals and horizontal channels. Other cross girders are located in the thirds of spans. Their spacing is compatible with transverse stiffeners. Headed stud connectors ϕ 19 x 150 mm were used for transmitting longitudinal shear between concrete and steel part of composite structure.

Figure 5. Cross-section of composite structure for the right express lane.

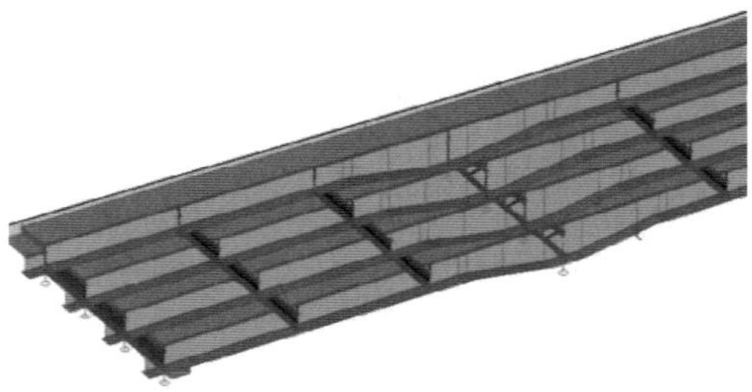

Figure 6. Selected viable finite elements structural modelling details.

Four trucks as testing load of the average weight around 41.15 tones with variation of 5 % were imposed in designed position, determined by calculation. Measurement of steel beams deflections and bearing settlements were completed in the two edge spans by strains measurements. Electrical resistant gauges were used in the mid-spans and above the adjacent piers.

3.2. Calculation models and testing

Three types of numerical models were used for structural analysis of the bridge. The first one was based on three-dimensional structural finite analysis. This model 1 has approximated concrete deck with plate elements of variable parameters. The steel girders as well as cross beams were meshed by beam elements, as it can be seen in Figure 6. The real structure curvatures and ways of bearing movement were implemented in this model, as well.

In the alternative grid model 2, the main girders consisted of steel-concrete composite_sections. The crossbeams and other bracing were connected to the girders, taking into consideration the vertical axe differences of the structural member. The slab strip 2.0 m wide simulated concurrent working of the concrete deck. The slight curvature of the structures was neglected in this model.

The third numerical approximation, specified as the model 3, might represent a quite simple approach. The longitudinally separated steel-concrete main girders without any interaction had to transmit an equivalent part of load applied on the deck, determined by conventional method supposing grid with an infinitely rigid cross-bracing.

Influence of concrete cracking on flexural stiffness modification may be considered by several methods. For this case, in the model 1, the simplified ω–method according to our research [5], introducing additional rotation due to crack development in hogging regions has been also used. The lines corresponding to this approach are labelled as model 1-ω.

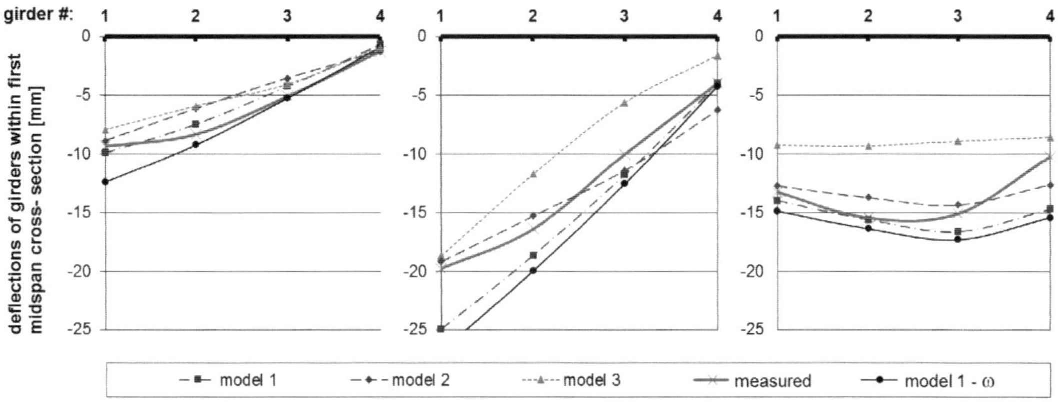

Figure 7. Girder deflections in the mid cross-section of the first span for the chosen testing load positions from I. to III.

Several locations of testing loads have been managed during the test. In Figure 7, deflection of all four girders in the middle of the first span caused by three load cases is presented. First load position (I.) represents testing loads unsymmetrically placed within the cross section in both adjacent spans (the first and the second). In the second load case (II.), the heavy weight lorries in the first span stayed in the same position, while the second span was unloaded. During the third load simulation, only the first span was loaded again, but the load was rearranged to a symmetric position within the bridge width.

The resulting deflections of girders illustrated by curves in Figure 7 are showing that the simplest model of separated beams may deliver somewhat inaccurate deformations of cross-sections. The grid model 2 and especially the complex tree-dimensional model 1 can reflect more properly flexural bridge behaviour. Anyway, consideration of concrete cracking did not lead into improvement of the results in all cases.

4. Steel-concrete composite highway bridge

4.1. Structural concept

Roadway deck of this highway bridge is supported by two girders and slim, rather high piers. Because of an angular crossover and arch curvatures of side road approaches, theoretical spans of main girders are 69.0 + 81.0 + 87.0 + 75.0 + 63.0 + 45.0 m. The roadway has a width of 11.5 m between barriers and the total deck width varies from 14.11 to 16.61 meters. Thus, this composite superstructure defines two-lane overcrossing over its whole 422.0 metres length [6]. Figure 8 is showing the bridge layout consisting of the concrete slab and only two built-up plate girders of I-section axially 6.7 m spaced.

The steel plate main composite girders were selected for the bridge having the constant depth of the slender web 4.94 m. Three longitudinal angle stiffeners welded to the outer sides of the webs were placed for meeting slenderness requirements and allowing developing shear capacity as well as bending resistance by preventing local buckling. The top flanges that act with the concrete slab are of the constant 900 mm width and proportioned by varying thickness from 25 to 40 mm. For increasing the flexural strength of cross-sections, the bottom flanges are 1 200 mm width, and vary in the thickness from 30 to 55 mm. The structural steels S355 and S420 as well as S460 have been used for the steel bridge structural parts. Truss cross-frames consisting of horizontal chords and diagonals made of HEB sections provide the lateral stability of the plate-girders and help to distribute vertical loads. The concrete class C 35/45 was used in the slab 250 mm thick with haunches increasing at 450 mm above the girders. Shear stud connectors Ø 19/150 from steel grade S235 should ensure a full composite action.

Solid wall piers of oblong section, both slender and aesthetically appealing provide vertical supports for spans at intermediate points. The seat-type abutments were constructed separately from the bridge superstructure.

4.2. Transformation model and testing under static loads

The concrete deck and steel main girders were approximated with shell finite elements. Varying thicknesses of elements were adopted for better account of variable deck as well as in fact fifteen different girder sections, as shown in Figure 9. Cross truss beams as well as support diaphragms were meshed by member elements.

Figure 8. Bridge structure during erection.

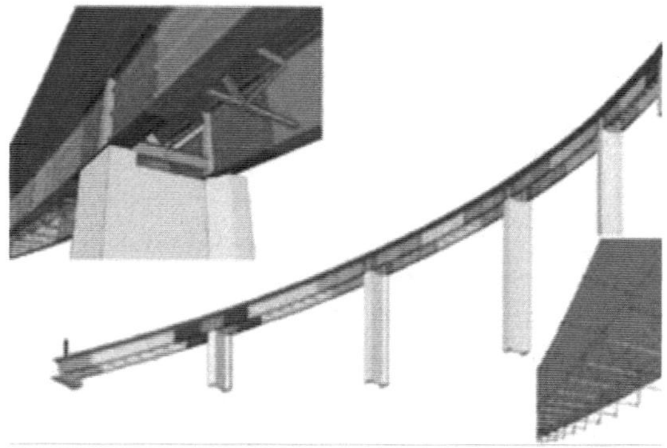

Figure 9. Bridge finite element modelling.

The efficiency factor as relation η between experimental deflection S_e = 24.6 mm produced by test loading and calculated values S_{cal} = 26.0 mm in the middle of the fifth span developed by code ideal load model LM1 [7], in the case of this bridge structure $\eta = 0.94$ imposed eight trucks, having the total weight of 42.0 tons each. This additional condition of the remaining deformations was also satisfied, since extreme ration $S_r/S_{tot} = 0.086 < 0.10$. This static testing has approved sufficient conformity of the behaviour of the structure with the design calculations.

4.3. Dynamic load testing

The bridge was tested also dynamically through passages of two fully loaded four-axle trucks, with a normal axle spacing of 4.5 m. The gross weight of the vehicle rested near the allowed limit of 42 tons as the maximum legal gross weight of vehicle. The test vehicles were driven at constant speed, along the longitudinal bridge axis and always in the same direction. The tests were begun with a speed of 5 km/h, which was then increased after every passage in steps of 5 to 10 km/h up to the maximum speed 50 km/h. Tests on the undisturbed pavement were repeated with a plank placed on the roadway in the main measurement cross section. The plank was approximately 60 mm thick and 500 mm wide.

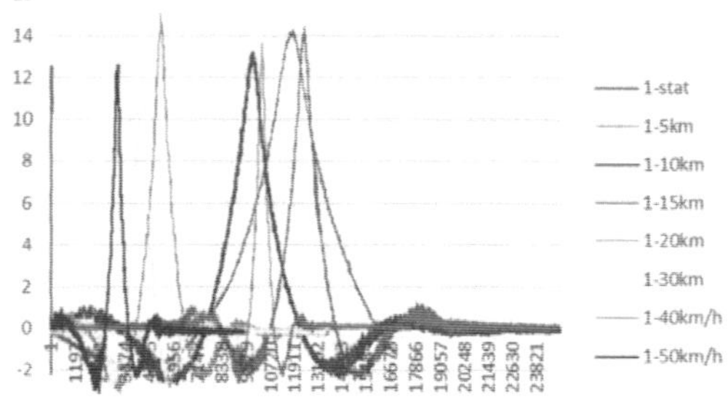

Figure 10. Strains of the midpoint of the third 87 m long span under the passage of vehicles traveling from 5 km/h to 50 km/h on the undisturbed pavement.

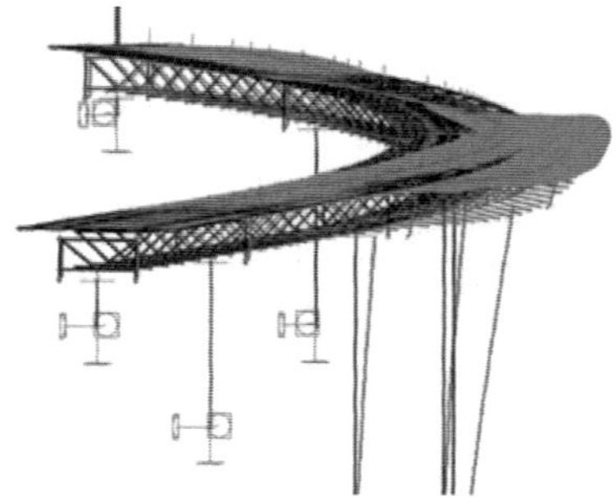

Figure 11. Second fundamental mode of bending and torsional oscillation.

Produced strains were measured at the mid-point of the second and third spans, as the bridge main characteristic measurement cross sections. The mechanical vibrations were recorded by two accelerators. The strain time dependent variations, transformed at electrical signals were registered by measurement accuracy apparatus Spider, controlled by a notebook.

From the registered dynamic signals showed as example in Figure 10, the fundamental frequency $^{obs}f_{(2)} = 1.05$ Hz, very close to theoretical value $^{theo}f_{(2)} = 1.03$ Hz was obtained for the second proper mode of the loaded bridge oscillations, as illustrated in Figure 11. Then logarithmic damping decrement, which required measurement of the magnitude of the first S_{fund} and last strain amplitudes S_i having the sane phase, could be identified. Considering the number of periods i, the damping was determined from $\upsilon = ln\ (S_{fund}/S_i)/i$, and going from 0.052 to 0.064. From the peak value S_{max} of the bridge response during passages of the test vehicles and strain observed under static loading with the same vehicles S_{stat} at Figure 10, the dynamic increment can be found, as $\phi = S_{max}/ S_{stat}$. Finally, the bridge critical velocity c =26 km/h was identified as corresponding to the maximum dynamic increment $\phi_{max} = 1.15$.

5. Concluding remarks

Generally, the actual stiffness of the tested structures turned out to be higher than the values predicted analytically. The primary reasons for this were the unintended composite action of non-structural members and partial fixity of supports. Thus, the bridge case studies in this paper might illustrate, that loading tests together with analytical modelling represent very powerful control mechanism. Practically, all relevant differences, defects or even static imperfections can be revealed before putting a bridge into exploitation. Therefore, experimental data and numerical models can present powerful tools for identification of real bridge behaviour, the integrity of the structure evaluation and validation and optimisation of the structural design.

Acknowledgements

The paper presents results of the research activities supported partly by the Slovak Cultural and Educational Grant Agency; grant No. 019ŽU-4/2016 and Science Grant Agency project No. 1/0343/18.

References

[1] Bujňák J and Odrobiňák J D 2014 The footbridge SO-206 at the highway D1 Dubná skala – Turany *Final report from proof load test* (University of Žilina, Faculty of Engineering, Slovakia)

[2] Slovak Institut of Technical Standardization (SUTN) 1993 Proof load tests of bridges, *Slovak Technical Standard (STN 73 6209)*

[3] European Committee for Standardization (CEN) 2005 Design of composite steel and concrete structures; General rules and rules for bridges *Eurocode 4, Part 2 (EN 1994-2)*

[4] Bujňák J and Odrobiňák J D 2015 The bridge SO-211 at the expressway R2 Pstruša – Kriváň *Final report from proof load test* (University of Žilina, Faculty of Engineering, Slovakia)

[5] Bujňák J and Odrobiňák J D 2005 Cracking of Concrete Deck in Composite Structures *Proc. Int. 4th European Conf. on Steel and Composite Structures "Research – Eurocodes – Practise"* (Maastricht) (Amsterdam: Elsevier) Volume B, pp. 4.2-15 – 4.2-22

[6] Bujňák J and Gocál J D 2017 The bridge SO-242 at the highway D3 in 25,140 km Svrčinovec – Skalité *Final report from proof load test* (University of Žilina, Faculty of Engineering, Slovakia)

[7] European Committee for Standardization (CEN) 2003 Actions on Structures; Traffic loads on bridges *Eurocode 1, Part 2 (EN 1991-2)*

IOP Conf. Series: Materials Science and Engineering **419** (2018) 012012 doi:10.1088/1757-899X/419/1/012012

Numerical simulation of rib distortion in orthotropic steel decks

Heng Fang[1], Evy Van Puymbroeck[1], Nouman Iqbal[1], Zain Ul-Abdin[1] and Hans De Backer[1]

[1]Ghent University, Technologiepark 904, Ghent 9000, Belgium

E-mail: Heng.Fang@UGent.be

Abstract. When adopting the trapezoidal section of ribs in orthotropic steel decks, the torsional rigidity of ribs losses radically because of rib distortion especially when the load is eccentric from the axis of the rib. Hence, the rib distortion causes high stress concentrations at the rib-to-crossbeam joint which makes this joint easily prone to fatigue. In this paper, the influence of 4 conventional design parameters on distortional behaviour is investigated by numerical simulation adopting the hot spot stress approach and the sub-model analysis technique. A refined finite element model of an orthotropic steel deck specimen with the dimension of 8.2 m × 4.1 m is built. Research results indicate that the distortional stress rises with the increment of the spacing of crossbeam or the weld length of rib-to-crossbeam joint. The thickness of rib is negatively correlated with the distortional stress. The influence of the spacing of rib on distortional stress depends on the relative size between the width of load area and the spacing of rib. The most unfavourable load position is determined by rib geometries, i.e. the thickness of ribs, whereas rest three parameters do not have obvious influence.

1. Introduction

The orthotropic steel deck is widely adopted in long-span bridges around the world since it has many merits e.g. high load carrying capacity, light weight, rapid construction, etc. However, stress concentrations caused by complex structure and the existence of numerous welds have impeded the further usage of orthotropic steel deck. According to previous research, there are four positions in total where fatigue cracks typically appear: rib-to-deck joint, splice joint of the longitudinal rib, deck plate at the position of crossbeam and rib-to-crossbeam joint [1]. Even though relevant research of these fatigue-prone details had been reported, the stress concentration at the rib-to-crossbeam joint caused by rib distortion was more or less ignored in previous research.

In orthotropic steel decks, one of the broadly adopted rib geometries is the trapezoidal section. Nevertheless, the torsional rigidity of rib is essentially reduced by distortion because of the existence of cut-outs in crossbeams, especially when the load is eccentric from the axis of the rib [2]. High stress concentrations are caused by the rib distortion at the rib-to-crossbeam joint. Moreover, the eccentric load that causes tension stress on one stem of the rib always causes compression stress on the other stem of the rib in the meantime. Thus, the rib-to-crossbeam joint carries positive and negative stress alternately when vehicles passing by, which makes the joint more vulnerable to fatigue damage. Unfortunately, no explicit recommendations on relevant parameters with respect to the influence of rib distortion are given in current major design codes.

This paper presents the influence of a number of classical design parameters on the distortional behavior. Firstly, the finite element model of an orthotropic steel deck specimen (8.2 m × 4.1 m) was developed. The sub-model including the investigated rib-to-crossbeam joint was also built to achieve a more accurate assessment. Then, the influence of the considered parameters was calculated and analyzed by getting influence lines of different parameters, which will provide the foundation for relevant experiments of the orthotropic steel deck specimen in the nearest future.

2. Numerical simulation of rib distortion

2.1. Finite element model

In order to investigate the influence of classical design parameters on the distortional behavior, parametric analyses were performed by ANSYS. The standard orthotropic steel deck is composed of 6 ribs, 3 crossbeams and 2 main girders. The dimensions of the longitudinal ribs are 300 mm high, 300 mm wide at the top and 125 mm wide at the lower soffit. Relevant geometries are shown in Figure 1.

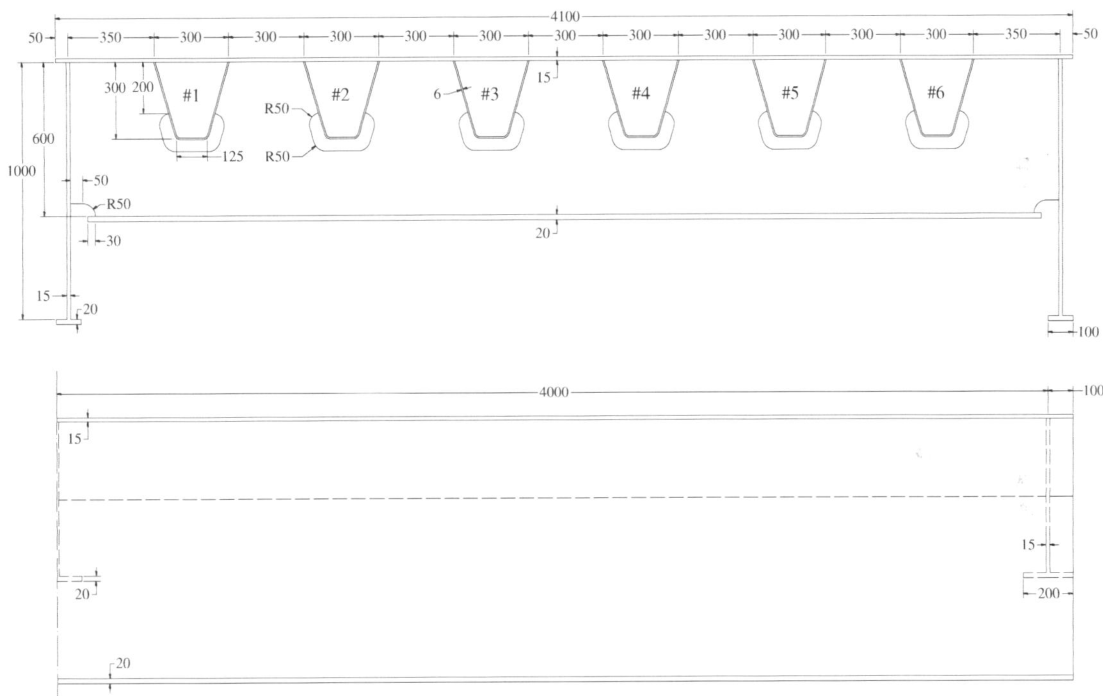

Figure 1. Geometries of the standard orthotropic steel deck (unit: mm)

Based on the geometries of the standard orthotropic steel deck, a global model was first built using shell element with relatively coarse mesh. Then, a sub-model was developed using solid element with the reference of the global model, which took the influence of welds into account. The sub-model was composed of segmental deck, crossbeam, rib and relevant welds as shown in Figure 2(b). The position of the sub-model was at the intersection of the third rib (see Figure 1) and the middle crossbeam, as shown in Figure 2(a). The cross section of weld in the sub-model was modelled as a triangular. The weld throat thickness of 4 mm was adopted.

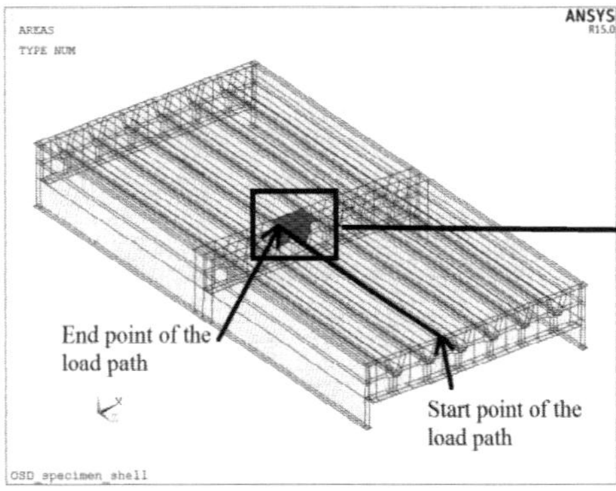

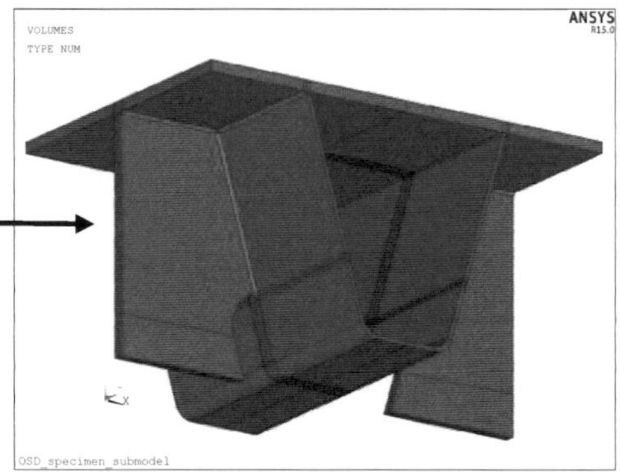

(a) global model (b) sub-model

Figure 2. Finite element models: (a) global model; (b) sub-model

The numerical simulation is based on linear elastic properties of the material. The modulus of elasticity is 210 GPa for steel and the Poisson's ratio in the elastic stage is 0.3. The structural hot spot stress approach is more appropriate for this joint according to previous research [3]. Meanwhile, the liner extrapolation method is also adopted. Based on the IIW recommendations [4], two reference points are at a distance of 0.4 times and 1.0 times the thickness of the rib separately. The mesh size of the elements at the investigated area is 0.1 times the thickness of the rib in order to fit the IIW recommendations. Figure 3 displays locations of two reference points and relevant mesh nearby.

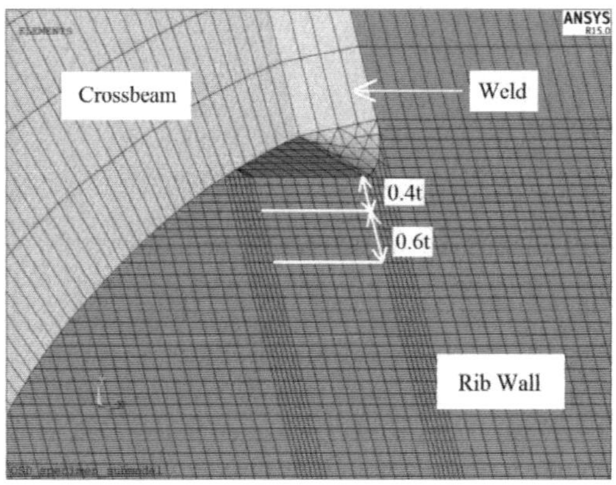

Figure 3. Reference points and mesh

2.2. Boundary conditions and load case

The boundary conditions of the global model are based on the real support conditions of the orthotropic steel deck specimen. Six bearings are situated at both ends and at the middle of two main girders separately. The contact area of a bearing is 80 mm × 80 mm. The boundary condition of the sub-model comes from the calculation result of the corresponding global model. Displacements and rotations of nodes along the cutting boundary between two models are placed on the sub-model as external loads.

In order to achieve the most unfavourable load position of distortional stress and to analyse the influence of different parameters, a movable unit load is considered to get the influence lines of distortional stress. The unit pressure of 1 MPa is adopted with the load area of 100 mm wide and 100 mm long and a load of 10kN. The load path is at the central line between rib 2 and rib 3 as shown in Figure 2(a). The interval distance between two loading points is 200 mm.

2.3. Results and discussion

In total 4 classical design parameters are considered, namely the thickness of rib, the transversal spacing of rib, the weld length of rib-to-crossbeam joint and the longitudinal spacing of crossbeam. Specially, the thickness of rib is determined by the stiffness of rib, which is normally selected in accordance with the traffic category, etc. The second moment of area of ribs acts as the constant reference when adjusting the thickness of rib. Relevant models with various parameters are shown in Table 1. The hot spot stress of following figures is the average of hot spot stress calculated from the two reference lines which are shown in Figure 3.

Table 1. Parameters of geometries

Model number	Height of rib (mm)	Upper width of rib (mm)	Bottom width of rib (mm)	Thickness of rib (mm)	Spacing of rib (mm)	Spacing of crossbeam (mm)	Weld length of rib-to-crossbeam joint(mm)
0(standard)	300	300	125	6	300	4000	200
1	275	-	-	8	-	-	-
2	255	-	-	10	-	-	-
3	-	-	-	-	200	-	-
4	-	-	-	-	375	-	-
5	-	-	-	-	-	2500	-
6	-	-	-	-	-	5000	-
7	-	-	-	-	-	-	150
8	-	-	-	-	-	-	240

2.3.1. Rib thickness

Figure 4 shows the influence lines of different rib thicknesses. Generally, the distortional stress decreases as the rib thickness increases. On the tension side, the maximum distortional stress decreases from 10.27 MPa to 5.27 MPa when the rib thickness increases from 6 mm to 10 mm. On the compression side, the maximum distortional stress decreases from 7.94MPa to 4.19 MPa as the rib thickness increases. Figure 4 also presents that the most unfavorable load position moves toward the middle crossbeam for about 400 mm when the rib thickness changes from 6 mm to 10 mm.

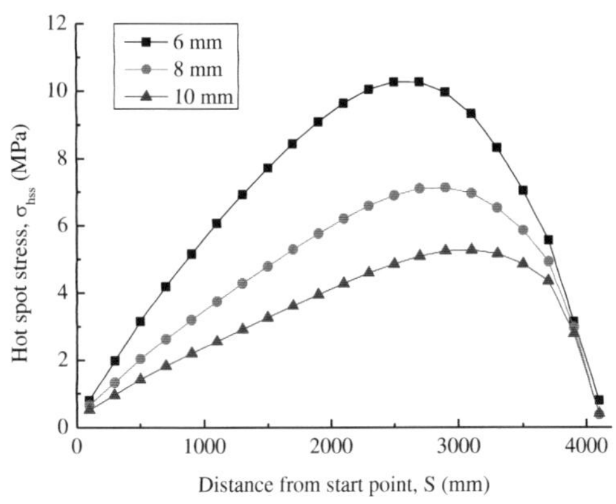

(a) tension side

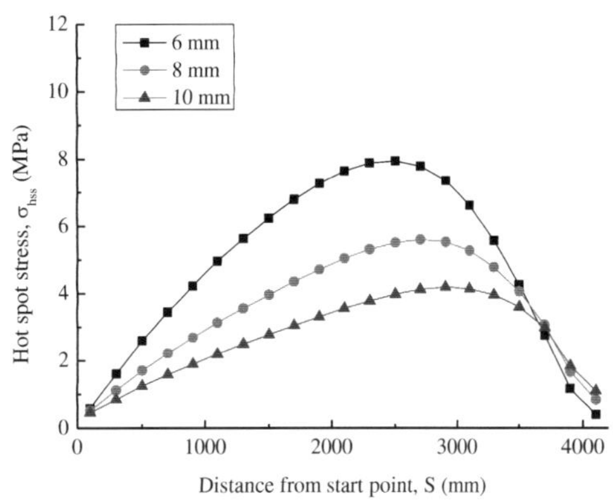

(b) compression side

Figure 4 Influence lines of different rib thicknesses: (a) tension side; (b) compression side

IOP Conf. Series: Materials Science and Engineering **419** (2018) 012012 doi:10.1088/1757-899X/419/1/012012

2.3.2. Rib spacing

Figure 5 displays the influence lines of different rib spacing. When the rib spacing decreases from 375 mm to 200 mm, the maximum distortional stress increases from 9.30 MPa to 10.18 MPa on the tension side and the maximum distortional stress increases from 7.07 MPa to 8.16 MPa on the compression side. According to previous research, when adopting a larger load area, the distortional stress would increase when the rib spacing increases [5]. Thus, the load dispersal is correlated with the relative size between the rib spacing and the width of load area. The most unfavourable load positions of both sides are not influenced by the changes of the rib spacing , which are about 400 mm away from the middle span in the direction to middle crossbeam.

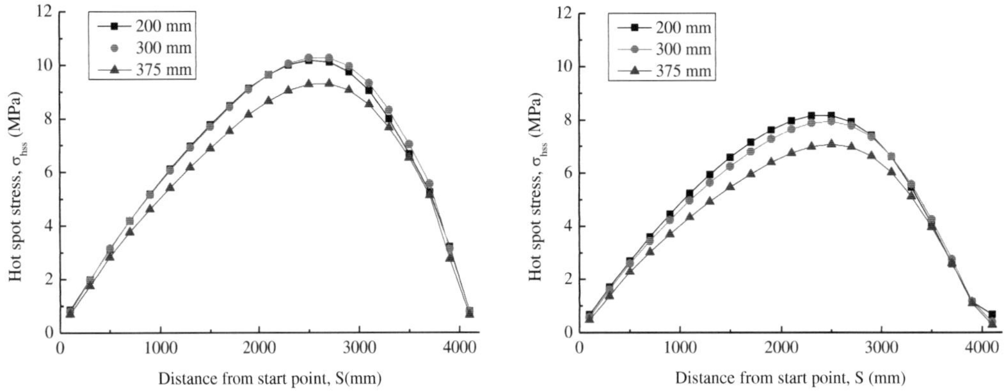

(a) tension side (b) compression side

Figure 5 Influence lines of different spacing of rib: (a) tension side; (b) compression side

2.3.3. Crossbeam spacing

Figure 6 presents the influence lines of different spacing. The distortional stress and the crossbeam spacing are also positively correlated crossbeam. On the tension side, the maximum distortional stress decreases from 11.57MPa to 6.90MPa as the crossbeam spacing decreases from 5000 mm to 2500 mm. Meanwhile, on the compression side, the maximum distortional stress decreases from 9.14MPa to 5.28MPa. The most unfavourable load position is around 1600 mm away from the middle crossbeam, which does not change with the crossbeam spacing.

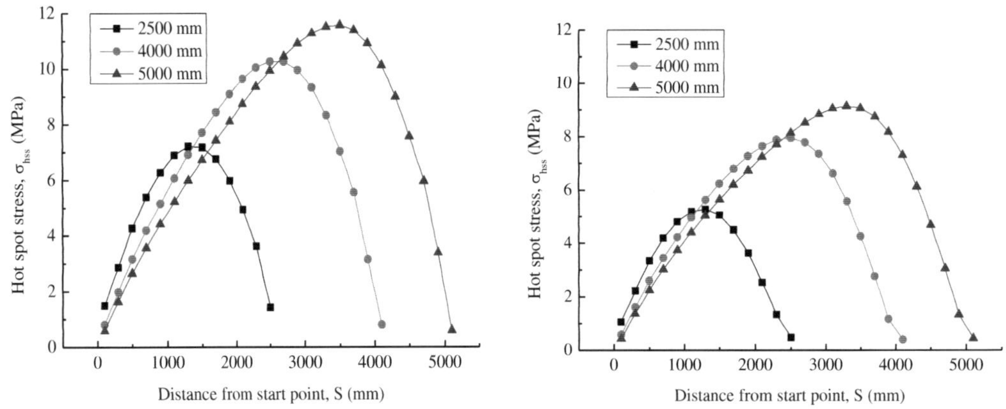

(a) tension side (b) compression side

Figure 6 Influence lines of different spans: (a) tension side; (b) compression side

2.3.4. Weld length of rib-to-crossbeam joint

Figure 7 shows the influence line of different weld length of rib-to-crossbeam joint. The distortional stress and the weld length of rib-to-crossbeam joint are positively correlated as well. On the tension side, the maximum distortional stress increases from 9.32 MPa to 11.18 MPa while the weld length of rib-to-crossbeam joint increases from 150 mm to 240 mm. At the same time, on the compression side, the maximum distortional stress increases from 6.20 MPa to 9.44 MPa. The most unfavorable load position is not affected by the weld length of rib-to-crossbeam joint which remains around 1600 mm to the middle crossbeam.

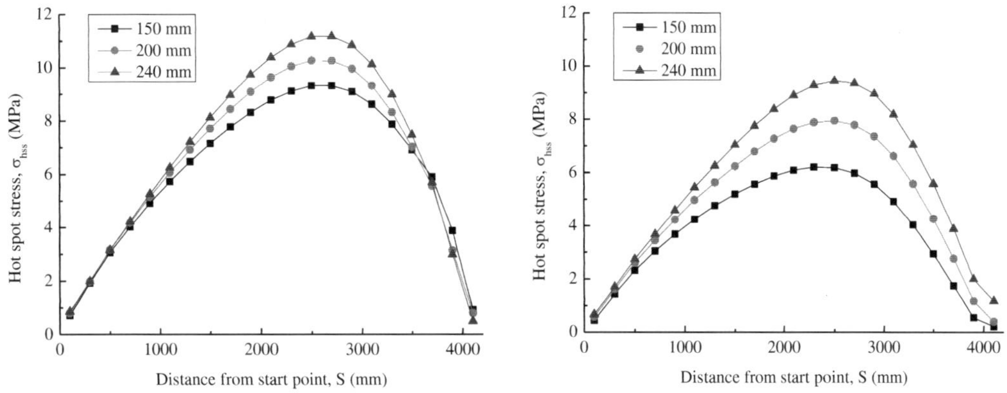

(a) tension side (b) compression side

Figure 7 Influence lines of weld length of rib-to-crossbeam joint: (a) tension side; (b) compression side

3. Conclusions

This paper presents the parametric analysis of 4 classical design parameters on the distortional behavior of ribs in orthotropic steel deck. Based on the presented results, the following conclusions can be drawn:

- Among these 4 parameters, the spacing of crossbeam and the weld length of rib-to-crossbeam joint have positive correlations with the distortional stress. The rib thickness has a negative correlation with the distortional stress. The influence of the rib spacing on distortional stress depends on the relative size between the width of load area and the rib spacing, which requires further research.
- The most unfavourable load position of distortional stress is determined by the geometries of ribs, whereas rest three considered parameters of this paper do not show obvious influence. In this case, the most unfavourable load position of distortional stress is at a distance of about 1600 mm away from the investigated crossbeam. When the rib thickness increases, the most unfavourable load position moves toward the investigated crossbeam.
- Analysis results will be verified by experiments in the nearest future. Simplified calculation method of distortional stress and possible alternatives to reduce the distortional behaviour require further research.

References

[1] Nagy W, Wang B and Culek B 2017 Development of a fatigue experiment for the stiffener-to-deck plate connection in orthotropic steel decks *International Journal of Steel Structures* 17 pp 1353-1364

[2] Connor R, Fisher J and Gatti W 2012 *Manual for Design, Construction, and Maintenance of Orthotropic Steel Deck Bridges* (Pittsburgh: HDR Engineering Inc.)

[3] Kolstein M H 2007 *Fatigue classification of welded joints in orthotropic steel bridge decks* (Delft University of technology, Faculty of Civil Engineering and Geosciences)

[4] Hobbacher A F 2015 *Recommendations for Fatigue Design of Welded Joints and Components*

(Switzerland: Springer International Publishing)
[5] Fang H, De Backer H 2018 Parametric analysis of rib distortion in orthotropic steel decks *9th International Conference on Bridge Maintenance, Saffety and Management* In press

Hybrid measurement techniques used to a study of historic cast iron suspension bridge

Wojciech Anigacz[1], Damian Beben[1] and Jacek Kwiatkowski[2]

[1]Opole University of Technology, ul. Proszkowska 76, 45-758 Opole, Poland
[2] J&L Consulting, ul. Obroncow Stalingradu 29, 45-594 Opole, Poland

E-mail: w.anigacz@po.opole.pl

Abstract. Traditional measurement techniques consist of discrete measurements and they are time-consuming. Moreover, information obtained by means of these methods is limited to over a dozen points located on the examined bridge. The latest measurement methods such as laser scanning and digital photogrammetry enable to obtain quasi-continuous surface of the tested bridge. It consists of many millions of points with known spatial coordinates. The tested structure is a historic suspension bridge in the town of Ozimek in Poland. Nowadays the bridge is intended for pedestrian traffic across Mala Panew river. It was built in 1827 and it is the oldest cast iron suspension bridge in continental Europe. In the analyzed case, the measurements were performed with the use of terrestrial laser scanning (a FARO Focus 3DX130) and high-resolution photogrammetry (Canon 650D camera with the lens of fixed focus of 85 mm and the GigaPano turntable). The obtained results indicate that the differences between the two methods are within a few millimeters, which should be considered satisfactory. The advantage of the presented measurement methods is minimizing time of fieldwork and obtaining a 3D model of the bridge with millimeter accuracy.

1. Introduction

In most cases control measurements are held on bridge structures to confirm design basis or as diagnostic checks to ensure proper maintenance. Inspection results of bridges constitute basic source of information about their technical condition and existing damages. In majority of countries in the world they are obligatory and are held on the basis of a systematic condition check of transportation infrastructure. Typical measurements and tests are usually executed by bridge inspectors with the use of traditional methods. The majority of maintenance or inspection works require direct contact with the structure, which, in some conditions, can cause problems with structure accessibility for tests and often requires using scaffolding or lifts. In many cases, especially historic bridges, design documents do not exist. This causes a major problem at renovation and/or maintenance works. Maintenance or inspection works are very time-consuming, especially when complex and complicated structures are concerned. Most of researchers [1,7,9] have a classic approach to the structure examinations, which means that they put a reference measurement grid outside the tested object and on the object itself. They mark or signalize points that are measured. Such approach is fully correct in the case of conducting long-term tests. The reference points outside the tested structure allow to determine changes of the bridge structure geometry in the absolute frame of reference comprised of the reference points, with coordinates determined, for instance, using the GPS method.

The method of laser scanning allows to obtain a cloud of points [4], and, as a result, it can be used with success for survey checks of various structures, especially ones with difficult access. Yu et al. [9] reported a very interesting comparison of the measurement methods, measurement accuracy, measurement type, bridge type and measurements costs. However, the presented information, e.g. with regards to the laser scanning, turned out to be too optimistic because obtaining accuracy below 1 mm is in practice limited to small bridges only (where the distance between the scanner and the bridge is small). While Tang et al. [8] compared potential of the tachymetry method and laser scanning in the measurements of bridges. They reported values of the measurement errors on real objects.

This article verifies the method of laser scanning on the example of a historic suspension bridge. Measuring capabilities of this method have been analyzed, i.e. the quality of inventory works was checked with use of a laser scanner on the historic cast iron bridge. Fast developing measurement techniques, especially geodetic, and in particular the mass measurements, in theory and practice, enable to obtain full information on the structure and changes in the bridge geometry [1]. The authors of the article have such technology, the latest equipment (often prototype-pre-production), the latest software, and adequate knowledge and experience [6]. Similar measuring techniques used for dynamic measurements of a technological suspension bridge are presented in papers [2,3]. It was found that one of the most important factors affecting the accuracy of the measurement is the distance of the scanner, camera or total station from the object being tested. This paper presents continuously modified measuring methods, equipment and software on the example of a relatively small bridge, but it is very interesting structurally and architecturally.

2. Description of bridge structure

The object is a historic bridge on Mała Panew river in the town of Ozimek, district Opole, in Opole Voivodship in Poland (Fig. 1). The parameters of the bridge are as follows: span – 27.7 m, axial spacing of pylons – 31.5 m, width of the bridge deck – 6.6 m (Fig.2).

a) b)

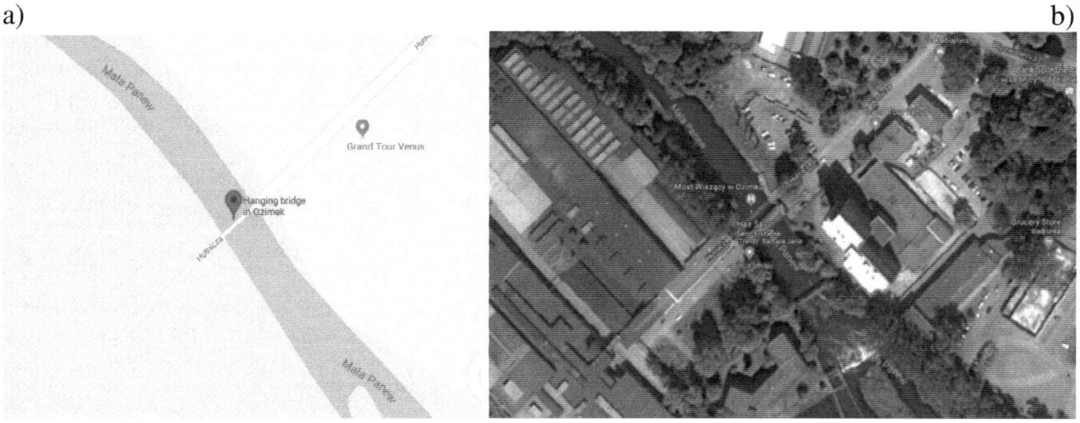

Figure 1. Location of the bridge: a) on a map, b) on aerial photograph (www.maps.google.pl).

Nowadays the bridge is intended for pedestrian traffic across the Mala Panew river. It was built in 1827 and it is the oldest cast iron suspension bridge in continental Europe. It is a suspension chain bridge completely made of cast iron. Following this parameter it should be acknowledged to be the oldest bridge of this type in the world existing till today. An older existing iron suspension bridge was built between 1819 and 1826 over the Menai Strait in Wales, however this one is suspended on stone pylons. Complete renovation of the bridge was carried out between 2009 and 2010. It was disassembled, renovated and reassembled. The bridge structure was strengthened with steel ropes. In the FOOTBRIDGE AWARD 2014 world competition the bridge won in the category of historic renovation/reuse of existing footbridges as the best renovated structure [5].

Figure 2. View of the historic bridge on Mała Panew river in Ozimek.

3. Methodology of research

The presented historic bridge on Mała Panew river in Ozimek is characterized not only by historical, technical and technological originality, but also by its design and cultural values. Due to the above conditions, it is desirable to use a wide range of measuring techniques to show not only the key structural elements, but also the aspects of aesthetics.

The method of bulk measurement, i.e. laser scanning and digital photogrammetry was used for survey check of the structure. In the presented case, it was intended to minimize preparatory works. A decision was made to use well-visible details of the bridge structure as measurement points. Vertical verification of selected points on the bridge was executed by means of spirit leveling with use of the patent [1]. To verify distance measurement of characteristic points of the structure by means of scanning and photogrammetric methods, total station TC2002 Leica was used.

3.1. Laser scanning method

Laser scanning method is a relatively new measurement method allowing to obtain a quasi-constant, spatial representation of the area of the examined structure seen in central view. The accuracy of the results depends mainly on distance between the scanner and the examined structure. In the last several years, fast development of laser scanning has been observed [2, 4]. A scanner uses a laser beam which, after being reflected from the measured structure, returns to the scanner. The distance is determined on the basis of the difference of phases of the generated and the reflected beam. During the measurement, scanner can be turned by 360 degrees horizontally. Horizontal and vertical angle of the scanner mirror set, information on the measured distance and quality of the reflected signal are calculated in real time into the coordinates.

The measurements were using a TLS (Terrestrial Laser Scanning) method by FARO Focus 3DX130 scanner. Scanning was executed in 13 positions (Fig. 3). Density of the registered cloud of points allowed for fully satisfactory identification of elements and even details of the bridge.

a)

b)

Figure 3. Distribution of scanner positions: a) a view from above, b) a side view.

In the case of the bridge the results clearly indicate that the distance to the bridge about 40 m allows to obtain satisfactory results, i.e. a full inventory of the bridge elements. For the bridge experimental examinations a FARO Focus 3D X130 scanner was used which measures up to 1 million of points per second and allows to create clouds of over a billion of points. The accuracy, declared by the accuracy of the measurement of the distance was ±2 mm, and angle resolution amounted at 0.009° (0.16 billion). The standard unit of resolution is dpi, i.e. a dot per inch of length. In technical applications where most frequently elements of the examined structure are located in different distances from the scanner, it is more convenient to use the angle resolution, and not the dpi which constantly changes as the distance changes. Data obtained by means of the scanner were then analyzed with help of a functionally advanced software Trimble Real Works Advanced Plant version 10.3.

3.2. Photogrammetric method

The second used method was a high-resolution photogrammetry. A Canon 650D camera with the lens of fixed focus of 85 mm and a GigaPano turntable were used to take photos (Fig. 4a). Nine 360°

a)

b)

Figure 4. Positions of the measuring equipment used: a) GigaPano turntable with Canon 650D camera, b) FARO Focus 3D X130 scanner.

panoramic photographs of high resolution were taken (Fig. 5). The aim was to ensure that the photos are taken from the exact scanning positions. (Fig. 4b).

Figure 5. The 360° panoramic image taken with use of Canon 650D camera from a position located within the half of the bridge span.

The level of spherical aberrations of the camera was confirmed in the laboratory on the test rectangle (Graph paper 420×297 mm), it did not exceed 3 columns of the matrix pixels, to be exact 5184×3456 pixels. To process the pictures Kolor AutoPano Giga 4.0 software was used. To calibrate the generated panorama through rescaling for defining the "size" of individual pixel, the Photoshop CC computer software was used. The Photoshop CC software enables linear rescaling of the pixel size. To scale the panorama of the bridge to the set pixel size, characteristic distances were used, which in the studied case were the distances between subsequent segments of the span (Fig. 6) that previously were measured with the tachymetry method. The photogrammetric method has the following advantages, among others: short time of making measurements for taking a picture and their sequences, possibility of visualization with the elements of spatialization, and possibility of multiple interpretations of the measurement. It is a relatively cheap method, and commonly available.

4. Analysis of the results

Figures 6 and 7 show location of control sections on photos and scans. Table 1 presents lengths of test sections together with differences in length.

Figure 6a. The photogrammetric rescaled image of the bridge in Ozimek.

Table 1. Tabular summary of lengths of control sections measured by means of the photogrammetry and scanning methods.

Bridge element	Dimensions obtained by means of: photogrammetric method [m]	Dimensions obtained by means of: laser scanning method [m]	Difference [m] (2)-(3)
(1)	(2)	(3)	(4)
Distance in abutments clearing	21.214	21.208	0.006
Spacing of pylons	31.470	31.477	-0.007
Height of the right-bank abutment	3.020	3.012	0.008
Height of the left-bank abutment	3.032	3.022	0.010
Distance between the middle hangers (at the top)	2.922	2.914	0.008
Distance between the middle hangers (at the bottom)	2.924	2.916	0.008
Length of the 4th hanger from the left bank	1.674	1.680	-0.006
Length of the 4th hanger from the right bank	1.672	1.678	-0.006

Figure 6b. Scan of suspension bridge in Ozimek made with use of FARO Focus 3D X130 scanner.

Figure 7a. Control sections shown on a section of 360° panoramic view.

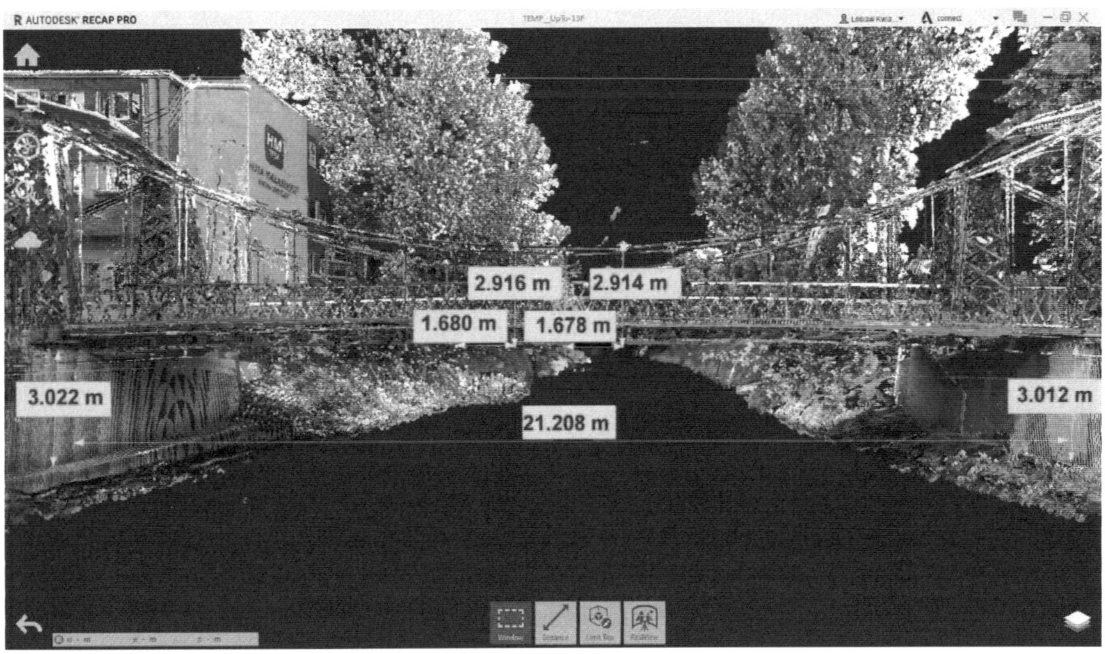

Figure 7b. Control sections shown on a section of the scan.

All the differences fall within the range of -7 to +10 mm. Laser scanning should be considered a more precise method. Individual images were recorded on a 24 Mpix matrix. Resolution is a function of the distance of the camera from the structure. The analyzed control sections were verified with rescaled resolution of the panorama to 1 pixel = 1 mm.

Such a resolution of a 360° panoramic image predisposes it to a highly photorealistic visual evaluation of the bridge and the surroundings. Differences in section length result from accuracy of binding the panorama by software and faithfulness of the parallel projection used i.e. projection of a sphere onto a plane. Measurement of distance on a photograph is made within the plane of the image, counting pixels and multiplying the result by pixel size in accordance with rescaled resolution.

5. Conclusions

As a result of measurements of historic cast iron suspension bridge conducted with use of two methods, i.e. photogrammetric and laser scanning, the following conclusions can be drawn:

1. The measurements unequivocally indicate that for conditions where distance to the bridge is up to about 40 m, the two methods allow to obtain fully satisfactory inventory results. The scans present structure details with sufficient resolution and accuracy.
2. In the photogrammetric method the camera's focus needs to be chosen in such a way as to ensure that the value of a pixel would amount at, for example, 1 mm. A higher scale of the photograms will allow to obtain additional information on the surroundings of the measured point, e.g. permanent deformations, cracks, etc.
3. The results of measurement of various bridge elements with use of two discussed methods confirm that both of them are appropriate for survey checks of bridge structure of complex geometry. The differences in results obtained with use of the respective methods fall within the range of a couple of millimeters.
4. The final analytical data illustrating the effects of the measurement session depend on three main factors:
   - resolution of the matrix of the digital camera,
   - quality of the optics (resolution, reduction of the spherical aberration),
   - quality of parallel projection.
5. Hybrid measurement techniques used to study the bridge are methods which allow to get external object topology and a high quality photorealistic image of the structure. It is of special importance when the object is a structure of historic value with no existing design documentation.

From the presented measurement methods, the equipment and software connected to the laser scanning is the most expensive. It would be ideal to use three methods at the same time for examination of bridges. Then the information about the bridge structure geometry would be a synthesis of all the methods used.

References

[1] Anigacz W, Partyka M A and Tomaszewski S D 1985 Adapter for a levelling rod, *Polish Patent Office* Utility model application number W-72405.

[2] Beben D, Anigacz W 2017 Examine changes in geometric parameters of a suspension bridge using various geodetic methods *Proc. 19th Int. Geodätische Woche Obergurgl 2017*, ed K Hanke and T Weinold (Berlin: Wichmamm) pp 21-30.

[3] Beben D, Anigacz W and Kwiatkowski J 2016 Analysis of suitability of new measurement technologies to examine changes in geometric parameters of a suspension bridge *Maintenace, Monitoring, Safety, Risk and Resilience of Bridges and Bridge Networks*, ed T N Bittencourt, D M Frangopol and A T Beck (London: Taylor & Francis Group) pp 2204-2211.

[4] Beshr A A A 2015 Structural deformation monitoring and analysis of highway bridge using accurate geodetic techniques. *Engineering* **7** pp 488-498.

[5] Jakiel P D 2014 Refurbishment of iron suspension bridge in Ozimek: general design, *Proc. 5th Int. Conf. Footbridge 2014- Past, Present, Future* (London: Hemming Information Services) pp 154-155.

[6] Kwiatkowski J L 2014 *Construction life cycles monitoring of industrial installations verified by hybrid bulk measurement techniques*. PhD thesis (Opole University of Technology Faculty of Civil Engineering) Supervisor: Anigacz, W, auxiliary supervisor: Beben, D), Opole, Poland.

[7] Schofield W and Breach M 2007 *Engineering Surveying* (Amsterdam: Elsevier) 6th ed.

[8] Tang P, Akinci B and Garrett J H 2007 Laser scanning for bridge inspection and management *IABSE Symposium Weimar 2007* (*Check Engineering and Monitoring—International Review on Quality Control Systems*) pp 17-24.

[9] Yu Y, Liu H, Li D, Mao X and Ou J 2013 Bridge deflection measurement using wireless mems inclination sensor systems. *International Journal on Smart Sensing and Intelligent Systems* **6**(1) pp 38-57.

Structural health monitoring for combined damage states

Martin A Butler[1], James A Swanson[1] and Gian A Rassati[1]

[1]University of Cincinnati, 2850 Campus Way, Cincinnati OH, United States

E-mail: butlerm3@mail.uc.edu

Abstract. The current state of the art in automated structural health monitoring (SHM) is to compare the results of a set of sensors to a set of previously established values, picking the most similar set. To identify a damage state, it must be envisioned upon the creation of the SHM system and analysed to determine the output. Following catastrophic events, bridges may be damaged in several ways in several locations, causing problems for the automated system in identifying the exact repairs that are necessary or whether sufficient strength remains to sustain reduced loading. An alternative method to modelling a massive number of damage states and memorizing them all is presented based on subdivided attractor networks. These artificial neural networks are similar to the feedforward networks that have found much use in industry. Instead of acting as an approximator for some function, these networks work as content addressable memory. Partitioning these networks allows different portions of the network to represent different portions of the structure, and to mix and match states that may be associated with each other. In this way a SHM regimen is produced that needs far fewer memorized states to recall all of the possible damage states.

1. Structural Health Monitoring System

Structural Health Monitoring (SHM) is used to maintain constant knowledge of a structures state, this may be necessary for particularly important pieces of infrastructure, or a bridge that is approaching the end of its design life that it may not be economical to replace. Current state of the art is to compare the results of a set of sensors to known sets of data that have been already established using some algorithm to determine the most similar damage state. Many different approaches have found success including effective independence method [1], Fourier component analysis [2], and time series analysis [3]. In order to do this, damage states must be thought of beforehand so that the structures state may be compared. Combinations of damage are difficult to discern, as they will resemble both of the states in some ways, and neither in some ways. Damage states in one locale may, through an alternative load path, result in a strain state that somewhat resembles damage to the subsystem that provides this alternative load path. The proposed SHM system is meant to avoid these problems, by producing a network of subsystems that communicate with each other regarding the state of each subsystem.

The proposed SHM system uses subdivided attractor artificial neural networks to identify the state of the structure. Specifically, strains are collected describing the bending and axial stresses at multiple members within the subsystem, these strains are fed through a feedforward artificial neural network (FF-ANN). The FF-ANN outputs a bottleneck image for the SA-ANN, a set of interface synaptic weights processes the bottleneck image to the full sized initial image. Finally, the SA-ANN converges to one of its stable neural states, describing the structure.

For this work strains are the inputs considered. Static behaviour is more sensitive to localized damage which is what is most important when identifying multiple damages in one structure [4]. Use of static behaviour has the added benefit of avoiding interference from local modes of vibration overlaid with global ones [2].

1.1. Pony Truss Bridge

The example bridge is a small pony truss bridge. A five panel bridge with a corrugated steel deck, seven rows of stringers, and skew of 20 degrees. The nonlinear behaviour of the connections was modelled using nonlinear links. The nonlinear behaviour is defined in the way suggested in [5]. The only loads applied to the structure were dead loads. Two kinds of damage were introduced to this structure: release and stiffness reduction. Release type damages include the introduction of a release on one of the ends of the frames. What releases and where they were applied depends on the type of member, truss members only experience axial releases, except for vertical members which are zero force members and are not considered for monitoring. Floor beams and stringers release moments, shears, and axial constraints at either end. The second damage type is property modifiers to the member, factoring by 0.3 the area or moment of inertia of the section. The releases and section reductions are shown in Table 1.

Table 1. Breaks and Softening states of the Pony Truss Example

Break Classification

	P_i	V_{2i}	V_{3i}	T_i	M21	M_{3i}	P_j	V_{2j}	V_{3j}	T_j	M_{2j}	M_{3j}
Bottom Chord	1	0	0	0	0	0	0	0	0	0	0	0
Top Chord	1	0	0	0	0	0	0	0	0	0	0	0
	1	0	0	0	0	0	0	0	0	0	0	0
Diagonals	0	0	0	0	1	1	0	0	0	0	0	0
	0	0	0	0	0	0	0	0	0	0	1	1
Verticals	0	0	0	0	0	0	0	0	0	0	0	0
Floor Beams	1	1	1	0	1	1	0	0	0	0	0	0
	0	0	0	0	0	0	1	1	1	0	1	1
Stringers	0	1	1	0	1	1	0	0	0	0	0	0
	0	0	0	0	0	0	0	1	1	0	1	1

Softening Classification

	A	As2	As3	J	I22	I33	Mass	Weight
Bottom Chord	0.3	1	1	1	1	1	1	1
Top Chord	0.3	1	1	1	1	1	1	1
	1	0.3	0.3	1	0.3	0.3	1	1
Diagonals	0.3	1	1	1	1	1	1	1
Verticals	1	1	1	1	1	1	1	1
Floor Beams	1	1	1	1	1	1	1	1
Stringers	0.3	1	1	1	1	1	1	1
	1	0.3	0.3	1	1	0.3	0.3	1

The subsystems considered were the two trusses and the floor system. The two trusses each have 53 damage states, they also have another 6 and 7 states that indicate that the substructure is not damaged, these either correlate to the whole structure being intact or other damage states existing in one of the other systems. The floor system is larger, with 82 states indicating the damage of the floor system, and a further 18 states indicating an intact structure or damage to other subsystems. The two trusses each had 11 states which induced collapse under dead load. These included eight of the top chord elements the diagonals at both ends, and one chord member. They were not included in training data.

2. Artificial neural networks as content addressable memory

SHM requires some degree of signal processing to discern what state the structure is in. For this work subdivided attractor networks are used, these are Hopfield networks of McCulloch-Pitts neurons that have been cleaved into several subnetworks [6, 7]. When not divided, attractor networks act as content addressable memory, converging to the most similar state to what has been initiated on the network [8]. This is done by connecting each neuron to each other neuron with a synaptic weight, the more often the neurons activate together (or are inactive together), the larger the weight, the more often one activates and the other does not the smaller the weight. McCulloch-Pitts neurons have values of +/-1, whether the value changes in an iteration is based on the signal given by the rest of the network; this signal is:

$$h_i(t) = \left(\sum_j^N J_{ij} s_j(t-1)\right) \quad (1)$$

Where: $s_i(t)$ is the state of the ith neuron at iteration t, J_{ij} is the synaptic weight between neuron i and j, and N is the number of neurons.

Other things may add to this signal field such as biases, new information, or noise meant to draw the network out of local minima. It is helpful to consider the negative of the signal field analogous to energy, with the attractor network falling into the lowest energy state. Simulated annealing is used, in this case a constant heat bath is used to allow the network to raise out of local minima.

2.1 Subdivided attractor networks

Subdivisions produce several smaller subnetworks, each with their own output called subdivided attractor artificial neural networks (SA-ANN) [7]. The synaptic weights between the nodes of the local subnetwork are treated normally, but the weights between the different subnetworks are reduced. The multiple networks identify different damage states that are non-orthogonal. Because the networks states are related to one another, an exchange of signals occurs. As an example, in identifying Superman as he flies overhead, he is typically misidentified as a bird or a plane. The movement type (flying) and the shape (man) are not typically combined, the movement type strongly suggest that what is seen is either a bird or a plane. The other subnetworks add a term to the signal, making it:

$$h_i(t) = \left(\sum_j^{N/q} J_{ij} s_j(t-1)\right) + \left(\sum_j^{N(1-q)/q} J_{ij} s_j(t-1)\right) \quad (2)$$

Where q is the number of subnetworks

For this work, we divide the structure into subsystems each with their own attractor network monitoring the state of the strains in members due to moment and axial loads of members. Damage to one substructure will affect the stress state of the rest of the structure, though not so much as the local damages will. The signal between subnetworks is of less import than internal signals, and in the case of a bridge with several spans, the import of the signal is a function of how close the subsystems are to each other. If the states are related, the subnetworks work together to determine the state of each network, in the case of a superman, this is the case of an odd looking bird being positively identified as a bird thanks to its movement, not merely its shape. However, because of the reduction in signal from outside the subnetwork, it is possible to remix the states of different subnetworks to those which are not usually associated.

Such an event would be damage present in multiple subsystems at once. If the full system is monitored this may produce states that are not recognizable as one pre-modelled damage state or another. Conversely, a regimen of independently monitoring subsystems may be able to identify combinations of damage states, but they will be more easily confounded by damages outside of the subsystem, likely requiring more inputs. The SA-ANN combines the benefits of both these methods while mitigating the drawbacks.

2.1. Feedforward Artificial Neural Networks

Feedforward artificial neural networks are a flexible way to categorize inputs or interpolate and extrapolate a function that is difficult to model directly [9]. They are by far the form of artificial neural networks to see the most application [7, 10]. In the proposed work, FF-ANN are used to categorize the state of a substructure based on a set of strain data gathered from that substructure. While in this work FF-ANN are used as an intermediate step in identifying the damage state of a structure, they have been used as SHM systems in their own right [11].

2.2. Bottleneck image

SA-ANN require a large number of neurons to function, far more than is necessary to describe as many states it memorizes in more conventional computing. The FF-ANN require a more complex training regimen than SA-ANN, so it is desirable to have fewer synaptic weights as long as there are enough to sufficiently approximate the output. The FF-ANN is set to output a bottleneck image of logistic function defined nodes that is smaller than the full subnetwork. This bottleneck image is multiplied by a set of one way synaptic weights created by Hebbian learning, in the same way as synaptic weights are defined for the SA-ANN. The reduction in size will necessarily cause some loss of fidelity relative to routing the FF-ANN directly onto the initial state of the attractor network. However the reduction in training time for the FF-ANN is worth it as the SA-ANN subnetworks are quite large. The damage to the initial network state caused by the bottleneck was examined as a function of the size of the bottleneck at several levels of damage to the bottleneck image. The results are shown in Figure 1. They indicate that even for a perfect image in the bottleneck, there is substantial damage to the initial network image. As the bottleneck approaches the full size of the network however, the interface between the two begins to act as an iteration of the SA-ANN and it begins to have lower damage to the initial subnetwork state than to the bottleneck state. The 50% size of bottleneck was selected for this work.

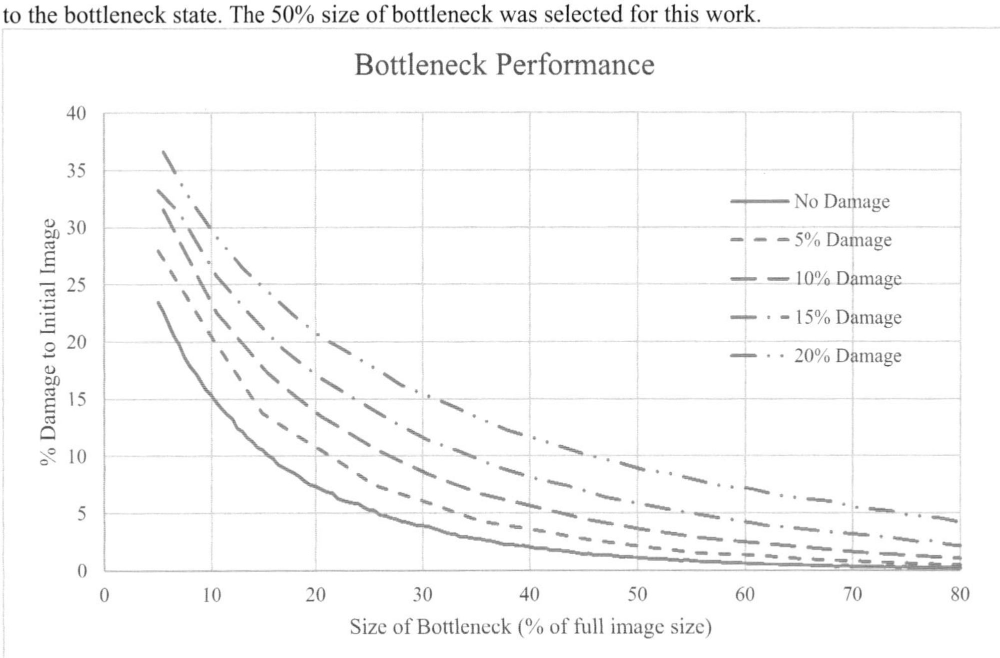

Figure 1. Averages of the effects of bottlenecks on 100 networks with 20 memorized states

2.3. Grouping

While each damage state considered requires its own memory state in its associated subnetwork, this is not the case for other subsystem damages. Each damage state is associated with a network state in each subnetwork. If the damage is not located in the subnetwork's subsystem, the associated subnetwork state will be shared with other less pertinent damage states. For example, in a long span bridge with many supports damage near one end of the bridge will have little effect on the other end. Part of this is covered by reducing the magnitude of the synaptic weights, but the physical state of the undamaged bridge end will be approximately the same, thus only a couple subnetwork states are memorized to cover all the damages to the other subsystems. This reduces the signal to the subnetwork, specifically the second term of equation 2. This reduction may not appear to necessarily happen when one looks at equation 2, however if the same subnetwork state in one subnetwork is associated with every single network state in another subnetwork the signal must necessarily go to zero from this term, just as if the whole world were painted green, the color green would not suggest that an object is a leaf rather than a piece of paper.

A method of grouping was developed based on the distance between states. The input vectors were standardized and principal component analysis was performed. The square distance between states was taken as:

$$D_{ij} = \sum_k^N (S_{i\,k} - S_{j\,k})^2 \tag{3}$$

Where D_{ij} is the standardized square distance between the ith and jth input vector and S_{ik} is the kth value for the ith input vector. The values in D_{ij} are then compared to each other; if they are less than some previously determined distance from each other they are placed into the same group. This does allow for a group "drift" as each new member slightly changes the bounds of what is considered similar to the group for the next vector.

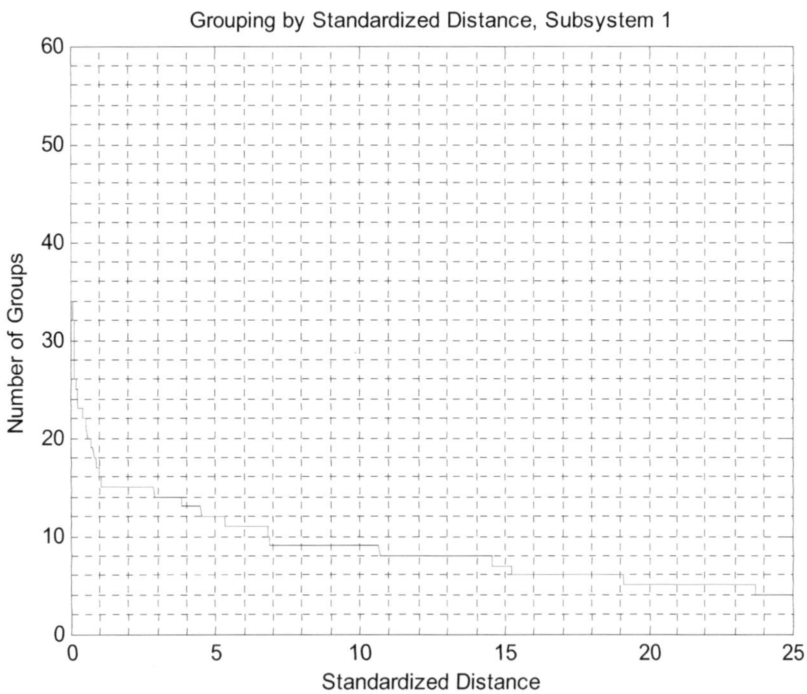

Figure 2. Number of groups of non-pertinent damage states based on normalized square distance

For the example pony truss bridge, subsystem 1 is a truss, it has 96 damage states to remember, there are 152 damage states for system 3 and another 96 damage states for the other truss, for a total of 248 non-pertinent states for this subnetwork. Initial increases in the acceptable square distance give precipitous decrease in the number of groups, but the decrease slows as the acceptable distance increases as shown in figure 2. It appears that grouping the non-pertinent input states into 5 to 9 states will be stable with substantial similarities between all group members. For the two trusses, (subsystems 1 and 2) we sort the non-pertinent groups into 7 and 6 groups respectively, for the floor system (subsystem 3) the non-pertinent states are grouped into 18 states. The floor system is closer to both the trusses and has substantially more sensors, allowing it greater granularity.

3. Methodology

The proposed SHM scheme was produced in Matlab and trained on strains from a nonlinear model produced in SAP2000. The nonlinear properties of the connections of the pony truss bridge were considered. Following this, some strain states were produced that the system was not trained to identify, including damage to multiple systems at once. These combined damage states were made up of damage states that the system was trained to recognize

The FF-ANN were defined using a standard backpropagation algorithm with momentum and regulation, as well as normally distributed noise on the inputs to prevent overfitting and ensure generality. Principal component analysis was used to reduce the inputs to the FF-ANN from 23 inputs for the trusses and 242 for the deck system to 13 and 65, respectively. The trusses each had 90 nodes in one processing layer, while the floor system had 250 nodes in its processing layer.

Results

The SHM system is able to identify the states it is trained for, but this is not all that is expected of this system. It was also subject to variety of damage states that it was not trained to recognize. The success of the network at identifying these untrained combinations is shown below in Table 2. All of the tested states are shown, failures of a subnetwork to converge to the correct neural state are shaded in. When the minor damage types such as merely reducing the moment of inertia or cross sectional area to 30% of what it is when intact, are paired with more substantial damage types such as truss member fracture, the system often fails to discern the more minor damage, either failing to converge to a state at all or converging to a state indicating that the subsystem is undamaged but the other subsystem is damaged. This second state at least enjoys some of the benefits of the subdivisions.

When significant damage is experienced in locations that are fully distinct in the geometry the system works at its best, such as when a stringer breaks off in the middle of the deck and a chord member completely fails. The physical system being a single span non-redundant structure may have reduced the success rate of this initial system, the systems are tightly associated with each other, and attempts to introduce multiple instances of fractures often result in collapse.

Table 2. Performance of SA-ANN on Combinations of Damage

TRUSS 1	TRUSS 2	FLOOR
bottom chord 1 axial loss	diagonal 16 RBS	intact
bottom chord 2 axial loss	diagonal 16 RAS	intact/damage elsewhere
bottom chord 3 axial loss	diagonal 15 RBS	intact/damage elsewhere
bottom chord 4 axial loss	diagonal 15 RAS	intact/damage elsewhere
top chord 6 axial loss	diagonal 14 RBS	intact/damage elsewhere
top chord 7 axial loss	diagonal 14 RAS	intact/damage elsewhere
diagonal 2 axial loss	diagonal 13 RBS	intact/damage elsewhere
diagonal 3 axial loss	diagonal 13 RAS	intact/damage elsewhere
diagonal 4 axial loss	diagonal 12 RBS	intact
diagonal 5 axial loss	diagonal 12 RAS	intact
diagonal 6 axial loss	diagonal 11 RBS	intact/damage elsewhere
diagonal 7 axial loss	diagonal 11 RAS	intact/damage elsewhere
bottom chord 1 RAS	diagonal 10 RBS	intact
intact/damage elsewhere	bottom chord 1 RAS	intact/damage elsewhere
intact/damage elsewhere	bottom chord 2 axial loss	intact/damage elsewhere
diagonal 8 RBS	bottom chord 3 axial loss	intact/damage elsewhere
diagonal 8 RAS	bottom chord 4 axial loss	intact/damage elsewhere
diagonal 7 RBS	top chord 6 axial loss	intact/damage elsewhere
diagonal 7 RAS	top chord 7 axial loss	intact/damage elsewhere
diagonal 6 RBS	diagonal 2 axial loss	intact/damage elsewhere
diagonal 6 RAS	diagonal 3 axial loss	intact/damage elsewhere
diagonal 5 RBS	diagonal 4 axial loss	intact
diagonal 5 RAS	diagonal 5 axial loss	intact
diagonal 4 RBS	diagonal 6 axial loss	intact/damage elsewhere
diagonal 4 RAS	diagonal 7 axial loss	intact/damage elsewhere
diagonal 3 RBS	bottom chord 1 RAS	intact
bottom chord 2 RAS	intact/damage elsewhere	FB 1 end 1 break
bottom chord 3 RAS	intact/damage elsewhere	FB 1 end 2 break
intact/damage elsewhere	diagonal 15 axial release	FB 2 end 1 break
intact/damage elsewhere	bottom chord 4 RAS	FB 2 end 2 break
top chord 1 RAS	top chord 5 axial release	intact
bottom chord 4 RAS	bottom chord 5 RAS	intact
bottom chord 5 RAS	intact/damage elsewhere	floor beam 1 release end 1
top chord 2 RAS	intact/damage elsewhere	floor beam 1 release end 2
bottom chord 2 axial release	intact/damage elsewhere	floor beam 2 release end 1
bottom chord 4 axial release	intact/damage elsewhere	floor beam 2 release end 2
top chord 7 axial release	intact/damage elsewhere	floor beam 3 release end 1
intact/damage elsewhere	bottom chord 1 axial release	floor beam 3 release end 2
intact/damage elsewhere	bottom chord 2 axial release	floor beam 4 release end 1
intact/damage elsewhere	bottom chord 3 axial release	floor beam 4 release end 2
intact/damage elsewhere	bottom chord 4 axial release	stringer 1 release end 1
intact/damage elsewhere	top chord 4 axial release	stringer 1 release end 2
intact/damage elsewhere	top chord 5 axial release	stringer 2 release end 1
intact/damage elsewhere	diagonal 3 axial release	stringer 2 release end 2
RAS = Reduced Axial Stiffness		
RBS = Reduced Bending Stiffness		

Conclusion

We propose a new method of SHM using SA-ANN, it allows for multiple damage states to be identified at once. The SA-ANN's ability to discern small damages alongside significant damages is marginal, showing the same weakness in discerning minor damages alongside large damages that is present in other approaches. In cases of damages of similar significance the combinations are determined more easily. The relationship between either of the trusses in the example problems and the floor system is closer than the relationship between the two trusses. The minor damages considered in the reduced area and bending stiffnesses, particularly in the bending stiffness of truss members are too small for this SHM system to identify in any but optimum conditions. This approach may have greater success with systems with more alternative load paths than a pony truss bridge, where the subsystems are less closely related.

References

[1] Yi, T.H., Li, H.N. and Gu, M., 2011. Optimal sensor placement for structural health monitoring based on multiple optimization strategies. The Structural Design of Tall and Special Buildings, 20(7), pp.881-900.

[2] Antonacci, E., De Stefano, A., Gattulli, V., Lepidi, M. and Matta, E., 2012. Comparative study of vibration-based parametric identification techniques for a three-dimensional frame structure. Structural Control and Health Monitoring, 19(5), pp.579-608.

[3] Sohn, H. and Farrar, C.R., 2001. Damage diagnosis using time series analysis of vibration signals. Smart materials and structures, 10 (3), p.446.

[4] Jenkins, C.H., Kjerengtroen, L. and Oestensen, H., 1997. Sensitivity of parameter changes in structural damage detection. Shock and Vibration, 4(1), pp.27-37.

[5] Butler, M.A., Swanson, J.A., Rassati, G.A. and Dues, E.F., 2018. High Resolution Modeling and Modeling of Connections in Pony Truss Bridges (No. 18-03496).

[6] Coughlin, J.P. and Baran, R.H., 1995. Neural computation in hopfield networks and boltzmann machines. (University of Delaware Press)

[7] Bar-Yam, Y., 1997. Dynamics of complex systems. (Reading, MA: Addison-Wesley)

[8] Amit, D.J., 1992. Modeling brain function: The world of attractor neural networks. Cambridge university press.

[9] Samarasinghe, S., 2016. Neural networks for applied sciences and engineering: from fundamentals to complex pattern recognition. CRC Press.

[10] Muttil, N. and Chau, K.W., 2007. Machine-learning paradigms for selecting ecologically significant input variables. Engineering Applications of Artificial Intelligence, 20 (6), pp.735-744.

[11] Bakhary, N., Hao, H. and Deeks, A.J., 2007. Damage detection using artificial neural network with consideration of uncertainties. Engineering Structures, 29 (11), pp.2806-2815.

Fatigue performance evaluation of steel arch bridge based on experimental tests in the light of increased operating loads

M Kużawa

Department of Civil Engineering, Wrocław University of Science and Technology, Poland

mieszko.kuzawa@pwr.edu.pl

Abstract. The paper presents a case study of experimentally supported fatigue performance evaluation of over 22-year old composite (steel-concrete) arch bridge. The bridge is a landmark road structure crossing the Narew River in Ostrołęka in Poland. It's main span is composed of through arch static system. The structure was designed to carry out the local town traffic. The bridge due to the emergency condition of the neighboring bridge located along national roadway was temporarily subjected to increased live loads. Thereby it became to play an important role in the transportation system of the whole region being intensively exploited by heavy vehicles in many cases neglecting compliance with restrictions for total weight of heavy trucks. Thus, taking into account limited load capacity and intensity of the traffic crossing the bridge, some doubts about the fatigue performance of bridge critical members were formulated. Fatigue performance evaluation of the steel superstructure was executed according to general assumptions of the Eurocodes and the *European Recommendations for Estimation of Remaining Fatigue Life of Existing Steel Structures*. Essential input data used for estimation of fatigue damage accumulation in critical bridge components as well as for prediction of residual lifetime of the whole structure was based on experimental vibration tests performed under operating loads.

1. Bridge structure and objectives of the study

The investigated structure called the Antoni Madaliński Bridge is located over the Narew River in Ostrołęka city. It consists of 4 spans with the lengths: 32 + 32 + 110 + 32 meters. Shorter spans are built of beam type superstructures, each constructed of 4 steel plate girders, 2,5 m high, composed with typical RC slab. The main span is the form of through arch static system constructed of composite–steel-concrete deck connected with the central steel arch by means of cable stays. The structure was designed to carry out the local town traffic. According to the system of the Polish national standards, which was used for the bridge design, the total weight of single vehicle crossing the bridge should not exceed 30 tonnes. The structural form of the investigated bridge is presented in figure 1 and figure 2.

In the years 1996-2016, due to its limited load capacity and in order to keep the heavy traffic out of town, the bridge was used only for carry out vehicles of total mass do not exceeding 3,5 tonnes. However, in the end of 2016 the neighboring bridge located along national road No 61 was closed for service and reconstructed due to its emergency condition. Thereby, the considered structure was subjected to increased live loads and become a very important element in the transportation system of the whole region. Thus, the bridge owner temporarily increased admissible weight of vehicles to 30 tonnes (designed load capacity). Taking into account intensity of the heavy traffic crossing the bridge – in many cases apparently neglecting compliance with weight restrictions for heavy vehicles introduced by bridge owner – some doubts about the fatigue performance of bridge critical members were formulated [1].

In September of 2017 intensity of the traffic on the bridge has been significantly decreased as admissible weight of vehicles was again limited to 3,5 tonnes when the alternative crossing was put to service.

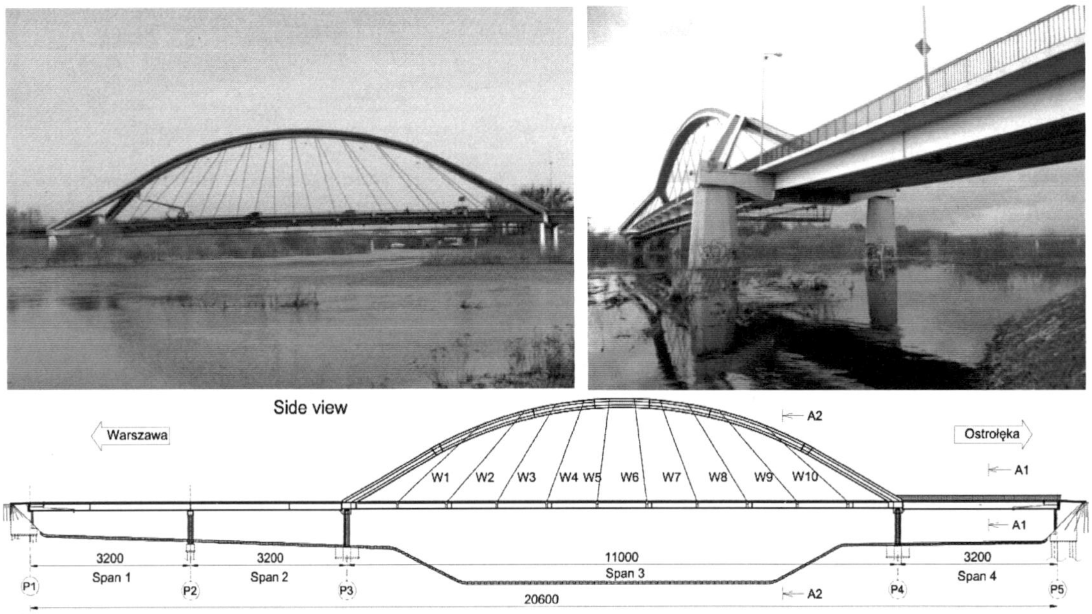

Figure 1. Structural form of the Antoni Madaliński Bridge over the Narew River in Ostrołęka.

Figure 2. Investigated bridge: a) bottom view of superstructure, b) inside view of the arch.

In the course of special inspection some defects of the bridge superstructure were observed, however fatigue cracks were not detected. Anyway, considering the intensity of heavy traffic crossing the bridge during the period of excessive use, some uncertainties about the fatigue performance of the bridge critical members appeared. Thus, the main goal of the presented investigation was to estimate the level of fatigue damage accumulated in the critical bridge components during the approximate 9 months period of increased heavy traffic as well as to predict the remaining life time of the whole structural system.

2. General procedure of fatigue performance evaluation

The fatigue evaluation of the Antoni Madaliński Bridge was performed following the general rules of the European design codes [2]-[3] and recommendations [4]-[5]. The assessment is aimed at providing evidence that the bridge, after a period of excessive use in 2017, will function safely over a specified residual service life. The applied approach is based on a three-step procedure described below.

Phase I – a preliminary evaluation – aimed at identification of critical members in the structure for which a fatigue hazard exists. It was performed by means of the Finite Element Method (FEM) model of the bridge developed on the basis of data coming from the structural drawings supplemented by a site investigations. Within the initial calculations the Fatigue Load Model 3 (FLM3) of the Eurocode 1[2] was applied and its effects in bridge steel members were referred to the S-N curve corresponding to relevant detail fatigue categories according to [3]. The initial evaluation of the fatigue safety level was carried out according to the formula:

$$\mu_{\mathrm{fat}} = \frac{\Delta\sigma_{\mathrm{C}}}{\gamma_{\mathrm{Mf}} \cdot \gamma_{\mathrm{Ff}} \cdot \Delta\sigma_{\mathrm{E,2}}} \geq 1 \tag{1}$$

where: μ_{fat} – fatigue safety level; $\Delta\sigma_{\mathrm{C}}$ – reference value of the fatigue strength corresponding to $N_{\mathrm{c}} = 2 \cdot 10^6$ cycles; $\Delta\sigma_{\mathrm{E,2}}$ – equivalent constant amplitude stress range related to $N_{\mathrm{c}} = 2 \cdot 10^6$ cycles; γ_{Mf} – partial factor for fatigue strength $\Delta\sigma_{\mathrm{C}}$; γ_{Ff} – partial factor for equivalent constant amplitude stress ranges $\Delta\sigma_{\mathrm{E,2}}$.

Rules defining values of the above mentioned parameters are given in [2]-[3]. The equivalent constant amplitude stress range $\Delta\sigma_{\mathrm{E,2}}$ can be calculated as follows:

$$\gamma_{\mathrm{Ff}} \cdot \Delta\sigma_{\mathrm{E,2}} = \lambda \cdot \Delta\sigma_{\mathrm{p}}(\gamma_{\mathrm{Ff}}, Q_{\mathrm{k}}) \tag{2}$$

where: $\Delta\sigma_{\mathrm{p}}(\gamma_{\mathrm{Ff}}, Q_{\mathrm{k}})$ – the maximum stress range generated by fatigue loads in an analysed bridge member; λ – total damage equivalent factor determined according to [3]; Q_{k} – characteristic value of a single variable action generated by FLM3 [2].

When the calculated fatigue safety level $\mu_{\mathrm{fat}} < 1$ for a structural component then such a member is classified as a critical element. Its safety needs to be further assessed in Phase II.

Phase II – a detailed fatigue performance investigation based on structure monitoring under increased live loads – is focused on estimation of the total fatigue damage $D_{\mathrm{d}}^{\mathrm{exc}}$ being accumulated during the 9 months period of excessive use in the critical bridge components. The main input data at this phase are results of stress/strain measurements carried out on site in bridge critical elements under increased traffic loads covering a representative (minimum 24-hours) period of the structure operation. The measured strain history in the selected steel components (supplemented, if relevant, by appropriate statistical and projected extrapolations) was used to create the real stress range spectrum representing bridge current, excessive operation conditions.

The measurement results were also used in the FEM model validation. In Phase II the accumulated fatigue damage in the selected structural members was calculated using the Palmgren-Miner damage summation rule, simply stated as follows:

$$D_{\mathrm{d}}^{\mathrm{exc}} = \sum \frac{n_{\mathrm{E}i}}{\mathrm{N}_{\mathrm{R}i}} \leq 1 \tag{3}$$

where: $D_{\mathrm{d}}^{\mathrm{exc}}$ – the total damage accumulated during the 9 months period of excessive use in an analysed spot of an evaluated steel member; $n_{\mathrm{E}i}$ – number of cycles occurring at stress range $\gamma_{\mathrm{Ff}}\Delta\sigma_i$ for band i in the factored spectrum; $N_{\mathrm{R}i}$ – the endurance (expressed in cycles) obtained from the factored $\frac{\Delta\sigma_{\mathrm{C}}}{\gamma_{\mathrm{Mf}}}$ - $\mathrm{N_R}$ curve corresponding to a stress range of $\gamma_{\mathrm{Ff}}\Delta\sigma_i$.

Phase III – a remaining fatigue life assessment after the period of excessive use – is aimed at estimation of remaining fatigue life T_{fat} considering all (past, present and expected future) periods of the structure operation, leading to damage accumulation. The prognosis of the fatigue damage $D_{\mathrm{d}}(t)$ in the next t years may be determined using formula:

$$D_{\mathrm{d}}(t) = D_{d}^{\mathrm{past}} + D_{\mathrm{d}}^{\mathrm{exc}} + D_{\mathrm{d}}^{\mathrm{fut}} \cdot \beta(t) \cdot t \tag{4}$$

where $D_{\mathrm{d}}^{\mathrm{past}}$ represents the estimated total damage accumulated during the bridge lifetime (excluding the period of increased traffic in 2017) considering natural (site specific) traffic conditions; $D_{\mathrm{d}}^{\mathrm{fut}}$ represents predicted damage accumulation for one year of bridge normal operation at future traffic conditions (admissible weight of vehicles is now again limited to 3.5 tonnes) and $\beta(t)$ is a time-dependent factor considering average changes in volume of freight vehicles crossing the bridge.

In presented case study both $D_{\mathrm{d}}^{\mathrm{past}}$ and $D_{\mathrm{d}}^{\mathrm{fut}}$ parameters were assumed to very small (close to zero) since for corresponding lifetime the bridge was subjected to lightweight traffic only which do not contribute to fatigue damages.

Finally, assuming in formula (4) $D_{\mathrm{d}}(t) = 1$ and $t = T_{\mathrm{fat}}$, the remaining fatigue life time T_{fat} can be obtained as:

$$T_{\mathrm{fat}} = \frac{1 - D_{\mathrm{d}}^{\mathrm{past}} - D_{\mathrm{d}}^{\mathrm{exc}}}{\beta(T_{\mathrm{fat}}) \cdot D_{\mathrm{d}}^{\mathrm{fut}}} \tag{5}$$

3. Assessment Phase I

For the purpose of the fatigue evaluation of the structure included in Phase I of the presented procedure a 3 different numerical FEM models representing separately bridge superstructures (beam type spans and arch spans were evaluated separately) were created (figure 3). Each model is composed of one- and two-dimensional finite elements: one-dimensional beam type elements are utilized for representation of all steel components including bracings and stay cables, while two-dimensional shell type elements are used for modelling of the RC deck slab.

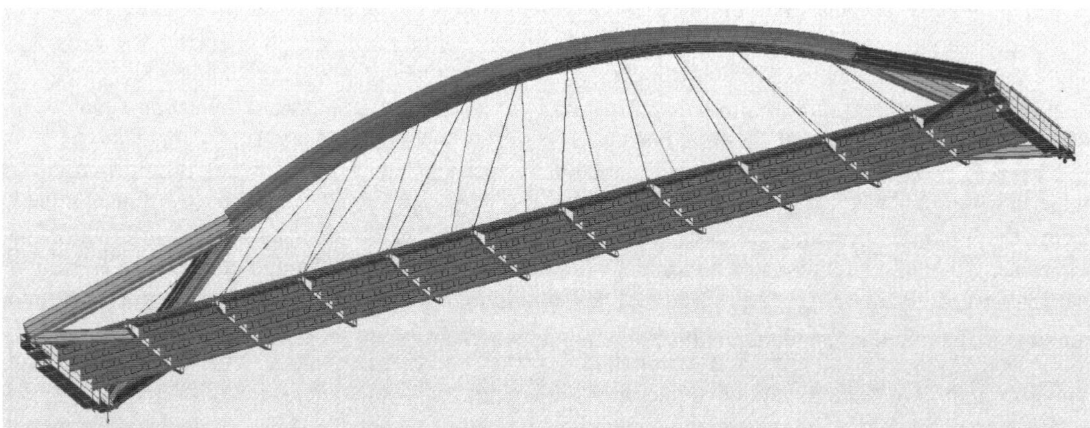

Figure 3. General view of FEM model of the main span of the Antoni Madalinski Bridge.

In Phase I of the bridge assessment presented FEM model was used for calculation of the stress ranges in steel members induced by the FLM3 vehicle of the Eurocode 1 [2] traveling along single traffic lane. The preliminary analysis pointed out critical elements of superstructure potentially exposed to fatigue phenomenon and made a detailed examination of these members necessary.

The identified critical elements (assuming $\gamma_{\mathrm{Mf}} = 1.35$) exposed to fatigue are:

- deck elements in span 3 in a place of connections of stringers with cross-beams – $\mu_{\mathrm{fat}} = 0.58$,
- main girders in span 4 in places where their cross-section changes – $\mu_{\mathrm{fat}} = 0.69$,
- in cable stays in span 3 – $\mu_{\mathrm{fat}} = 0.90$.

As it can be seen especially the deck elements in span 3 show very low fatigue safety level μ_{fat}. Therefore the detailed evaluation in Phase II was focused on estimation of damage accumulation in these elements.

4. Assessment Phase II and Phase III

4.1. Methodology of experimental test

On site testing of the bridge in Ostrołęka provided information on current live loads and rate of damage accumulation in bridge critical spots during the period of increased live loads. The experimental fatigue performance investigation was based on measurements of structure response in selected points to various vehicles. Choice of the points' location was based on numerical analysis in Phase I using the described FEM models of the bridge spans which indicated the members with the lowest fatigue safety level μ_{fat}. Dynamic tests of the structure were executed according to recommendations and experience from previous similar tests, e.g. [5]-[8].

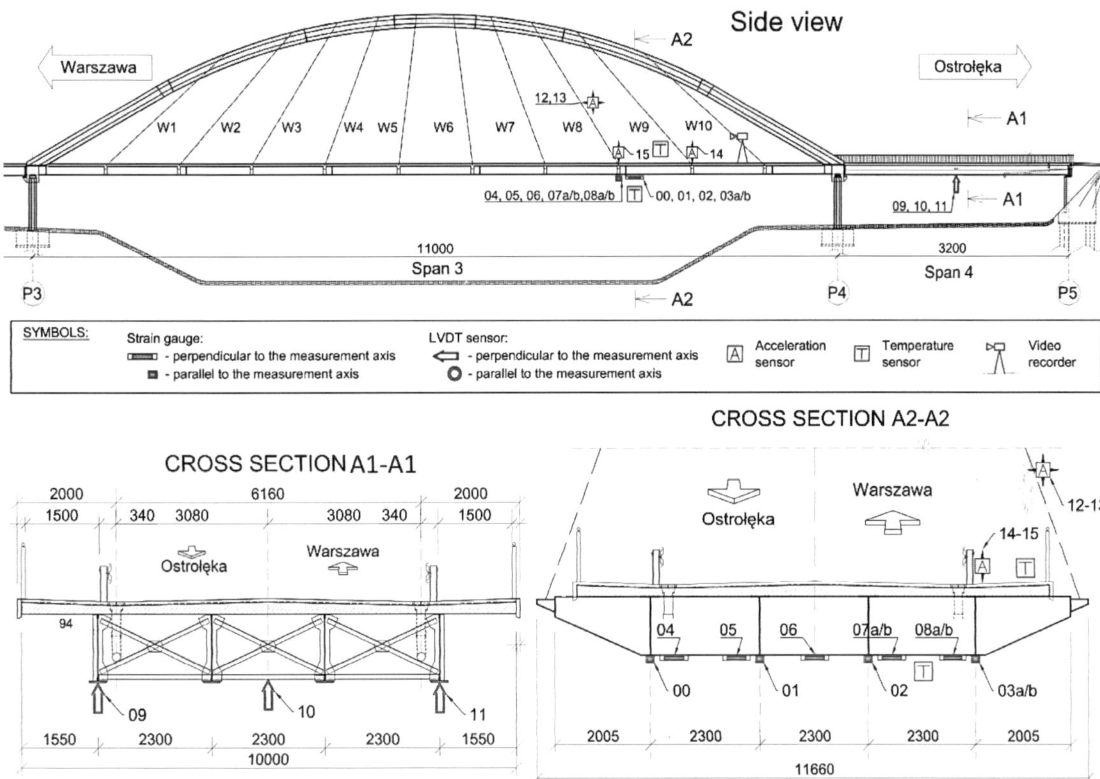

Figure 4. Arrangement of gauges during the tests: a) in side view, b) in cross sections.

The 24 hours special testing of the structure response to live loads included measurements of:

- strains in steel members of the deck within span no. 3, i.e. a stringers (points 00-03) and cross-beam (point 04-08),
- vertical displacements (points 9-11) in middle section of span no 3,
- vertical accelerations (points 12-15) of deck and stay cable in span 3.

During 24-hours measurements all vehicles crossing the bridge were registered using the camera and digital video recorder (DVR), for matching type of vehicles with the corresponding measured physical values. The temperature was controlled continuously in two specific areas of the structure.

Location of the measurement points and types of gauges applied in tests is given in figure 4. The measurements were carried out:

- under controlled, quasi-static and dynamic load scenarios executed by means of utilization of special vehicle of designed and controlled parameters (i.e. axle loads and spacing, vehicle velocity and location),
- during a 24-hours session under increased operation conditions.

The obtained results were used threefold. Firstly, they enabled initial calibration of the numerical model applied in Phase I of the assessment procedure. Secondly, the measured response provided information on real stress ranges within the structural members and in this way gave a direct input to calculation of fatigue damage being accumulated during the period of structure exploitation under increased live loads. Thirdly, it was also possible to identify level and distribution of approximate total vehicles weights corresponding to fatigue progress in critical bridge elements.

4.2. Results of experimental test under controlled load scenarios

Tests under dynamic loads of designed and controlled parameters consisted of a series of passages of a single 40-tonne truck along the bridge at the following speeds: 5, 20, 40, 60 km/h. Parameters of the applied truck (see figure 5a) have been selected in such a way to determine the highest level of effects in the analyzed bridge components induced by the heavy trucks meeting the regulatory requirements for admission to traffic in Poland.

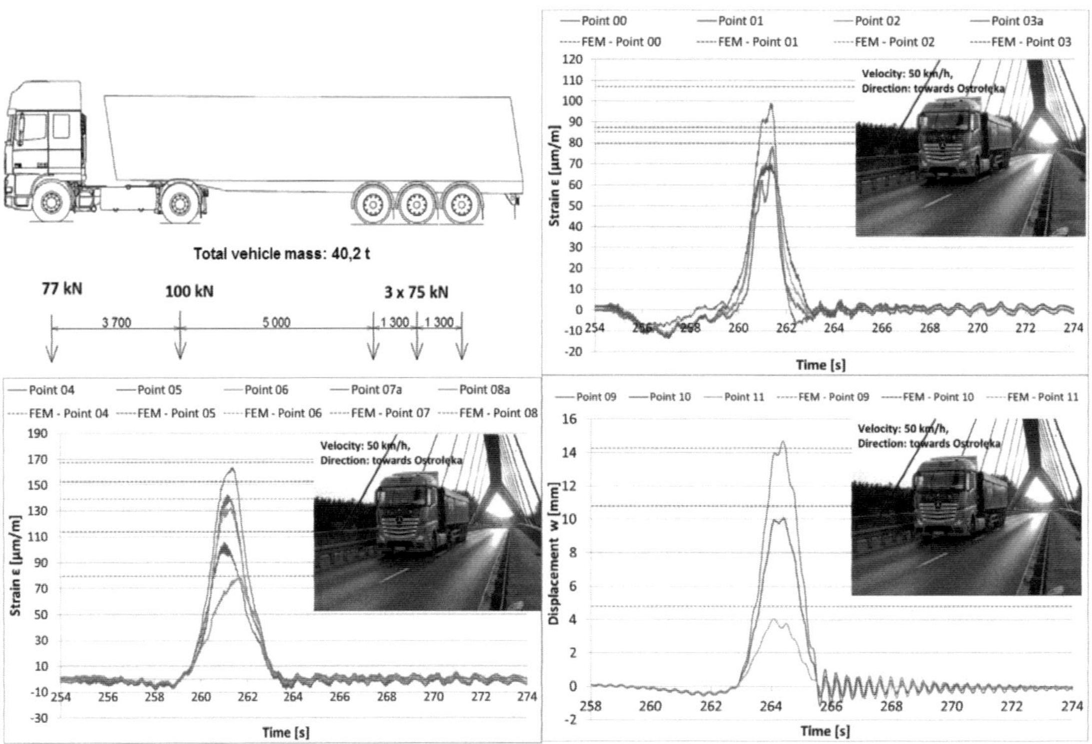

Figure 5. History of strains and displacements induced in selected points by a single 40-tonne, 5-axle vehicle traveling at a speed of 50 km/h towards Ostrołęka: (a) vehicle parameters, b) strains in stringers of span 3, c) strains in cross-beams of span 3, d) deflections of span 4.

Exemplary history of strains and displacements induced in analyzed points of the bridge by the applied vehicle (figure 5a) crossing the bridge at speed of 50 km/h is shown in figure 5b-d. In the diagrams the results of strain and displacement measurements were also compared with extreme static effects of loading with a moving theoretical vehicle of the same type determined by means of the applied numerical model. As it can be seen the experimentally determined peak values of strains related to real vehicle load proved to be consistent with the results of the theoretical analysis. The theoretical strains based on FEM analyses are very close to the experimentally identified values. The differences between the experimental and theoretical results are relatively small and do not exceed 10% on average. Therefore it can be concluded that the obtained results confirm high precision of applied numerical model as well as a proper modelling approach utilized in Phases I and II of the theoretical fatigue evaluation procedure.

During the model validation process significant (due to possibility of fatigue phenomenon) horizontal bending of steel deck elements of span 3 was revealed. Therefore, for capturing more realistic peak values of stresses in the vicinity of potential cracks at connections of stringers with cross beams an additional strain measurement points (03b, 07b and 08b) were installed before executing of 24-hours session under operational conditions. This was of great importance for determination of more precise strain ranges induced in critical spots by real live loads as the extreme strains recorded at "b" points were significantly greater than recorded at the corresponding "a" ones.

4.3. Results of experimental test under operational loads

Measured quantities caused by the real traffic provided data on extreme real strain/stress ranges occurring in critical components of the bridge superstructure induced by heavy traffic crossing the bridge during 9 months period of excessive use.

Maximum values of strains registered during measurements were relatively high and amounted to ε_{max} = 148.41 μm/m (point 03b) in stringers and ε_{max} = 283.74 μm/m (point 07b) in cross beams (figure 6). It was concluded that recorded peak values of strains, significantly larger than obtained by means of special truck use for model validation, were induced by passage of about 50 tonne truck.

Dynamic testing of the bridge superstructure under real operating conditions aimed also at identification of the contemporary loads which contribute the most to fatigue durability of bridge critical components. It was found that the highest values of measured physical quantities at present time of bridge operation are essentially related to the passing of 4-axle or 5-axle trucks transporting goods with total weight usually varying between 30-40 tonnes. What is more, in many cases estimated total weight of vehicles was varying between 41-50 tonne. Crossings of such vehicles of excessive total mass, in comparison with bridge load capacity, significantly reduce fatigue life.

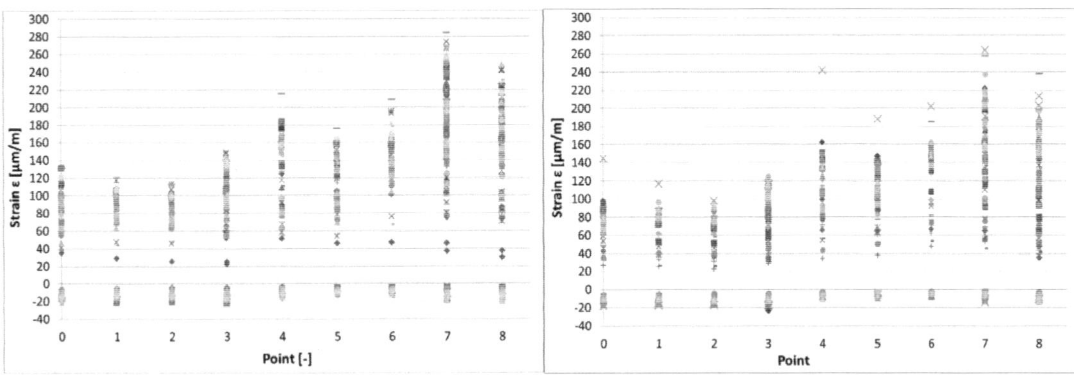

Figure 6. Extreme values of strains induced by live loads during consecutive measurements (10 minutes each): a) in the daytime: 7 a.m.–7 p.m., b) at night: 7 p.m.–7 a.m.

4.4. Fatigue performance assessment

Measurements during a 24-hours session were carried out to get real-time information on the amount of damage accumulated during the period of increased operating loads.

The measured history of strains caused by the actual traffic constituted essential input for the prediction of the bridge remaining fatigue life. The rainflow counting algorithm was used in order to reduce a spectrum of varying strains into a set of simple stress reversals. Histograms presented in figures 7 show number of cycles for 50 selected threshold values of stress ranges occurring in measurement points during 24-hours of bridge operations conditions under increased live loads. Assuming detail category of 50 MPa (defined in [3]) for span 3 welded grillage system components the corresponding fatigue cut-off limit (indicated also in figure 7) below which micro-damage does not accumulate in material is: 15.20 MPa considering safety factor γ_M = 1.35 and 17.84 MPa assuming safety factor γ_M = 1.15. Meanwhile the experimentally obtained maximum values of stress ranges caused by the real traffic was equal to $\Delta\sigma_{\max}^{\mathrm{EXP}}$ = 61.17 MPa (point 07b) which confirmed, according to the current knowledge (e.g. [4]), a risk of fatigue failure in the studied structural members.

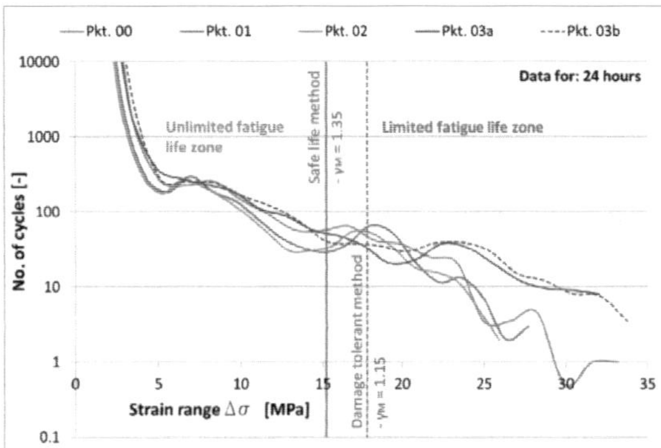

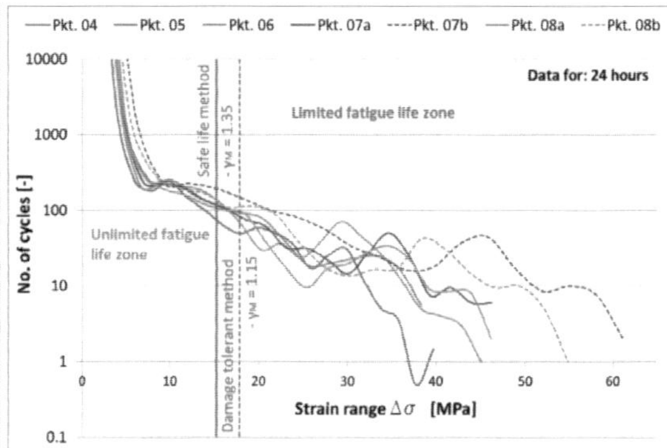

Figure 7. Stress range [MPa] spectrum calculated for 24- hours of bridge operation on basis of experimental tests results: a) points at stringers, b) points at cross beams

In Phase II, based on results of 24-hours measurements, the calculated total fatigue damage $D_{\mathrm{d}}^{\mathrm{exc}}$ being accumulated during the 9 months period of excessive use in the critical bridge components was less than 1 for all investigated members. The maximum estimated level of fatigue damage $D_{\mathrm{d}}^{\mathrm{exc}}$ of the deck elements of the span 3 is diverse and ranges from 0.012 (stringers, point 03b) to 0.062 (cross beams, point 07b).

In Phase III, considering all stages of bridge exploitation, the maximum estimated total damage accumulated during the elapsed 22 years bridge lifetime in analysed spot of the evaluated critical members was assumed to be around 0.132 (point 07b). This value was assessed assuming $D_{\mathrm{d}}^{\mathrm{past}}$ at level 0.069 corresponding to the traffic volume during 22 years bridge lifetime equal to 5% of traffic measured during the tests. This means that the fatigue life of critical components of the bridge has not been completely exhausted so far.

Similar assumptions were made for representing predicted damage accumulation for one year of bridge normal operation at future traffic $D_{\mathrm{d}}^{\mathrm{fut}}$. Finally assuming in formula (5) $D_{\mathrm{d}}^{\mathrm{past}} = 0.069$, $D_{\mathrm{d}}^{\mathrm{exc}} = 0.062$, $D_{\mathrm{d}}^{\mathrm{fut}} = 0.003$ and $\beta = 1.5$ the remaining fatigue life time T_{fat} was assessed to be equal 180 years.

It was concluded that when the limitation related to the total weight of a single vehicle crossing the bridge is respected by bridge users than the fatigue damage does not accumulate in critical bridge members and in that case there is no risk of superstructure failure associated with it.

5. Summary

The case study on fatigue evaluation performance of 22 years old, steel-concrete, through arch bridge in Ostrołęka in Poland was described in this paper. The applied procedure of structure evaluation was performed following the general assumptions of the European design codes [2]-[3] and recommendations [4]-[5].

The basic input data used for bridge assessment were based on 24-hours, experimental dynamic tests carried out using real live loads of designed and controlled parameters as well as under natural (random) operation conditions.

Taking into account the results of the theoretical analysis and experimental load tests the following general conclusions may be formulated:

- Applied numerical modelling and analysis approach of composite, steel-concrete superstructure of the Antoni Madaliński Bridge enabled precise representation of strain/stresses in all structural members taking into consideration both, global and local effects of live loads.
- Fatigue evaluation supported by results of the experimental vibration tests provided information on actual loads acting on the structure and enabled refinement of FEM model of bridge superstructure, thereby allowing a more detailed assessment of the level of fatigue damage of the bridge critical members.
- The estimated, current maximum level of fatigue damage is $D_{d,\ max}(t=0) = 0.132$ and actual total weight of vehicles crossing the bridge is now limited to 3,5 tonnes. This indicates that the remaining life of the critical structural components has not yet been fully exhausted and bridge fatigue performance should be satisfactory throughout its design life.
- Overall maintenance of the considered bridge is satisfactory. However, localised corrosion pits in any members need a special attention because simultaneous acting of corrosion and fatigue will dangerously accelerate degradation mechanism.

The applied method of testing planning and execution as well as measuring system confirmed practicability of the solutions in identification of bridge characteristics as well as in determination of real load spectra acting on a structure. The proposed methodology can be efficient in condition evaluation of other bridge structures.

References

[1] Kużawa M and Bień J 2017 Experimental tests and analyses of the Antoni Madaliński bridge over the Narew River in Ostrołęka in Poland, Scientific Report of Civil Engineering Department of Wroclaw University of Science and Technology SPR Series No. 27/2017, Wrocław (in Polish)

[2] EN1991-2 2003 Eurocode 1: Actions on Structures - Part 2: Traffic Loads on Bridges, European Standard (Brussels: CEN)

[3] EN1993-1-9 2003 Eurocode 3: Design of Steel Structures – Part 1-9: Fatigue, European Standard (Brussels: CEN)

[4] Kühn B, Lukić M, Nussbaumer A, Günther HP, Helmerich R, Herion S, Kolstein MH, Walbridge S, Androic B, Dijkstra O and Bucak Ö 2008 Assessment of existing steel structures: recommendations for estimation of remaining fatigue life, In: *Joint Report Prepared under the JRC-ECCS cooperation agreement for the evolution of Eurocode 3 (programme of CEN/TC 250)*

[5] Kużawa M, Kamiński T, and Bień J 2018 Fatigue assessment procedure for old riveted road bridges, *Arch. Civ. Mech. Eng.* 18 (4)

[6] Bień J, Kużawa M and Kamiński T 2015 Validation of numerical models of concrete box bridges based on load test results, *Arch. Civ. Mech. Eng.* 15 (4)

[7] Kużawa M, Bień J, Rawa P, Kamiński T and Zwolski J 2012 Static and dynamic load tests of cable-stayed bridge over Odra River in Wrocław, In: *Proc. 6th Intern. Conf. on Bridge Maintenance, Safety and Management "IABMAS"*, Stresa

[8] Bień J, Kużawa M, Zwolski J, Banakiewicz A, Rabiega J, Rawa P and Adamcewicz S 2009 Fatigue life assessment of Grunwaldzki Bridge based on experimental vibration tests, In: *Proc. 3rd Intern. Conf. Experimental Vibration Analysis for Civil Engineering Structures "EVACES'09"*, Wrocław

IOP Conf. Series: Materials Science and Engineering **419** (2018) 012016 doi:10.1088/1757-899X/419/1/012016

Advanced solutions with hot-rolled sections for more economical bridges

Wojciech Lorenc[1], Maciej Kożuch[1], Dennis Rademacher[2] and Wojciech Ochojski[3]

[1]Wrocław University of Technology, Faculty of Civil Engineering, Wybrzeże Wyspiańskiego 27, 50-370 Wrocław, Poland
[2]ArcelorMittal Europe – Long Products, 66, rue de Luxembourg, 4221 Esch-sur-Alzette, Luxemburg
[3]ArcelorMittal Commercial Long Poland, Al. Piłsudskiego 92, 41-308 Dąbrowa Górnicza, Poland

E-mail: dennis.rademacher@arcelormittal.com

Abstract. The use of composite steel structures has gained significant market shares in road bridges across Europe over the last several decades. In particular, when high-strength steel is used, the total weight and the construction depth of the bridge deck can be significantly reduced. This also means lighter substructures and – if necessary – shorter approaching ramps to the bridge. Current developments of the market force designers, producers and fabricators to go new ways and to invent continuously new efficient solutions to save money and time. One big advantage can be achieved when using hot-rolled sections, which are available in higher strength S460 as standard and allow for simpler and more economical construction due to lower fabrication costs. Sophisticated designs with rolled sections allow not only small and medium span lengths, but also longer ones, e.g. in arch bridges with span lengths over 100m. This paper will show new trends in Europe using hot-rolled sections in steel and composite bridges.

1. Introduction

On the basis of hot-rolled steel sections there do exist several composite solutions, which are well known all over Europe. Multiple I-girder composite decks and filler beam decks are standard bridge designs since decades; the latter exists since more than 100 years. New and alternative methods of corrosion protection, such as self-weathering steel and hot-dip galvanizing, offer simple and economical solutions so that the steel part of the composite bridge can endure the designated lifetime without any need of maintenance or refurbishment. However, new market demands force designers and producers, as well as fabricators, to go new ways and to invent continuously new solutions, such as PreCoBeam bridges, where halved rolled sections are used in a more efficient way than full I-sections. About 35 PreCoBeam bridges have been built since 2003. Further optimization of this construction type allows to build bridges in medium span length range at lowest costs.

For longer bridges other advanced composite bridge type solutions were developed in Poland in the last years and successfully implemented in the market with first built examples. Here worth mentioning are the tandem solution and network arch bridges with extremely low steel consumption and manufacturing costs, due to the use of hot-rolled sections in high strength steel.

2. New PreCoBeam system using composite dowels for span length about 35m

A new system of PreCoBeam, dedicated to medium span lengths, was used for the first time for the "Elbląg" bridge (L=~38m). The system is quite aesthetic, see general view in Figure 1. The principles of the system (Figures 2-3), which are presented in detail in [2], were developed in 2015 in Poland [1].

Figure 1. General view of the "Elbląg" bridge in Poland finished 2016.

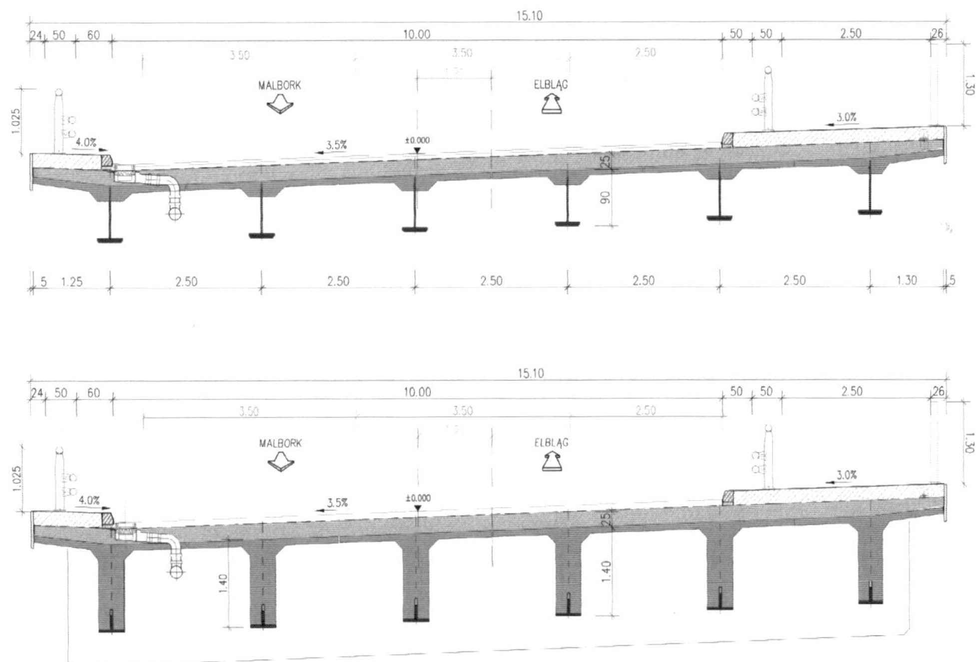

Figure 2. Cross-section of the bridge in the midspan and at the support [2].

The cross-section of this structure adapts significantly better to the distribution of forces in the bridge girder than a cross-section with a concrete girder of constant height. In this advanced design, the height of both the steel girder and the concrete web alters over the girder length. In support regions, where negative bending moments lead to compression in the lower part of the cross section, the concrete web goes down to the lower flange. The steel profile acts as external reinforcement. Towards midspan the compression moves to the upper part of the cross section. Thus, the concrete web is reduced and the steel web is enlarged and goes up to the concrete slab. The web of the steel profile is then in the tension zone. For large spans with corresponding beam depths there is the advantage that no tensile stresses arise in the concrete which lead to a crack pattern. At the same time, the weight of the beams is reduced considerably.

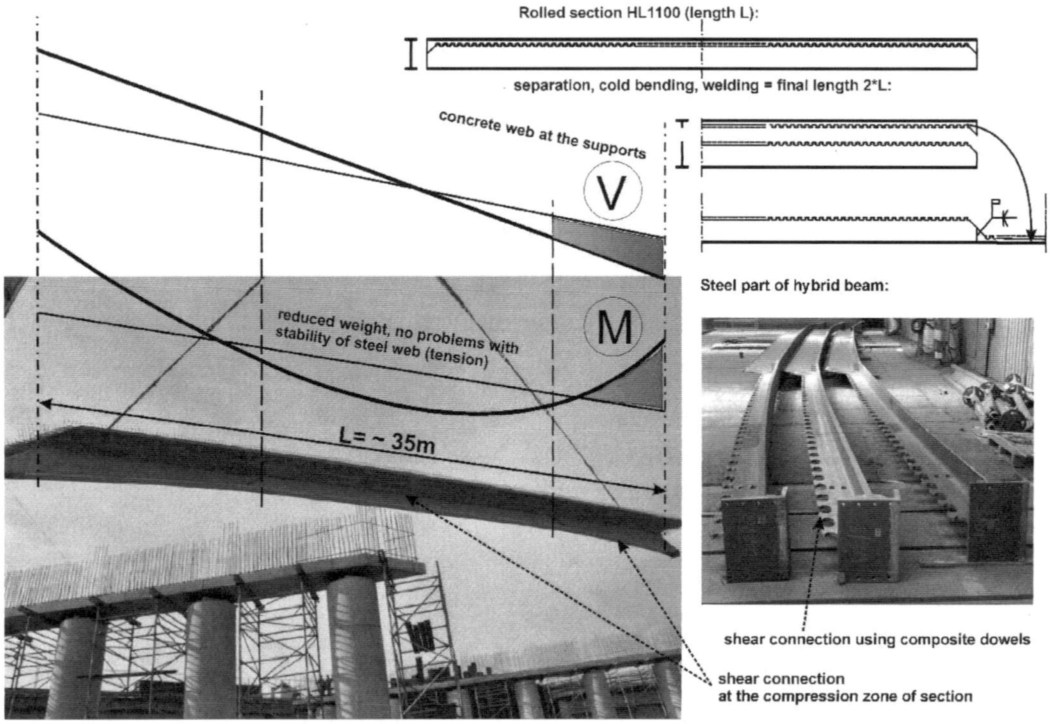

Figure 3. Principles of the new kind of PreCoBeam system (road viaduct WA-352, picture of prefabricated element from http://www.a1-rzasawa-blachownia.pl).

Figure 4. Steel part of PreCoBeam for WA-352, made from rolled sections HL1100.

Figure 5. WA-352 under construction, January 2018 (http://www.a1-rzasawa-blachownia.pl).

Although the solution is dedicated to rolled sections, the first bridge was constructed using plated elements because of specific contract conditions (the main contractor was a company with own steel production plants producing structures from welded plates). For the second bridge, a road viaduct in Poland with a larger skew than the first one (Figure 5), the initially intended hot-rolled sections of the HL1100 series (Figure 4) were used. The bridge WA-352 (Figure 6) transfers the highway A1, km 417+402,48 over the railway line no. 146 Częstochowa Wyczerpy – Chorzew Siemkowiced. The highway is still under construction.

Figure 6. Road viaduct WA-352, bottom view.

The solutions constructed in Poland and presented in this paragraph are a new kind of general composite forms. The basis for design are presented in [4]. They are possible to implement and develop due to the advanced shape of shear connection called "composite dowels". The implementation of composite dowels in a new CEN Technical Specification will make clear design rules widely available all over Europe. Appropriate works are ongoing in the frame of the CEN working groups for Eurocode 4. At the moment, only national guidelines and approvals are existing.

3. New system for spans at range 50-60 m.

The new system for spans at range 50-60 m was introduced in Poland for the first time. It is based on using rolled I-sections in midspan and T-sections with additional plates in between at support regions, see Figure 7. In addition, double composite action is applied (Figure 8). The system was developed as a result of R&D works of ArcelorMittal. The development and design of the system is described in detail in the free available report [5]. The 2017 finished bridge constructed in city of Opole in Poland (Niemodlińska street) is the first implementation [6], see Figures 9-12. The main span of the bridge is 54 m. First dynamic load tests proved good dynamic behavior of the bridge despite of its big slenderness.

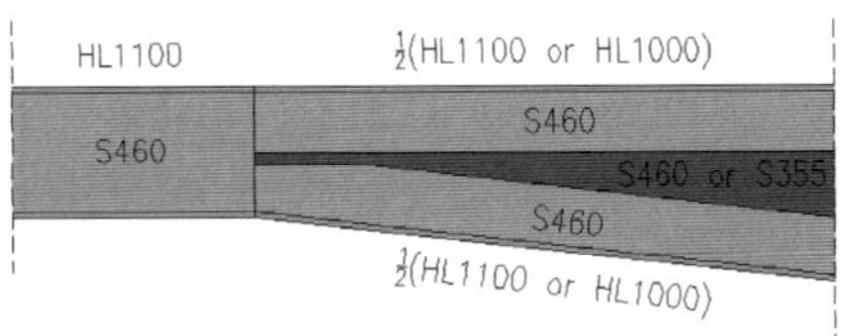

Figure 7. Assumption for steel beam of the new system [5].

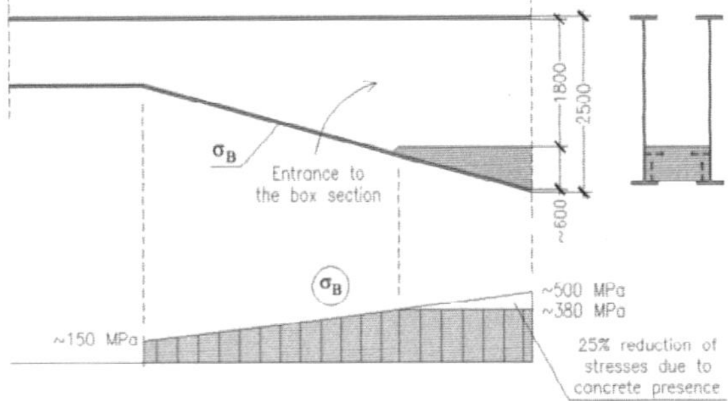

Figure 8. Assumption for double composite action of the new system [5].

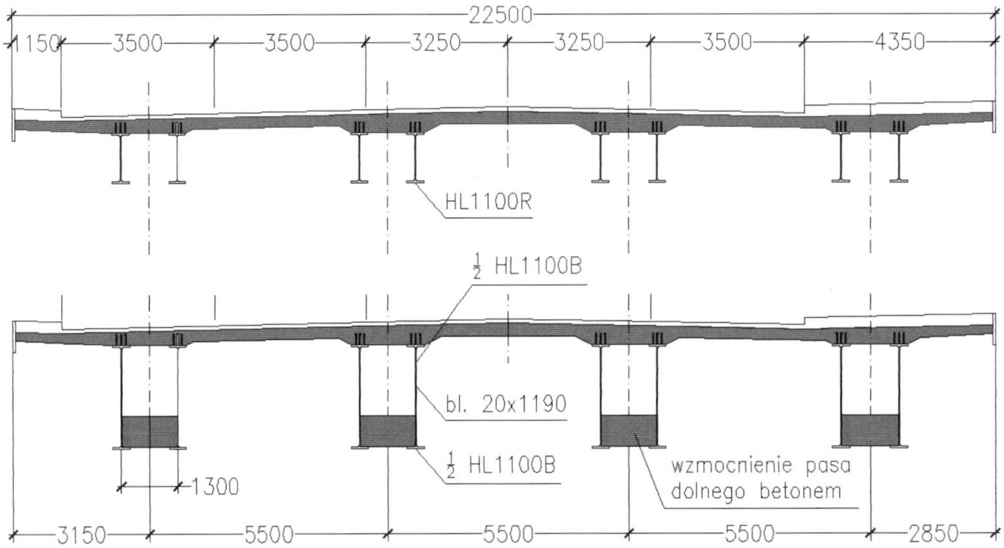

Figure 9. Cross sections of composite bridge at Niemodlińska street in Opole (top: midspan, bottom: supports).

Figure 10. The bridge in Niemodlińska street in Opole during load test.

Figure 11. Steel girder using two T-sections and welded-in plate of bridge in Niemodlińska street in Opole.

Figure 12. Bridge in Niemodlińska street in Opole: bottom view after construction of concrete slab.

The steel beams of the bridge were produced and delivered by ArcelorMittal. The steel structure was welded and the whole bridge was constructed by Banimex. More information and pictures from the construction can be found on their company website "www.banimex.pl". A detailed calculation report of a very similar structure (span 50 m) is included in the R&D report [5].

In the same year (2017) the construction of a second bridge with a similar solution started (Figures 13-14) next to Rzuchów in Poland. The span lengths are: 24.5 + 39.7 + 68.0 + 39.6 + 25.0 + 24.8 m. The new system is dedicated to bigger spans and it assumes the separation of HL sections at midspan as well. Thereby an asymmetric I-section is used at midspan and its web thickness differs along the web height. The main span is constructed using this new system with partly double composite action (4 girders connected by bottom concrete slab). For the other spans standard rolled sections are used.

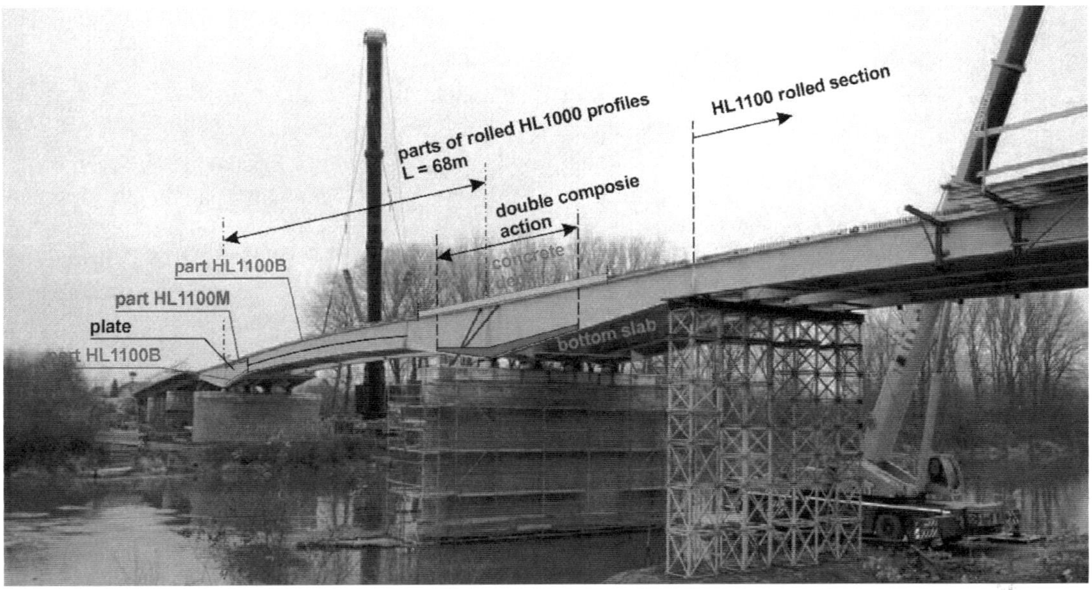

Figure 13. New system presented on the basis of the bridge in Rzuchów in Poland. Picture from http://www.nowiny24.pl/

Figure 14. Bridge in Rzuchów in Poland: erection of steel structure, Source: http://www.nowiny24.pl/

4. Net-arch bridges using HD-sections

Network arch bridges with the use of rolled I-beams in their arch structure (Figure 15) have been constructed for some time in Poland [7]. The main reason to use rolled sections instead of traditional welded box sections are cost savings and shortening of the construction time. Due to the availability of S460 as standard and a wide range of heavy rolled sections (HD profiles up to 140 mm flange thickness and with maximum weight of 1299 kg/m), less steel is needed to resist the high compression forces in the arches. Additionally, rolled sections are less costly in terms of fabrication and the connection of the hangers is much easier. Gusset plates can easily be welded into the chamber of the HD sections (Figure 16).

Figure 15. Net-arch bridge with span length 75m using rolled I-sections in western Poland (WD-57 bridge, S5 express road Poznań – Wrocław, part Kaczkowo-Korzeńsko) [8].

Figure 16. Detail: Gusset plate to connect hanger with arch of HD section.

Figure 17. Weld preparation of bended beams at ArcelorMittal Beam Finishing Center.

Figure 18. Prepared beam with weld access holes.

The positive experiences gained with this construction method and the cost-effectiveness have allowed this solution to spread and its significance is increasing now. Currently a network arch road bridge with 120 m span length is under construction (Figure 19). For this bridge HD400x744 and HD400x1086 profiles in HISTAR460 steel grade were used (Figure 17). Since the arches could not be delivered in one piece to the job-site, the construction had to be designed considering welded site joints. Therefore, a certain weld preparation (Figure 18) was applied to the beam segment ends to allow for a residual stress reduced joint.

Figure 19. Construction of MS-15 bridge with 120 m span length in northern Poland, road S51 Olsztyn-Olsztynek, part Olsztyn East – Olsztyn South [8].

5. Conclusions

The here shown projects can be considered as good examples of modern design of medium span bridges, and even longer ones. Thanks to the use of high strength steel sections and due to the latest developments of composite technologies such as PreCoBeams the efficiency of bridge constructions can be optimized. This makes bridge construction fast, efficient, sustainable and cost saving.

References

[1] Stacewicz M., Hukowicz A., Windorpski H., Kołakowski T., Kosecki W., Lorenc W., Nowatorska konstrukcja zespolona przęseł nowego wiaduktu ciągu DK nr 22 nad linią PKP koło Elbląga, Mosty 2017-1, pp. 57-62.

[2] Lorenc W., Kołakowski T., Hukowicz A., Seidl G. *Verbundbrücke bei Elbląg: Weiter-Entwicklung der VFT-WIB Bauweise.* Stahlbau. 2017, Jg. 86, H. 2, pp. 167-174.

[3] Lorenc W., Seidl G., *Anwendung von Verbunddübelleisten im Brückenbau.* Münchener Stahlbautage 2017

[4] Lorenc W., *Model for a general composite section with dowels.* Steel Construction, 2017 vol. 10, nr 2, pp. 154-167.

[5] Lorenc W. Kożuch M. Innovative design of composite bridges for medium spans based on rolled sections Project PFR338-14, 2016. (http://sections.arcelormittal.com/library/steel-research-reports/bridges.html)

[6] Kożuch M., Lorenc W., Rabiega J., Ochojski W., *Technologia budowy mostu drogowego o rozpiętości przęsła powyżej 50 m z dźwigarów walcowanych.* Seminarium Naukowo-Techniczne Wrocławskie Dni Mostowe, Wrocław, 28-29 listopada 2017

[7] Kaczmarek T., Radziecki A., Sygit M. *Network arch bridges over motorways in Poland – selected design and construction problems.* ARCH 2013

[8] Kaczmarek T., Galewski T., Topolewicz K., Sęk R., Ochojski W., Lorenc W., Kożuch M., *Mosty network arch z zastosowaniem dwuteowników HD giętych na zimno (Network arch bridges using cold bended HD sections)* MOSTY 3/18, 2018.

The particular challenges of expansion joint installation in steel bridges – Case studies

David Jelínek[1], Niculin Meng[2] and Colm O'Suilleabhain[2]

[1]Mageba CS s.r.o., Novoměstská 1c, 621 00 Brno, Czech Republic
[2]Mageba SA, Solistrasse 68, 8180 Bulach, Switzerland

E-mail: cosuilleabhain@mageba.ch

Abstract. How a bridge's expansion joints are connected to the main structure is an important detail, with relevance not only for ease of installation and replacement but also for durability and long-term performance. In the case of steel bridges – or more accurately expressed, in the case of connection to steel surfaces – the challenges tend to be greater than with concreted connections. With steel connections, special attention is required in the design of the expansion joints, and considerably more effort and precision is required on site during installation work. This paper shall describe the particular challenges, and illustrate these, and ways of addressing them, with reference to steel bridge construction and renovation projects.

1. Introduction

Expansion joints are critically important bridge components, and how they are installed in a structure is important in a number of ways. Of course, the suitability of the connection as designed is of primary importance in ensuring that loads are properly transferred, but the quality of the connection, as designed and executed, is also of great importance in ensuring the durability and long-term performance of the components and the structure as a whole. Connection of expansion joints to concrete bridges is relatively straightforward, as concrete can very often be poured around the joints' anchorages, allowing a high degree of flexibility and tolerance without any impact on load transfer or long-term durability. In the case of steel bridges, however, the connection of a steel expansion joint directly to a steel structure often allows little or no such tolerance. Steel surfaces must generally be perfectly flat and flush with each other to ensure even transmission of loading without undesired moment forces or distortion, and in the case of bolted/screwed connections, anchors/bolt holes must be perfectly aligned with each other across each connection interface. While expansion joints may be specially designed and detailed to accommodate greater tolerances as specified, such measures require considerable additional design and fabrication effort, impacting on supply costs.

These challenges, and how they may be addressed, are described and illustrated below with reference to various steel bridge construction and renovation projects.

2. Connection of expansion joints to bridge structures – general

In principle, an expansion joint should be installed in such a way that all its parts are properly supported and will not be subjected to any unnecessary forces. Its presetting, or gap width at the time of installation, must be appropriate for the gap width of the structure at that time, considering the prevailing structure temperature etc., with allowance for the future opening and closing movements that the joint must accommodate. And any designed pre-tensioning within the joint should be as designed, without increase or decrease due to lack of proper levelling. This will ensure that the joint's durability and performance will be unimpeded by installation-related deficiencies, enabling it to perform well for many years.

Design and fabrication of an expansion joint for installation in a concrete bridge deck (i.e. with concrete connection) generally allows great installation tolerance, with the expansion joint simply lifted into a block-out / recess that can be over-sized as desired to ensure the joint does not clash with the superstructure. As long as any reinforcement steel (which can generally be bent out of the way if necessary) is not in the way, the expansion joint can then be freely, easily and precisely positioned in the correct position, at the correct height and with the correct longitudinal and transverse gradients. Furthermore, since the edges of the joint can be individually freely positioned for connection at both sides of the bridge movement gap, the expansion joint can be installed with a complete lack of undesired distortion or twist, and adjustment of presetting is easy and precise. The installation tolerance offered by concrete connection is illustrated in Figure 1, for two common types of expansion joint.

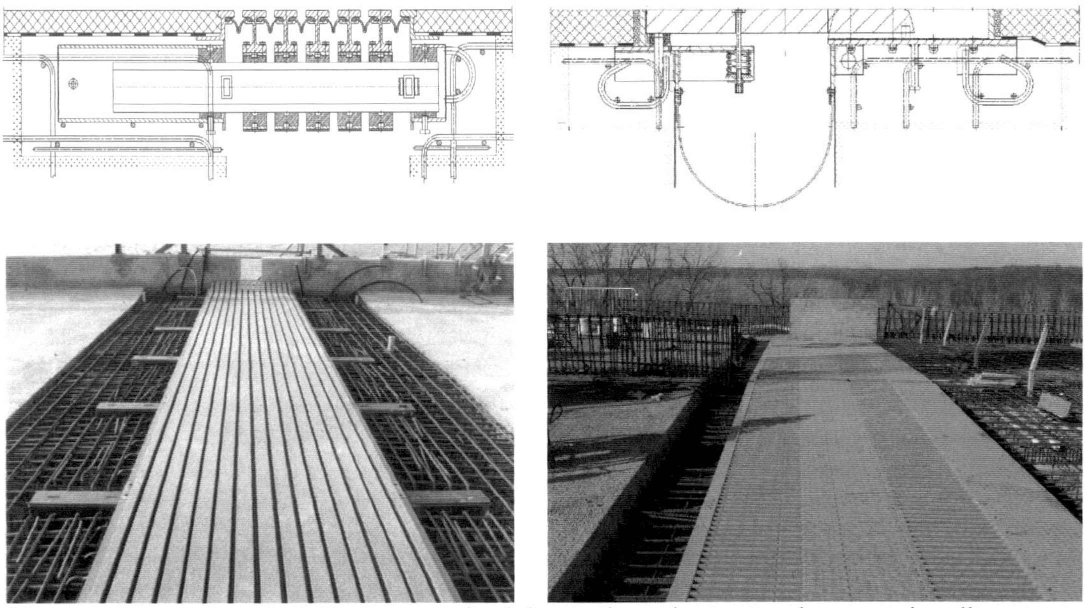

Figure 1. Connection of a bridge's expansion joints to the main structure by concreting allows a great degree of tolerance in terms of positioning, with the joint simply lifted into an oversized recess, as shown by these cross section sketches and photos (Left: Modular joints. Right: Sliding finger joints).

3. Connection of expansion joints to steel bridge superstructures – typical challenges

Design and fabrication of an expansion joint for installation in a steel bridge deck (i.e. with steel connection) generally requires more attention to detail and greater effort than would arise if the deck was of concrete. And installation of the expansion joint on site then also typically requires more effort and care.

Connection of expansion joints to steel structures may often be achieved by welding or bolting. Bolting offers advantages in terms of installation (especially, for example, in the case of galvanized steel) and replaceability, but this approach requires yet more precision / allows yet less tolerance than welding, and bolting generally requires significantly more space to transmit a specified load than welding – making bolting impractical in some circumstances. And as a rule, achieving a desired level of quality control is considerably more laborious with welded connections than with bolted ones.

The particular challenges of steel connections include:

3.1. Uniqueness
Steel connections are always non-standard from an expansion joint design perspective, and generally unique, so they require special solutions to be developed and implemented for every structure / joint.

3.2. Load transmission to the main structure
Without the load-distributing benefits of an intermediate "layer" of concrete between the expansion joint and the main structure to which it is connected, load transmission at connection points tends to be much more highly focussed as point/line loads, increasing design and execution challenges locally in these areas. This is especially true if connecting steel surfaces are not perfectly flat and parallel with each other, resulting in yet higher point loading and undesired torsions or bending moments.

3.3. Accommodation of tolerances
The installation tolerance described in Section 2 above associated with concreted connections does not apply in the case of steel connections. Not only does this increase the challenge of installation on site, and the difficulty of constructing the steel superstructure to receive the expansion joint exactly as designed, it also increases the effort required to fabricate the expansion joint precisely as designed, with a complete lack of distortion.

3.4. Transportation and installation frames
Connection of transportation and installation frames, which must often be designed to facilitate on-site adjustment of presetting (depending on the bridge structure's temperature at the time of installation, etc.), is less straightforward in the absence of anchor loops/studs of the type typically available for this purpose on a joint designed for concrete connection. In the case of concreted connections, such frames can simply be welded to the anchor loops/studs, and cut off at the time of installation – but in the case of steel connections, connection points are more limited, and must be either precisely designed for bolted connections or, where simpler welded connections are used, must be smoothened and treated with corrosion protection following installation. Furthermore, steel connections often require specially designed installation/positioning frames, tailored to the particular steel structure to which the joint is to be connected – a considerably more demanding aspect than arises with a concrete connection.

3.5. Accessibility during installation
Connection to a steel superstructure, especially by welding, typically requires a much higher degree of access for installation workers than does connection by concreting. Such access must be ensured in the design of the superstructure and of the expansion joint – if, indeed, it can be accommodated at all.

3.6. Distortion due to welding
Care must be taken when installing an expansion joint by welding to avoid distorting the joint or the steel structure to which it is connected, and to avoid damaging adjacent corrosion protection etc.

3.7. Subsequent application of corrosion protection to welded areas
Since steel which is to be welded must be clean, without any corrosion protection applied to the welding area, corrosion protection must subsequently be applied to the entire welded area following completion of welding. Not only does this require effort, it also introduces a new durability / quality control issue, since site-applied corrosion protection may not offer the same long-term protection as the factory-applied equivalent.

4. Examples of steel-connection expansion joints in major bridge structures
Some of these challenges are illustrated by the following images showing examples of installation of large expansion joints with steel connections in major bridge structures around the world.

Figure 2. Installation of 7-gap and 14-gap *Tensa-Modular* expansion joints, side by side, during renovation of the Lillebaelt Bridge, Denmark (2002).

Figure 3. Installation of a 15-gap *Tensa-Modular* expansion joint during construction of the Guangan Bridge, Busan, South Korea (2002).

IOP Conf. Series: Materials Science and Engineering **419** (2018) 012017 doi:10.1088/1757-899X/419/1/012017

Figure 4. Installation of a 24-gap *Tensa-Modular* expansion joint during construction of the Incheon Grand Bridge, South Korea (2008).

Figure 5. Installation of a 12-gap *Tensa-Modular* expansion joint with noise reducing "sinus plates" on its surface on the Köhlbrand Bridge, Hamburg, Germany (2016).

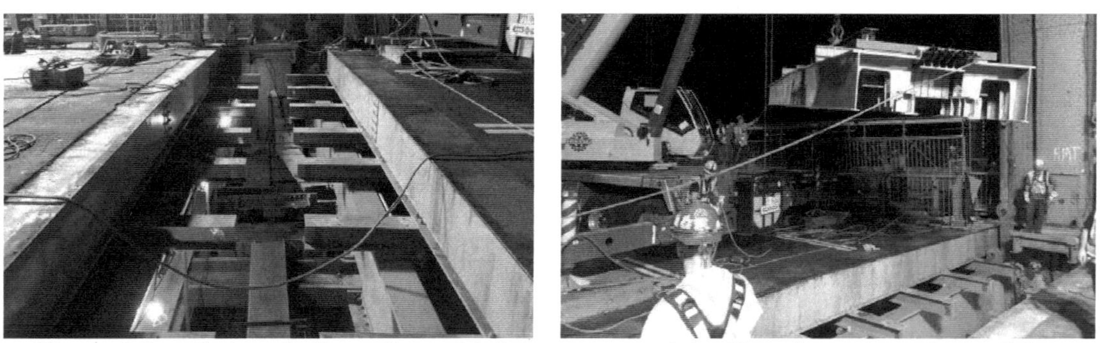

Figure 6. Installation of a 7-gap *Tensa-Modular* expansion joint (featuring noise-reducing "sinus plates" on its surface during renovation of the Angus L. Macdonald Bridge in Halifax, Canada (2017).

Figure 7. Replacement of *Tensa-Finger* sliding finger expansion joints with movement capacities of up to 2700 mm in the Verrazano Narrows Bridge, New York, USA (2017).

5. Expansion joint solutions that can avoid the need for steel connections on steel bridge decks

In some cases, the use of a particular type of expansion joint solution can avoid steel connection difficulties and challenges of the types described and illustrated above. A number of such solutions are presented below.

5.1. Expansion joint solutions that are designed for easy future replacement/renewal, retaining the existing connections to the main structure

A number of expansion joint solutions have been developed which enable the main dynamically-loaded part of an expansion joint to be replaced without replacing the parts which are connected to the main structure at each side and not subjected to dynamic loading, and which therefore will likely not need to be replaced. Such solutions can be particularly beneficial in steel bridges, as it is these connections to a steel bridge that are especially challenging and problematic.

5.1.1. The "Box-in-Box" solution for modular joint renewal

Where a modular joint on a bridge requires to be replaced, the "Box-in-Box" method may be used in certain circumstances. This method involves cutting short sections out of the existing joint's edge beams at each support bar, and cutting the top plate off the boxes at each end of each support bar. This enables the joint's mechanical structure, consisting primarily of its centerbeams and support bars, to be lifted out and replaced with a similar construction which also features new boxes on the ends of the support bars. When the new construction is lifted into place, the new boxes are secured inside the existing boxes, hence the method's name. This method requires that the existing joint is suitable in certain ways – in particular, with boxes that are large enough to receive fully functional new boxes. The method is illustrated in Figures 8 to 10, and described further by Spuler et al [1].

Figure 8. Renewal of a *Tensa-Modular* expansion joint by the "Box-in-Box" method: Removal of rubber seals and steel centerbeams; Cutting of edgebeams; and removal of transverse support bars.

Figure 9. "Box-in-Box" method (cont.): Lifting in of new structure consisting of centerbeams and support bars; Welding of new box inside old box; Concreting of spaces between new and old boxes.

Figure 10. "Box-in-Box" method (cont.): Reinstatement of edgebeams where cut (with recesses for rubber seals); Insertion of rubber seals; Reinstatement of road surfacing – if needed, one lane at a time.

5.1.2. The "Quick-Ex" solution for modular joint renewal

The "Quick-Ex" (or "quick-exchange") solution for replacement/renewal of a modular expansion joint is similar to the "Box-in-Box" method described above, but far easier to carry out when the time comes to replace the joint thanks to the designing of the existing joint to facilitate the renewal process. All parts that will need to be removed to enable the joint's mechanical structure to be replaced are bolted/screwed in place and can thus be easily removed, without cutting of steel or damaging the bridge's road surfacing or waterproofing. Then, the new mechanical structure can be lifted into place and secured by bolting, without any need for welding or concreting. The impact on the bridge and on traffic are thus kept to an absolute minimum. The approach is illustrated in Figure 11, and described in more detail by Adam et al [2].

Figure 11. A Tensa-Modular expansion joint as specially designed for easy future renewal by means of the "Quick-Ex" method – maintaining the joint's connections to the main structure.

5.2. *Expansion joint solutions that do not have steel at connection interface*
Connection of an expansion joint to a steel structure can be relatively unproblematic if the joint does not have steel at the connection interface – for instance, as illustrated by the following low-movement solutions that may be applied to any type of structure, especially during renovation work.

5.2.1. *Flexible plug expansion joints*
Modern flexible plug joints, such as the *Polyflex* joint (Figure 10) which primarily consists of a poured polyurethane compound, are far superior to the traditional asphaltic type as noted by Moor et al [3].

5.2.2. *Single-gap joints with polymer concrete forming high-strength bond to main structure*
Single-gap joints that have their steel edge profiles anchored in freshly poured quick-drying polymer concrete offer great tolerance in positioning the steel elements of the joint and avoid any need for welding or bolting. An example of such a solution is the *Tensa-Crete* joint shown in Figure 11, which has numerous benefits when used to replace old existing joints as described by Spuler et al [4].

Figure 12. Poured polyurethane flexible plug joints such as the *Polyflex* joint offer many advantages over the traditional asphaltic type.

Figure 13. Single-gap joints with polymer concrete anchorage, such as *Tensa-Crete* joints, offer many benefits for joint replacement works.

6. Conclusions
For reasons as mentioned in Section 3 above, and as illustrated in Section 4, the challenges arising when expansion joints are to be connected to a steel bridge are very varied, and the situations arising are very often unique. While certain types of expansion joint do not have steel on their connection surface, which can reduce connection effort and introduce a significant element of tolerance/flexibility for fabrication/construction and installation work, these tend to have limited applicability – particularly in relation to the magnitude of movements these components can accommodate. Where steel connections are required, great care and attention to detail are needed in developing and implementing the solution, as are extensive knowledge of the technical challenges arising and in-depth experience in the development of suitable solutions. The potential for problems to arise during installation on site should be recognised, with adequate time allowed for installation, and all appropriate measures taken, in consultation between the bridge designer, the bridge constructor and the expansion joint supplier, to minimise the risks and ensure a satisfactory long-term solution.

References
[1] Spuler T, Moor G and O'Suilleabhain C 2013 Renewal of modular expansion joints - an innovative approach that minimises impacts on traffic and on the main bridge structure *Proc. IABSE Conf. on Assessment, Upgrading and Refurbishment of Infrastructures* Rotterdam

[2] Adam S, Spuler T and Hoffmann S 2018 The quick-exchange expansion joints of the Köhlbrand Bridge in Hamburg *Proc. 40th IABSE Symposium* Nantes, France

[3] Moor G, Baillés B and Pakniat P 2018 Modern flexible plug expansion joints with polyurethane surface – far superior to the traditional asphaltic type *Proc. 10th International Conference on Short and Medium Span Bridges* Quebec City, Quebec, Canada

[4] Spuler T, Moor G and O'Suilleabhain C 2012 Renewal of small movement expansion joints with minimum break-out and time requirements *Proc. 6th International Conference on Bridge Maintenance, Safety and Management (IABMAS 2012)* Stresa, Italy

IOP Conf. Series: Materials Science and Engineering **419** (2018) 012018 doi:10.1088/1757-899X/419/1/012018

Experience of duplex stainless steels as structural materials for bridges

Sukanya Hägg Mameng[1], Andrew Backhouse[2], Jonathan McCray[3], Graham Gedge[3]

[1]Outokumpu Stainless AB, Avesta Research Centre, Avesta SE77244, Sweden
[2]Outokumpu Stainless Ltd, Europa Link, Sheffield, S66 1EH, United Kingdom
[3] Arup, Fitzroy Street, London, W1T 4BQ, United Kingdom

E-mail: andy.backhouse@outokumpu.com

The use of duplex stainless steel (DSS) in bridge construction has increased significantly in the last twenty years because of the material's high strength and good corrosion resistance. The high corrosion resistance of DSS results in a low maintenance requirement and hence low life cycle costs. There is now a range of DSS grades available and it is possible during the design phase to undertake a material selection exercise for DSS to align the required material performance with the bridge environment. Lean alloyed and less costly DSS grades can be specified in less harsh environments. This paper presents the results from inspection of various existing road, rail and pedestrian bridges. In each case a DSS was used for the main structure and each bridge illustrates the performance of the material in a representative European environment. The environments were retrospectively classified in accordance with the recently introduced EN 1993-1-4 Annex A and a comparison of actual performance made with the guidance given in this annex. Previous high-level cost studies have indicated that durable, low-maintenance bridge designs using DSS are achievable without incurring excessive initial construction costs, and that significant life cycle cost savings can be made over the life of the structure. The present work has confirmed that, when appropriately selected and used, DSS can be anticipated to give long service lives in bridge structures, greatly reducing corrosion problems that may necessitate re-painting and remediation of carbon steel structures.

1. Introduction

The structural use of duplex stainless steels (DSS) in bridge construction has increased significantly in the last twenty years. DSS are ideal for bridges because of their high strength, good corrosion resistance, very low need for maintenance and hence low life cycle costs. The use of stainless steel for structural applications is now recognized in design codes such as EN 1993-1-4 (Eurocode 3) and AISI Steel Design Guide 27[1,2]. Three DSS grades suitable for use in bridges are EN 1.4462 (UNS S32205), 1.4362 (UNS S32304) and 1.4162 (UNS S32101). All DSS have high strength, good fabrication characteristics and good weldability. In recent years the lean alloyed DSS 1.4162 has often been selected in preference to DSS 1.4462 and 1.4362 since it combines high strength with lower cost[1,4]. It is important to select the DSS grade that is appropriate for the intended service environment, bearing in mind that appropriate use of a leaner alloy can reduce costs. EN 1993-1-4 Annex A [1] sets out a methodology for selecting an appropriate grade of stainless steel for the service environment, using five Corrosion Resistance Classes (CRC I to V). The CRC are assessed in terms of three components: risk of exposure to chlorides from salt water or deicing salts (distance from the sea and roads), risk of exposure to sulfur dioxide and cleaning regime including natural washing by rain which has a cleaning effect. The standard suggests appropriate grades for the CRC classes. Different elements of the same structure may have different exposure conditions, i.e. one part may be fully exposed to rain and another part fully sheltered from it. However, there is little documented information available about the service performance of actual structures such as bridges made of DSS. The main objective of this paper is to present such information for seven DSS bridges.

2. Inspection procedure, results and comments for each bridge inspection

Seven different DSS bridges were inspected to assess durability in their actual service environment. The methodology used involved selecting bridge sites, visual inspection conducted at each site and later analysis in the laboratory. The criteria used for selecting the sites were: the type of DSS used for the main structure, the environmental conditions and the age of the bridge. The environment was classified in terms of the CRC class [1]. The selected bridges were between 4 and 12 years old at the time of inspection. All bridge structures are intended to have a service life and economic life longer than this. The age of the bridge structures inspected was considered to be adequate for making assessments of the suitability of the DSS selected for the service environments. The list of the DSS bridges inspected, including the DSS grade used, the type of DSS elements, the construction year, the CRC class and the estimated C-Class (Corrosivity class) [5] are given in Table 1. At each inspection site visual inspection was used to identify the condition of DSS elements and digital photographs used to record the condition of each element. Where access was possible, a hand-held X-ray spectroscopic analyzer was used to confirm the grade of steel used. Surface replicas were taken for surface roughness measurements in the laboratory. When any surface contamination was observed, the contamination was transferred to an adhesive scanning electron microscope (SEM) stub for subsequent analysis by SEM and Energy Dispersive X-ray Spectroscopy (EDS). For sites close to bodies of water, chloride ion (Cl^-) levels in the water were checked using a chloride test kit. A water sample was also taken for subsequent laboratory analysis of Cl^- content. In the laboratory surface roughness measurements were taken on the replicas using a confocal microscope. Any contamination was examined by SEM/EDS to identify the nature and chemical composition. The chloride ion level from water samples was analyzed by chromatography.

2.1 Road Bridge in Orrhammarvägen, Flen, Sweden – 1.4162, CRC II

The Orrhammarvägen Road Bridge in Flen, Sweden (Figure 1) is a composite bridge with concrete deck and abutment, supported by 1.4162 DSS beams and bearing plate. the environment was classified as CRC II with a low risk of exposure to Cl^- (>10 km from Baltic Sea) and SO_2, and the DSS 1.4162 elements were in a sheltered condition. The inspection results showed that the performance of 1.4162 was very good with no sign of corrosion observed after six years exposure in service. Some accumulation of dirt was found on the underside DSS surfaces due to the sheltered position and lack of natural cleaning by rain. A superficial orange stain was noted at regular intervals along the bottom face of the bottom flange of one beam. A sample of the contamination was taken and analyzed by

Table 1. List of the DSS bridges inspected, the DSS grade used, the type of DSS elements, the year of construction, the CRC class[1] and the estimated Corrosivity C-Class[3]

Steel Grade	Bridge name (Location)	Year (Age[1])	DSS elements	CRC	Exposure[2] to Cl	SO_2	Exposed condition	Estimated C-Class[3]
1.4162	Orrhammarvägen Road Bridge (Flen, Sweden)	2009 (6 y)	Bridge beam, bearing plate	II	L	L	Sheltered	C2
1.4162	Añorga Railways Bridge, (San Sebastian, Spain)	2011 (4 y)	Arch, beam	II/III	M	L	Open/ sheltered	C3
1.4162	Road Bridge (Nynäsham, Sweden)	2011 (4 y)	Bridge beam, Abutment plate	IV	V	L	Sheltered	C3
1.4362	Puerto Arrupe Pedestrian Bridge (Bilbao, Spain)	2003 (12 y)	Box girder	II/III	M	L	Open/ sheltered	C3
1.4362	Celtic Gateway pedestrian Bridge (Holyhead, UK)	2006 (9 y)	Arch	III/IV	V	L	Open/ sheltered	C3
1.4462	Pedestrian Bridge (Zumaia, Spain)	2008 (7 y)	Bridge beam, balustrade	III/IV	H	L	Open/ sheltered	C4
1.4462	Cala Galdana Road Bridge (Menorca, Spain)	2005 (10 y)	Arch, deck, balustrade	IV/V	V	L	Open/ sheltered	C5

Notes: 1 = Age of the bridge at the time of inspection
2 = Risk exposure to chloride Cl^- and sulfur dioxide SO_2 : L = Low, M = Medium, H = High, V = Very high
3 = Estimated corrosivity class in accordance with ISO 9223.

SEM-EDS. The result showed the stain was contamination by iron particles from carbon steel or a similar source. Due to the regularity of the spacing of contamination it was suspected that the beam had been resting on steel trestles at some point during its fabrication, transportation or installation. The contamination was superficial and had not affected the structural integrity of the beam.

Figure 1. Orrhammarvägen Bridge, Flen, and close up of underside periodic contamination

2.2 Añorga Railway Bridge in San Sebastian, Spain – 1.4162, CRC II/III

The main structural material of the railway bridge in Añorga, San Sebastian, Spain was DSS 1.4162. This bridge had replaced an existing carbon steel structure that had required intensive maintenance over its lifespan. The bridge is located about 3 km from the Bay of Biscay and has areas both with and without rain washing. The CRC was therefore II and III due to the moderate risk of exposure to chlorides.

wt %	O	Mg	Al	Si	K	Ca	Fe
AVG	48.4	1.1	7.9	20.5	3.2	16.0	3.4

Table 2: SEM-EDS result of the contamination on the DSS 1.4162 surface

Figure 2. Añorga Railway Bridge (DSS 1.4162) in San Sebastian, Spain

The inspection results showed that the DSS 1.4162 was performing well, with no sign of corrosion. All stainless steel surfaces had a fine covering of dirt and dust. A sample of this showed that iron (Fe), aluminium (Al), calcium (Ca), magnesium (Mg) and Silicon (Si) were present on the DSS surfaces, Table 2. These deposits were probably from the cement industry which is very near to the inspected site. This surface deposit was not causing any corrosion of the DSS.

2.3 Road Bridge in Nynäshamn, Sweden – 1.4162, CRC IV

The Nynäshamn bridge is a concrete composite bridge with DSS supporting beams, and is in a sheltered inlet off the Baltic Sea adjacent to an oil refinery. In accordance with the standard EN 1993-1-4 Annex A, which was published after the bridge was constructed, the environment condition is CRC IV, due to being a distance of less than 0.25 km from the coast and the absence of rain washing. The standard recommends that a high alloyed steel such as DSS 1.4462 material is required in this environment, but the bridge is constructed of DSS 1.4162, a lower alloyed grade. The results showed that the DSS 1.4162 performed well and appeared to be in good condition, as seen in Figure 3. Some cosmetic surface corrosion attack in the form of micro-pitting was observed on the beam. This is possibly due to the presence of traces of chloride found on the surface in combination with no washing by rain. However, these isolated micro-pits were superficial and not causing any deterioration to the structural performance. It is clearly seen that DSS 1.4162 is performing adequately in this location, despite being outside the recommendation given in EN1993-1-4 Annex A.

Figure 3. General appearance of the composite bridge in Nynäshamn, and isolated pits on the surfaces

2.4 Puerto Arrupe Pedestrian Bridge in Bilbao, Spain – 1.4362, CRC II/ III

The DSS Bridge in Bilbao is located 10km from the Bay of Biscay, and acts as a link between the university and Guggenheim museum. The environment was classified as CRC II and III. The DSS grade 1.4362 was used for the support structure and external parts.

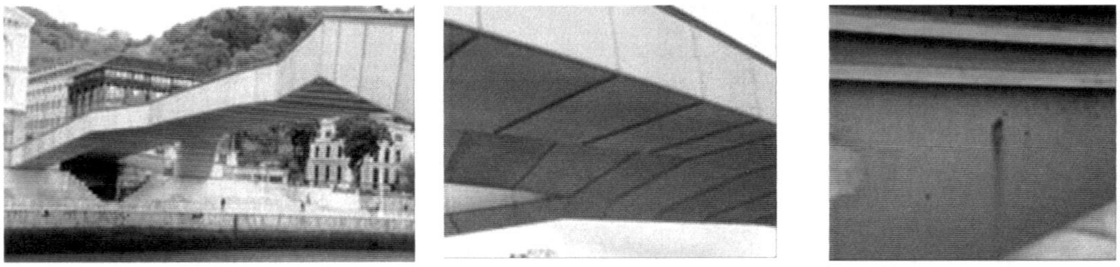

Figure 4. Puerto Arrupe bridge overall design, and close-up view of sides and sheltered underside

The inspection results indicated very good performance of DSS 1.4362 and no signs of corrosion were observed, see Figure 4. However, some discoloration of panels and spots of contamination were occasionally seen at the bridge box surface. Discoloration is likely due to limited washing by rain or from fabrication or transportation. The contamination spots were probably from fabrication processes - either grinding or spatter from welding. One panel was painted on the internal side surface at pedestrian level in order to cover graffiti. This is not normally a recommended method for dealing with such issues with DSS, and a more appropriate solution should be used.

2.5 Celtic Gateway Bridge, Holyhead, Pedestrian Bridge – 1.4362, CRC III/IV

The bridge in Holyhead (Figure 5) is adjacent to an ocean harbor, so it may be considered CRC IV due to salt exposure from the ocean. However, the harbor below the bridge is well sheltered, and therefore wave conditions are likely to remain calm. There is no high-speed water traffic passing under the bridge to generate splashing seawater. Consequently, the position of the 'coast' could be interpreted as being at the harbor mouth and that the bridge is therefore 800–900m from the coast in a CRC III environment. The Celtic Gateway Bridge spans over Victoria Road in Holyhead. The road is likely to be de-iced in winter by using grit/salt, and this would also result in a CRC III classification. In a CRC IV environment DSS 1.4462 is recommended by EN 1993-1-4. For CRC III the recommendation is DSS 1.4162 or 1.4362. This bridge is constructed of 1.4362, although it was found to have a particularly smooth surface finish (Ra=0.4-0.5 μm) which is known to be beneficial in improving corrosion resistance of stainless steels.

Figure 5. Celtic Gateway Bridge in Holyhead, UK; close ups of good condition, and localized run-off

The inspection result demonstrated that the arch surfaces and detailing made of DSS 1.4362, were in generally good condition as shown in Figure 5. Some lime was deposited and appeared to be from water run-off from the concrete deck onto DSS surfaces. This was not causing any damage to the DSS.

2.6 Pedestrian bridge in Zumaia, Spain – 1.4462, CRC III/IV

The first pedestrian bridge constructed of a combination of DSS and glass fiber reinforced plastic is in Zumaia, Spain over the Narrondo River (Figure 6). The marine climate with high risk of exposure to chloride (about 1km from Bay of Biscay) required a highly corrosion resistant grade such as 1.4462.

Figure 6. Pedestrian Bridge (DSS 1.4462) in Zumaia, Spain, with DSS elements performing well. One incidence of a balustrade endplate with orange-brown deposit of carbon steel contamination

The inspection result show that DSS 1.4462 was performing very well in this highly corrosive environment without any sign of corrosion to the DSS itself. The only exception was that some grey staining and discoloration was observed on the surface. The main cause of the damage found on this bridge was contamination. One specific area of balustrade top surface and end plate showed an orange-brown stain, as seen in Figure 6. It was speculated that this orange stain could be from the fabrication (grinding or cutting) of a carbon steel fence that is adjacent to the bridge. It is usually possible to remove this type of contamination by use of dilute acid-based cleaners, and restore the appearance of the DSS to be similar to the majority of the bridge.

2.7 Road Bridge in Cala Galdana, Menorca, Spain – 1.4462, CRC IV and V

This bridge in Menorca was the first stainless steel road bridge in Europe. The DSS bridge replaced an existing concrete bridge that was only 30 years old but exhibited significant degradation due to the very corrosive marine environment. The environment condition for this bridge was classified as CRC IV and V due to very high risk to chloride (<0.25 km from the sea) and elements exposed with and without rain washing. A suitable grade for this class is DSS 1.4462. The site X-ray spectroscopy found that this bridge uses two stainless steel grades for the major components. DSS 1.4462 is used for the bridge arches and balustrades (Figure 7).

Figure 7. Cala Galdana Bridge (DSS 1.4462) in Menorca, Spain, with close up of arch and underside

The austenitic stainless steel grade 1.4401 (ASTM 316), which has a lower corrosion resistance than 1.4462, is used for a tubular barrier between vehicle and pedestrian areas of the deck. Both alloys are finished with a similar surface roughness (Ra 3.5-4.0μm), however the performance is noticeably different. The 1.4401 is covered with orange-brown staining, resulting from a corrosion reaction with the coastal atmosphere, whereas the DSS 1.4462 has a good appearance. The beams (DSS 1.4462) located under the bridge deck were also in a good condition, despite being used in a position with no rain washing. Some white staining appeared to be deposits of carbonate washing out of the concrete bridge deck. This has not adversely affected the stainless steel.

3. Guidance on use of DSS as a construction material in bridges for specific environments

Seven DSS bridges have been inspected in environments which were classified in accordance with EN 1993-1-4 Annex A. The inspection results show that correct selection of a stainless steel grade for the actual service environment is the most significant factor affecting the long term durability of the bridge structure. The surface condition is also of importance, as illustrated by the DSS 1.4362 Celtic Gateway Bridge in Holyhead, UK where the good performance may be partially attributed to the smooth surface finish as well as grade selection.

Table 3. Summary of the inspection results

Bridge name	Designation EN (UNS)	DSS elements	Surface roughness Ra μm	Cl- Level in water ppm	Cause of damage	Loss of structural integrity
Orrhammarvägen Road Bridge (Flen, Sweden)	1.4162 (S32101)	Bridge beam, bearing plate	4.0-6.0	~10	Contamination, transportation	0*
Añorga Railways Bridge, (San Sebastian, Spain)	1.4162 (S32101)	Arch, beam	~8.0	Not tested	Contamination	0
Road Bridge (Nynäsham, Sweden)	1.4162 (S32101)	Bridge beam, Abutment plate	6.0-8.0	~20	Micro-pitting	1
Puerto Arrupe pedestrian Bridge (Bilbao, Spain)	1.4362 (S32304)	Box girder	3.0-5.0	~2500	Contamination	0*
Celtic Gateway pedestrian Bridge (Holyhead, UK)	1.4362 (S32304)	Load bearing arch	0.4-0.5	Not tested	n/a	0
Road Bridge (Zumaia, Spain)	1.4462 (S32205)	Bridge beam, balustrade	4.0-6.0	~15000	Fabrication	0*
Cala Galdana Road Bridge (Menorca, Spain)	1.4462 (S32205)	Arch, deck, balustrade	3.5-4.0	~17000	n/a	0

Note: Loss of structure integrity: 0 = no corrosion, 1 = cosmetic surface micro-pitting, 2 = corrosion that might require repair for reasons of structural integrity, 3 = general attack by the environment with loss of structural integrity.
* Cosmetic staining observed that was attributed to contamination with mild steel

Minor problems found during inspection were generated by contamination during fabrication, installation or transportation. The results obtained in this work can be used as examples of successful application of DSS structures. They can also be used to provide data for future revisions of EN 1993-1-4 Annex A, relating to materials selection. For example, the bridge in Nynäshamn, Sweden and the Celtic Gateway, Holyhead, UK, indicate that the CRC may be too conservative in some cases. An overall summary of results is given in Table 3. A study of construction costs using a reference composite road bridge design has been previously undertaken[6]. This high-level cost study demonstrated that similar build costs in DSS 1.4162 compared to using S355 carbon steel bridge beams could be achieved if the design of the DSS beams was optimized to utilize the higher strength ($Rp_{0.2}$ = 460MPa) of DSS. A further study[7] examining the impact on life cycle costs concluded that direct savings of up to EUR 140K (discounted to present day value and not including costs for disruption and loss of amenity during maintenance) could be anticipated over a 60-year life through the reduction in repair and re-painting costs. The hypothesis in this earlier work was that little maintenance of DSS would be required, and this assumption has been shown to be valid by the inspection work presented in this paper. DSS has been used in bridge structures built in the last 20 years, hence it is not possible to check older structures to confirm the very long term performance, although it is noted that design lives are often up to 120 years. Nevertheless, it is generally known that the corrosion performance of stainless steels in the built environment is stable over the long term, and that if significant corrosion problems will occur, the onset of such problems will most likely be easily identifiable within the first 6 – 36 months of service.

4. Conclusion

DSS offers a long-term cost effective option for bridge construction with minimal requirement for maintenance throughout the life of the structure. In this investigation, the performance of DSS for existing bridges under specific service environments was good. Occasional minor problems were cosmetic, with surface discoloration generally caused by contamination from carbon steel during fabrication, transportation or in-service. These had not resulted in loss of structural integrity, and can be eliminated . The results obtained from this exercise can be used as illustrative examples of successful application of DSS bridge structures and used to provide data for future design standard revisions of EN 1993-1-4 Annex A relating to materials selection. The present study examining the performance of bridges in service has confirmed that a high corrosion resistance can be expected of DSS, when appropriately selected for the particular environmental conditions. The hypothesis that a low level of maintenance is required for duplex stainless steel bridge elements has been validated. It would be valuable to carry out repeat inspections of these structures after a further 10 – 20 years of service to confirm the ongoing performance.

5. References

[1] European Committee for Standardization, *Eurocode 3-Design of steel structures- Part 1-4: General rules - Supplementary rules for stainless steels*, 2015.

[2] American Institute of Steel Construction, *Design Guide 27: Structural Stainless Steel, 2013.*

[3] Gedge G, *Structural uses of stainless steel-buildings and civil engineering*, Journal of Constructional Steel Research, 2008; 64: 1194–1198.

[4] Outokumpu Handbook of Stainless Steel, First edition; 2013.

[5] ISO 9223: *Corrosion of metals and alloys – Corrosivity of atmospheres–Classification, determination and estimation*. 2012

[6] Davis, I, Arup, *Stainless Steel Composite Bridge Study Report*, October 2011 (available on request from the authors of this paper)

[7] Gittens, R, Arup, *Life Cycle Cost for Maintenance Study Report*, August 2012 (available on request from the authors of this paper)

IOP Conf. Series: Materials Science and Engineering **419** (2018) 012019 doi:10.1088/1757-899X/419/1/012019

Qualification of the welding technology of the structural steel S355J2G3

V Lazić[1], D Arsić[1], I Ilić[2], S Aleksandrović[1], R Nikolić[3], M Djordjević[4] and B Hadzima[3]
[1]Faculty of Engineering, University of Kragujevac, Sestre Janjić 6, 34000 Kragujevac, Serbia
[2]Wacker Neuson Kragujevac Ltd, Milice Miljojković 13, 340000 Kragujevac, Serbia
[3]Research Center, University of Žilina, Univerzitna 8215/1, Žilina, Slovakia
[4]Faculty of Technical Sciences, University of Priština, Knjaza Miloša 7, 38220 Kosovska Mitrovica, Serbia

E-mail: dusan.arsic@fink.rs

Abstract. The selection procedure and qualification of the optimal welding technology for the structural steel S355J2G3 are presented in this paper. The considered steel is widely applied for manufacturing the supporting structures in various areas, especially in civil and mechanical engineering. It is delivered mainly in form of plates of various thicknesses and solid profiles of various diameters. Considering that welding is mostly used in manufacturing of structures, it is very important to be aware of possibilities for this steel welding, as well as of problems that may arise during that process. Here are presented: a calculation procedure of the base metal's weldability, selection of the filler metal and the welding technology, with results of experimental testing of the base metal and welded samples. The following experimental investigations were performed: tensile, bending and impact toughness test, metallographic investigations and hardness measurement. All the samples were cut out from the welded plates. Obtained results have shown that the structural steel S355J2G3 is suitable for welding, since there is no danger of appearance of cracks or other defects. With adequate choice of the welding technology the mechanical properties of the welded joint can be comparable to those of the base metal.

1. Introduction

This paper deals with investigation of weldability of the structural steel S355J2G3 and possibilities for its application for building the welded structures. The S355J2G3 steel belongs into a group of unalloyed (carbon) structural steels with the yield stress of 355 MPa and it is aimed for manufacturing the responsible support structures in various industries like bridge building, ship building, manufacturing of industrial equipment, metal working industry, machine building industry, etc. The steel is delivered in form of bars of various diameters and in form of plates of various thicknesses. The latter form of steel is the most frequently aimed for welding, thus its possibility for being welded must be known.

Here is presented an analysis of the base material and the welded samples by the destructive tests, which included the following: tensile tests, bending tests, metallography and hardness measurements. Besides that, the prescribed welding technology is presented, as well. The presented analysis can be considered as a kind of qualification of the welding technology of the S355J2G3 steel. Authors of this paper were already dealing with similar analyses of the welding technologies of other types of steel, like the high-strength steels S690QL [1-2] and HARDOX [3], tool steels for hot working [4-5], high-speed cutting steels and carbon steels [6-7], etc.

2. The base metal

As stated earlier, the paper presents analysis of the base metal and the welded joint of the S355J2G3 whose chemical composition, application and mechanical properties are given in Tables 1 and 2 [8].

Table 1. Chemical composition of the S355J2G3 steel.

Chemical composition, %													
C	Mn	Si	P	S	Al	Cu	Cr	Ni	Mo	Ti	V	Nb	N
0.16	1.36	0.021	0.013	0.005	0.04	0.02	0.02	0.01	0.004	0.01	0.005	0.028	0.007

Table 2. Different notations and mechanical properties of the S355J2G3 steel (EN 10025) [8].

Standard			Mechanical properties					
EN 10027-1	EN 10027-2	DIN 17100	R_m, MPa	R_{eH}, MPa				
			Thickness 3-100 mm	<16	16-40	40-63	63-80	80-100
S355J2G3	1.057	St 52-3N	490-630	355	345	335	315 325	305 315

The steel hardness in the annealed condition is about 230 HBS, while in the drawn condition it is about 260 HBS. The elongation of the steel is higher than 20 % for the plates of thickness up to 100 mm, while for the thicker plates the guaranteed elongation is 18 %. The similar is valid for the impact toughness. Though the minimal toughness, prescribed by the standard, is 27 J at – 20 °C, what can be also seen from the steel's notation, for plates of thickness < 100 mm that value is higher than 35 J.

The S355 steels are being delivered in qualities JR, J0, J2 (J2G3, J2G4), K2 (K2G3, K2G4). The difference in quality is related to weldability and guaranteed impact toughness at lower temperatures. Letter J denotes guaranteed impact toughness of 27 J, letter K of 40 J and letter L of 60 J at lower temperature, which is denoted by a number behind one of those letters. Thus, steel S355J2G3 has the guaranteed toughness of 27 J at – 20 °C. G3 denotes supply condition - normalized or normalized rolled plate. Authors of [9] have shown that (even highly alloyed) steel's toughness can be positively influenced by adding some micro-alloying elements.

Besides the earlier mentioned areas of application of this steel, it is also used for manufacturing the cold-rolled thin sheets and pipes. It is frequently used for manufacturing the shafts, axles, gears, spring carriers, screws, covers, casings, etc. Considering the mechanical properties, the structural steels must possess high yield stress, sufficient plasticity, high creep strength and strength at elevated temperatures, satisfactory toughness and dynamic strength. Besides those, the structural steels have to be resistant to wear and corrosion, machinable by particles detachment (cutting), weldable, convenient for cold forming (bending, punching, deep drawing), etc. Generally, the structural steels are divided into:

- Carbon steels for supporting (carrying) structures and
- Steels for machine building.

Taking into account that the chemical composition is not guaranteed for this group of steels, due to the higher presence of impurities with respect to other steels, the generic structural steels are not predicted for the heat treatment. Steels for the supporting structures are mainly grouped according to the value of the yield stress and impact toughness. Marking of the structural steels is done according to the

EN 10027-1 standard. They possess the yield stress within range 190 - 370 MPa, with elongation of 10 - 28 %. The yield stress is higher with the increased carbon content and share of pearlite in the microstructure. These steels are mainly used within the temperature range – 40 - +50 °C.

3. The welding technology

Prior to prescribing the welding technology, the calculation of the base metal weldability was done, according to standard formulas for the carbon equivalent for the low-alloyed steels [3]:

$$CE = C + \frac{Mn}{6} + \frac{Cr + Mo + V}{5} + \frac{Ni + Cu}{15}, \ \% \tag{1}$$

$$CE = C + \frac{Mn}{6} + \frac{Si}{24} + \frac{Ni}{40} + \frac{Cr}{5} + \frac{Mo}{4} + \frac{V}{14}, \ \% \tag{2}$$

$$CE = C + \frac{Mn}{20} + \frac{Ni}{15} + \frac{Cr + Mo + V}{10}, \ \% \tag{3}$$

The steels that have *CE* less than 0.45 % are considered as well weldable. The steels with the higher values of the carbon equivalent are considered as the conditionally weldable, since they must be subjected to special regimes: preheating, additional heating and/or post-welding heat treatment. Results obtained are as follows: according to formula (1) *CE* = 0.395 %; according to formula (2) *CE* = 0.393 % and according to formula (3) *CE* = 0.235 %. It can easily be concluded that the S355J2G3 steel belongs into a group of well weldable steels and that the preheating prior to welding is not necessary.

Taking that into account, the solid electrode wire G3Si1 (according to standard ISO 14341-A) of diameter Ø1 mm was selected as the filler metal, while the GMAW (135) procedure in the protected gas mixture Ar + 18 % CO_2 was selected for executing the welding. The welding parameters were I = 140 A, U = 23 V with welding speed of 300 mm/min. Welding was executed in the horizontal position. Between the two welds the break of 3 to 5 min was prescribed to avoid the plates' deformations.

Within the samples preparation, welding of plates of two thicknesses of 8 and 30 mm was performed. The 8 mm thick plates were welded in the V groove and 30 mm thick plates were welded by the "double V" or the X groove. Samples for bending and hardness measurements were prepared from the thinner welded plates (8 mm), while samples for tensile test and impact toughness test were prepared from the thicker welded plates (30 mm).

4. Experimental procedure and discussion of results

4.1. The tensile test

This experiment was done at the room temperature in the static loading conditions until the tearing, in series of 3 samples. The shape, dimensions and actual appearance of a sample are presented in Figure 1. Samples were made of the considered carbon steel S355J2G3, cut out from the welded plate in such a way that they contained the welded joint, while there were also samples prepared without the welded joint portion, i.e. of the base metal only.

Tensile tests were conducted in the specialized laboratory on the testing machine with the loading range 1 to 100 kN. The deformation rate was 10 mm/min. Obtained results are presented in tables 3 and 4, while the summary diagrams for the tensile test of the base metal and the welded joint are shown in Figures 2 and 3, respectively.

IOP Conf. Series: Materials Science and Engineering **419** (2018) 012019 doi:10.1088/1757-899X/419/1/012019

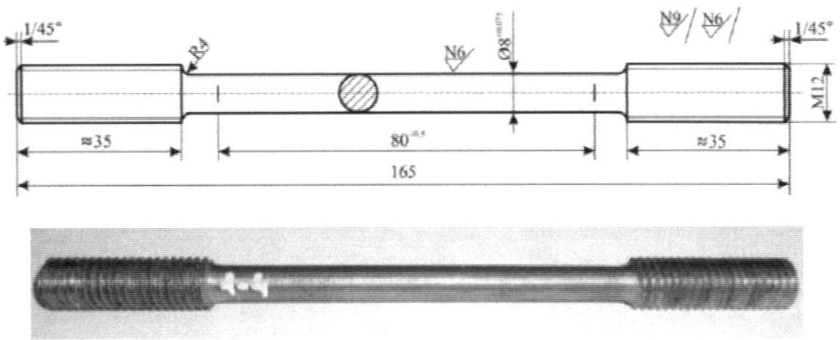

Figure 1. Drawing and actual appearance of the tensile test samples.

Table 3. Experimental results of the S355J2G3 steel tensile test – base metal.

Sample number	S_0, mm^2	L_0, mm	$R_{p0.2}$, MPa	R_m, MPa	A, %
1	49.02	140	522.70	633.11	5.37
2	49.02	140	537.86	638.21	4.82
3	49.02	140	538.54	643.78	4.93

Table 4. Experimental results of the S355J2G3 steel tensile test – welded joint.

Sample number	S_0, mm^2	L_0, mm	$R_{p0.2}$, MPa	R_m, MPa	A, %
1	49.02	140	522.70	633.11	5.37
2	49.02	140	537.86	638.21	4.82
3	49.02	140	538.54	643.78	4.93
4	44.18	90.00	411.15	535.97	14.24
5	44.18	90.00	420.26	546.79	16.79
6	44.18	90.00	405.96	526.92	13.98
7	44.18	90.00	405.77	530.69	16.60
8	44.18	90.00	399.93	532.80	18.46
9	45.36	90.00	393.49	524.93	13.44
10	45.36	90.00	406.62	535.46	18.16
11	44.18	90.00	414.84	531.80	13.41

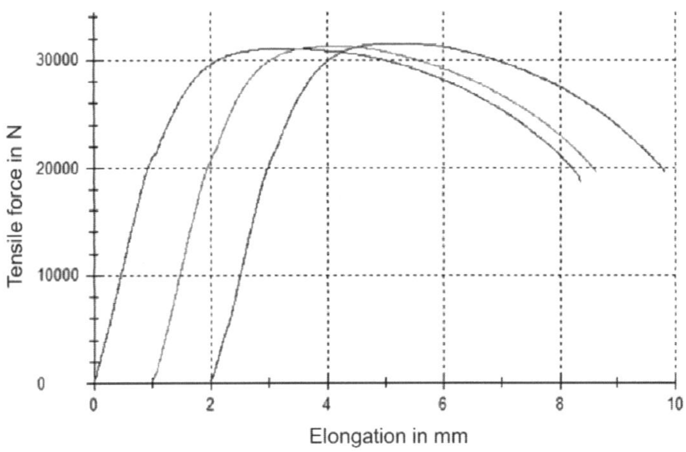

Figure 2. Summary diagram of the base metal tensile test.
(Curves in order from left to right correspond to samples 1 to 3 in Table 3).

The tearing point in all the tested samples was outside the welded joint zones, i.e. in the base metal, what is a good indicator that the applied welding method and welding technology parameters and conditions were properly selected and applied.

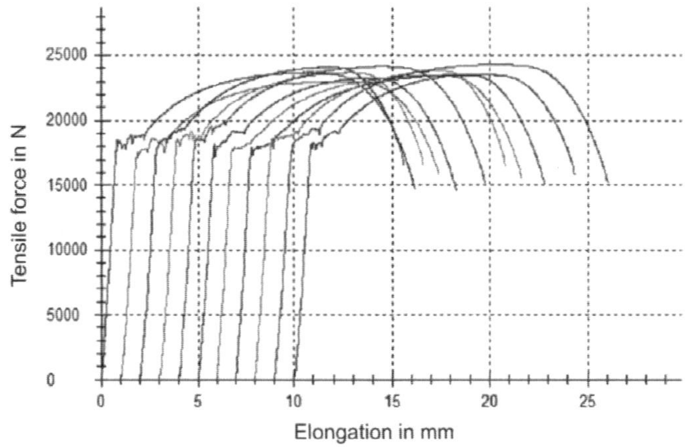

Figure 3. Summary diagram of the welded joint tensile test.
(Curves in order from left to right correspond to samples 1 to 11 in Table 4).

4.2. The bending test

The bending test is often used in practice. Samples for the bending test are usually of the circular or rectangular cross-section; in this particular case the cross-section was rectangular. The welded samples were, after the adequate preparation, subjected to the bending force, Figure 4. If during the bending test, the sample has reached the required value of the bending angle without appearance of cracks, the test result is considered as positive. The results of the bending test serve for determination of the welded joint plasticity. If the small cracks are noticed during the test, the result is negative, under the condition that the bending angle is smaller than the required one.

Figure 4. The bending test

The objective of the bending test was determination of the deformation capability of the welded joint and the heat affected zone of the weld. The test was performed in the standardized conditions, by continuous bending with the loading rate of 10 mm/min. The plasticity measure is the bending angle. Results of the bending test are shown in Tables 5 to 7.

IOP Conf. Series: Materials Science and Engineering **419** (2018) 012019 doi:10.1088/1757-899X/419/1/012019

Table 5. Experimental results of the bending test of the base metal – BM.

Sample number	Temperature °C	Maximum force range F_{max}, daN	v_{def}, mm/min	F_{max}, daN	Bending angle α, °
1				165	80
2	20	10000	10	165	89.5
3				160	89

Table 6. Experimental results of the bending test of the weld's face – WF.

Sample number	Temperature °C	Maximum force range F_{max}, daN	v_{def}, mm/min	F_{max}, daN	Bending angle α, °
1				167	90.5
2	20	10000	10	165	88
3				129	92,5

Table 7. Experimental results of the bending test of the weld's root – WR.

Sample number	Temperature °C	Maximum force range F_{max}, daN	v_{def}, mm/min	F_{max}, daN	Bending angle α, °
1	20	10000	10	155	89

4.3. The impact toughness test

This test was conducted on samples made from the base metal, as well as those with the welded joint, executed by the MAG procedure. The samples were tested on the Charpy pendulum in the accredited laboratory according to the ISO TC/164/SC 4 standard, on samples with dimensions 55×10×10 mm. The notch – groove for testing the impact toughness was first cut along the axis of the welded joint, then in the melting zone and, finally, in the heat affected zone, both at the weld's face and the weld's root sides. The welded joint characteristic zones were chemically etched by the 4 % solution of the nitric acid, what made them visible for precise notching the groove for the impact test in the desired zones.

The technical drawing of the sample for the impact toughness tests, as well as appearance of the actual samples, are presented in Figure 5. Results of the impact toughness tests of the base metal (BM), weld metal (WM), melting zone (MZ), HAZ and at the weld's root (WM-R) and the face (WM-F) side are presented in Tables 8 to 13, respectively, and in Figure 7, while the appearance of the sample from the weld metal, before and after the test, is shown in Figure 6.

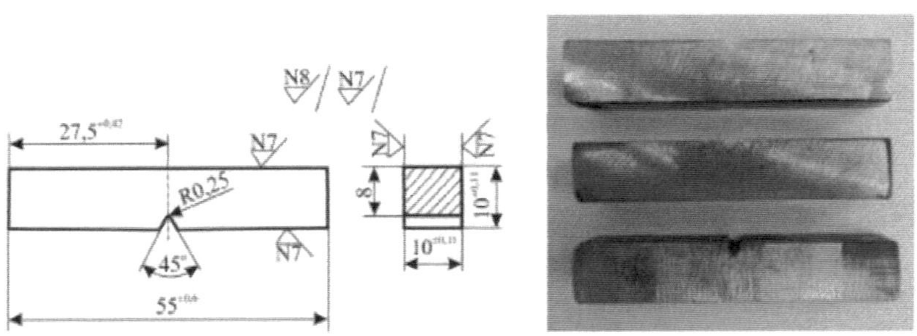

Figure 5. Technical drawing and actual appearance of samples for the impact toughness tests.

Table 8. Experimental results of the impact toughness test – BM.

Sample number	Temperature, °C	Absorbed energy A_0, J	Average value, J
1		180	
2		178	
3	20	179	179
4		182	
5		178	
6		178	

Figure 6. Appearance of the weld metal samples before and after the impact toughness tests.

Table 9. Experimental results of the impact toughness test – WM.

Sample number	Temperature, °C	Absorbed energy A_0, J	Average value, J
1		146	
2	20	158	151.3
3		150	

Table 10. Experimental results of the impact toughness test – MZ.

Sample number	Temperature, °C	Absorbed energy A_0, J	Average value, J
1		168	
2	20	184	167.3
3		150	

Table 11. Experimental results of the impact toughness test – HAZ.

Sample number	Temperature, °C	Absorbed energy A_0, J	Average value, J
1	20	152	152

Table 12. Experimental results of the impact toughness test – WM-Root.

Sample number	Temperature, °C	Absorbed energy A_0, J	Average value, J
1		120	
2	20	90	110
3		120	

IOP Conf. Series: Materials Science and Engineering **419** (2018) 012019 doi:10.1088/1757-899X/419/1/012019

Table 13. Experimental results of the impact toughness test – WM-Face.

Sample number	Temperature, °C	Absorbed energy A_0, J	Average value, J
1		164	
2	20	140	148
3		140	

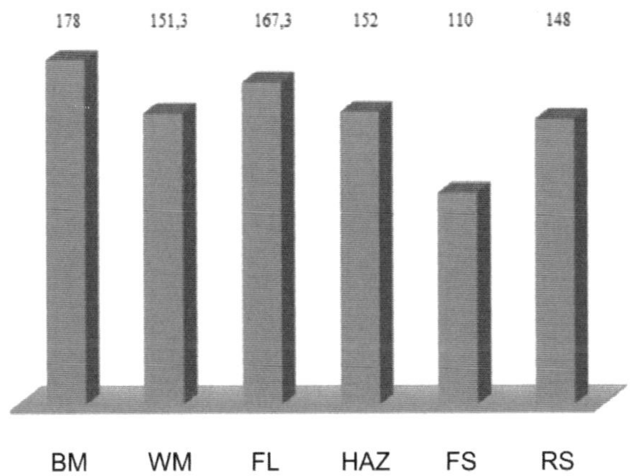

Figure 7. Histogram presentation of the absorbed fracture energy for various positions of the notch; BM, WM, FL (fusion line), HAZ, FS (face side), RS (root side)

Experimentally determined values of the fracture energy exhibit very small deviations; the force vs. time dependencies exhibit the similar character. Thus, the obtained results can be considered as relevant for the further analysis.

Based on the fracture energy, obtained for the various positions of the notch at the welded joint's face, one can conclude that the highest values of the fracture energy are obtained for the samples made of the base metal, somewhat smaller values were for the samples with the notch in the melting zone, HAZ and the weld metal. The smallest values were obtained for the samples with the notch in the weld's root.

Obtained results indicate the drop of the toughness of the welded joint zone with respect to the base metal. Such a case is unfavorable from the homogeneity aspect, as well as the safety of the welded joint, since appearance of the zones of the lower impact toughness may be the places where the stress concentration may arise, causing micro cracks that could lead to fracture of the welded structure. The very significant indicator of the material's toughness is also the share of the brittle and ductile fracture. The larger the share of the brittle fracture the higher is the possibility for appearance of the brittle fracture, and vice versa. What concerns the base metal, its characteristics related to toughness are exceptionally favorable at the room temperature.

4.4. Hardness measurements and metallographic investigations

This part of investigation assumed measuring of the micro-hardness in three different, mutually parallel directions, on the sample with the welded joint, executed by the MAG procedure.

For the 8 mm thick plate the measurement direction I-I includes the base metal, the heat affected zone and the weld metal, while the direction II-II is placed along the root pass in the weld metal zone. Measurement was executed according to the SRPS EN ISO 9015-1:2013 standard. Appearance of the sample and the measurement directions are shown in Figure 8, while the results of the hardness measurement are given in Table 14.

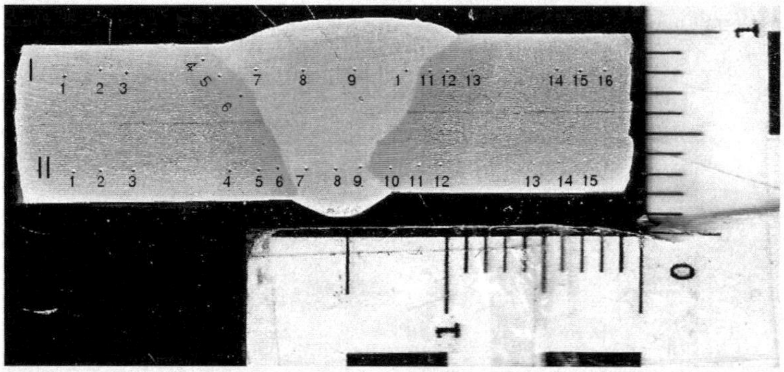

Figure 8. Micro-hardness mesurements points on the metallographic sample - 8 mm thick plate

Table 14. Micro-hardness mesurements results for the sample made from the 8 mm thick plate [HB].

	Mesurement points														
Measurement direction	1	2	3	4	5	6	7	8	9	10	11	12	13	14	15
I	170	176	176	183	192	192	189	185	189	199	199	189	170	170	170
II	170	176	176	177	177	189	185	185	191	192	192	185	170	176	176

For the 30 mm thick plate the measurement direction I-I includes the base metal, the heat affected zone and the weld metal. The II-II direction is selected along the middle of the joint in the zone where the mixing of the root pass and the cover filler metal occurred. The III-III direction was placed on the root pass side in the welded joint zone where the dominant influence is exhibited by the way of the weld execution, parameters and the filler metal of the root pass. The micro-hardness measurement points on the sample, as well as the sample appearance after the executed micro-hardness measurements are presented in Figure 9, while the obtained results are given in Table 15.

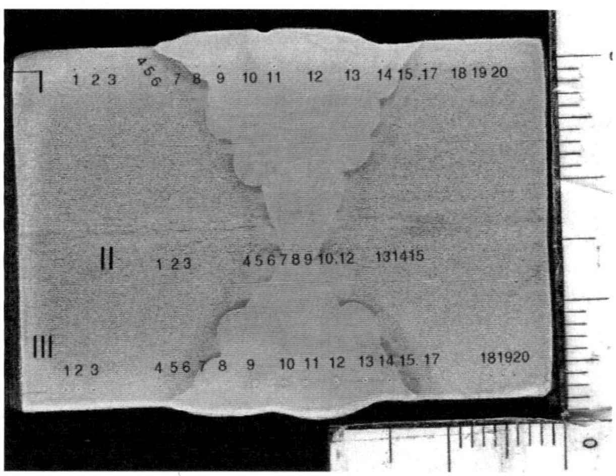

Figure 9. Micro-hardness mesurements points on the metallographic sample - 30 mm thick plate

Table 15. Micro-hardness mesurements results for the sample made from the 30 mm thick plate [HB].

	Mesurement points														
Measurement direction	1	2	3	4	5	6	7	8	9	10	11	12	13	14	15
I	156	206	199	213	322	206	206	193	199	206	193	236	245	193	156
II	165	185	199	228	213	206	206	221	199	187	170	165	165	165	-
III	160	199	245	297	206	206	206	193	199	199	193	185	254	254	213

Results of the metallographic analysis are shown in Figure 10. They pointed to conclusion that the structure was homogeneous, pearlite-ferrite, without appearance of the brittle zones that should be pointed at, which could lead to catastrophic damages [10-11]. The reason for that is the very good weldability of the base metal, as well as the adequately selected welding parameters and the filler metal. In addition, no geometric defects, like scratches, dimples, dross inclusions etc., which could influence the quality of the welded joint, were noticed on any of the samples. Besides that, the filling of the groove was homogeneous and sufficient, since in the case of the too high overfill of the weld, the dynamic strength of the welded joint would be compromised [12].

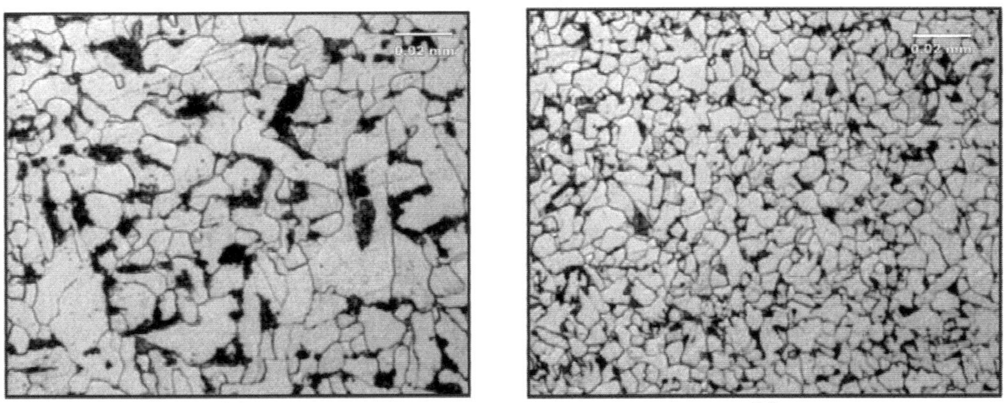

(a) (b)

Figure 10. Microstructure: (a) the base metal – pearlite-ferrite structure
(b) the heat affected zone – the normalization zone with ferrite-pearlite structure.

The objective of the whole technology qualification, which is proven by all the tests performed on the welded joints, as well as on the base metal, is thus fulfilled. This particular steel can be used as a part of the bridge constructions and can be welded by the proposed technology. Furthermore, the verified technology can be applied for welding of other types of steels, as well, with necessary adjustments regarding the different material's mechanical properties, weldability and hardness. In any case, additional tests would have a good guidance with the procedure here presented.

5. Conclusions

Based on the conducted experiments and certificate by the adequate accreditation body, the so-called qualification and verification of the selected welding technology was performed.

Based on the experimental investigations, the following conclusions were reached:

- The tensile tests have shown that the most important properties of the welded joint (strength and deformation) are approximately the same or even higher than those of the base metal;
- The impact toughness tests of the welded joint individual zones led to establishing the fact that there were no large deviations with respect to impact toughness of the base metal and that no brittle fracture was registered;

- The conducted bending tests have shown the exceptionally high plasticity of the base metal and the welded joint (at the sides of the weld's face and root) and that there are no major differences in obtained results;
- The macro-graphic tests have shown that the multi-pass weld was homogeneous, of the regular shape and without noticed defects;
- The hardness measurements in the individual welded joint's zones (in the three selected directions) have shown that there were no significant deviations in the measured values, namely that there are no present brittle zones, since all the measured values were lower than the critical one (350 HB).

Based on all the above one can conclude that the selected welding technology was qualified and verified (certified by the authorized accreditation body) and that it can be successfully applied in manufacturing the real structures made of the S355J2G3 steel.

The technology thus qualified is ready for welding of any structural part of the bridge constructions and can be, with necessary adjustments be applied for other types of steels welding.

Acknowledgements
This research was partially financially supported by European regional development fund and Slovak state budget by the project "Research Centre of the University of Žilina" and by the Ministry of Education, Science and Technological Development of Republic of Serbia through grant TR35024.

References

[1] Arsić D, Lazić V, Nikolić R, Aleksandrović S, Hadzima B and Djordjević M 2015 The optimal welding technology of high strength steel S690QL *Materials Engineering - Materiálové inžinierstvo* **22** (1) 33-47

[2] Lazić V, Aleksandrović S, Arsić D, Sedmak A, Ilić A, Djordjević M and Ivanović L 2016 The influence of temperature on mechanical properties of the base material and welded joint made of steel S690QL *Metalurgija - Metallurgy* **55** (2) 213-216

[3] Lazić V, Arsić D, Nikolić R, Djordjević D, Prokić-Cvetković R and Popović O 2017 Application of the high strength steel HARDOX 450 for manufacturing of assemblies in the military industry *Key Engineering Materials* **755** 96-105

[4] Arsić D, Lazić V, Samardžić I, Nikolić R, Aleksandrović S, Djordjević M and Hadzima B 2015 Impact of the hard facing technology and the filler metal on tribological characteristics of the hard faced forging dies *Tehnički Vjesnik – Technical Gazette* **22** (5) 1353-1358

[5] Arsić D, Lazić V, Sedmak A, Nikolić R, Aleksandrović S, Djordjević M, Bakić R 2016 Selection of the optimal hard facing (HF) technology of damaged forging dies based on cooling time t8/5 *Metalurgija - Metallurgy* **55** (1) 103-106

[6] Ratković N, Arsić D, Lazić V, Nikolić R, Hadzima B, Palček P and Sedmak A 2017 Influence of friction welding parameters on properties of the Al-Cu joint *FME Transactions* **45** (1) 165-171

[7] Ratković N, Lazić V, Arsić D, Nikolić R, Meško J and Nigrovič R 2016 Influence of the friction time on the shape and microstructure of the mixing zone of the friction welded joint *Manufacturing Technology* **16** (6) 1355-1359

[8] European Committee for Standardization (CEN) 2011 Hot rolled products of structural steels: Technical delivery conditions for non-alloy structural steels *Eurocode 3, Part 2 (EN10025-1:2011)*

[9] Todić A, Čikara D, Todić T, Pejović B, Čamagić I and Vukojević V 2017 The Influence of the Vanadium Content on the Toughness and Hardness of Wear resistant High-alloyed Cr-Mo Steel *FME Transactions* **45** (1) 130-134

[10] Nourya P and Eriksson K 2017 Failure analysis of martensitic stainless steel bridge roller bearings *Engineering Failure Analysis* **79** 1017-1030

[11] Bošnjak S, Arsić M, Savićević S, Milojević G and Arsić D 2016 Fracture analysis of the pulley of a bucket wheel boom hoist system *Eksploatacja i Niezawodnosc – Maintenance and Reliability* **18** (2) 155-163

[12] Djoković J, Nikolić R, Bujnak J and Sedmak A 2016 Influence of welding geometry on cylindrical welded joint life *Structural Integrity and Life* **16** (1) 35-38

IOP Conf. Series: Materials Science and Engineering **419** (2018) 012020 doi:10.1088/1757-899X/419/1/012020

Influence of High Frequency Impact Treatment (HiFIT) on fatigue strength of welded joints of high-strength steel S700MC for bridges applications

Krzysztof Krasnowski

Instytut Spawalnictwa, Błogosławionego Czesława Street 16-18, Gliwice 44-100, Poland

E-mail: krzysztof.krasnowski@is.gliwice.pl

Abstract. The paper deal with experimental fatigue test of arc welded joints of thermomechanicaly rolled structural steel S700MC, uses for heavy duty application like steel bridges structures. The aim of the work was to compare the fatigue strength of welded joints in two conditions, i.e. as-welded and after HiFIT process as a post-weld treatment and set the influence of HiFIT process on fatigue strength improvement. For this purpose three different welded joints have been prepared. The experimental test series were consist of butt welded joints, T-joints with longitudinal and transverse stiffeners. In the paper the fatigue characteristics S-N of welded joints have been presented as well as calculated fatigue category FAT. The study shows that the fatigue strength of welded joints of S700MC steel grade after HiFIT treatment is much more higher in each cases than the fatigue strength of non-treated welded joints. The obtained results can be used by engineers at the design stage or estimating the remaining fatigue life of the welded steel structure made of similar high-strength steel used for bridges application.

1. Introduction

Numerous investigations in recent years show that the fatigue strength of welded structures can be increased by different post weld processes that mainly rely on residual stress and weld geometry improvement methods. One of this method, which belongs to group of high-frequency mechanical impact methods (HFMI), is the High Frequency Impact Treatment (HiFIT) for post-weld fatigue strength improvement technique of existing and new welded steel structures exposed to variable loading conditions. The general idea of HFMI treatment is to employ alternate power sources, for example, ultrasonic piezoelectric elements, ultrasonic magnetostrictive elements or compressed air which accelerate cylindrical steel indenter or set of intenders against a component or structure with high frequency (>90 Hz). The impacted material is highly plastically deformed causing changes in the material microstructure and the local geometry as well as the residual stress state in the region of impact [1-4]. The area of application of the HiFIT includes steel bridges, load suspension means, cranes, wind power plans, building machines and many other steel structures [5].

In the literature there are many names to describe the devices used in HFMI treatment, for example ultrasonic impact treatment (UIT) [6], ultrasonic peening (UP) [7], ultrasonic peening treatment (UPT) [8] [9], high frequency impact treatment (HiFIT) [5], pneumatic impact treatment (PIT) [10] and ultrasonic needle peening (UNP) [11] [12]. In comparison to conventional hammer peening, the HFMI operation is considered to be more user-friendly and due to smaller spacing between alternate impacts the final surface is more smooth. While HFMI can be considered as environmental friendly, safe and relatively easy to apply, operators must still exercise safe work practice and understand the equipment and the nature of the post-weld operation which is being imparted to a welded structure.

2. Material and experimental procedure

2.1. Material

The material used in tests was thermo-mechanically rolled steel S700MC in the form of 10 mm thick plates. The chemical composition of S700MC steel is presented in table 1 and basic mechanical properties in table 2.

Table 1. Chemical composition of S700MC steel plate (acc. to certificate).

C %	Si %	Mn %	P %	S %	N %	Al %	Cu %	Cr %	Ni %	Mo %	Nb %
0,07	0,18	2,0	<0,012	0,001	0,005	0,049	0,03	0,06	0,16	0,12	0,08

Table 2. Mechanical properties of S700MC steel plate (acc. to certificate).

Yield stress $R_{e,}$ MPa	Tensile strength $R_{m,}$ MPa	Elongation $A_{5,}$ %	Impact energy KV_2 at -40°C, J
686	762	21	107

2.2. Test joints

Welded joints were produced with 10mm-thick thermo-mechanically rolled steel S700MC. The test joints were produced by means of semi-automatic MAG welding method (135) with 1.2 mm-diameter solid filler wire BÖHLER X 70-IG and by use of KEMPPI ProMIG 500 welding machine. As a shielding the gas M21(82% Ar +18% CO_2) was used.

The test joints were made in three following configurations:

- Butt welded joins with full penetration,
- T-joints with double-sided fillet welds,
- T-joints with longitudinal rib with fillet and peripheral fillet weld.

The details of welding parameters are summarized in table 3.

Table 3. Technological parameters of welding process.

Type of joint	Butt joint	T-joint with double sided fillet weld	T-joint with longitudinal rib and peripherial fillet weld
Joint design and welding sequences	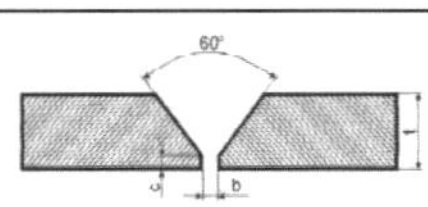 t = 10 mm; b = 2 - 3 mm; c = 1 – 2 mm 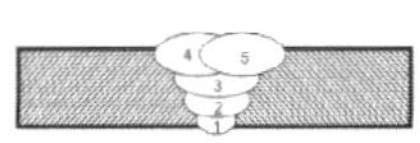	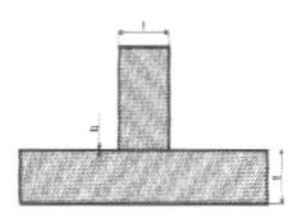 t = 10 mm; b = 0 - 1 mm 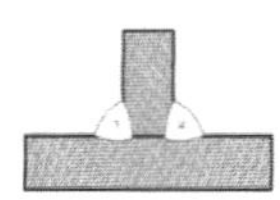	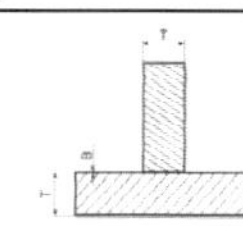 t = 10 mm; b = 0-1 mm 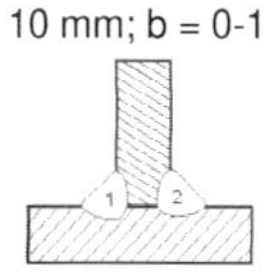
Welding method:	135	135	135
Size of filler material:	Ø1,2 mm	Ø1,2 mm	Ø1,2 mm
Current:	112-121 A (run 1) 240-254 A (runs 2-5)	230-244 A	248 A
Voltage:	16 V (run 1) 26 V (runs 2-5)	26-26,5 V	26,8 V
Travel speed:	138 mm/min (runs1) 350 mm/ min (runs 2-3) 400 mm/min (runs 4-5)	380 mm/min	380 mm/min
Wire feed speed:	6,5 m/min	6,5 m/min	6,5 m/min

2.3. HiFIT process of test joints

Proper impact machining (hammering) using the HiFIT method requires the selection of appropriate parameters of this process. The selection of such parameters was carried out in an experimental manner, in accordance with the device operating instructions.

The relevant parameters are selected by setting the correct working pressure, selection of the intensity of hammering by turning the screw causing the piston stroke to change and proper pin angle to obtain a smooth surface without burrs in the work area and to get the processing depth within 0.2-0.35 mm.

The HiFIT was carried out at the station equipped with HiFIT device with standard pin with a tip diameter D=3 mm, flexible air joint and air filter with pressure gauge to adjust the supply pressure The test joints to be treated were fixed to the table using conventional joinery clamps (Fig. 1) and the HiFIT process was carried out according to parameters as presented in table 4.

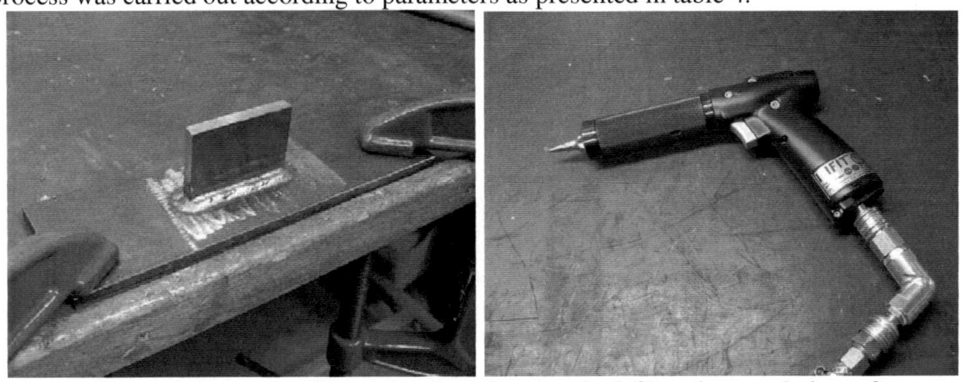

Figure 1. Example of test joint clamping (on the left) and general view of HiFIT system device (on the right).

Table 4. HiFIT process parameters.

Working pressure (bar)	Impact intensity	Pin movement speed, (mm/s)	Pin angle against treated surface (°)	Pin angle against movement direction (°)
4,8-5,0	Small (high frequency)	3,0-5,0	60-80	70-90

2.4. Experimental fatigue test procedure

Fatigue tests of three type of welded joints in "as welded" and after HiFIT treatment conditions were conducted with testing machine MTS 810. Fatigue tests of each lot of samples were performed for several ranges of stresses $\Delta\sigma$ with stress ratio R=0.2 ($R=\sigma_{min}/\sigma_{max}$) and changing load frequency of 20 Hz until reaching the moment of the test sample failure. The number of samples in each lot amounted to 10, which enabled the determination of S-N curves and calculation of fatigue category FAT pursuant to the guidelines of the International Institute of Welding (IIW) [13].

3. Results and discussion

The goal of these investigation was to verify whether the HiFIT can be used for effective fatigue life improvement of most popular types of welded joints made of S700MC steel grade and to estimate the efficiency of this modification. The main work was focused on experimental fatigue test (axial tension) of the joints performed in two different conditions, i.e. as welded and after HiFIT process. The test results as a fatigue curves are presented in figures 2-4. The blue lines represents S-N curves

IOP Conf. Series: Materials Science and Engineering **419** (2018) 012020 doi:10.1088/1757-899X/419/1/012020

while the red triangle represents fatigue lives of each tested specimens. In table 5 the list of calculated FAT classes and HiFIT process efficiency for each type of joints are presented.

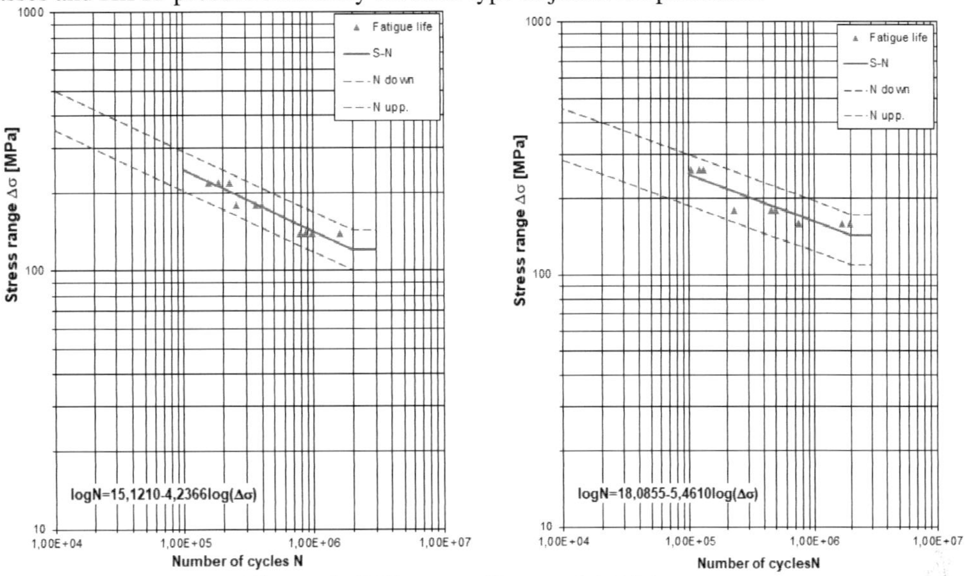

Figure 2. S-N curve of butt welded joints;
a) in "as welded" condition, b) after HiFIT process.

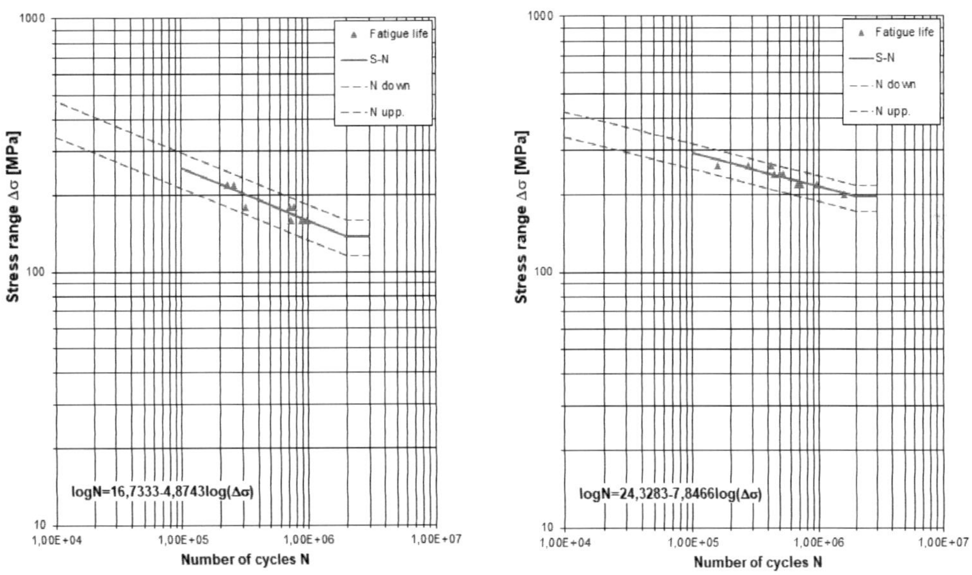

Figure 3. S-N curve of T-joints with double fillet welds;
a) in "as welded" condition, b) after HiFIT process.

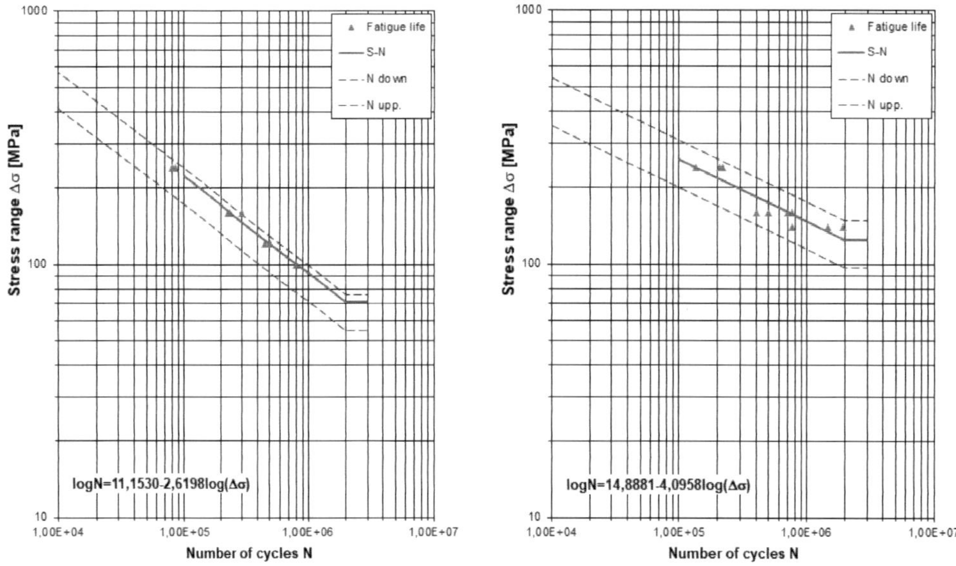

Figure 4. S-N curve of T-joints with longitudinal rib and peripheral fillet weld; a) in condition "as welded", b) after HiFIT process

Table 5. Comparison of FAT classes for welded joints with and without HiFIT process

Type of joint	Joint sketch	Fatigue class FAT, MPa: As welded	Fatigue class FAT, MPa: After HiFIT	HiFIT efficiency, %
Butt joint		72	85	**18%**
T-joint	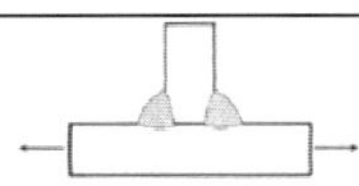	98	159	**62%**
T-joint with longitudinal rib	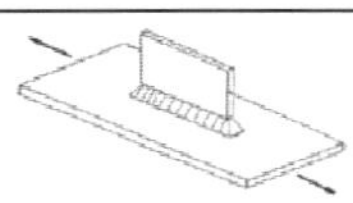	30	71	**137%**

Analyzing the results, it can be seen that in every type of welded joints of S700MC steel, the use of HiFIT process resulted in the increase of fatigue strength. For butt welded joints the increase in fatigue strength expressed in the FAT class was 18%, which is the smallest increase among the tested joints in this work. In the case of T-joints with double-sided fillet welds, the benefits of HiFIT reach a 62% increase by comparing fatigue categories of joints before and after HiFIT process. The highest increase in fatigue strength after the application of HiFIT process was obtained for joints with an longitudinal rib with a circumferential fillet weld, for which the fatigue category of FAT is higher by as much as 137% in comparison to fatigue category FAT joints in "as welded" condition.

The test results confirm the principle that if the notch is bigger (the more significant its effect), the increase in fatigue strength due to the HiFIT impact treatment is also bigger. Regarding the subject of the research, the biggest notch was the longitudinal rib, and for these joints the effect of HiFIT process is the highest, while the mildest notches are undoubtedly the reinforce of face and root in butt welds, therefore the impact of HiFIT process on fatigue strength increase is also lower.

In conclusion, and referring to the purpose of the work, it should be stated that the application of HiFIT process for welded joints of S700MC steel is justified in order to increase the fatigue strength of welded steel structures. Obtained results of research can be an unquestionable argument for the environment of both designers and manufacturers of steel structures of high strength steel, which are designed taking into account their exposure to fatigue.

4. Conclusions

Based on the conducted tests and the analysis of results, the following conclusions are drawn:

1. It is possible to modify the fatigue strength of welded joints of S700MC thermomechanical steel by means of local impact machining of welds using HiFIT process.
2. The degree of increasing the fatigue strength of FAT after applying HiFIT process for a selected steel grade depends on the type of joint and is connected with the sharpness of the notch of a given welded joint.
3. The increase in fatigue strength of S700MC welded joints after the HiFIT process is 18% for butt joints, 62% for T-joints with double-sided fillet welds and 137% for T-joints with longitudinal rib and peripheral fillet weld.
4. HiFIT device for impact treatment of joints is easy to use and offers the possibility of conducting an efficient process of modification of fatigue strength of welded joints.

References

[1] Marquis G, Barsoum Z 2014 Fatigue strength improvement of steel structures by HFMI: Proposed procedures and quality assurance guidelines *Welding in the World* **58** pp 19-28

[2] Haagensen P J, Maddox S J, 2013 *IIW Recommendations on methods for improving the fatigue strength of welded joints* ISBN:9781782420644 (Woodhead Publishing)

[3] Barsoum Z, Jonsson B 2011 Influence of weld quality on the fatigue strength in seam welds, *Engineering Failure Analysis* **18** pp 971-979

[4] Yildirim H C, Marquis G B 2012 Overview of Fatigue data for high frequency mechanical impact treated welded joints *Welding in the World* **57** 7/8

[5] http://www. hifit.de

[6] http://www.appliedultrasonics.com

[7] http://itlinc.com/

[8] http://www.lets-global.com

[9] Huo L, Wang D, ZhangY 2005 Investigation of the fatigue behavior of the welded joints treated by TIG dressing and ultrasonic peening under variable-amplitude load *International Journal of Fatigue* **27** pp 95-101

[10] http://www.pitec-gmbh.com

[11] http://www.sonats-et.com

[12] Bousseau M, Millot T 2006 Fatigue life improvement of welded strucrtures by UNP compared to TIG dressing *International Institute of Welding* Document XIII-2125-06

[13] Hobbacher A F 2016 Recommendations for Fatigue Design of Welded Joints and Components *International Institute of Welding* DOI 10.1007/978-3-319-23757-2

IOP Conf. Series: Materials Science and Engineering **419** (2018) 012021 doi:10.1088/1757-899X/419/1/012021

New Modular Bridges Solutions

A André[1], J Fernandes[1], I Ferraz[1]; P Pacheco[1]

[1]MBS, BERD, Edifício Olympus, Av. D. Afonso Henriques, 1462 – 2º, 4450-013 Matosinhos, Portugal

Abstract. The present paper describes new modular bridges solutions developed by MBSbyBERD, with higher range of spans and load capacity, fast assembly and smart packing. As an upgrade of the Bailey bridges type, MBSbyBERD's developed new structural solutions thought to be launched by increment launching.
Natural disasters, conflicts between peoples and the aging of existing infrastructures have been the main causes for the search of modular bridges as a solution for the rapid reestablishment of interrupted roads. MBSbyBERD solutions aims to accelerate the reestablishment of the connections comparing with the existing market solutions. Designed to be pre-assembled and containerized in order to minimize assembly time, 4 different models of modular bridges have been developed, optimized to cover a range of spans from 15m to 120m. The new range of spans up to 120m offered by BERD, open new possibilities of bridging further margins reducing distances between populations and so social and economic impact.

1. Introduction
The modular bridges Bailey type were developed during World War II to be used as logistical bridges with a recognized structural efficiency that has allowed them to be durable with slight geometric adjustments and improved class of materials and accessories. However, these bridges have limitations for spans greater than 60m, namely load capacity, high deformability and assembly time. Nowadays modular bridges are used to respond as a rapid reestablishment solution of disrupted roads in several situations such as natural disasters, conflicts or aging of existing bridges. MBSbyBERD developed new structural solutions that aim to accelerate the reestablishment of the disrupted connections comparing with the existing modular bridges solutions on the market and overcome spans up to 120m, fulfilling the current load parameters for this type of bridge - HL-93 (AASHTO) and MLC80T / 110W (STANAG). Designed to be pre-assembled and containerized in order to minimize assembly time, 4 different models of modular bridges have been developed and optimized to cover a range of spans up to 120m.

2. Modular bridges – conventional solutions

2.1. Background
The prompt reestablishment of connections is often ensured through the installation of temporary bridges. Mostly of these solutions are based in Bailey Bridge - modular panel bridge – invented during the World War II by Donald Bailey and widely tested and implemented all over the world.

Fig. 1. Bailey bridge - incremental launching during WWII [1] [2];

Fig. 2. Bailey bridge for route reestablishment during the collapse of a section of Highway A14 – Figueira da Foz/ Coimbra, Portugal (2016);

Bailey bridge type is based in two trusses constituted by modular panel pinned between each other. For longer spans (up to 70m) or heavy loads, truss reinforcement can be reached simply by joining more layers or levels of modular panels. The solution achieved, more than 70 years ago, is still being applied. However, the response capacity of the existing solutions is currently limited - in many situations the installation of a modular bridge is not a viable alternative to minimize the impact caused by the lack or loss of circulation. This is due to the limitations of the existing solutions, within which were identified as the most relevant [3]:

- For spans longer than 60m, the existing solutions impose incrementally operative constraints (vehicle weight and traffic speeds);
- There are no solutions for spans larger than 90m;
- For long spans (80 to 90m), assembly times are very long and need very significant human and mechanical means;

2.2. Logistic and bridge assembly

Modular panel bridges are usually transported in containers. As so panel measures (10feet long by 7feet height) were limited to fit in containers since their invention. If by one hand light panels were achieved allowing the bridge assembly without mechanical and/or lifting equipment, by the other hand results in a limitation of stiffness that for long spans (>50m) is not solved with the two level panel truss girder.

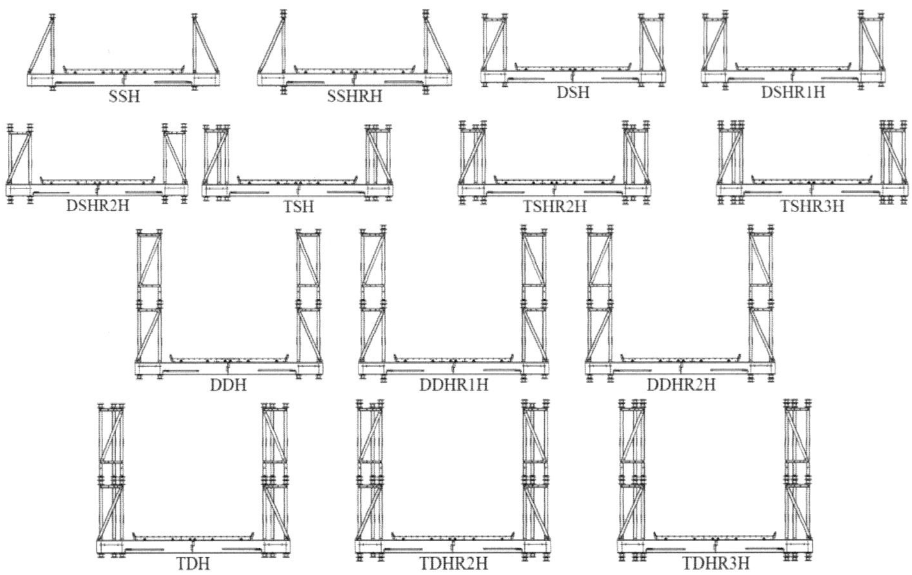

Fig. 3. Bailey bridge – cross sections types – panel combinations [3];

Modular bridges are assembled in one of the margins and then placed using cranes (allowable for spans up to 30m) or using incremental launched construction process. For the launching operation is used a lighter structure "launching nose" to reach the opposite margin. During the bridge launching pavement panels are located in the back of the bridge. This permits to have a lighter structure during launching operation and the needed counterweight to keep the static equilibrium.

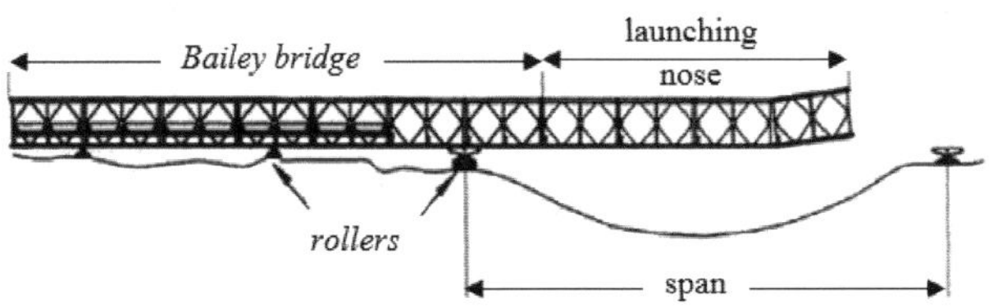

Fig. 4. Bailey bridge - incremental launching [3];

3. Design criteria for modular bridges

3.1. General parameters

The design of modular bridges is strongly conditioned by a set of parameters, namely geometric requirements (span, number of lanes, width of the deck), assembly (topographical conditions and accessibility of the margins), transport, connections and modularity, among others. The deck width is standard and depends from the number of lanes: 1 lane normal width - 3.15m; 1 lane extra width - 4.20m; 2-lanes - 7.35m.

Since modular bridges had their backgrounds in the military area, the main guidelines to design are also military. The Trilateral Design and Test Code for Military Bridging (TDTC) [4], developed by the US, UK and Germany, provides recommendations on the several parameters that should be considered in the design of temporary bridges, where modular bridges can be included.

3.2. Loads

Apart the conventional loads normally considered in bridge design, in the case of vehicle loading there are some particularities that depend of the bridge application - civil or military. It has been found that in general the main manufacturers of modular bridges (Mabey, Acrow, Wagner-Biro among others) use the load model defined by AASHTO (USA) - HL-93 (or previous versions) for civilian applications and

STANAG2021 (NATO) for military applications. In the case of civilian application, the vast majority of countries that regularly acquire modular bridges, namely Peru, Colombia, Chile, Indonesia, among many others, have based their codes in AASHTO, being therefore deprived the EC1, even by the European manufacturers.
In the case of military actions, NATO developed the STANAG standard on the military load class of bridges, barges and vehicles. The NATO classification system considers 32 classes (MLC) of vehicle type - VT (16 wheeled vehicles and16 tracked vehicles): 4, 8, 12, 16, 20, 24, 30, 40, 50, 60, 70, 80 , 90, 100, 120 and 150 [5]. Besides vehicle loading defined by STANAG, TDTC introduces other loads that should be considered during modular bridges design.

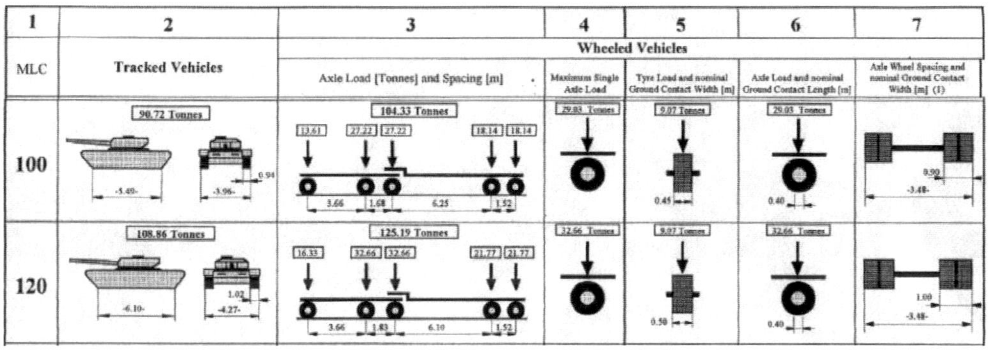

Fig. 5: Characteristics of the vehicles used in the MLC category system of STANAG (extract for MLC100 and MLC120) [5]

In case of bridges to be used temporarily in the construction of other infrastructure works, namely works involving the movement of large volumes of land in places with crossings of valleys and / or water courses, design must be done considering the specific vehicle to be used in that work, like high load capacity dumpers.
Vehicle loads should also be increased by impact factors - 15% for bridge and 20% for access ramps in the case of TDTC and 1.33 in the case of AASHTO [4].

4. New modular bridge solutions

4.1. General

The main challenge associated to the development of a modular bridge is the combination of the conditions that must be met in the replacement or materialization of a connection between two margins in one of the scenarios identified above.
In addition, to the necessary study of the structure of the modular bridge and the particular aspects of this type of bridges, namely the constitution of the modular elements with fast connections to one another, ease of transportation, new configurations, material efficiency, etc. there are important challenges associated with the construction/ assembly of a modular bridge. In case of underdeveloped countries, the reestablishment time of a collapsed or damaged bridge is relevant but might have a low social-economic impact. In developed countries the time for a bridge reestablishment have a high social-economic impact. In order to respond to these issues, MBSbyBERD developed new modular bridge solutions that were designed considering several core aspects, namely:

- Load capacity required by main standards;
- Service limitations defined by standards (for example deflection);
- Enable the pre-assemblage of the most part of the pieces in factory and the final assembly in the site become much faster than the traditional solutions;
- The pre-assembled module is containerized (20 or 40feet);

- The "wrap effect" condition the geometry of longitudinal truss. Geometry is also conditioned by internal forces.;

MB60, MB80 and L-MB120 (spans up to 120m) are bridges with faster assembly[2] combining modularity with high load capacity ASSTHO HL-93 and STANAG MLC110W/80T. The modular bridges use common material (steel S355), pinned and bolted connections and the same construction process – increment launching, as Bailey bridges type.

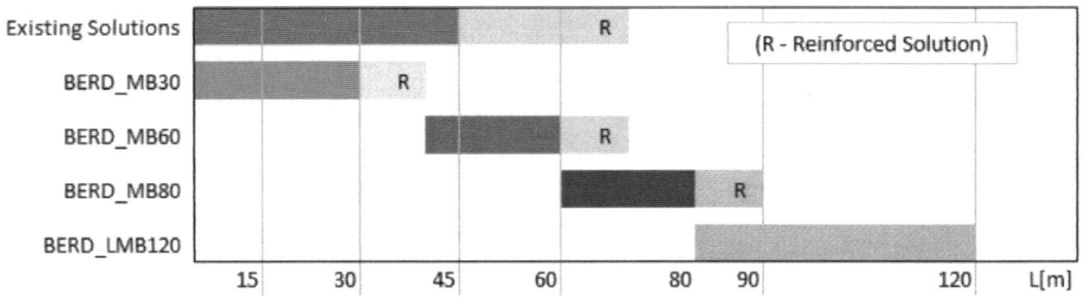

Fig. 6. Bailey bridge - incremental launching [6];

4.2. Solutions to spans up to 60m

For spans up to 60m two solutions were developed, the MB30 optimized for 30m spans, and the MB60 for 60m spans. In the case of the MB30, it was decided to place the deck on the top of the truss beam thus taking advantage of the transversal beams and deck panels in controlling the instability of the upper chord under compression. This solution allows you to adjust the load capacity as well as the width of the deck by adding new planes of longitudinal trusses. MB30 can also be assembled without the use of mechanical lifting equipment like the Bailey bridges type.

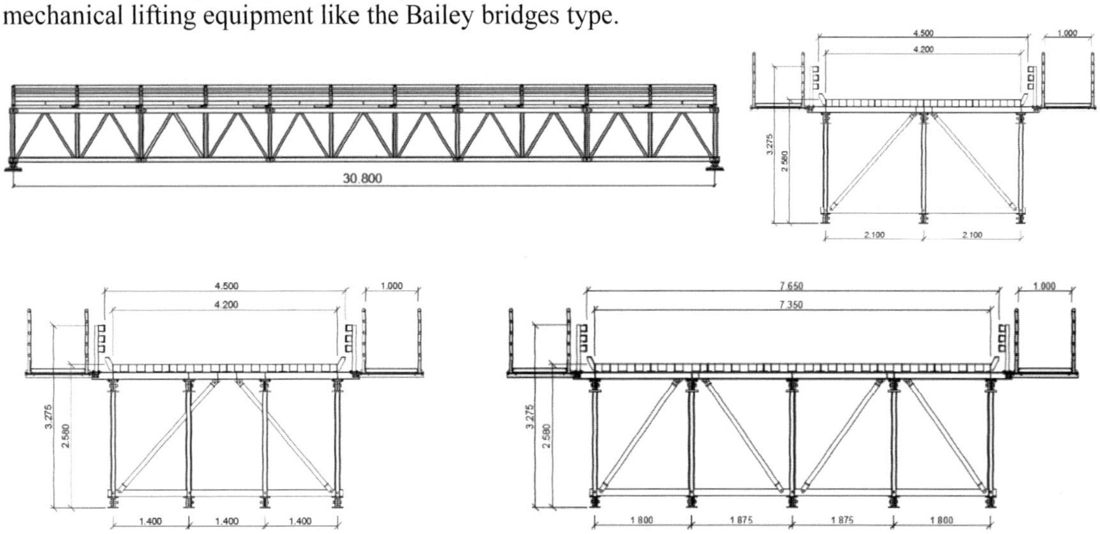

Fig. 7: MB30 – Elevation and cross sections corresponding to different deck widths and load capacity.

2 When compared with main market players (Mabey and Acrow) with similar spans (60m and 90m). No cases are known of modular bridges for spans longer than 90m

The MB60 has a high load capacity and meets the deflection limits of AASHTO. To do so, the height of the trusses was increased to 4.1m which implied the impossibility of producing welded modules by containerization reasons. Thus, a new concept was developed of articulated modules, pre-assembled in the factory and optimized for containerization and rapid assembly on site.

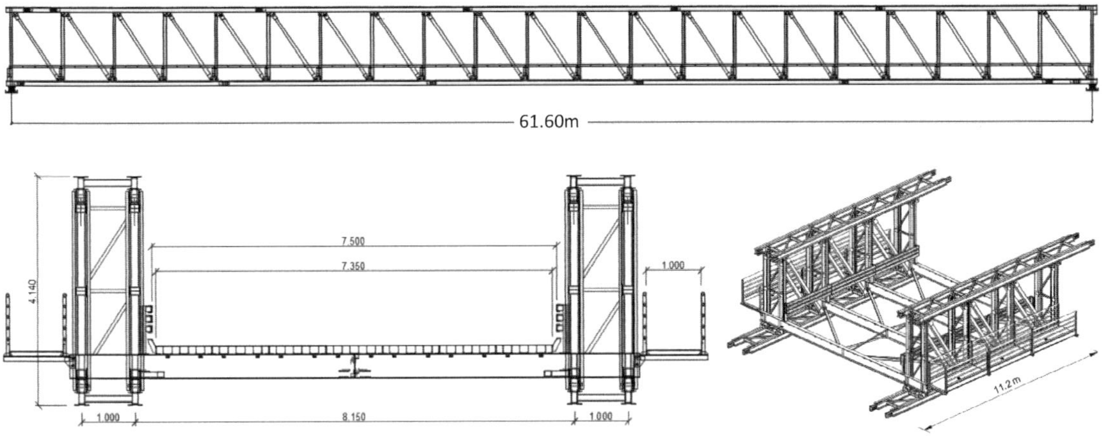

Fig. 8: MB60 – Elevation, cross sections and perspective for two lanes solution

4.3. Long span solutions (up to 120m)

For spans longer than 60m, the modules developed for the MB60 have limitations, mainly in the deflection. The possibility of applying reinforcements in the truss chords allows some flexibility for spans up to 70m. However, for longer spans these reinforcements are ineffective. Thus, it was designed specific modules for this range of spans (MB80 and LMB120), with a height of 8.84m, which from 80m are reinforced with the inclusion of an upper bow-shaped chord. The significant height of the modules allows creating a closed cross-section with clear benefits in controlling instability phenomena. These modules are also articulated to be containerized.

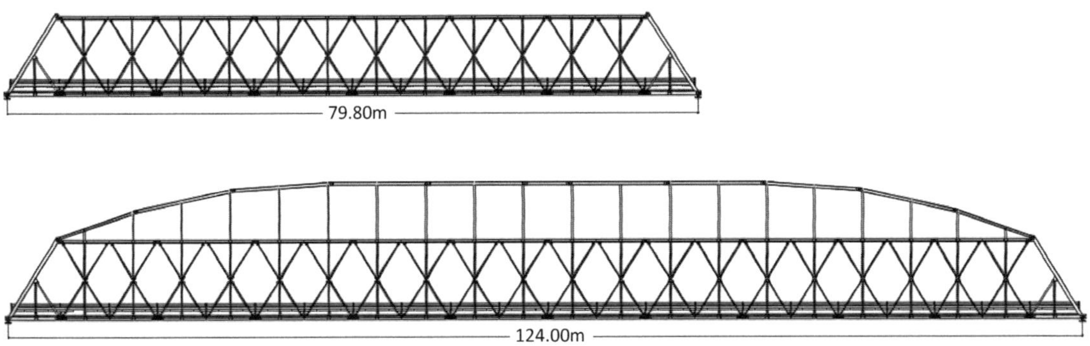

Fig. 9: MB80 and LMB120 elevations

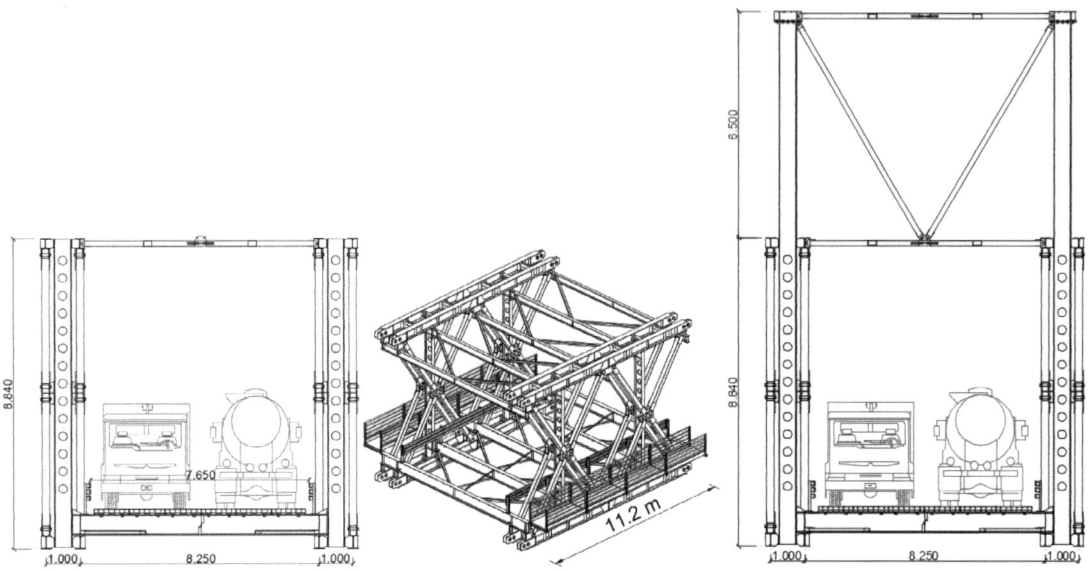

Fig. 10: MB80 and LMB120 cross sections

4.4. Logistic and bridge assembly

New bridge solutions have modules that are packed for being easily removed from containers and are partially pre-assembled in order to accelerate assembly process (see Fig. 5 and 6). The reduction of number of assembly connections on site and longer modules (5.6m or 11.2m) than Bailey bridge type (3.05m- 10feet) are crucial aspects to reduce the global duration assembly time.

MB60 and MB80/LMB120 modules types are longer and heavier and common pulling and lifting equipment is required on assembly operations. Although Bailey bridge modules are lighter, the use of mechanical equipment on lift and launching operations is also common.

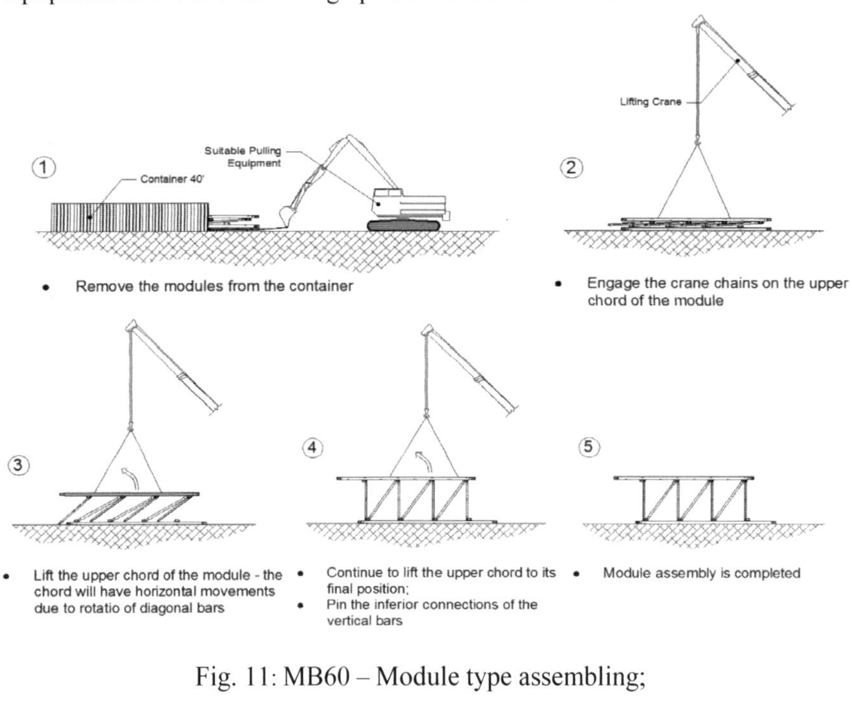

Fig. 11: MB60 – Module type assembling;

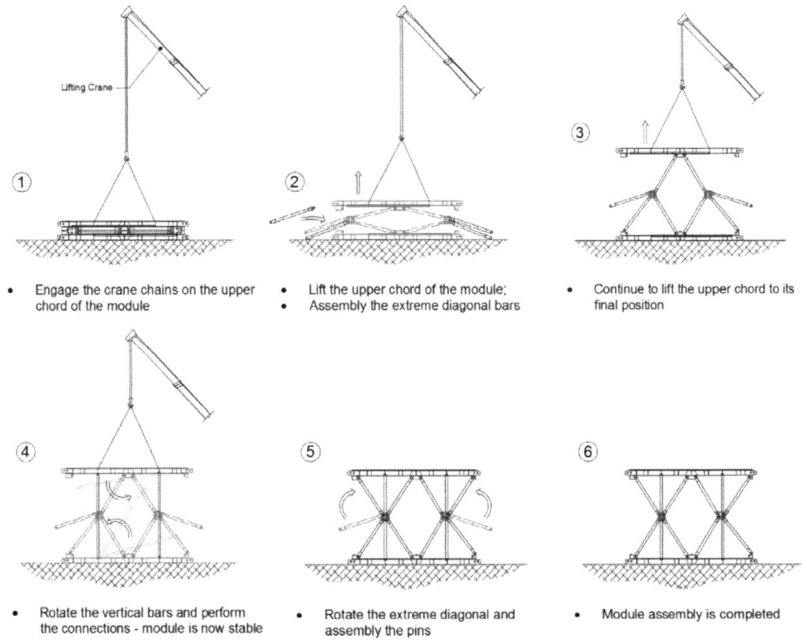

Fig. 12: MB80/LMB120 – Module type assembling;

Both module types were designed considering the scissor opening movement.

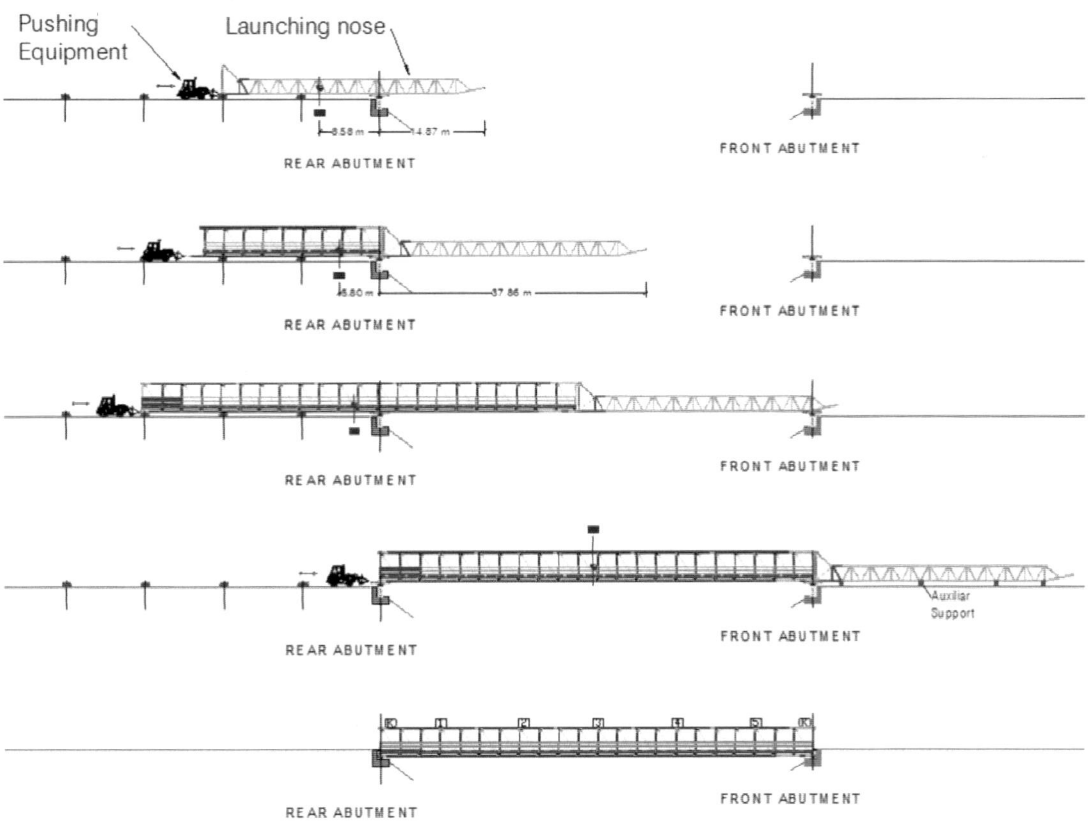

Fig. 13: MB60 – Incremental launching;

IOP Conf. Series: Materials Science and Engineering **419** (2018) 012021 doi:10.1088/1757-899X/419/1/012021

As already mentioned before, MB60 and MB90/LMB120 are installed in the final position using incremental launching method, similar to the common modular bridges installation. MBSbyBERD kinematic studies point out to a global assembling time reduction up to 40% [6].

5. Conclusions

Modular bridges are being used as fast reestablishment solutions for routes disruption and to the reduction of social-economic impact. However, this impact can be even more reduced if the installation of the modular bridge becomes faster and if longer spans are available. As so, MBSbyBERD developed new modular bridge solutions to accelerate the reestablishment of routes in different scenarios and for different levels of need.

6. References

[1] https://en.wikipedia.org/wiki/Bailey_bridge#/meia/File:%22Don%27t_tell_me_there % 27s_any thing_the_engineers_can%27t_do._We_built_bridges_where_bridges _couldn %27t_ be_built. We_built..._-_NARA_-_535980.tif.

[2] Beck, Alfred M., et al 1985 The Corps of Engineers: "*The War Against Germany, Center of Military History*", U S Army, 1985, p. 524.

[3] André A., 2016. Estudo da Aplicação de Pré-esforço Orgânico em Pontes Provisórias, PhD Thesis, Faculty of Engineering of University of Porto, Portugal.

[4] Hornbeck, B., Kluck, J., Connor, R. *"Trilateral Design and Test Code for Military Bridging"*. Tardec Bridging, 2005.

[5] STANAG 2021. Agreement, N. S. Ed.6 "*Military load classification of bridges, ferries, rafts and vehicles*", NATO, 2006.

[6] André A., Carvalho D. and Pacheco P. 2017. Novas soluções de pontes modulares para vãos até 120m, Proceedings of XI Congress of Steel and Composite Steel and Concrete Construction, Coimbra, Portugal.

Barkarby Bridge

Manuel Biedma García[1], Mayra Toledo Serrano[1], F. Javier Martínez López[1] and José Manuel González Parejo[2]

[1]ICCP. Acciona Ingeniería. Centro de Negocios Albatros Edificio D.
C/Anabel Segura, 11. Alcobendas (Madrid) 28108, Spain
[2]ICCP. Cuerpo de Ingenieros de Caminos, Canales y Puertos del Estado.
Confederación Hidrográfica Guadiana. Spain

E-mail: franciscojavier.martinez.lopez@acciona.com

Abstract. Barkarby Bridge is a bridge designed to take part of the Barkarby-Kallhäl section in the Sundbyberg-Kungsägen highway line project in Stockholm, Sweden. This bridge allows the road traffic to cross over the Sundbyberg-Kungsägen railway line and over the Västeras-Stockholm highway. This is a singular bridge because is intended to be used as a road bridge in a first phase, and as a combined pedestrian-road bridge in a second phase. In addition, it will provide access to the pedestrians to the new railway station. These factors have implied to design a non-usual trapezoidal and variable deck. The deck typology corresponds to a composite steel/concrete type.

1. Introduction

Barkarby Bridge is located in the Barkarby-Kallhäl section in the Sundbyberg-Kungsägen highway line project in Stockholm, Sweden. This bridge allows the road traffic to cross over the Sundbyberg-Kungsägen railway line and over the Västeras-Stockholm highway.

This bridge was intended to replace the old Barkarby Bridge which was unable to adequately make the traffic compatible with the new uses needed: the bridge had to be connected to the new railway station. In addition, the old bridge was affected by several typical pathologies of old concrete bridges in city environments. Hence, Trafikverket, the Swedish road authority decided to replace the old bridge and construct a new one adjacent to the old one which was finally demolished.

This is the first bridge that Acciona Infraestructuras Swedish Filial has constructed in Sweden as an EPC contractor. Acciona Ingeniería carried out the detailed engineering of the project and was in charge of the works assistance during its execution. Acciona Infraestructuras itself executed and delivered the project.

2. Description of the bridge

Barkarby Bridge presents a total length of 250.25 m developed in a straight line in plan view except for the last 17 m where follows a circular route with a radius of 500 m. The total length is distributed over 9 spans with a maximum clear span of 32.40 m. It is modulated in three different decks. The first deck is comprised by the first span only. The second deck is comprised by the following three spans and the third deck is comprised by the last 5 spans. See figure 1.

The first and the third deck are 8.00 m wide and are designed with the same deck typology: twin I-cross section steel girders of constant depth of 900 mm separated 3.46 m at the bearing level and are aligned in a parallel way to the centerline of the deck. See figure 2. The singularity of these girders is that their webs are inclined: the planes that contain them are not vertical. The second deck has a variable width from 11.10 m to 13.80 m, and it consists of three I-cross section steel girders of constant depth of 900 mm and inclined webs as well. See figure 3.

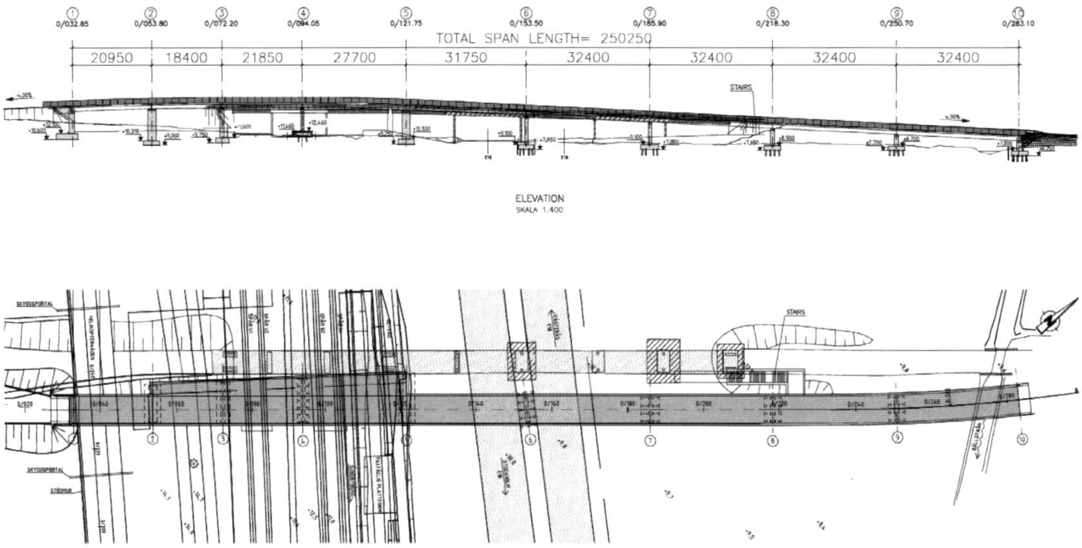

Figure 1. Plan and elevation view.

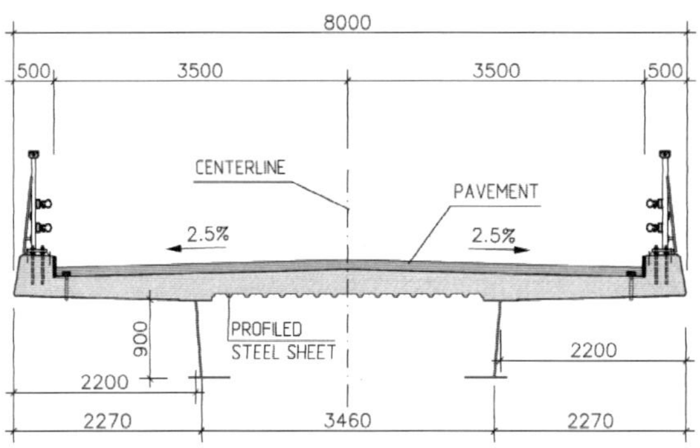

Figure 2. Twin-girder deck cross section.

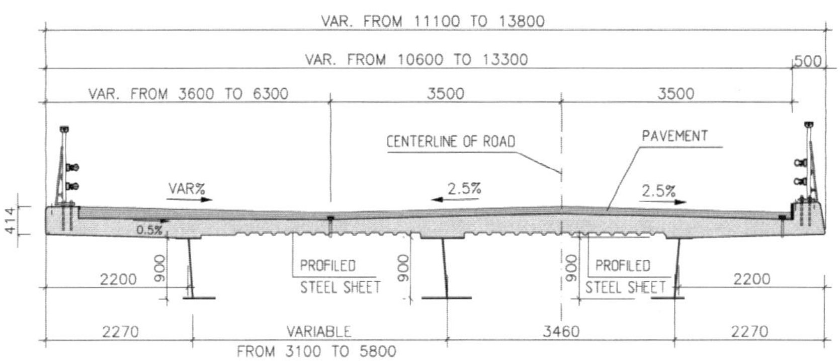

Figure 3. Three-girder deck cross section.

A diamond-type bracing was designed between the lower flanges of the steel girders to withstand the torsional forces that could act on the deck. K-bracing intermediate stiffeners have been located each 3.459 m to control de distortional effect of the eccentric loads that act on the deck. The diaphragms are located on the piers and the abutments. See figure 4.

Figure 4. Diaphragm, intermediate stiffeners and diamond-bracing.

On top of the steel girders a concrete slab of 28.2 cm maximum depth was casted by means of temporary hanging formworks in the cantilevered zones of the decks and permanent profiled steel sheets in the central zones. The top concrete slabs have a constant cross slope of 2.5% in the first and third decks, which implies a minimum depth of 21.6 cm at the edge of the deck, but varies in the second deck. Hence, the decks are composite steel/concrete decks. Each cantilever of the slab is 2.20 m long. The decks accommodate two road lanes of 3.50 m wide each one, and 0.50 m at both sides to locate the lateral barriers.

All of the exposed steel surfaces, including the profiled steel sheets, were applied an anticorrosion protective system C5M-H. The preparation grade of welds, edges and other areas with surface imperfections was P3 [1] according with the strict Swedish standards.

The pavement is 11 cm thick and is located between the inner sides of the lateral barriers.

At both ends two abutments have been designed. All of the intermediate supports are reinforced concrete wall piers and their mean height equals to 6.00 m, with maximum height of 7.90 m. The support 6 is designed parallel to the highway direction, so its axis is skew with an angle of 7.639g. The support 4 is located in the middle of the railway platform. As a result, due to the few hours available for the construction of this pier, the pier has been designed with three steel profiles columns in a first phase. On top of these three columns, a precast concrete pier cap is located. The dimensions of this pier cap are 1.00 m thick, 9.760 m wide and 1.00 m long. The pier was finally casted with the rest of the concrete in a second phase. See figure 5.

Figure 5. First stage of support 4 and unshored erection of deck.

The embankment behind abutment 10 is designed with cellplast (extruded polystyrene, or XPS) in order to reduce the weight transmitted to the soil due to the low bearing capacity of the latter. In addition, the foundation soil of this abutment needed to be treated to improve its bearing capacity by means of gravel cement columns. See figure 6.

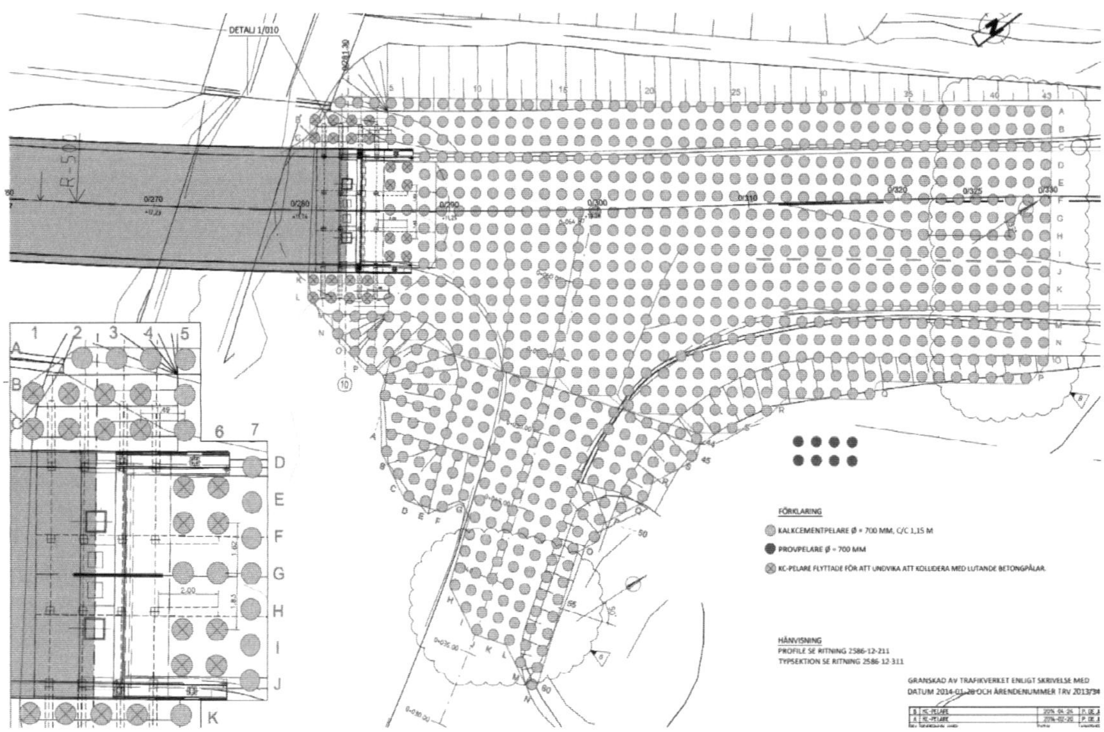

Figure 6. Treatment of embankment behind abutment 10.

IOP Conf. Series: Materials Science and Engineering **419** (2018) 012022 doi:10.1088/1757-899X/419/1/012022

Supports 1, 2, 3, 5 are designed with spread foundations. The rest of supports are designed as deep foundations with precast square piles of 275×275 mm, except for support 4, where the piles are designed as steel pipe piles Ø220×10 mm.

The bearings are comprised by laminated elastomeric bearings which although widely spread in Europe, they are not of common use in Sweden.

3. Erection of the bridge

The foundation was constructed using classical construction methods for the spread foundations and for the deep foundations as well.

The piers and the abutments were carefully casted to account for the inner and surface temperature of the young concrete in order to limit the surface cracking of these elements. It is really common in Sweden, due to the low temperatures during the winter and the hydration heat of the concrete, to partially isolate the formwork to limit the temperature gradients.

Simultaneously, the steelwork of the deck was fabricated in Spain by the workshop of Acciona Infraestructuras. See figure 7. Then the steel components were transported from Spain to Sweden by road and sea. Due to transport dimensional limitations the steelwork of the deck was broken down but keeping in mind to minimize the need of field splices on site. These splices were executed on site by welding.

All of the construction stages of the bridge have been designed in order to avoid road and railway traffic disruption. As a result, the construction process is a relevant part of the bridge design. So, it was decided to erect the decks in an unshored way, just reducing the need of shores to the main welded girder splices only.

Once the steel girders and the diamond-type bracing were completely erected, the top concrete slab formwork was erected on the girders on site by means of conventional mobile cranes. See figures 8 and 9. The formwork was removed as soon as the top slab concrete strengthened and reused in other sections of the deck.

The final stages of the bridge erection consisted of the erection of the barrier, light poles, pavement, road signs, electric heating pipes and urban services, etc. See figure 10.

Figure 7. Fabrication at workshop.

Figure 8. Unshored formwork of the concrete slab.

Figure 9. Several erection stages, simultaneously.

Figure 10. General view of the bridge.

4. Design and calculations

In essence, the decks of the bridge act as a box girder deck: the lower diamong-type bracing simulates the lower plate of a box girder deck and the top concrete slab acts as the top plate of a box girder deck once the concrete strengths. However, detailed calculations were carried out to assess the risk of lateral-torsional buckling during the casting of the top concrete slab because in that very moment the box girder deck mechanism is not completely closed. See figure 11.

All of the construction stages have been taken into account in the calculations by means of the computer program RMBridge. This computer program allows to easily accumulate the tensional state of all of the most significant cross sections of the decks as the erection progress.

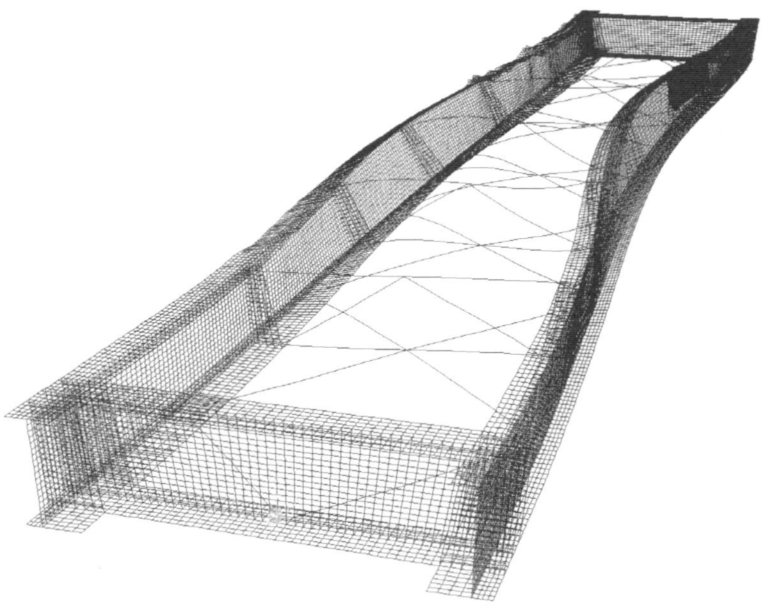

Figure 11. Lateral-torsional buckling assesment.

IOP Conf. Series: Materials Science and Engineering **419** (2018) 012022 doi:10.1088/1757-899X/419/1/012022

Acknowledgements
The authors wish to thank the hard work done by all Acciona Infraestructuras professionals who have made possible the design and fabrication of this bridge, the first bridge built by Acciona Infraestructuras in Sweden.

References
[1] International Organization for Standardization (ISO) 2006 Preparation of steel substrates before application of paints and related products. Visual assessment of surface cleanliness. Part 3: Preparation of grades of welds, edges and other areas with surface imperfections *ISO 8501-3*

Long-Span Orthotropic Steel Deck Bridges of Turkey

Nurdan M. Apaydin [1] and Selcuk Bas[2]

[1]Head of Structures Department, General Directorate of Turkish State Highways, 06100 Ankara, Turkey
[2]Department of Civil Engineering, Faculty of Engineering, Bartin University, 74100, Bartin, Turkey

E-mail: napaydin@kgm.gov.tr

Abstract. Long-span bridges are the key structures for transportation systems and they are considerably needed in Turkey since it is a transit country between Europe and Asia Continents. Besides, Turkey has many large reservoir dams, so these kinds of bridges are essential to provide transportation. Long-span orthotropic steel deck bridges are the main and the most important links in the transportation network of the country. This study presents the general features of the long-span orthotropic steel deck bridges in Turkey, namely The July 15 (the 1st Bosphorus) Martyrs Bridge, the Fatih Sultan Mehmet Bridge (the 2nd Bosphorus Bridge), the Yavuz Sultan Selim Bridge (the 3rd Bosphorus Bridge), the Osman Gazi (İzmit Bay) Bridge and the Canakkale 1915 (Dardanelles) Bridge. The importance behind the selection of long-span bridges in Turkey is explained and detailed information for recently ongoing project of the Canakkale 1915 Bridge is given. The outcomes from the current study are aimed to provide important vision for the construction of new long-span bridges in the world.

1. Introduction

Of all the civil infrastructures, long-span cable-supported bridges are the lifeline structures for the transportation network in a country/state. These bridges are not only a substantial component in transportation system, but also offer a respectable contribution to economic and cultural growth of their regions [1].

Transportation infrastructures play a critical role in Istanbul and Marmara Region, Turkey. İstanbul is located approximately in the middle of the Marmara Region and the city links between two continents of Asia and Europe. As a result of the increase in population and consequently in traffic of Marmara Region (25% of total population of Turkey), the importance of long-span bridges raises significantly and rapidly in Turkey. More specially, the city of İstanbul with a population of 18 million has a strategic position holding three Bosphorus Bridges and being close to the Osman Gazi and Canakkale 1915 Bridges.

Along with the existing 1st and 2nd Bosphorus Bridges (the Fatih Sultan Mehmet), numerous efforts have been continuously made to meet all needs for the transportation service with the newly completed long-span bridge projects of the 3rd Bosphorus Bridge (the Yavuz Sultan Selim Bridge) and the Osman Gazi Bridge located on Izmit Bay relatively close to Istanbul. The Osman Gazi Bridge is one of the bridges that were opened to traffic in the soonest time, in July 2016. The bridge, a 2.682-meter long-structure, is said to be the 4th longest suspension bridge in the world and the 2nd longest in

Europe. The importance of this structure is to ensure the integrity of transport between the two sides of İzmit Bay, the country's western provinces and consequentl reducing travel time between İstanbul. With this project, the average travel time between İzmir (a province of the Aegean region) and İstanbul was reduced from ten hours to approximately four hours [2, 3].

It was decided to construct the Canakkale 1915 Bridge (named in memory of Turkey's historic victory in the region during World War I) with the idea of connecting Thrace to the Aegean and the Western Mediterranean via Canakkale. Upon completing this project, the bridge will have been the longest (32 meters longer than the world's longest bridge, the Akashi-Kaikyo Bridge) bridge between two shores in the world. Besides, a significant support will be provided to solve the traffic problem in İstanbul, especially on the Bosphorus Bridges. It will be the 5th suspension bridge in the sea of Marmara Region, will contain six traffic lanes, three in each direction and will reduce the Canakkale crossing time from 45 minutes to 4 minutes. The bridge will be built on approximately 200 kilometers away from the southwest of İstanbul and it is expected to open in 2022. The locations of the bridges on the Turkey Map are shown in Figure 1.

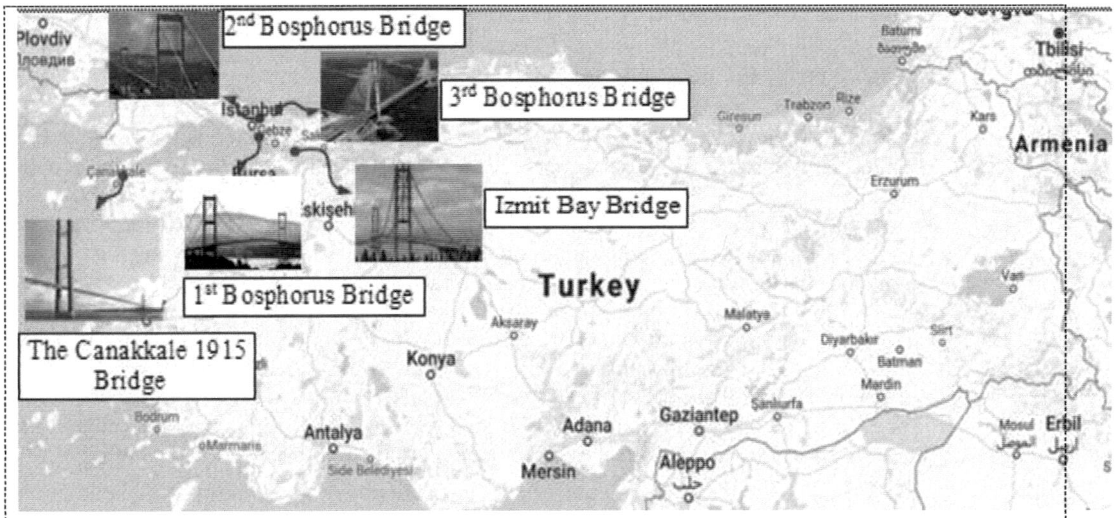

Figure 1. The locations of the long-span orthotropic steel deck bridges in Turkey.

2. The First Bosphorus (15th July Martyrs) Bridge

The history of bridge projects on the Bosphorus Straits dates back ancient times, to 522 BC - 485 AC. However, the first decision to build a long-span bridge over the Bosphorus River was in 1957. The British firm Freeman Fox & Partners designed the bridge in 1968. After its construction started in 1970, the bridge was commissioned in 1973 corresponding to the 50th anniversary of the Republic of Turkey. With its main span, the bridge was the longest suspension bridge in Europe until the completion of the Humber Bridge in 1981. It was also the 4th longest bridge in the world upon completion in 1973. The First Bosphorus Bridge is a steel suspension bridge with three standard traffic lanes. Its structural type and design considerations are mostly similar to that of the Humber and Severn Bridges, which are the first modern bridges in Europe with the aerodynamic box deck section [3, 4].

The First Bosphorus Bridge is a key civil infrastructure in transportation networks of Istanbul, with one of the beltways between the two continents designed after the construction of the bridge. Therefore, the bridge currently serves as significant link for the city of Istanbul. As indicated in Figure 2, the First Bosphorus Bridge is a gravity-anchored type of long-span steel suspension bridge. The aerodynamic box deck section and the towers with tapered box section are the key indicators of a modern bridge. The bridge has the main span length of 1074 m and two approach-spans with the length of 231 m at the European side and 255 m at the Asian side, respectively. Only the main span is suspended with the suspenders; nevertheless, the approach spans are supported at the base instead of the suspender elements. The width of the deck is 33.40 m and its height is approximately 3.00 m [3].

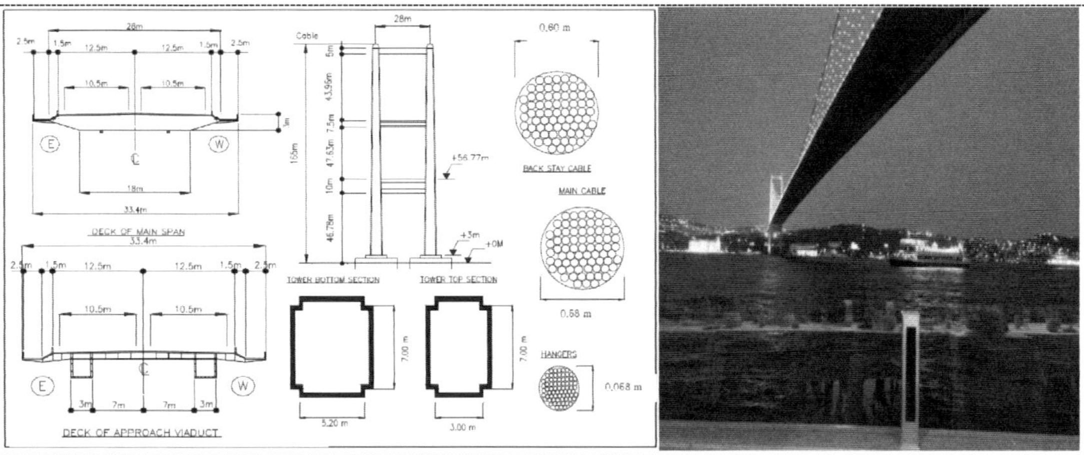

Figure 2. Dimensions and general view of the First Bosphorus Bridge [1].

3. The Second Bosphorus Bridge (The Fatih Sultan Mehmet Bridge-FSM)

Based on the decision on new beltway (O-2) in Istanbul called Trans-European Motorway (TEM), constructing the 2nd Bosphorus bridge was inevitable. Accordingly, the British firm Freeman Fox & Partners prepared the second project for the long-span bridge over the Bosphorus Straight. The construction was started in 1986 and the bridge was completed in 1988 in which the bridge was the 5th longest suspension bridge in the world; it is now ranked 21st. The Second Bosphorus Bridge located on the north side of the First Bosphorus Bridge was named the Fatih Sultan Mehmet Bridge (FSM) [5, 6].

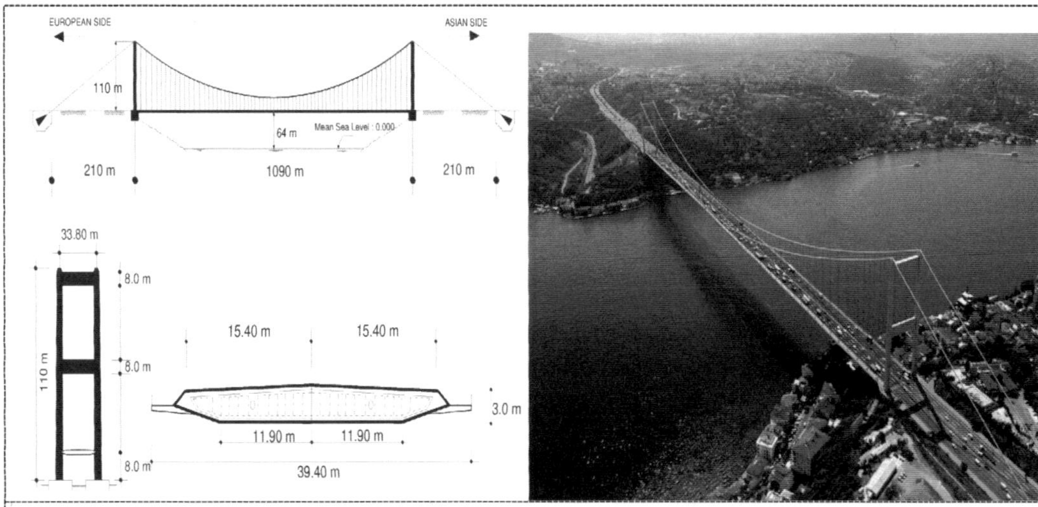

Figure 3. Dimensions and general view of the Fatih Sultan Mehmet Bridge [1].

As shown in Figure 3, the Fatih Sultan Mehmet Bridge is a steel long-span suspension bridge. The towers of the bridge are supported at the ground level, which means that the ends of the deck are the level of the tower base. Therefore, the bridge has no approach span. The main span length of the bridge is 1090 m and the deck with aerodynamic box section has a width and height of 39.40 m and 3.00 m, respectively. The tower of the bridge is a rectangular box section and the total height of the tower is 110 m. The deck of bridge at the mid-span is 64 m high from the mean sea level. With the side span length of 210 m at the two continents, total length of the bridge is 1510 m. Since the TEM highway is the only route allowed for truck and other heavy vehicles, the bridge is subjected to various heavy traffic loads [6].

4. The Third Bosphorus Bridge (The Yavuz Sultan Selim Bridge-YSS)

The construction of the Third Bosphorus Bridge was started in 2013 and it was finished within three years. The Third Bosphorus Bridge called the Yavuz Sultan Selim Bridge, which is a hybrid cable-stayed-suspension structural system, is located at the entrance of the Bosphorus River to the Black Sea, which is approximately 5.0 km north from the First Bosphorus Bridge. With the 322 m height of tower and approximately 59 m width of the deck, it is the widest and the second tallest suspension bridge in the world. The YSS Bridge is designed not only as a motorway with four lanes but also as a railway with one lane in each direction. Therefore, the bridge is the world's longest suspension bridge with railway system. As indicated in Figure 4, the main span length of the bridge is 1408 m and total length is 2164 m. The towers with triangular hollow section are designed as reinforced concrete, but the other elements are made of structural steel [1].

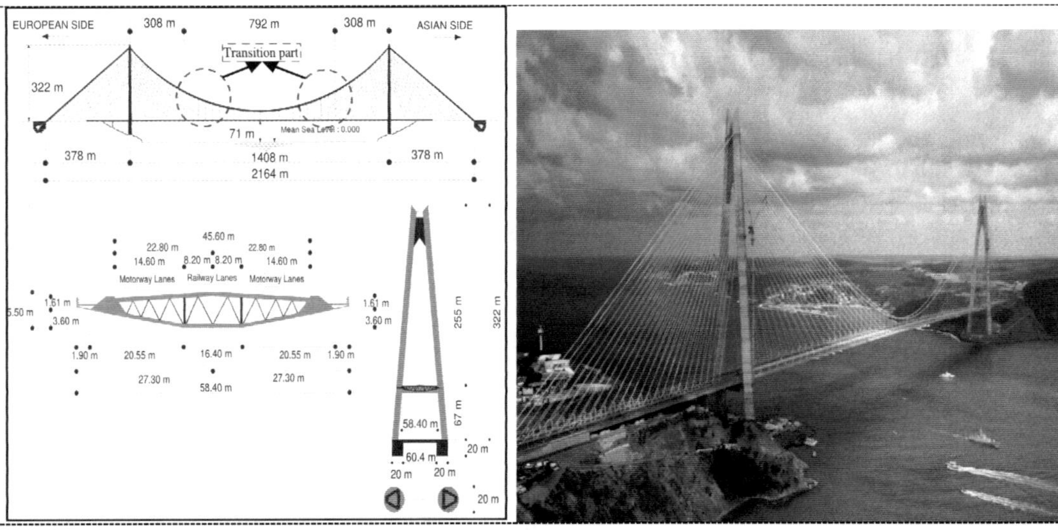

Figure 4. Dimensions and general view of the Yavuz Sultan Selim Bridge [1].

5. The Izmit Bay Bridge (The Osman Gazi Bridge-OGB)

Other than the Bosphorus Bridges in Istanbul, the first long-span suspension bridge is constructed for the Izmit Bay at the eastern side of the Marmara Sea, very close to the city of Istanbul. The long-span suspension bridge is planned as a significant link for the Istanbul-Izmir Motorway to considerably reduce the distance between two cities. The construction of the Osman Gazi Bridge was started in 2013 under the build-operate-transfer contract, and the bridge was opened to traffic in 2016. According to the main span length of 1550 m, the bridge is the fourth longest suspension bridge in the world.

General arrangement of the bridge is shown in Figure 5. The bridge has three spans: two approach spans with the length of 566 m and one main span. However, there is also transition span with the length of 120 m at the anchorage blocks at each side of the bridge. Accordingly, the total span length of the bridge is 2907 m. The height of the tower with box section is 252 m and towers are supported by concrete foundations submerged into the sea. The main deck with the width of 30.10 m and the height of 4.75 m was designed for the three standard traffic lanes in each side. The deck has also aerodynamic box section braced by the internal truss elements and diaphragm [1].

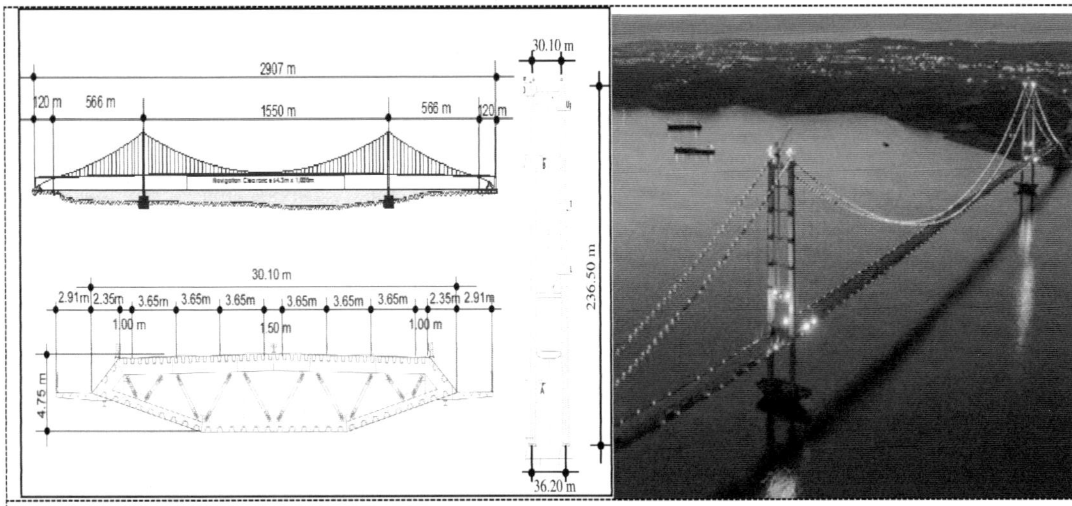

Figure 5. Dimensions and general view of the Osman Gazi Bridge [1].

6. The Canakkale 1915 (Dardanelles) Bridge

Another important link to Istanbul and the Marmara region where the majority of the population of Turkey is concentrated is the Canakkale Straits. Due to this crucial geographic location of the Canakkale, It was decided to construct the Canakkale 1915 Bridge (named in memory of Turkey's historic victory in the region during World War I) with the idea of connecting Thrace to the Aegean and the Western Mediterranean via Canakkale.

As shown in Figure 6, the bridge is connected with a new highway project of Malkara-Canakkale to be completed at the same time with the Canakkale 1915 Bridge. This highway project is planned to make a connection of Istanbul to Canakkale and Northern Aegean location. This new connection including the suspension bridge project also provides an effective solution to link to many harbors, railways and air transportation systems between Aegean and Marmara regions that are the most important locations of Turkey. Besides, general dimensions of the bridge in plan are summarized in Figure 6 with the main span length of 2023 m and the side span length of 770 m [7].

Figure 6. The location of the Canakkale Bridge and its transportation networks.

The bridge has a tapered rectangular hollow section towers with a total height of 318 m from the sea level. The tower foots are casted on the sea bottom at 40 m below from the sea level. As depicted in Figure 7, dimensions of the base and top tower sections are 11.00mx10.50m and 8.00mx7.50m, respectively. The main span deck has an orthotropic steel deck with the height of 3.50 m and the width of 45.06 m. Three lanes are considered for each direction. The side span of the bridge also follows the approach viaduct supported at the base for each side as indicated in Figure 7. With these properties, the bridge is will be the longest suspension bridge in the world upon constructed as revealed in Figure 8.

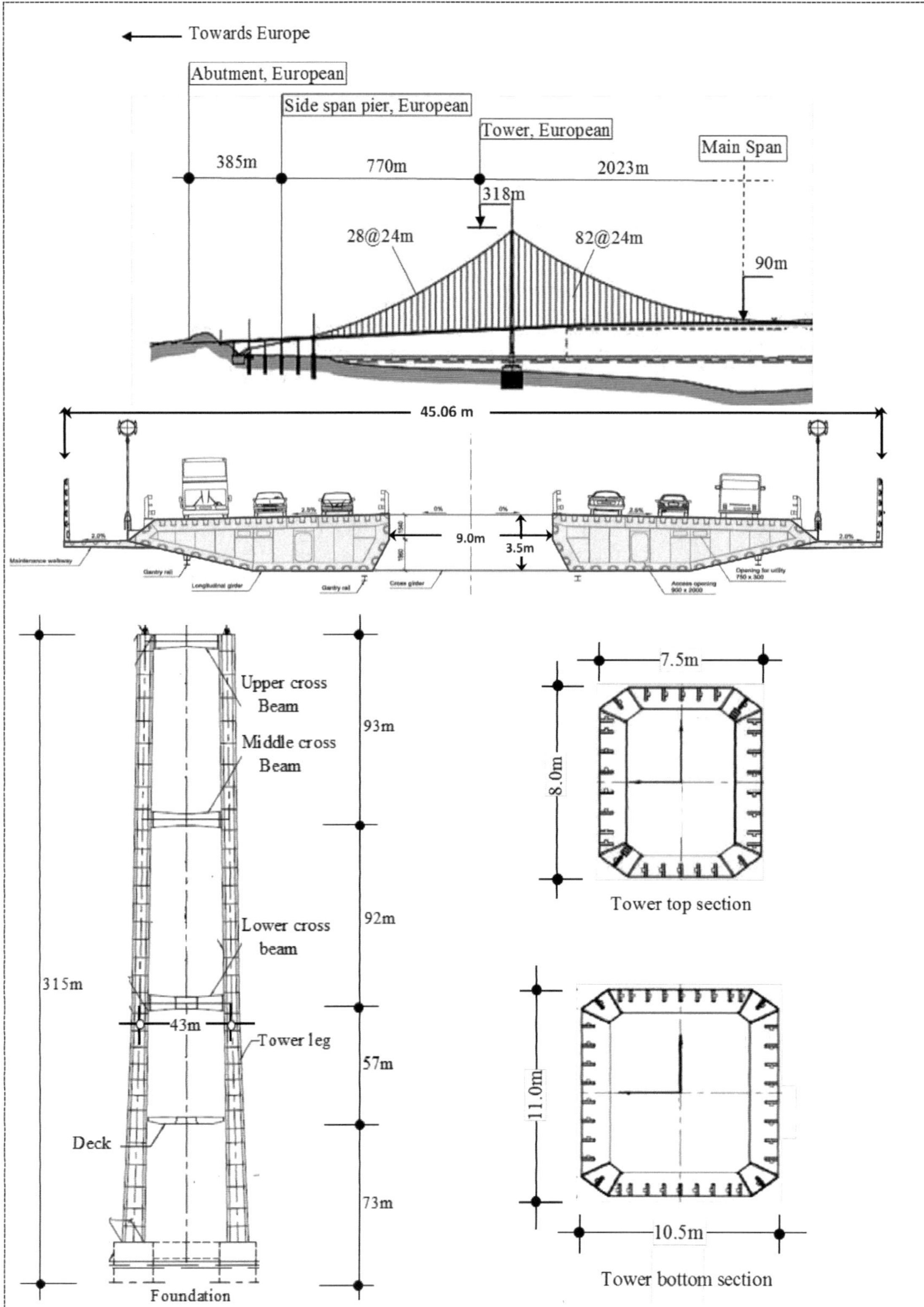

Figure 7. Structural properties of the Canakkale 1915 Bridge [7].

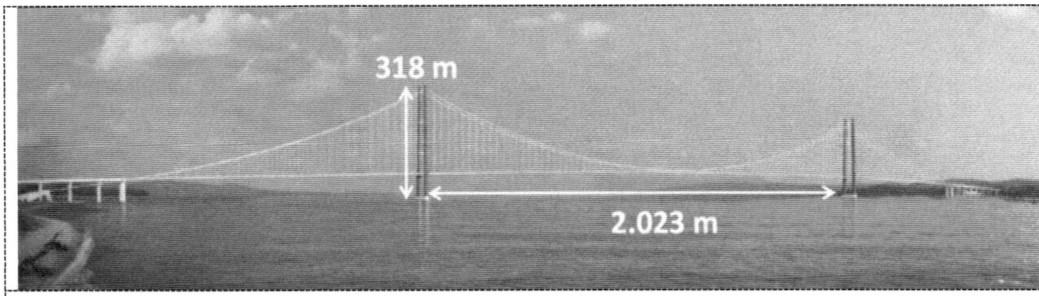

Figure 8. General view from the completed version of the Canakkale 1915 Bridge.

7. Conclusions

The most important regions of Turkey, Aegean and Marmara Regions, have a natural straits and rivers more suitable to build landmark civil infrastructures. With the rapid increase in the population of these regions, the need for the long-span bridges that provide critical connections is indispensable to establish effective transportation networks in Turkey. Due to these properties of Turkey, long-span orthotropic steel bridges are the key infrastructures in its transportation system. In this paper, general structural properties of the existing long-span orthotropic suspension bridges, the 1st Bosphorus (15 July Martyrs), the 2nd Bosphorus Bridge (the Fatih Sultan Mehmet), the 3rd Bosphorus Bridge (the Yavuz Sultan Selim) and the Osman Gazi Bridge, and the ongoing bridge project of the Canakkale 1915 (Dardanelles) Bridge are presented. Based on the critical properties of the bridges, the following points are underlined,

- Considering certain challenges and disadvantages during the construction of 1st and the 2nd bridges, the new completed Yavuz Sultan Selim Bridge (YSS) and Osman Gazi Bridge (OGB) effectively designed, which means that maintenance, management and safety requirements are estimated appropriately in the bridge.
- The outputs and experiences obtained from the new bridge project of the YSS and OGB provide an opportunity to specifically determine project considerations of the Canakkale 1915 Bridge and to make an effective solution to design and construction processes of the bridges.
- The experiences from the long-span orthotropic steel deck bridge projects indicate that Turkey is one of the best example of implementing landmark civil infrastructures projects and thus have a number of experiences for both project and construction works.
- With the last bridge project of the Canakkale 1915 Bridge, the longest suspension bridge with orthotropic steel deck in the world according to its main span length of 2023 m will be constructed on the Dardanelles River.

Considering these important conclusions on the long-span orthotropic steel deck bridges in Turkey, it is clear that civil infrastructure technologies and resilience considerations should be developed for the effective post design strategies of maintenance, management, safety and life extension of these bridges.

References

[1] Bas S, Apaydin NM, Ilki A and Catbas FN 2018 Structural health monitoring system of the long-span bridges in Turkey *Structure and Infrastructure Engineering*, **14**(4) pp 425-444.
[2] Bas S, Apaydin NM, Ilki A and Catbas FN 2016 Structural health monitoring system-SHMs of the long span bridges in Turkey *Maintenance, Monitoring, Safety, Risk and Resilience of Bridges and Bridge Networks* ed T.N. Bittencourt *et al* (London: Taylor & Francis)

[3] Bas S 2017 *An Investigation on structural identification (St-Id) of a long-span bridge for performance prediction* (Istanbul Technical University, Faculty of Civil Engineering, Structural Engineering)
[4] Apaydin, N, Bas S and Catbas, N 2015 Long-span bridges: behavior under operational and extreme wind loads *The Third International Conference on Smart Monitoring, Assessment and Rehabilitation of Civil Structures*-SMAR201 (Antalya)
[5] Apaydin N 2002 *Seismic analysis of the Fatih Sultan Mehmet Suspension Bridge* (Bogazici University, Kandilli Observatory and Eearthquake Research Institute)
[6] Apaydin N, Bas S and Harmandar E 2016 Response of the Fatih Sultan Mehmet Suspension Bridge under spatially varying multi-point earthquake excitations *Soil Dynamics and Earthquake Engineering* **84** pp 44-54
[7] General Directorate of Turkish State Highways (KGM) 2017 A preliminary technical report for The Canakkale 1915 Bridge

Damage or change detection in a small scale model of steel bridge deck under static loading by extensometery

S Iglouli, N Boumechra and K Hamdaoui

Civil Engineering Department, EOLE Laboratory, University of Tlemcen, Algeria

E-mail : sidali.iglouli@student.univ-tlemcen.dz

Abstract. In order to monitor the health state of a bridge structure and to ensure that it still fulfills its functions, an identification method for possible changes or damages has been developed by their locations and the assessment of the damage or change rate. The method is based on a comparison between two structural responses (displacements or stresses) under static loadings. The first is obtained by numerical modeling of the healthy structure, where the second by experimental measurements on the real structure under its operational conditions using sensors placed on structural elements. The two responses are analyzed by an algorithm (IAFEM) based on the inverse analysis method by computing the structure's eventual change or damage factors. The method was initially tested experimentally on a small scale model of a steel bridge deck, built in laboratory, and by analyzing the displacement and stress responses of the structure.

Keywords: bridge, damage identification, inverse analysis, static, sensors, extonsometry

1. Introduction

The health of civil infrastructures (bridges, buildings, etc.) is directly related to the citizen's life security on the one hand and the economic development of the countries on the other hand, especially the old structures that are still operational. Considering the importance of the impact of a possible degradation, it is therefore essential to monitor them to ensure that they still fulfill their functions or require repair.

Structural health monitoring is the design of a process that will enable to evaluate the integrity of a structure and will define any changes (damage) that may affect it and thereby disrupt its normal functioning [1].The performance of such a system is determined by the level of accessibility to the information collected on these changes (detection of the damage, its location, its quantification and the estimation of the damaged structure's residual life [2].

According to Worden K and Dulieu-Barton JM [3] there are two complementary approaches for damage identification:

The first approach considers it to be a classification problem, in which pattern recognition algorithms are used to group in damage classes the different characteristics extracted from the measurement data of the system. The second approach regards damage identification as an inverse problem and generally adopts a reference model (that of healthy structure), and attempts to relate the changes obtained in the measurement data to those obtained by simulation to damage , so the gap is produced due to damage. The algorithms used for this approach are mainly based on linear algebra or optimization theory. The main existing methods based on the exploitation of the dynamic characteristics according to the two approaches as well as their applications can be consulted in the work of Fritzen CP and Kraemer P[4]; Doebling SW and all [5] and Friswell MI [6].

We are interested in this work to the method mentioned in Boumechra N [7] which is based on an inverse analysis of the static displacement responses of one or more points of the structure subjected to a moving load in order to define the real stiffness and therefore a possible change of stiffness, compared to a reference numerical model developed in finite elements, of the healthy structure. The real stiffness determined could then be greater or less than an initial value assumed by reference to the numerical model. This will allow us first of all to check the behavior of the real structure as well as its composition with respect to this design model and in a second place to detect if there are damages in the case where this stiffness would be smaller.

The aim of our work is to evaluate by an inverse analysis the real stiffness of a steel structure composed of frame elements, realized in the laboratory, with respect to displacement and stress measurements using displacement sensors and strain gauges respectively.

2. Methodology

2.1. Method principle

The principle is the initial definition of a structure's geometrical dimensions, physical properties, and supports conditions. Subjected to a movable static load, this structure is deformed so we can record displacements or stresses at several points (static response). From these results, a research to identify the real stiffness of each structural element will be done. The real stiffness thus determined may be greater or smaller than the initial or assumed value of this one. If this stiffness is smaller, then it is said that there is damage. The mathematical method developed is an inverse analysis of those displacement and / or stress responses of one or several points of the structure. This inverse analysis developed on structure's finite element numerical model allows the evaluation of each finite element's real stiffness Boumechra N [7].

The assumptions of the developed method are as follows: the geometrical and mechanical characteristics of the undamaged structure are assumed to be known, the structure is elastic and linear, the intensity of the moving load is known and the dynamic effect of the moving load is neglected.

The studied structure is discretized into n finite elements. For each stiffness $[K_i]$ of the finite element *(i)* corresponds a correction coefficient α_i. If this coefficient is positive, it is said that there is an additional stiffness. If this coefficient is negative, it is said that there is damage on the finite element and is thus called reduction coefficient. From this, we develop the system's static equilibrium equation which is defined in linear analysis as:

$$[K]\{\delta(x_0)\} = \{F(x_0)\} \tag{1}$$

The structural stiffness matrix $[K]$ is the assembly of the stiffness matrices of each finite element. The load vector $\{F(x_0)\}$ varies according to the position of the moving load (x_0). So from the global displacement vector $\{\delta(x_0)\}$, we can determine the movement of the section $w_S(x_0)$ by the matrix [N] defining the displacement of the studied section S.

$$w_s(x_0) = [N]\{\delta(x_0)\} = [N][K]^{-1}.\{F(x_0)\} \tag{2}$$

After the assembly of the structure's global stiffness matrix, we have:

$$[K] = [K_0] + \sum_{i=1}^{n} \alpha_i.[K_i] \tag{3}$$

The global stiffness matrix $[K_0]$ and the elementary stiffness matrices $[K_i]$ of the assumed healthy structure are defined from the initial information about the structure's geometrical dimensions and material's mechanical characteristics. Therefore, the matrix $[K]$ is function of variables (α_i) that are unknown and to be determined. Because of the importance of the system, Neumann series are used to inverse the global stiffness matrix $[K]$:

$$[K]^{-1} = \left([K_0] + \sum_{i=1}^{n} \alpha_i.[K_i]\right)^{-1} = \left(\sum_{q=0}^{T}\left(-[K_0]^{-1}.\left(\sum_{i=1}^{n}\alpha_i.[K_i]\right)\right)^q\right).[K_0]^{-1} \tag{4}$$

Where T is the maximum order of the Neumann series. It must be chosen so great enough to ensure convergence of the series. So the inverse matrix of $[K]$ is a multivariable polynomial matrix (α_i $i=1, n$) of a degree T. It is a matrix whose elements are multivariate polynomials of degree T in n variables α_i. The Neumann series converges to the following condition:

$$\left\| [K_0]^{-1}.\left(\sum_{i=1}^{n}\alpha_i.[K_i]\right)\right\| < 1 \tag{5}$$

This indicates that the difference of stiffness, due to a reduction or an excess, must be small. After substitution of Equations (2) and (4), the displacement $w_S(x_0)$ is rewritten as follows:

$$w_S(x_0) = [N].\left(\sum_{q=0}^{T}\left(-[K_0]^{-1}.\left(\sum_{i=1}^{n}\alpha_i.[K_i]\right)\right)^q\right).[K_0]^{-1}.\{F(x_0)\} \tag{6}$$

Equation (6) defines an algebraic relationship with n unknowns (α_i). It is a multivariate polynomial equation of (α_i $i=1, n$) variables and of T degree. A set of m loading vectors ($\{F\}_i = \{F(x_0)\}$) can be developed corresponding to m positions of the moving load P. For these m loading vectors correspond m displacements of the selected point ($w_{S,i} = w_S(x_0)$). The data of the static displacement response thus allows the development of a system of m nonlinear equations (polynomial) and n unknowns, which is as follows:

$$f_d(\alpha_i)_{i=1,m} = w_{S,i} - [N].\left(\sum_{q=0}^{T}\left(-[K_0]^{-1}.\left(\sum_{i=1}^{n}\alpha_i.[K_i]\right)\right)^q\right).[K_0]^{-1}.\{F\}_i = 0 \tag{7}$$

Normal Stress in a given section point or fiber defined by its local coordinates y_L and z_L, is computed from the vector of the local element forces defined by its two ends nodes i and j.

$$\sigma_s(x_0) = \frac{N}{A} + \frac{M_y}{I_y}.z_L + \frac{M_z}{I_z}.y_L \tag{8}$$

The local solicitations vector is written as follows:

$$\{soll\} = [N_i, V_{yi}, V_{zi}, M_{xi}, M_{yi}, M_{zi}, N_j, V_{yj}, V_{zj}, M_{xj}, M_{yj}, M_{zj}]^T \tag{9}$$

$$\{soll\} = [K_{L,i}]\{\delta_{L,i}\} = [K_{L,i}][Tr_i][N_{L,i}]\{\delta_G\} = [K_{L,i}][Tr_i][N_{L,i}][K]^{-1}.\{F\}_i \tag{10}$$

$$\{soll\} = [K_{L,i}][Tr_i][N_{L,i}].\left(\sum_{q=0}^{T}\left(-[K_0]^{-1}.\left(\sum_{i=1}^{n}\alpha_i.[K_i]\right)\right)^q\right).[K_0]^{-1}.\{F\}_i \tag{11}$$

$[K_{L,i}]$, $[Tr_i]$ et $[N_{L,i}]$ are respectively, local stiffness matrix, transformation matrix from the global coordinate system to the local coordinate system of the corresponding finite element and the nodes' degrees of freedom extraction matrix of the corresponding finite element.

Thus, in the same way as equation (7), we can develop m nonlinear relations of corresponding stresses of the section ($\sigma_{S,i}=\sigma_S(x_0)$) and n unknowns, which is the following:

$$f_\sigma(\alpha_i)_{i=1,m} = \sigma_{S,i} - \sigma_S(x_0) = 0 \tag{12}$$

The correction coefficients resolve requires the solving of nonlinear equations system (7) or (12). It consists of an optimization problem. The Levenberg-Marquardt algorithm is the most efficient technique for solving this type of non-linear functional.

From this mathematical development, a Matlab code has been written for the inverse analysis of those displacement and/or stress responses on a finite element model: frame element with 6 degrees of freedom per node, and thus the computing of the correction coefficients (α_i). These coefficients allow the detection of the eventual damage rate or excess of stiffness in the structure's elements.

2.2. Experimental process and structure description

The structure to be studied is a small scale model of a steel bridge deck, composed of two chords (L40x40 x 4 mm^2), 7 floor beams (L40x40 x 4 mm^2), and two stringers (40 x 3 mm^2) (figure 1a), resting on two hinged supports and two movable supports, with a length of 2.4m, an width of 0.5m, gussets of 2.5 mm of thickness with several shapes and dimensions (in plan) are fastened by welding to the chords, the connections between chords and floor beams are bolted and those between floor beams and stringers are welded.

A tensile test was performed to measure the elasticity modulus of the system's steel material. It is equal to 210000 N / mm^2 with a Poisson coefficient of 0.30. A variable loading of 2x1kN was applied on 5 different longitudinal positions at the stringers–floor beams intersections.

Displacement measurements were performed at the ends of the floor beams by HBM inductive displacement sensors (WA-50mm-T) (figure 1b) [8]. The stress measurements were carried out by HBM strain gauges (LY11-10 / 120 ohms) on three points chosen on the chords near the mid-span (figure 1c) [9]. It was mounted in a half bridge circuit and using the multiple wires technology (5 wires) in order to increase the sensitivity of the signal and compensate for the effects of connection cables (voltage drop, zero drift) and temperature variation. A 4800 HZ carrier frequency power supply, shielded and properly grounded wiring were used to reduce electromagnetic noise and thermoelectric voltages according to Keil S [11] and Hoffmann K [12]. The data acquisition was done using the Quantum MX410 amplifier [10], the visualization and computer data processing by Catman software (HBM).

2.3. Structure modeling (Numerical reference model)

The finite element model is built by the frame finite elements with 6 degrees of freedom per node. It is performed by a Matlab code: 92 nodes, 81 frame finite elements, and 552 degrees of freedom (figure2).The model have two hinged supports (65; 86), two movable supports (71; 92) which are leaning on the brick walls (figure 1a). 28 Rigid links, are assigned to connection points to take into account the elements superposition like those between stringers and transvers beams or between main beams and transvers beams. In this numerical model, the existing gussets plates are neglected.

2.4. Static displacement and stress responses

Two parallel moving static loads of 1kN are applied at stringers–floor beams intersections. The loading step is 0,40m. Thus the displacement responses at the nodes (66; 67; 68; 69; 70; 87; 88; 89; 90 and 91) and stress responses at the nodes (15; 22 and 38) according to the different positions of the moving load (5 positions) have been recorded using displacement sensors and strain gauges: the total of 50 data from vertical displacements and 15 from normal stresses. The inverse analysis concerns a numerical model composed of 81 finite elements from which 81x4 = 324 unknowns or correction coefficients. The stiffness correction coefficients are presented with respect to cross section A and moments of inertia Ix, Iy and Iz.

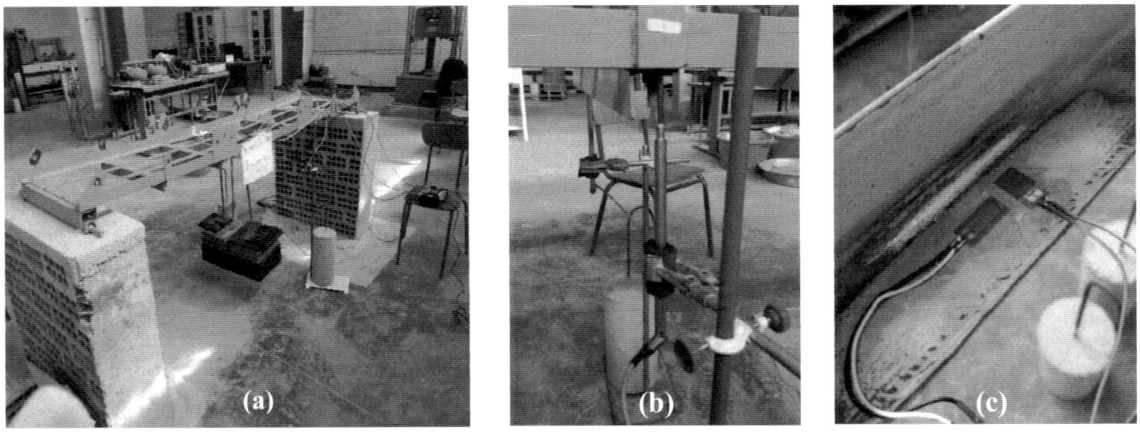

Figure 1. Experimental device for measuring displacements and stresses on the reduced model.

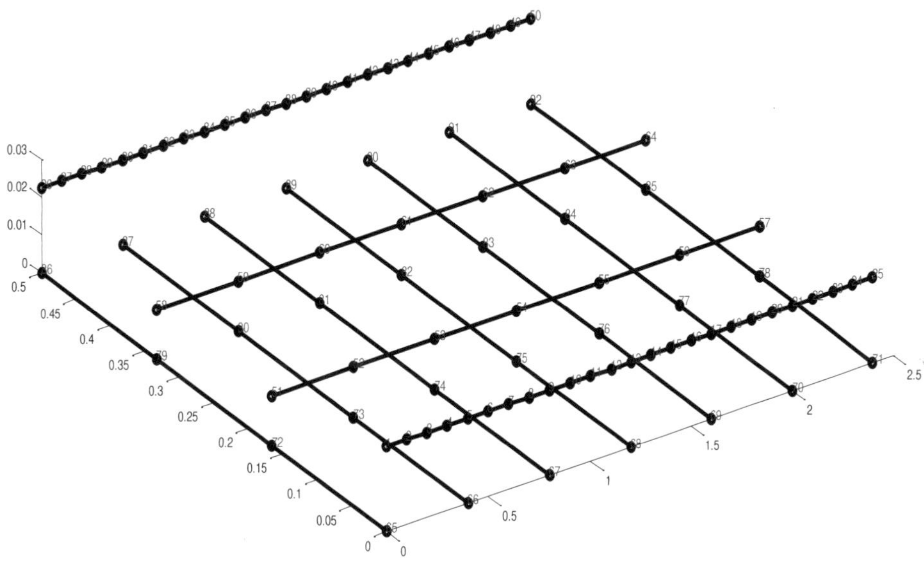

Figure 2. Numerical reference model of the structure (graphic presentation by Matlab).

3. Results and discussion

The inverse analysis to evaluate the real stiffness of the metallic system was made in three steps: that of displacements, stresses then displacements + stresses.

The results of the inverse analysis of the displacement responses recorded for different positions of the point loads, are illustrated in (figure 3).The axial stiffness with respect to the cross section A and the torsional stiffness with respect to torsional moment of inertia are variable between +0.2 and -0.23 in the chords (main beams), negligible on the stringers and quite variable between +0.00 and -0.23 for floor beams. For the flexural stiffness relative to the moment of inertia I_y, the correction coefficients of the main beams are close to 0.80 except finite elements close to the supports while the flexural stiffness corresponding to the moment of inertia I_z, their participations are insignificant. This is true considering the structural system's geometry which is horizontal and the loads are perpendicular thereto.

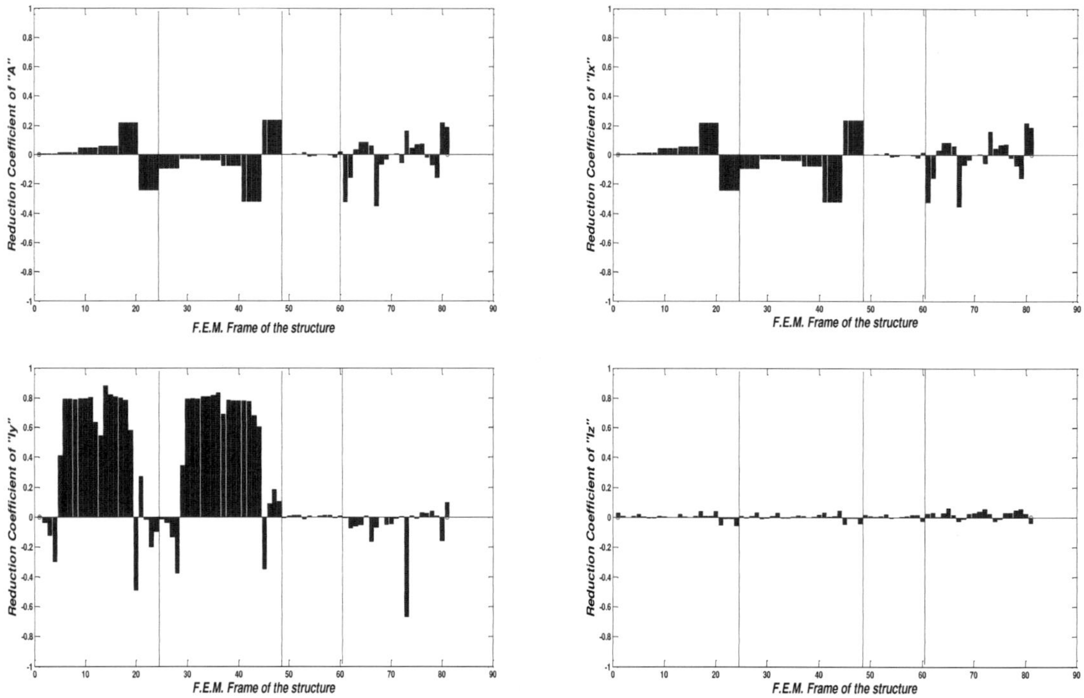

Figure 3. Inverse analysis results corresponding to the displacements, with respect to the cross section *A*, the torsional moment of inertia I_x and the flexural moments of inertia I_y, I_z respectively.

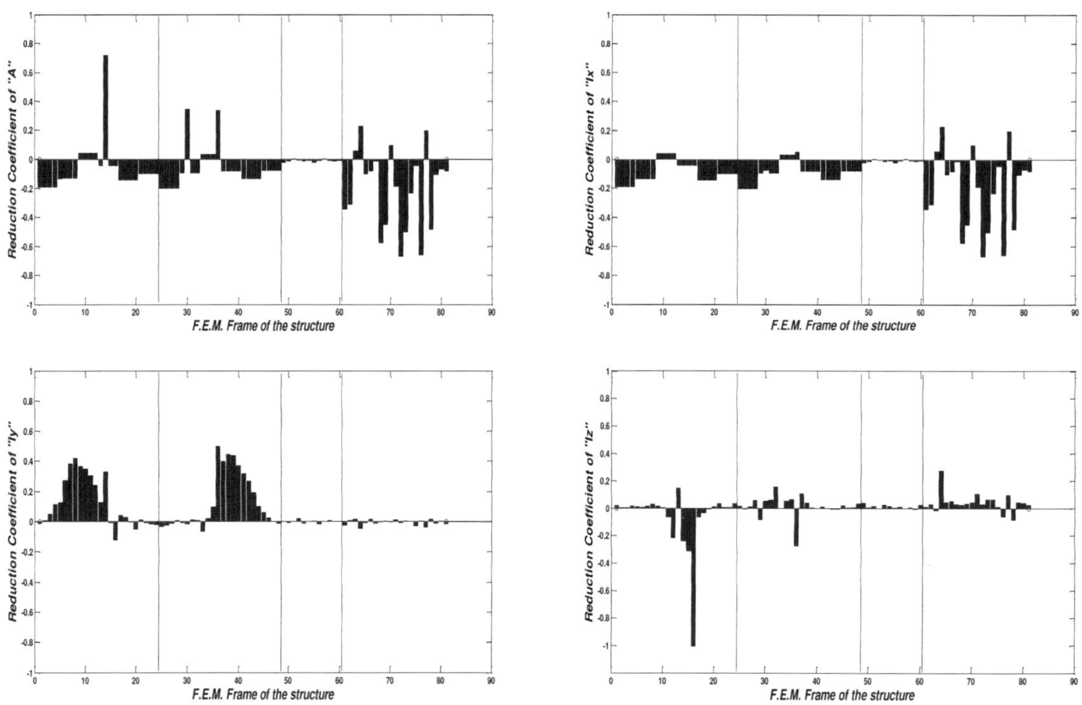

Figure 4. Inverse analysis results corresponding to stresses, with respect to the cross section A, the torsional moment of inertia Ix and the flexural moments of inertia Iy and Iz respectively.

Inverse analysis of the stress responses which is illustrated in, (figure 4) is roughly similar to (figure 3). The axial stiffness with respect to the cross section A and these relative to torsional moment of inertia are generally variable between 0.00 and -0.21 in the main beams, negligible on stringers and quite variable for the floor beams. For the corresponding flexural stiffness at the moment of inertia Iy, the correction coefficients are variable between 0.00 and +0.40 and lower when approaching the supports. For the flexional stiffness corresponding to the moment of inertia Iz, their participations are negligible.

When the displacement and stress measurements were taken together in the inverse analysis, the main remark is that the flexural stiffness /Iy of the longitudinal or principal beams are generally increased from + 0% to + 80% except those near the supports (figure 5). For axial and torsional stiffness, they vary from -23% to + 20%. The participation of the stringers is negligible and also the flexural stiffness / Iz.

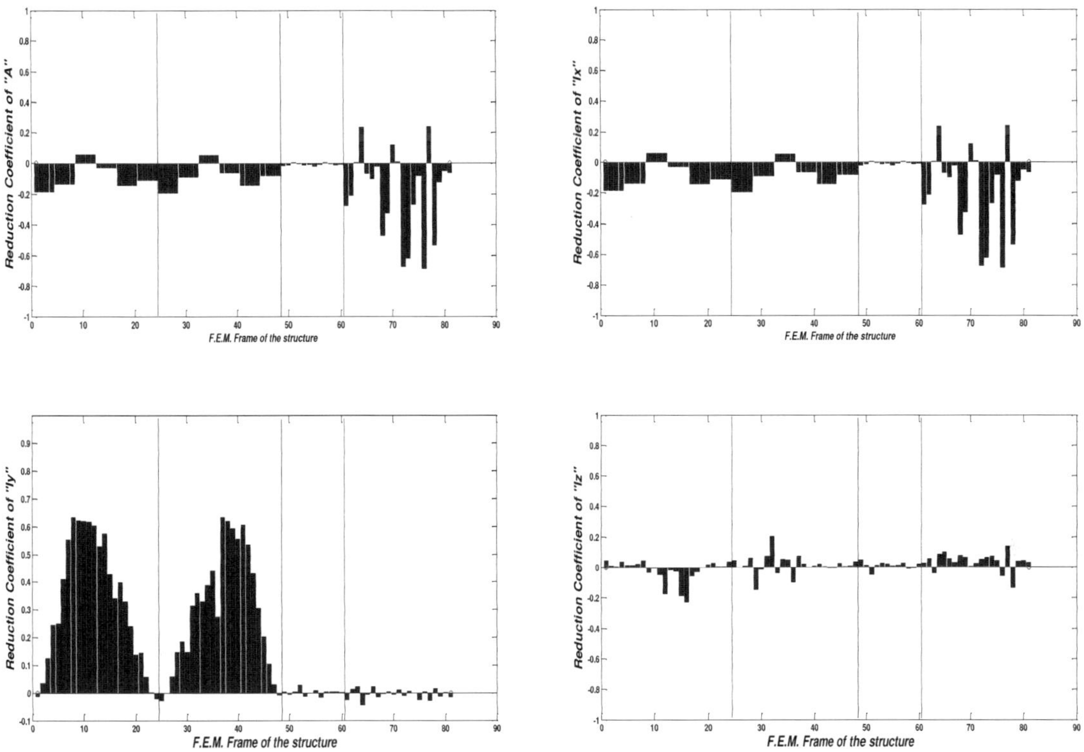

Figure 5. Inverse analysis results corresponding to the displacements + stresses, with respect to the cross section A, the torsional moment of inertia Ix and the flexural moments of inertia Iy, Iz.

It is worth noting that these results were based on the processing, by the IAFEM code, of 50, 15 and 65 pieces of information respectively to determine 324 unknowns or correction coefficients. It has been shown, in Boumechra's work [7], that it is necessary to have a greater number of information in order to have a better convergence.

Considering the design of the reduced model of the bridge deck studied in the laboratory and the results obtained by the developed inverse analysis code, it is noted that the addition of gussets plate on the main beams increased the flexural stiffness of this one. This result can be supported by a more approximate numerical model using plate or shell finite element (figure 6) and also frame finite element model with only the flexural stiffness correction in accordance with the results of the inverse analysis. The (Table 1) confirms that the method developed makes it possible to evaluate the real stiffness of any structural system.

The difference found between the numerical and experimental values of node 22's stress is due to the position of the strain gauge: very close to the bolt (4cm from the bolt) and therefore the effects of the stress concentration around the bolt's hole.

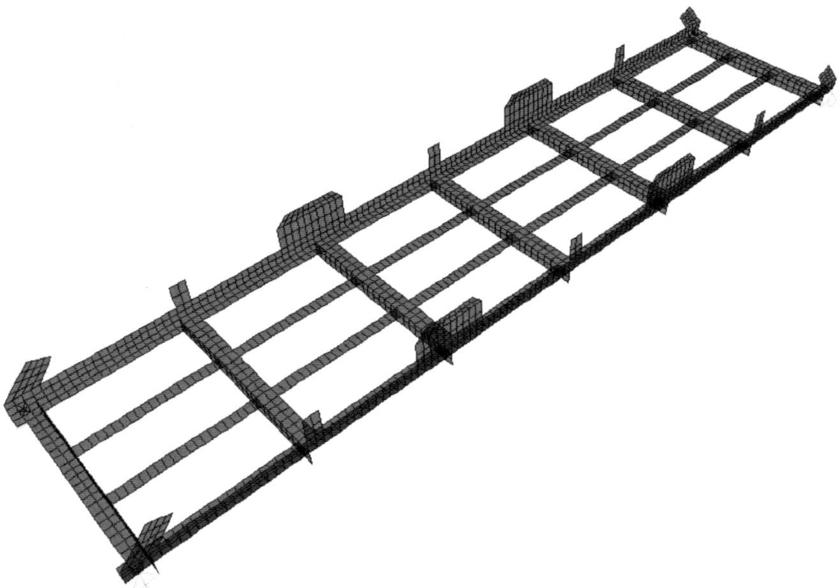

Figure 6. Plate finite elements model of the small scale steel bridge deck (Sap 2000).

Table1. Comparison between the different experimental and numerical models (Mid-span loading case).

	Experimental model	Plate finite element model	Frame finite element model (+80% chords stiffness)	Frame finite element model (without stiffness change)
σ (node 15)	57.99 N/mm²	57.73N/mm² (-0.5%)	60.8 N/mm² (+5%)	62.8 N/mm² (+8%)
σ (node 22)	96.75 N/mm²	117.15N/mm² (+21%)	108.4 N/mm² (+12%)	119.7 N/mm² (+23%)
σ (node 38)	33.7 N/mm²	31.202N/mm² (-7%)	34.7 N/mm² (+3%)	37.2 N/mm² (+10%)
δ_v (node 66)	9.5 mm	9.73mm (+3%)	10.0 mm (+2%)	14.4 mm (+47%)
δ_v (node 67)	16.27mm	16.76mm (+3%)	17.4 mm (+7%)	25.5 mm (+47%)
δ_v (node 68)	19.31 mm	19.91 mm (+3%)	20.1 mm (+5%)	30.0 mm (+56%)

4. Conclusion

Structure and its components' stiffness assessment is a method that may interest engineers to verify the conformity of any structural systems between those conceived in a design office and those determined by its real behavior or its response under different loading. It is also interesting to check numerical models or computation numerical approaches. The evaluation of stiffness also allows the engineer to detect possible damage on his structure. It will be more interesting to check this method on real bridges and under dynamic loads in the near future.

IOP Conf. Series: Materials Science and Engineering **419** (2018) 012024 doi:10.1088/1757-899X/419/1/012024

References

[1] Farrar CR and Worden K 2010 An introduction to structural health monitoring J. *New Trends inVibration Based Structural Health Monitoring*. ed Deraemaeker A and Worden (Vienna:Springer Vienna) pp 1–17

[2] Rytter A 1993 *Vibrational based inspection of civil engineering structures* Dept. of Building Technology and Structural Engineering (Aalborg University)

[3] Worden K and JM Dulieu-Barton 2004 An overview of intelligent fault detection in systems and structures *J. Structural Health Monitoring*. **3(1)** pp 85–98

[4] Fritzen CP and P Kraemer 2009 Self-diagnosis of smart structures based on dynamical properties. *J.Mechanical Systems and Signal Processing*. **23(6)** pp1830–45

[5] Doebling SW et al 1996 Damage identification and health monitoring of structural and mechanical systems from changes in their vibration characteristics: a literature review Los *Alamos National Lab* NM (United States)

[6] Friswell MI 2007 Damage identification using inverse methods *J.Philosophical Transactions of the Royal Society of London A: Mathematical, Physical and Engineering Sciences*.**365(1851)** pp 393–410

[7] Boumechra N 2017 Damage detection in beam and truss structures by the inverse analysis of the static response due to moving loads *J.Structural Control and Health Monitoring*.**24(10)** Doi:10.1002/stc.1972

[8] https://www.hbm.com/en/3059/wa-t-inductive-displacement-transducer-probe/

[9] https://www.hbm.com/en/4561/ly-linear-strain-gauges-with-1-measurement-grid/

[10] https://www.hbm.com/en/2133/quantumx-mx410b-4-channel-highly-dynamic-amplifier/

[11] Keil S 2017 *Technology and Practical Use of Strain Gages: With Particular Consideration of Stress Analysis Using Strain Gages* (John Wiley & Sons)

[12] Hoffmann K 2012 An introduction to stress analysis and transducer design using strain gauges *J.HBM Test and Measurement*.

IOP Conf. Series: Materials Science and Engineering **419** (2018) 012025 doi:10.1088/1757-899X/419/1/012025

Comparison of FEM Analysis And Experimental Investigation for A 168m Long Steel Double Twin Truss Launching Gantry Subjected To Different Load And Support Conditions

Doncho Partov [1], Lazar Georgiev [2]

[1] Prof. D-r Eng., Department of Mechanics and Mathematics, University of Structural Engineering & Architecture (VSU) "Lyuben Karavelov" - Sofia, 175 Suhodolska Street, 1373 Sofia, Republic of Bulgaria, partov@vsu.bg;
[2] Assoc. Prof. D-r Eng., Roads and Transport Structural Facilities Department, FTE, University of Architecture Civil Engineering and Geodesy, Hristo Smirnenski Blvd, Sofia 1046, Republic of Bulgaria, ldg_77@abv.bg, ; LAZAR_FTE@uacg.bg;

E-mail: ldg_77@abv.bg

Abstract. This paper describes the construction and behavior of a large steel double twin truss launching gantry with total length 168m, designed by Dr. Eng. Borislav Bankov (1912÷1992), used for erection of prestressed concrete bridge girders with span up to 60m (weight up to 220ton) along the Hemus highway, Republic of Bulgaria. It consists of two twin frame steel girders and rails on the top and on the bottom flanges for moving the special girder carrying devices. Experimental investigations are performed by the Road Agency and vertical deflections of the mentioned twin truss under different loads and support conditions are measured. 3D FEM of the existing steel twin truss launching gantry considering and neglecting the the increased axial stiffness of the lower chord at truss joint zones are developed by the authors to investigate its behavior under different load and support conditions. Results from these FEM are analyzed, compared with the test results and discussed..

1. Introduction.

At the 1980's the Hemus highway connecting Sofia and Pravets (Republic of Bulgaria) was constructed. It crosses hard for construction mountain terrain and high number of viaducts and tunnels were realized along the highway. Significant part of the viaducts were constructed with simply supported prestressed reinforced concrete (RC) bridge girder superstructures (GT) with continuous RC slab. The mountain terrain leading to the necessity of high piers (up to 120m for the Bebresh viaduct) and inconvenient terrain conditions under a lot of the bridge structures were critical for the choice of the erection technology. It was decided to build the bridge superstructures containing GT RC bridges using steel gantry cranes. For the erection of the 60m long GT prestressed concrete girders Dr. Eng. Borislav Bankov and his team in 1977 designed a unique steel double twin truss launching gantry with 168m length – Fig 1.1. It consists of two particular main spatial longitudinal parallel identical frame or trussed girders axially spaced in 5,6 m. Each of them has a total length of 168 m, 4,25 m in height, 1,2 m axially in width and its weight is of 210 t (Figs. 1.1, 1.2, 1.3). Every girder is composed from 15 separate prefabricated units (mounting blocks) in length - 13 inner uniform with a length of 11,0 m and a weight of 18 t and 2 outer non-uniform (the upper and the lower chords are inclined as a belaying cleat or as an fork arm) of 12,5 m long and of 21 t weight. The truss gantry moves on special trucks which can roll by 8 lower wheels on the two parallel cross double-track underlying railways – Fig 1.5. The rod splicing is made by weld junctions. Bolt-type joints are used within the units and also to link the adjacent units (Fig. 1.3, 1.4) on the side close to the bridge.

Figure 1.1. Double twin truss launching gantry used for positioning of the 220ton, 58m span RC girders of Bebresh viaduct along Hemus highway, Bulgaria [2].

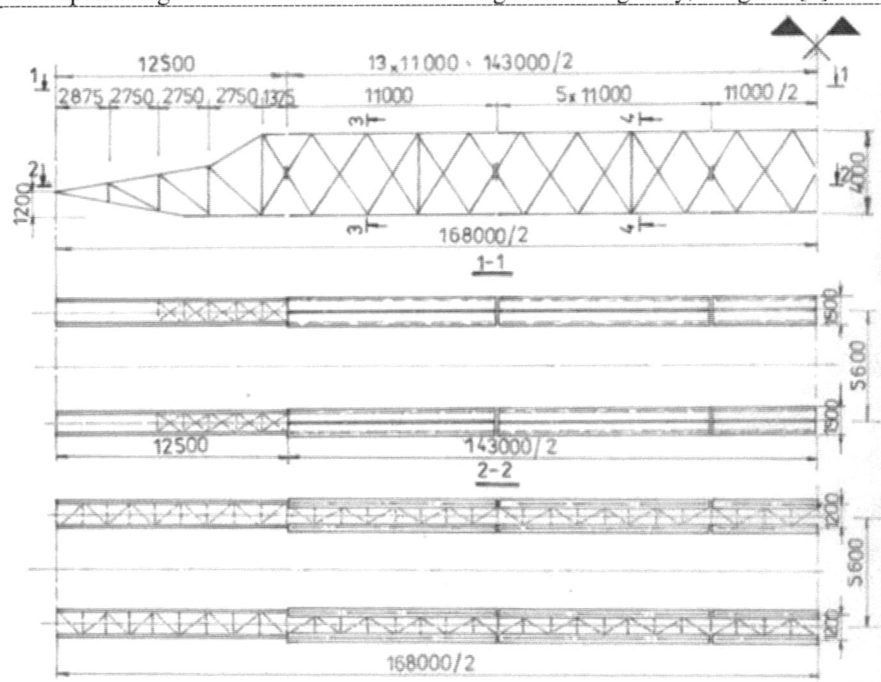

Figure 1.2. Double twin truss launching gantry schemes.

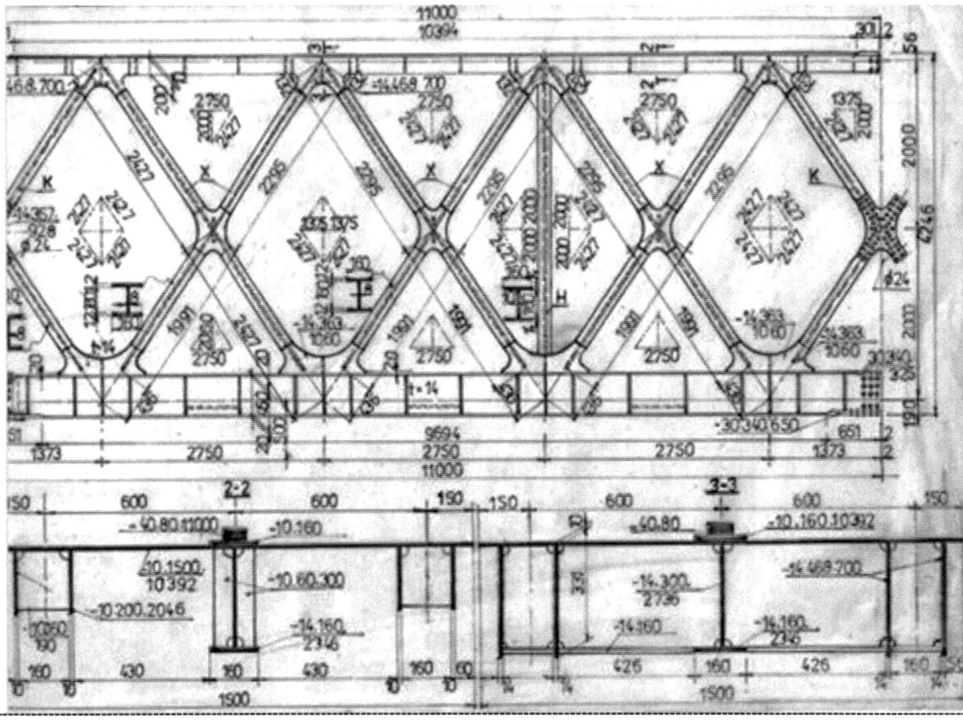

Figure 1.3. Longitudinal view of a fragment and cross section of the upper chord of the double twin truss launching gantry.

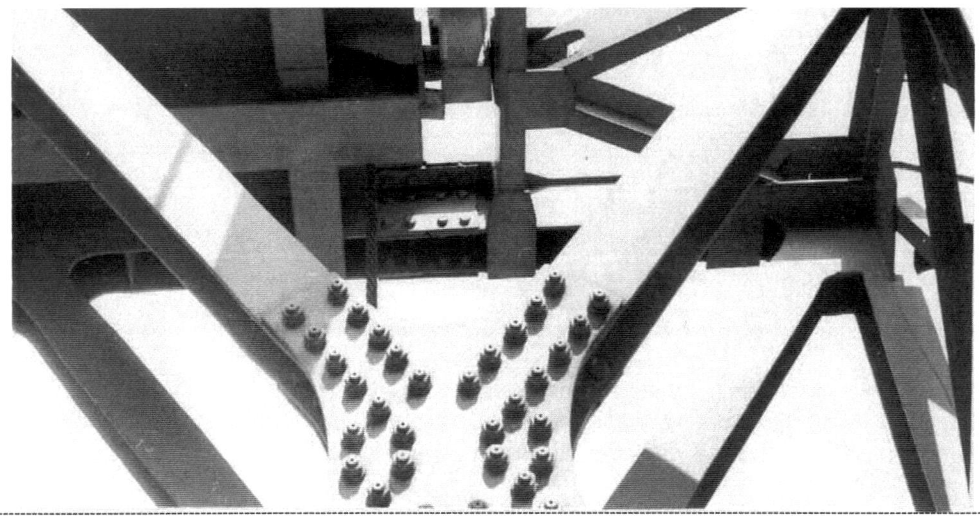

Figure 1.4. High-Strength Bolt joint between two mounting blocks of the gantry truss.

Figure 1.5. Double twin truss launching gantry – Bebresh viaduct [2].

2. Describtion of the developed fem and performed by the Bulgarian road agency tests of the analysed double twin truss gantry crane

Before using the gantry truss for transporting the 60m bridge girders with 220ton weight Bulgarian Road Agency[2] performed series of load tests to prove its load bearing capacity and compare the structural behavior with the project developed. Under load and support conditions in accordance with schemes SH 1.1, SH 2.1, SH 2.2 and SH 2.3 vertical displacements of the particular measuring points are measured by means of Maximov's deflectometer and with precise geodetic level measurer "KONI". The symbol of the vertical displacements considered in tests are as follows: f_p - total displacement, $f_{p,cor}$ - total displacement, corrected with supports settlement, f_{el} - elastic displacement, $f_{el,cor}$ - elastic displacement, corrected with supports settlement. The theoretical displacements according to the project are marked as f_t (neglecting the influence of transverse connections and section increase at the joints due to gusset plates).

Figure 2.1. 3D FEM of the double twin gantry truss with 168m length.

IOP Conf. Series: Materials Science and Engineering **419** (2018) 012025 doi:10.1088/1757-899X/419/1/012025

For the comparison of the results with the developed by the Bulgarian Road Agency tests and calculations a 3D FEM[1, 3-6] of the gantry truss is developed with SAP2000 software – Fig.2.1. The connection between elements is assumed as rigid.

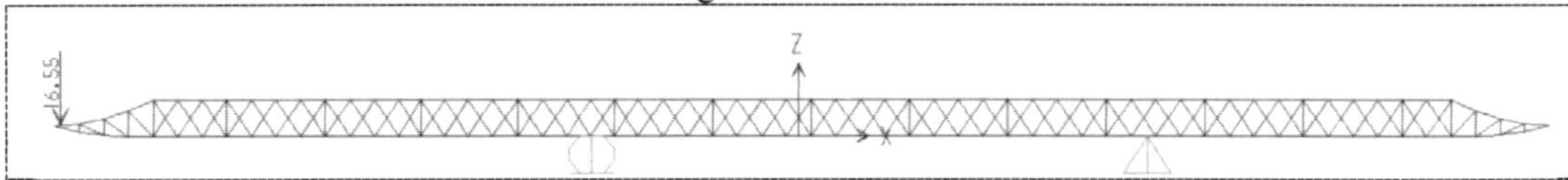

Figure 2.2. Support conditions and loads for scheme SH 1.1. (F=16.55kN).

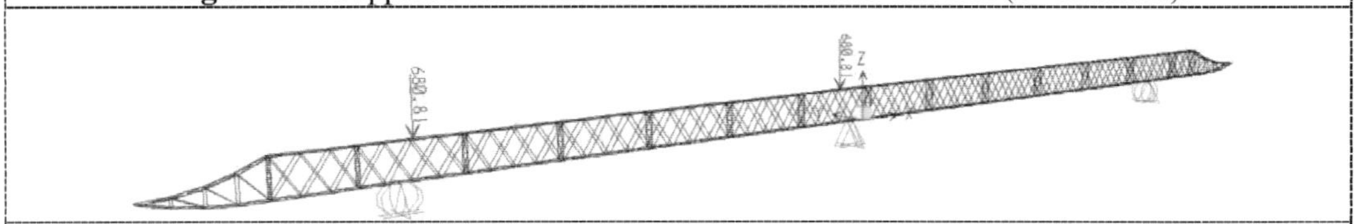

Figure 2.3. Support conditions and loads arrangement for scheme SH 2.1. (F=680.81kN).

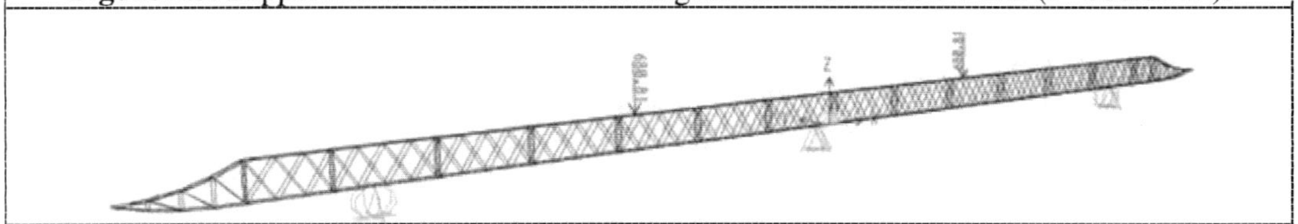

Figure 2.4. Support conditions and loads arrangement for scheme SH 2.2. (F=680.81kN).

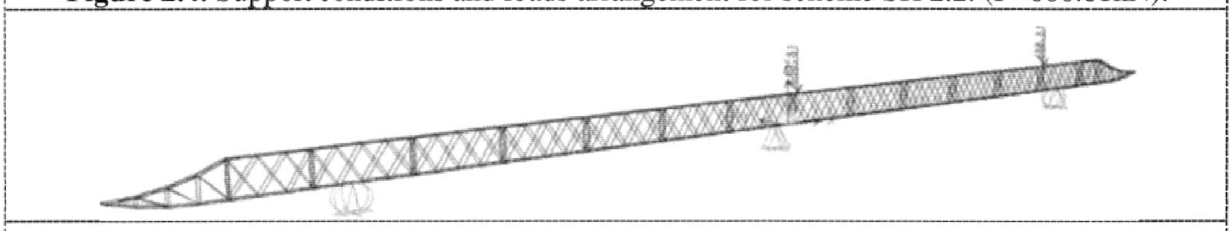

Figure 2.5. Support conditions and loads arrangement for scheme SH 2.3. (F=680.81kN).

For the FEM marked with index RJC the increased axial stiffness of the lower chord at truss joint zones resulting from gusset plate participation in the cross section is taken into account. For the 3D FEM this effect is considered using effective axial stiffness for the lower chord elements:

$$(EA)_{eff} / l_e = k_{eff} EA_e / l_e \tag{2.1}$$

$$k_{eff} = l_{tot} \left[l_1 \frac{A_e}{A_1} + \left(l_{tot} - l_1 - l_2 \right) + l_2 \frac{A_e}{A_2} \right]^{-1} = 1{,}05136 \tag{2.2}$$

Where $l_e = 2750mm$ is the axle distance between lower chord truss joints signed as 1 and 2, $A_e = 22200mm^2$ is the cross sectional area of the lower chord, $l_1 \sim 400mm$ and $l_2 \sim 400mm$ are the length near truss joints 1 and 2 where the area of the lower chord cross section is increased to $A_1 = 26680mm^2$ and $A_2 = 26680mm^2$ respectively. Cases with E=205GPa, E=210GPa and E=215GPa are considered in RJC FEM models for the determination of the vertical displacements.

3. Comparison of the results

At Fig 3.1 and Fig 3.2 and Tab.3.2 are presented measured and calculated vertical displacements for SH 1.1 (SH 1.1 3D FEM are for 3D frame model with rigid joints and E=210GPa, SH 1.1 3D FEM RJC E210GPa is the model considering the axial stiffness of the lower chord at truss joint zones resulting from gusset plate participation in the cross section and SH 1.1 3D FEM RJC E215GPa SH 1.1 3D FEM RJC E205GPa are for the same model but with E=215GPa and E=205GPa) of the truss in accordance with load and support scheme SH 1.1. Results for the fixed detecting points are presented in Table 3.1. As it is seen the difference between SH 1.1: 3D FEM RJC E210GPa model and the axial $f_{el,cor}$ is only 2,9% which is a good agreement. For smaller as an absolute value displacements differences are higher up to 6,4%. As it is seen from Fig3.2 the difference between 1.1: 3D FEM RJC E205GPa model and the measured $f_{el,cor}$ is 0,6% which is negligible.

At Fig 3.3, Fig3.4 and Fig 3.5 are presented vertical displacements measured and calculated for SH 2.1, SH 2.2 and SH 2.3 respectively. It is seen that for SH 2.1 and SH 2.3 where the absolute values of vertical displacements are relatively small (under 20mm) the difference between $f_{el,cor}$ and SH 2.i: 3D FEM for E210GPa (i=1;3) is relatively high.

For SH 2.2 as it is seen from Fig.3.4 the vertical displacement diagram f_t is symmetric which do not correspond to the unsymmetrical support conditions. For measuring points 2 and 5 the difference between $f_{el,cor}$ and SH 2.2: 3D FEM RJC (E210GPa) is 0,04÷5,65% and the difference between $f_{el,cor}$ and SH 2.2: 3D FEM RJC (E210GPa) is 1,4÷7,3% - Tab 3.1. As it is expected the results from SH 2.2: 3D FEM RJC fits slightly better to the measured in the test displacements but the difference between is SH 2.2: 3D FEM RJC and SH 2.2: 3D FEM low – 1,4÷1,5%.

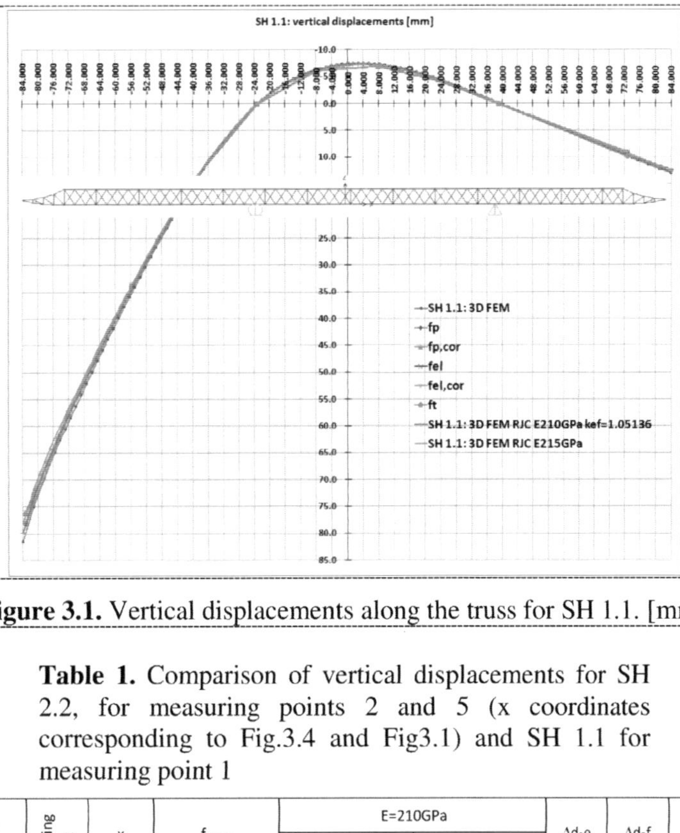

Figure 3.1. Vertical displacements along the truss for SH 1.1. [mm].

Table 1. Comparison of vertical displacements for SH 2.2, for measuring points 2 and 5 (x coordinates corresponding to Fig.3.4 and Fig3.1) and SH 1.1 for measuring point 1

SCHEME	Measuring point	x	$f_{el,cor}$	E=210GPa		Δd-e	Δd-f	Δf-e
				SH 2.2: 3D FEM	SH 2.2: 3D FEM RJC			
		[m]	[mm]	[mm]	[mm]	[%]	[%]	[%]
a	b	c	d	e	f	g	h	i
SH 2.2	2	-31.080	25.30	25.67	25.31	1.4	0.04	1.41
SH 2.2	5	29.420	28.20	30.25	29.79	7.3	5.65	1.52
SH 1.1	1	-83.240	76.00	79.73	78.20	4.9	2.90	1.96

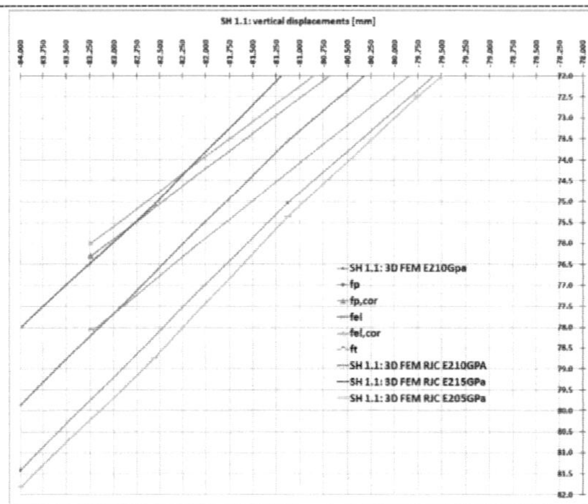

Figure 3.2. Vertical displacements for SH 1.1. at the end of the external load application [mm].

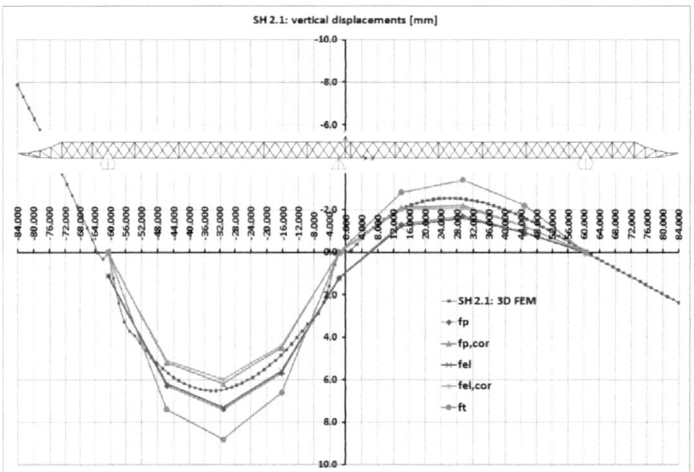

Figure 3.3. Vertical displacements along the truss for SH 2.1. [mm].

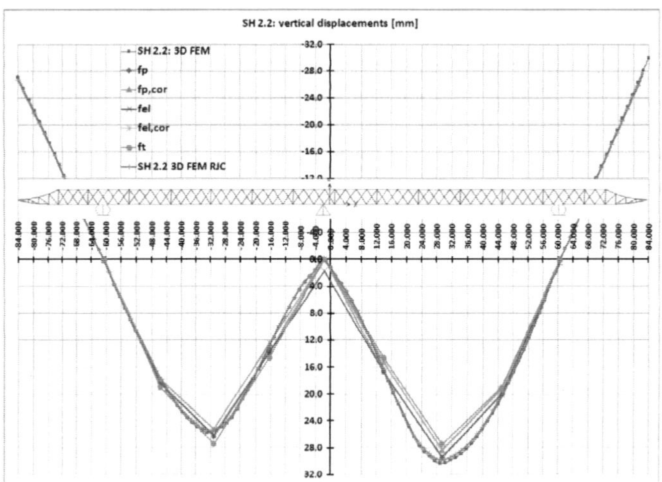

Figure 3.4. Vertical displacements along the truss for SH 2.2. [mm]

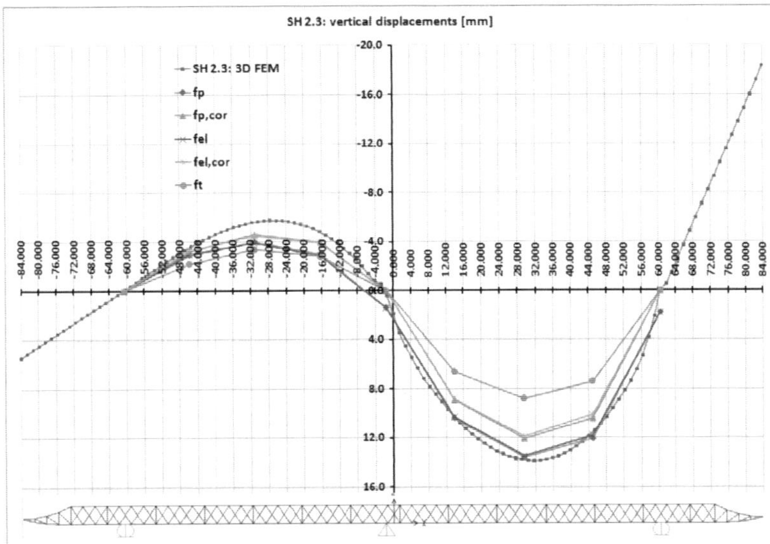

Figure 3.5. Vertical displacements along the truss for SH 2.3. [mm].

4. Conclusion

Described unique steel double twin truss launching gantry served for placing large number of RC prestressed bridge girders for bridges along Hemus highway, Republic of Bulgaria. Its structural behavior under different loads and support conditions (SH 1.1, SH 2.1, SH 2.2 and SH 2.3) are investigated in this paper and compared with corresponding test results obtained by the Bulgarian Road Agency tests. The influence of a slight difference of the modulus of elasticity (205GPa, 210GPa, 215GPa) and increased axial stiffness of the lower chord at truss joint zones resulting from gusset plate participation in the cross section are investigated. For 3D FEM with E=210GPa the difference between vertical displacements considering and respectively neglecting the increased linear stiffness of the lower chord at truss joint zones resulting from gusset plate participation for investigated points varies between 1,5÷2,0% which is practically negigible. For vertical displacements in SH 2.1 and 2.3 because of the low absolute values the percent difference between measured elastic vertical displacements and calculated with different 3D models are significant. For SH 1.1 and SH 2.2 the measured corrected with support settlements elastic vertical displacements and calculated with 3D FEM models are in good agreement (about 0÷6% percent difference). The conclusion is that developed 3D FEM considered correctly with relevant accuracy the structural behavior of the double twin truss launching gantry. This launching gantry designed by D-r Borislav Bankov is a part of the Bulgarian bridge cultural heritage from the history of the Hemus highway building but also after adequate rehabilitation it can be used in current practice.

References

[1] Bathe, K.J., "Finite Element Procedures" Prentice Hall New Jersey, 1996
[2] Hristov,G., "Static tests for gantry crane 220/65 for installation of 58m long bridge girders", Bulgarian Road Agency, Sofia, 1983
[3] Kazakov, K., „Finite Element Methods for Structural Analysis, University publishing House, VSU "L.Karavelov", 3rd Edition, Sofia, 2011
[4] Němec,I., Kolař,Vl., Ševčik, I., Vlk, Z., Blauwendraat, J., Buček, J., Teplý, B., Novák,D., Štembera, V., Finite Element Analysis of Structures, Shaker Verlag, 2009
[5] Ravinger. J., Numerical methods in Theory of Structures, SNTL, Bratislava, 2014
[6] Spyrakos, C. C., Raftoyiannis, J., Linear and Nonlinear Finite Element Analysis in Engineering Practice. Algor Inc., Pittsburgh, PA, 1997

IOP Conf. Series: Materials Science and Engineering **419** (2018) 012026 doi:10.1088/1757-899X/419/1/012026

The residual lifetime of steel bridges under the action of fatigue and corrosion effects.

Martin Macho[1], Pavel Ryjáček[1] and José António Campos e Matos[2]

[1]Faculty of Civil Engineering, Czech Technical University in Prague, Thákurova 7/2077, Prague, 166 29, Czech Republic

[2]Department of Civil Engineering, University of Minho, Campus de Azurém, 4800-058 Guimarães, Portugal

E-mail: martin.macho@fsv.cvut.cz

Abstract. Decades-old steel bridges have to face the unfavourable environment that causes the ageing and deterioration of material properties. Moreover, especially railway bridges, which are famous for permanent resistance to heavy movable load, are exposed to large number of stress ranges that may lead to fatigue damage. Combination of both phenomenon has significant impact on the condition of bridges and their remaining service life. The subject of the proposed paper is observation the response of steel bridge structures containing construction elements weakened by corrosion on cyclic load. For the purpose, the series of laboratory fatigue tests were performed on samples of riveted connections. Specimens with various level of corrosion weakening had been taken from historical railway bridge and were loaded by cycling force in tension subsequently. The laboratory tests were supported by numerical analysis. This paper presents knowledges gained from the study that was focused on how to take into account the combined effect of degradation process and fatigue. The evaluation of the tests indicates that the service life of members may be significantly reduced due to fatigue. The developed relationship between corrosion loss and reduced detail category is stated.

1. Introduction

Metallic bridge structures built many decades ago represent historical and cultural significant structures. Nevertheless increasing demands on transportation in combination with unfavourable environment have negative effects especially on the state of existing railway steel bridges. Civil engineers reassessing such existing structures face several sources of doubt, especially if the structure is deteriorated by a combination of corrosion and fatigue load.

The effects of corrosion weakening on the behaviour of steel elements has been the subject of many research projects. A set of corroded steel riveted stringers was subjected to laboratory tests to find the influence of material loss on the remaining bearing capacity of these bridge structure elements [1], [2]. In several cases [3], [4], the influence of corrosion on the fatigue behaviour of corroded steel plates was investigated. The origin and propagation of fatigue cracks on corroded elements were also studied [5]. The response to cyclic loading of steel webs and flanges with material loss due to corrosion was the topic of another study [6]. In the case of riveted connections, the effect of rivet head corrosion on the pre-stressing of the rivet was investigated by laboratory tests and numerical simulations [7]. From the point of view of fatigue, the shear category of rivets for riveted shear splices is considered to be category 100, which is equivalent to splices jointed with non-preloaded bolts [8].

The studies mentioned above were based largely on the behaviour of simple steel plates. The research presented in our paper focuses on elements taken from real structures that had been in service for more than 100 years. The experimental tests performed on these samples best depict the behaviour of the kinds of structures that civil engineers usually meet with. Moreover, the material loss of all tested samples is a result of the natural corrosion process, so the shape of the corroded surface and the irregularities fully corresponds to reality.

The paper presents observations from laboratory tests of riveted bridge components and remaining fatigue life, in the presence of natural corrosion weakening at various levels and with various distributions.

2. Laboratory tests on fatigue

Tested specimens for fatigue tests were taken from the bottom flange of the truss girder of the old railway bridge in Holubov (see Fig. 1). The samples were chosen to focus on details susceptible to fatigue and, at the same time, to focus on parts that will be deteriorated by corrosion. In addition, the samples were chosen from parts where the impact of the cumulated damage is negligible or below the limit of the permanent fatigue life. The aim of the experimental tests was to find how the remaining fatigue life of the steel bridge components is affected by various levels of corrosion weakening.

2.1. Bridge description

The Holubov bridge is a steel truss riveted railway bridge with an upper deck that was built in 1892 (see Fig. 1). From the static point of view, the structure acts as a continuous beam with 2 spans (2x41,9=83,30 m). As was mentioned above, this structure was chosen because of its age and because there was significant corrosion damage. In addition, the bridge was under reconstruction.

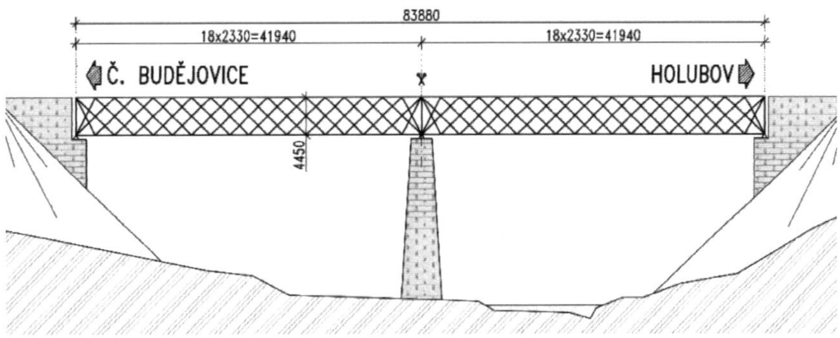

Figure 1. Holubov Bridge.

2.2. Tested samples

All specimens for laboratory tests were cut from the bottom flange of the riveted truss girder. The most significant corrosion weakening was found on the flange angles connecting the flange to the web. For this reason, the riveted joints of the sheets forming the bottom flange with the angle mentioned above were chosen as the experimental specimens (see Fig. 2). The samples were cut manually by a band saw. The final precise shape was done by milling. Due to this procedure, the temperature influence of samples was avoided.

Figure 2. Tested sample (photo).

The size of the tested samples is based on the geometry of flange angle L100x100x12 and position of rivets. Each specimen consists of 4 steel sheets 12+9+9+8 mm in thickness, which are connected by 3 rivets with a shank diameter of 20 mm. The number of 3 rivets was selected with respect to adequate total length of specimen. The axial distance between rivets is 116,5 mm, and the total length of the tested specimen is 626 mm. The width of the specimens is variable, and changes from 80 mm at both ends to 60 mm in the middle of the span. The total number of specimens was 13. All tested samples were divided into 4 groups on the basis of the corrosion level of the upper sheet. The corrosion level was determined as the change in cross-section area due to uniform corrosion in the place of the first rivet, and can be determined according to Eq. (1):

$$d_A = \frac{\Delta A}{A_0} = \frac{A_0 - A_1}{A_0} \quad [-] \tag{1}$$

- where: A0 is the cross-section area of the original uncorroded sheet in the position of the 1st rivet; A1 is the cross-section area of the corroded sheet in the position of the 1st rivet.

The general characteristics of each group are as follows:

- 1st group: Without corrosion - samples: S1.1, S1.2, S1.3, S1.4
  - characteristics: no measurable corrosion weakening, a change in surface colour only;
- 2nd group: Medium level of corrosion - samples: S2.1. S2.2, S2.3
  - characteristics: a medium level of corrosion loss - especially the upper sheet and the heads of the rivets on the upper surface, medium deep (0,5-1,5 mm) corrosion pits on the upper sheet; a range of corrosion thickness loss of about 1-2 mm;
- 3rd group: High level of corrosion - samples: S3.1. S3.2, S3.3
  - characteristics: a high level of corrosion loss - especially the upper sheet and the heads of the rivets on the upper surface, deep corrosion pits (2-3 mm) on the upper sheet; a range of corrosion thickness loss from about 2-4 mm;
- 4th group: Samples with missing rivet - samples: S4.1. S34.2, S4.3
  - characteristics: no measurable corrosion weakening, a change in surface colour only, the middle rivet is missing.

All tested specimens were sandblasted to remove the corrosion layer before installing the measurement devices. The corroded surface of samples was sandblasted to the smoothness Sa2,5 by the very fine siliceous sand. In order to prevent any impact on the fatigue behaviour, the very fine siliceous sand instead of traditional steel balls was used with the small jet device. The sandblasting of surface was necessary to do because of the installation of strain gauges and to measure the real geometry and observe the cracks on the surface. Subsequently, the photogrammetry method with sufficient accuracy ± 0,5 mm was performed to obtain a 3D model and to find the exact geometry, including corrosion pits, rivet damage and other irregularities. The depth of the corrosion pits was in the range of 0,5 – 1,5 mm, and for the samples with the highest level of corrosion, the depth reached a maximum value of 2 – 3 mm.

A preliminary numerical models of tested specimens have been created to see the behaviour of samples and find out the optimal values of loading force (see Fig. 3). Models were created in Dlubal RFEM software. All plates were modelled as 3D solid members, except fixing device.

A total of 8 small samples were taken from the bottom flange of the truss girder to find the material properties of the steel. According to the results, the material of the bridge components is a mild steel with an average yield strength value of 263 MPa and an average modulus of elasticity value of 207,8 GPa.

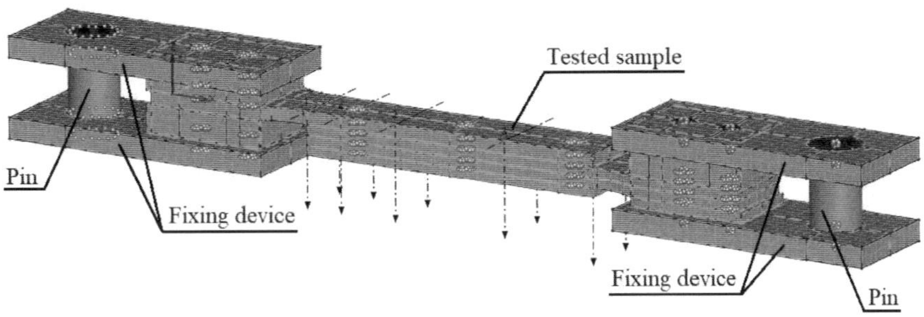

Figure 3. Numerical model of tested sample.

2.3. Indicators measured during the tests
All tested samples were divided into two groups. The first group of experimental specimens contained 2 samples that were equipped with a larger number of measurement devices. A total of 10 strain gauges were arranged on each of them to measure the relative strain during the fatigue tests (Fig. 4). The measured data will help to verify the numerical models of the tested samples. Each of these tested specimens was also equipped with a couple of displacement sensors, which measured the movement of individual sheets.
HBM strain gauges were used (1-LY11-1.5/120, 1-LY11-10/120) together with displacement inductive sensors (LVDT) Micro Epsilon of type DTA-10C-CA with nominal displacement ± 5 mm.

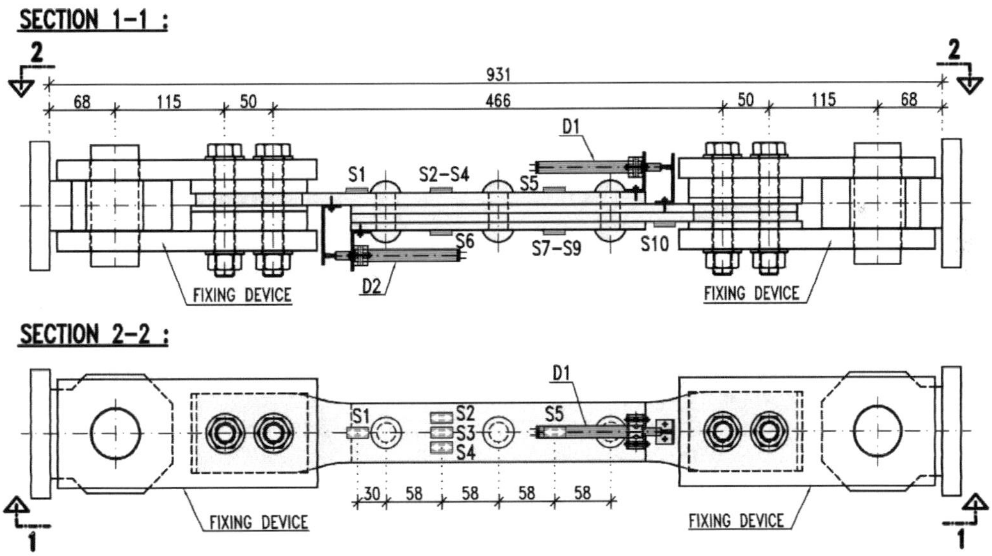

Figure 4. Tested specimen except the sensors and the fixing device (drawing).

All the other specimens were categorized as 2nd phase samples. Unlike the previous group, in this case just 2 strain gauges were installed in the middle of the gap between rivets on the bottom and upper side of the sample (S1+S2). The number and the position of the displacement sensors correspond fully to the previous case.

2.4. Process of the fatigue tests
The tested construction detail was considered to be a bending detail category 71 [9]. The tested sample simulated the connection of the truss members, for example: the connection of the bracing diagonal and the main girder, or the connection of the diagonal and the web of the flange of the truss riveted girder.

Due to unsymmetrical specimen there was a small effect of bending moment that caused additional normal stresses. However, the shape of testes samples was chosen to minimize this effect of bending moment. For this reason the bottom plate was shorter. Also the fixing of samples was designed to minimize the additional stresses caused by bending. As the result, the whole cross-section of sample is loaded by tensile stresses.

It was assumed that a crack occurs on the upper single sheet in the cross-section containing a hole for a rivet. The fatigue test was considered to be ended when:

1. a fatigue crack initiated and had grown to the critical length – in our case, the critical length was considered to have been reached at the moment when the tested sample failed,
2. there was a large displacement (found by the displacement sensors), or
3. when the rivet was damaged.

The values of measured data (relative strain and displacement) were observed continuously in a real time and recorded in period of each 10000 cycles in the phase before the crack initiation and then continuously until the collapse of sample occurred. The frequency of record was set to 100 Hz. Type of the measuring device was Dewetron Dewe 5000.

A special device was manufactured for fixing the samples during the experimental fatigue test (see Fig. 5).

Figure 5. Arrangement of the fatigue test.

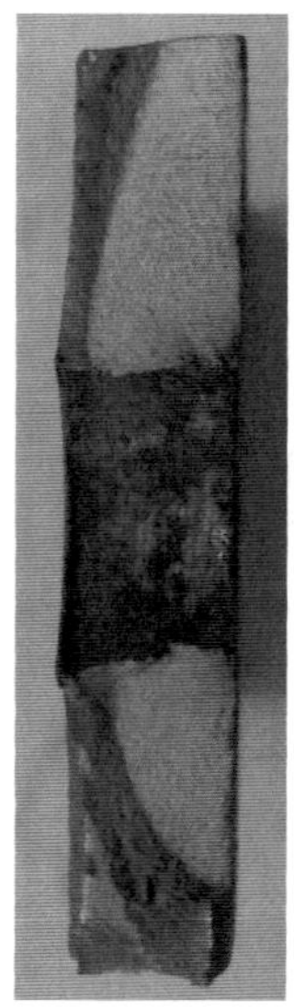

Figure 6. Fatigue damage to tested sample.

2.5. *Evaluation of the tests*

The values of the stress range were almost the same for all samples - nearly 180 MPa. The assumption of the bending detail category for an uncorroded tested sample was 71 MPa, which corresponds to 123000 cycles for stress range 180 MPa. The measured number of cycles was higher for uncorroded samples, and the ratio of measured-to-expected was equal to 1,5-2,0. The number of cycles for samples with a medium level of corrosion and with a high level of corrosion was also variable, but was generally lower. The ratio for medium level was: 1,3-1,8, and for high level the ratio was: 0,5-1,5.

The failure mode of the tested samples was as had been assumed. The first crack in all cases occurred in place of hole for the first rivet, in weakened cross-section area. The crack propagation was in direction perpendicular to main stresses. The samples became damaged after the crack reached its critical length and the weakened cross-section was not able to transfer the loading force. An example of a damaged specimen is shown in Fig. 6.

Figure 7 summarizes the number of cycles that were needed for the initiation of a crack, the number of cycles after which damage to the samples occurred, and the number of cycles that were necessary for crack growth. Each column shows the average number of cycles for one set of specimens (1 – samples without corrosion, 2 – samples with a medium level of corrosion and 3 - samples with a high level of corrosion). It can be seen that there is a downward tendency in the number of cycles for crack initiation. The number of cycles for crack initiation decreases slightly with the increase in corrosion weakening. However, this effect is not so evident for individual samples, because of the fluctuation in the numbers in each set. Because of the small dimensions of tested samples, the number of cycles for crack initiation prevails over the number of cycles for crack growth.

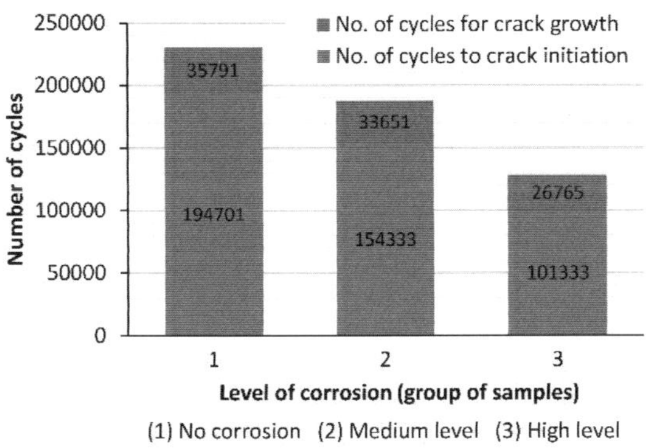

Figure 7. Number of cycles for crack initiation and crack growth.

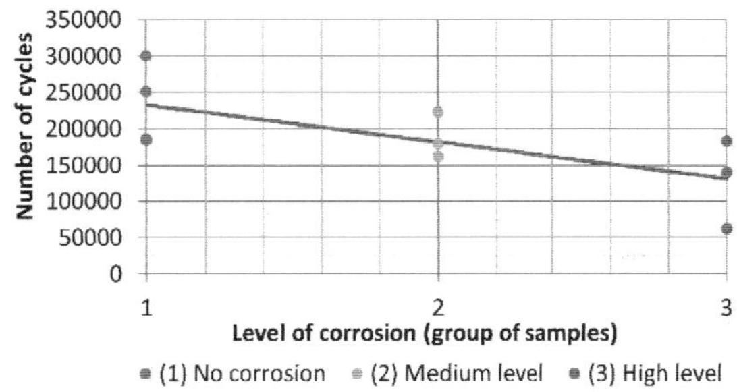

Figure 8. Dependence of the number of cycles on the level of corrosion.

This type of laboratory test was performed mainly to find out whether or not there is any change in the bending detail category when there is corrosion weakening. In most cases, corrosion weakening consists of a uniform part and an irregular part. The irregular part means a locally increased depth of corrosion in the form of corrosion pits. The results of the test are shown in Fig. 8 in the form of the relationship between the number of cycles and the set of specimens. There is a downward trend in number of cycles, accompanied by an increasing level of corrosion.

Table 1. Bending detail category of the tested samples

No.	Sample	Cycles Ni	Bending detail category $\Delta\sigma_{c,R}$ [MPa]		
1	S1.1	184452	79.2	average:	95% quantile:
2	S1.2	251053	84.5		
3	S1.3	186320	77.7	82.8	75.0
4	S1.4	300140	89.6		
5	S2.1	179633	77.7	average:	95% quantile:
6	S2.2	161769	75.4	78.9	73.3
7	S2.3	222551	83.6		
8	S3.1	139313	70.9	average:	95% quantile:
9	S3.2	62003	55.6	67.9	52.9
10	S3.3	182980	77.3		
11	S4.1	104000	62.0	average:	95% quantile:
12	S4.2	171843	72.5	76.3	54.1
13	S4.3	313553	94.5		

IOP Conf. Series: Materials Science and Engineering **419** (2018) 012026 doi:10.1088/1757-899X/419/1/012026

The measured number of cycles until the failure was higher than the assumed number, which means that the real bending detail category must also be higher. The real parameters (geometry, load) were therefore considered for a calculation of the real bending detail category of the tested riveted joint (see Table 1). Detail category was based on the number of cycles that corresponded to total failure of the sample. In this case, too, a downward tendency can be observed – the higher the corrosion loss, the lower the bending detail category, in terms of average values. The evaluation was based on the fatigue curve taken according to [9].

Neglecting tested sample S3.2, the bending detail category value of which was deviated most from the other values, and using regression analysis, the dependence between the category of the detail and the change in cross-section area can be expressed by a linear relation as (2) and (3) (see also Fig. 9):

$$\Delta\sigma_{c,R} = \Delta\sigma_c \cdot c_R \qquad [MPa] \qquad (2)$$

$$c_R = (1 - 0{,}4874 \cdot d_A) \qquad [-] \qquad (3)$$

- where: $\Delta\sigma c,R$ is the reduced bending detail category; $\Delta\sigma c$ is the original bending detail category of an uncorroded sample; dA is the level of corrosion determined as the change in cross-section area for members loaded by axial force. In case of members loaded by bending moment or loaded by combination of bending moment and axial force, the level of corrosion is considered as change in cross-section area of assessed part of member.

Coefficient dA takes into account the change in the cross-section area as a consequence of the uniform part of the corrosion. The influence of irregularities of the corroded surface are not included in this calculation. However, the authors are currently working on this issue, and the effects will be stated in a later stage of our investigations. Surface irregularities may have negative effects on the course and the distribution of the stresses in a member. If the corroded surface is rough, the influence of irregularities is higher than in case of smooth surface. Then the initiation of fatigue crack might become earlier, after lower number of cycles.
Since the stated formulas are derived from the experimental tests introduced above, their applicability is limited to narrow members, i.e. to steel plates not more than 60 mm in thickness.

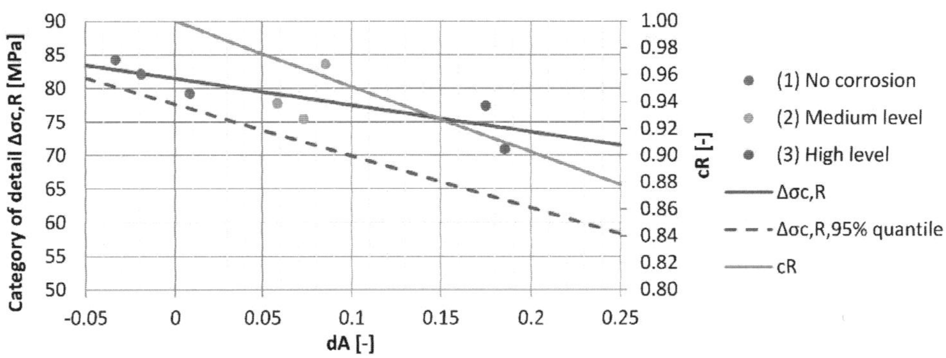

Figure 9. Relation between bending detail category and level of corrosion weakening.

To find out the accuracy of the formula that has been developed, the Root Mean Square Error (RMSE) was used to calculate how much error there is between the predicted values and the observed values. For the bending detail category the RMSE = 0,527 [-]. This error is due to the dispersion of the results from the experimental test, together with the fact that the combined effects of corrosion and fatigue phenomena are very complex. The value is therefore acceptable.

3. Conclusions

Laboratory tests on real steel bridge components led to several significant conclusions:

The bending detail category of tested samples without corrosion weakening corresponds to category 75 MPa in the 95% quantile. This means that the assumption of use category detail 71 was on the safe side. In terms of the number of cycles, the measured number is 1,5 to 2 times higher than the number expected for category 71.

Degradation in the form of corrosion has a negative impact on the fatigue behaviour of steel members. The irregularities of the corroded surface cause changes in the stress distribution. Fatigue cracks generally occur earlier, and the fatigue life is shorter for the same value of the nominal normal stress, in comparison with uncorroded elements.

The change in the length of the fatigue life can be described by the change in the fatigue curve in dependence on the corrosion level. The relation between the bending detail category and the corrosion loss may be expressed by a linear function, and the new detail category for narrow members can be calculated using reduction coefficient cR. However, it is not always the rule that the higher the level of corrosion, the lower the fatigue life. The roughness of the degraded surface also plays an important role, but is not included in the cR factor.

If the corrosion phenomena investigated here are not taken into account when evaluating the current and future state of existing structures, their remaining lifetime may be overestimated.

Acknowledgements

The research reported in this paper was supported by the Grant Agency of the Czech Technical University in Prague, grant No. SGS17/126/OHK1/2T/11 "Changes in fatigue behaviour of steel bridge structures weakened by corrosion" and by COST CZ (LD) project of the Ministry of Education, Youth and Sports (No. LD15127).

References

[1] Ryjáček, P., Macho, M., Stančík, V., and Polák, M. Deterioration and assessment of steel bridges. *8th International Conference on Bridge Maintenance, Safety and Management IABMAS 2016*. 2016; pp. 1188-95.

[2] Macho, M., and Ryjáček, P. The impact of the severe corrosion on the structural behavior of steel bridge members. *Advances and Trends in Engineering Sciences and Technologies: Proceedings of the International Conference on Engineering Sciences and Technologies, ESaT (2015)*. 2015; pp. 123-128.

[3] Xu, S., and Qiu, B. Experimental study on fatigue behavior of corroded steel. *Materials Science and Engineering: A*. 2013; **584**(Supplement C): pp. 163–169.

[4] Xu, S., and Wang, Y. Estimating the effects of corrosion pits on the fatigue life of steel plate based on the 3D profile. *International Journal of Fatigue*. 2015; **72**(Supplement C): pp. 27-41.

[5] Zhang, X.-Y., Li S.-X., Liang R., and Akid, R. Effect of corrosion pits on fatigue life and crack initiation. *13th International Conference on Fracture*. Beijing, China. 2013.

[6] Rahgozar, R., and Sharifi, Y. Remaining fatigue life of corroded steel structural members. *Advances in Structural Engineering*. 2011; **14**(5): pp. 881-890.

[7] Heinemeyer, C., and Feldmann, M. The influence of rivet corrosion on the durability of riveted connections. *Steel Construction*. 2011; **4**(3): pp. 188-192.

[8] Pipinato, A., Pellegrino, C., Bursi, O. S., and Modena, C. High-cycle fatigue behavior of riveted connections for railway metal bridges. *Journal of Constructional Steel Research*. 2009; **65**(12): pp. 2167-2175.

[9] UIC Leaflet 778-2 R. Recommendations for determining the carrying capacity and fatigue risks of existing metallic railway bridges.

IOP Conf. Series: Materials Science and Engineering **419** (2018) 012027 doi:10.1088/1757-899X/419/1/012027

Factors Influencing the Design Life of Old Steel Bridges

Nor Ashikin Muhamad Khairussaleh[1], Gerald A.R. Parke[2] and Boulent Imam[3]

[1]Universiti Malaysia Pahang, Fac. of Civil Engineering & Earth Resources, Gambang, 26300, Malaysia
[2,3]University of Surrey, Dept. of Civil & Environmental Engineering, Guildford, GU2 7XH, United Kingdom

E-mail: aasyiq@ump.edu.my

Abstract. When deciding about the specific design life for bridge structures, care should be taken to ensure that the structure fulfils all the fundamental requirements of structural reliability in terms of robustness, safety and serviceability in order to achieve the service life of the bridge. Load and material properties, cross-section and system geometry are the basic variables or parameters that are being used when considering the design life of bridge structures. The design life of bridge structures as specified in the Eurocodes and British Standards is 100 and 120 years respectively while in AASHTO-LRFD it is for 75 years. However, in all these codes of practice, there are no specifications or provision or guidelines related to sustaining the service life of bridge structures (Bartholomew, 2007 & 2009). This is because the service life depends on the durability of the structure which is heavily influenced by several factors such as fatigue, corrosion and changes in superimposed loads. Therefore, with an increase in traffic loading together with climate changes, the demand to ensure service life is very acute because of the importance of bridge structures in the infrastructure network, which is especially true for long span bridges. Also, with different load applications of heavy good vehicles (HGVs) occurring during the design life, this will affect the structural integrity of old steel bridges. For example, old bridges in many countries including Malaysia were designed using British Standard compliances. Since the Eurocodes have been widely practising nowadays, the estimation of loading applied might be different with the old code of practice. Plus, with the traffic increasing every year without any control, the maintenance for the old bridges especially should be more frequent as these bridges may not have benefited to the remedial measures to improve the fatigue performances. Therefore, the actual service life may not reach the expected service life as the actual service life depends on the exposure condition of the structure, quality of materials, design and construction and also the level of maintenance performance. In addition, with the increase in traffic load and frequency, this could seriously jeopardise the integrity of old bridges to meet their actual service life. This paper discusses the issue concerning the design and remaining service life estimation used when designing and appraising steel bridges.

1. Instructions

Bridges in many countries including Malaysia were designed using British Standard compliances. There are more than 7000 bridges in Malaysia excluding Borneo island. Nearly 10 percent of these bridges used steel as the primary material. This total numbers of bridges do not include modern, long-span bridges, such as the Penang Bridges, Langkawi Sky Bridges, and the Muar Second Bridge. For a steel bridge, the demand to ensure its service life is achieved is very acute especially for long-span bridges due to their importance in the infrastructure network. This is because steel bridges are prone to fatigue damage due to cyclic loading caused by moving traffic loading. According to Miki (2004), there are a large number of steel bridges that have been suffering fatigue cracking and several of them have collapsed due to fatigue damage since 1960. A recent incident related to the Fourth Road Bridge in Scotland where it was closed to traffic in December 2015 because of a sudden element failure due to fatigue. For all bridges, the actual service life may not reach the expected design life. This is because the actual life depends on the exposure condition of the structure, the quality of the materials used, the design and construction and also the level of maintenance received. On the contrary, the expected service life is based on the owner's desires and expectations through the design life. As specified in the British Standards and Eurocodes, the design life for bridges is 120 and 100 years respectively while for AASHTO – LFRD it is only 75 years. However, where the number of years has been set as 'mandatory' for bridges to fulfil their service life, this does not necessarily mean that the structure is able to complete its expected service life especially with the status of the current highway system which is both overloaded and overcrowded?

1.1. Fatigue phenomena

Traffic volume can expose individual members and connections in structures to fatigue damage. According to Gabler (2015), increasing traffic volume is one of the major factors contributing to fatigue crack initiation. This phenomenon can deteriorate the performance of a structure as it may result in unexpected failure. Defects or damage due to fatigue failure may occur and develop progressively in the structure but it may not be detected visually. Fatigue usually causes a very fine crack to develop in the steel member. This initial crack is then followed by crack propagation leading to complete fracture of the member after the structure undergoes a sufficient number of critical load cycles. These critical load cycles for a bridge may occur from traffic and/or wind loading. Also, a breakdown of the coating system used for the structural system will increase the damage level. This increases the phase propagation of any cracks. A deterioration mechanism such as chloride and carbonation attack can dissipate into the structure and induces corrosion to the structure. In addition, if corrosive water leaks into the structure, fatigue corrosion may also occur and this speeds up crack propagation. For example, leaks at the cable anchorage points in a cable-stayed bridge during the grouting process contribute to premature fatigue and fatigue corrosion due to the infiltration of moisture (Tabatabai *et. al.*, 1995).

1.2. Codes of practice used

Starting from March 2010, it was recommended that all newly-procured structural design work in the UK complied with the Eurocodes (ADEPT, 2010). Therefore, the assessment of fatigue for bridges should also be based on the guidance given in the Eurocodes. However, the Eurocodes use a slightly different approach for the applied loading compared with the British Standard. Furthermore, the increase in traffic loading each year could affect the loading application that has been used in designing bridges using British Standards. Also, both the British Standards and the Eurocodes do not state clearly the traffic load increments to use which results in the designer using their own experience and knowledge to overcome this shortcoming. However, the Eurocodes introduce five different types of fatigue load namely, fatigue load model 1 (FLM 1) to fatigue load model 5 (FLM 5) which gives the opportunity for engineers to design the bridge using an accurate traffic model (BS EN 1991-2). These fatigue load models are used under different conditions and consist of a set of equivalent lorries that simulate the traffic, producing fatigue damage which is equivalent to that due to the actual traffic. Engineers prefer to use FLM 3 in bridge design because it reduces some of the calculation procedure, yet it is accompanied with some shortcomings (Muhamad Khairussaleh, 2016). In addition, the concept and application of this practice are easy and nearly similar to the standard fatigue vehicle (SFV) given in the British Standard BS5400 Part 10. On the contrary, two other types of the fatigue load models given in the Eurocodes; FLM 4 and FLM 5 provide the more realistic condition of typical traffic. This is because the determination of the stress history can be achieved with the application of these fatigue load models as they simulate the actual traffic flow or by taking the exact data from the traffic movement on the bridge (in situ traffic). The complex stress ranges can be generated which are needed in the fatigue assessment through the use of the appropriate S-N curve. Therefore, FLM 4 and FLM 5 are the most accurate basis for predicting fatigue life.

2. Estimation fatigue damage using different codes of practice

Each code of practice uses its own distinctive heavy good vehicle (HGV) and conditions of traffic flow. There are two types of traffic load conditions recommended in the Eorocodes, FLM 4, namely traffic alone and traffic in a convoy. These types of load model simulate traffic where each lorry is expected to transit both alone and in a convoy along the bridge. In addition, this model classifies traffic loading into three types of traffic namely; *long distance, medium distance* and *local traffic* (BS EN 1991-2). Although there is no specific explanation in the Eurocodes but the Institution of Highway & Transportation, Department of Transport (1987) proposed the magnitude of these traffic loading should be chosen based on the design needs on the bridge one by one. The total load of the smallest lorry allowed to cross in the traffic alone condition is the lorry with two axles weighing 200kN (20 tonnes) while the total load of the heaviest lorry with a set of five equivalent axle loads is 490kN (49 tonnes). For the traffic in convoy condition, five equivalent lorries should run in convoy with a 40m spacing between each vehicle. This causes a total load of a set equivalent axle loads running in convoy to be equal to 1840kN (184 tonnes). This is quite a large traffic load being applied on the bridge for the fatigue assessment. In the AASHTO-LFRD code, the loading for the fatigue assessment is based on a single design truck with a total axle load of 320.3kN (72kip). This total axle load is 169.7kN lower compared to the biggest equivalent lorry load used with the traffic alone condition given in the Eurocodes and nearly five sixth lower than the total equivalent axle loads that are used with the traffic convoy condition in the Eurocodes. Table 1 shows the summary of the critical loading of HGV applied on a bridge from different codes of practice used for fatigue appraisal.

Table 1. Highest Axle Load recommended for Different Codes of Practice

Traffic Condition	EUROCODES FLM 4	EUROCODES FLM 3	BS 5400	AASHTO LRFD
		kN		
Traffic Alone	490	480	320	320.3
Traffic in Convoy	1840	516	-	-

Based on Table 1, it is evident that for the traffic alone model the Eurocodes provide nearly 53% or 1.5 times greater load compared with the AASHTO load value. Furthermore, in the convoy traffic condition, the load in Eurocodes is 5.75 times greater compared to the AASHTO traffic alone condition. However, BS 5400 provides the same magnitude of axle load as that given by AASHTO. Therefore, it is expected that the stresses that will be generated from AASHTO and BS 5400 will be substantially lower and the potential for fatigue damage to occur in the vicinity that is vulnerable to fatigue using these two codes of practice will be less than that expected from using the Eurocodes loading. It is important to remember that each code of practice also recommended a different service design life. That is why in order to achieve this sustained fatigue design working life up to 120 years, the Eurocodes provide a more stringent procedure compared to that adopted by the AASHTO code.

3. Potential factors affecting design life

Fatigue actions are determined according to the requirements of the fatigue assessment based on the statistical evaluation presented in S-N curves but fatigue appraisals are different from the ultimate and serviceability limits state verifications (Bijlaard, 2007). Therefore, any fatigue crack that develops

IOP Conf. Series: Materials Science and Engineering **419** (2018) 012027 doi:10.1088/1757-899X/419/1/012027

during the service life does not necessarily mean the end of the service life for the structure because the crack would be repaired (Bijlaard, 2007). This is to avoid any fatigue crack introducing more severe notch conditions that could shorten the service life of the structure even further.

3.1. Residual stress in the welded structure

The problems with welded structures are the presence of high tensile residual stresses occurring as a result of the welding process. This is because with high-stress concentrations forming in the vicinity of the connections and the occurrence of tensile residual stresses influences the fatigue life of welded structures. Therefore, the possibility of fatigue damage occurring in a welded structure is high even with an applied external compressive stress. Steel bridges generally contain a large number of welds and consequently high tensile residual stresses which may govern the fatigue life of the bridge. Although studies have been carried out to investigate ways of improving fatigue performance for weld connections (Lee *et al.*,2012; *Wang et al.*,2012; Fu *et al.*,2014; Zhao *et al.*,2016) old steel bridges can still be exposed to these high tensile residual stress fields. This is because only relatively modern bridges have benefited from these techniques used to improve the fatigue performance of the connections but not for old bridges. Therefore, steel bridges which have achieved a service life of more than 50 years should have a more frequent inspection for fatigue damage.

3.2. Traffic flow increasing and type of traffic characteristic

In December 2015, the Fourth Road Bridge in Scotland had major repair work undertaken after a crack was found in one of the support brackets for the truss carrying the bridge carriageway. The 20mm wide crack was not found during the last regular six months of maintenance regime work. Fatigue and overloading occurring over the bridge's 51-year lifespan have been reported to be the reason for the crack problem (BBC, 2015). According to Russell (2016), this structure carries 78,000 vehicles per day with 9% of them are heavy goods vehicles (HGVs). In addition, the occurrence of the fatigue crack was identified to be partly due to the increasing traffic load, which was not fully anticipated in the initial design. Most codes of practice do not indicate the level of traffic increment expected for bridge design. Gabler (2015) also mentioned that the weight of the average truck, loaded to its full capacity, has also increased steadily along with the general increase in traffic density. Therefore, both of these events will adversely affect the strength of the bridge and will generate bigger stress ranges at the critical or vulnerable areas prone to fatigue damage. Hence, old steel bridges are also exposed to the same problems if there is any traffic load increment crossing over these bridges. Thus, in order to determine the condition of the structural elements prone to fatigue, cycle loading analysis has been conducted for a cable anchorage block on a cable-stayed bridge by comparing the traffic characteristic applied as shown in Figure 1.

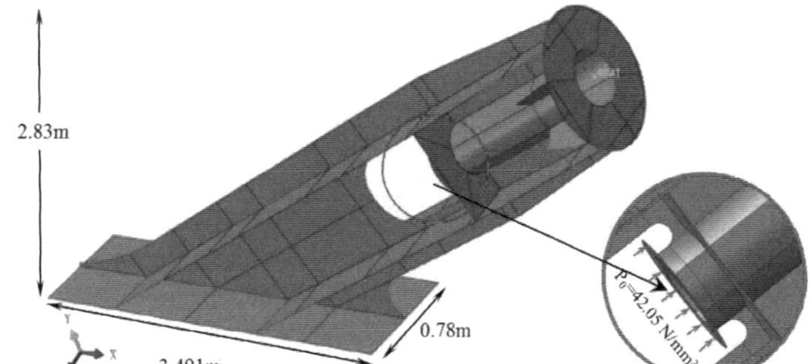

Figure 1: A 3D model of the anchorage block with the pressure applied to the bearing plate.

Table 2 present the total fatigue damage that could occur on the cable anchorage block with different types of traffic characteristics.

Table 2: Total damage occurring at the cable anchorage block with a different type of traffic characteristic

Type of traffic characteristic	Top Gusset (a)	Bottom Gusset (b)
Local Traffic	0.325	0.035
Medium Distance	0.847	0.087
Long Distance	1.270	0.113

From the top gusset and bottom gusset of a particular cable anchorage block have different fatigue damage with reference to the type of traffic characteristic. Based on Miner's rule, the damage accumulation (D_{max}) should not exceed 1.0 (BS EN 1993-1-9; Palmgren – Miners' Rule). This value means that with proper maintenance, the bridge can sustain to its design service life. However, the International Institute of Welding (IIW, 2008) and BS 7608 (2014) recommended the fatigue damage accumulation should not exceed 0.5 due to a large number of uncertainties present in the fatigue phenomenon. According to Nussbaumer *et al.* (2011), the use of D_{max} = 0.5 is needed for verifications under both proportional and non-proportional multi-axial stress cases. Therefore, based on this fatigue guidance rule, it reveals that both gussets will not have an expected fatigue failure during their service design life but only for local traffic and medium distance traffic. However, the top gusset will experience fatigue failure if long distance is used as the traffic characteristic in the analysis.

3.3 The hot-spot stress approach

Using the hot-spot stress method (Maddox, 2011) for steel bridges for fatigue evaluation has become important nowadays. This is because the actual stress value used in estimating fatigue damage can be difficult to determine. Note that fatigue is a local failure and assessing the local stress concentrations occurring at the critical location is very critical. It is very important to determine high-stress intensities resulting at any particular connection detail. Therefore, the concept of a hot-spot stress has been introduced in order to provide an appropriate stress for fatigue assessment. This approach has been used successfully for welded details especially in offshore structures but it is still limited for bridge design engineering (Bhargava, 2009). Determining the hot-spot stress cannot be directly acquired from a finite element analysis. The derivation and guidance on how to determine the hot-spot stress have been outlined in the codes of practice (Niemi, 1995; Niemi *et al.*, 2006; DNV - RP-C203; 2011; IIW, 2008; BS 7608, 2014). Basically, two stresses values located at a defined distance from the critical point will be used to extrapolate the hot-spot stress. This extrapolation is based on the finite element mesh density, element type, size, and reference point to the highest node stress concentration all of which has been recommended in the codes of practice. Even though this method has a high accuracy in assessing fatigue, the use of this method by bridge designers is still not widely adopted.

4. Conclusion

Bridges in Germany have to accommodate increasing loads due to the large increase in truck traffic and this was one of the primary factors in the initiation of a crack in the main girder of the Leverkusen Bridge (Gabler, 2015). The same factor was the cause of fatigue damage to the Fourth Road Bridge in Scotland causing it to close in December 2015 for a major repair. This increase in traffic density and vehicle loads which recently caused problems in two different countries alarming engineers as a significant increase in traffic volume over the past 50 years has not been properly considered in bridge design. This enormous growth in both traffic volume and weight will influence greatly the design life of old bridges. Moreover, with the different codes that have been used in the past for design, have resulted in different values for bridge service life. In addition, with the general lack of detailed information particularly regarding welded details on old bridges and the type of traffic characteristic used during design, old steel bridges are likely to have a high exposure to fatigue failure which will consequently be reflected in a shortened design life. Therefore, the maintenance schedule for old bridges should be clear in identifying critical details and these need to be appraised on a frequent basis to ensure a continuity structural integrity of the structures.

Acknowledgment
Research Grant of Universiti Malaysia Pahang RDU 1703202 tittle *Analysis of Fatigue Life of Steel Bridges.*

References

[1] ADEPT (Association of Directors of Environment, Economy, Planning and Transport) (2010) Guidance Document on the Implementation of Structural Eurocodes

[2] AASHTO (2012) *AASHTO LRFD Bridge: Design Specifications.* American Association of State Highway and Transportation Official.

[3] AASHTO (2002) *Standards Specifications for Highway Bridges. 17th edn. American Association of State Highway and Transportation Official.*

[4] Bartholomew, M. (2007) How Long Will Your Bridge Last? – The need to Predict Service Life of Bridge Components. *Western Bridge Engineer's Seminar.* Boise, Idaho, 23 - 26 September 2007.

[5] Bartholomew, M. (2009) Design for Service Life, Bridge Birth Certificate and Concrete Structures Managements Concepts. AASHTO Bridge Sub-Committee Meeting T-9 Technical Committe for Bridge Preservation. New Orleans,LA, 6 July 2009.

[6] BBC News *Fourth Road Bridge to be Closed Until New Year. BBC News 4th December 2015.* Available at: http://www.bbc.co.uk/news/uk-scotland-35001277

[7] Bhargava, A. (2009) *Fatigue Analysis of Steel Bridge Details: Hot Spot Stress Approach.* PhD. George Washington University.

[8] Bijlaard, F.S.K. (2007) *'Eurocode 3 Part 1-9: Fatigue' Paper of the Seminar on Eurocodes.* Oslo, Norway, 9 May 2007.

[9] BSI (2005) *BS EN 1993 Part 1-9: Design of Steel Structures: Fatigue.* London: British Standards Institution.

[10] BSI (2004) *BS EN 1991-2 Part 2: Traffic Load on Bridges.* London: British Standards Institution. London: British Standards Institution.

[11] BSI (2014) *BS 7608 : Guide to Fatigue Design and Assessment of Steel Products.* London: British Standards Institution.

[12] BSI (1980) *BS5400 Part 10: Steel, Concrete and Composite Bridges: Code of Practice for Fatigue.* London: British Standards Instituition.

[13] DNV - RP-C203 (2011) *Fatigue Design of Offshore Steel Structures.*

[14] Fu,G., Lourenco, M.I., Duan, M. and Estefan, S.F. (2014) 'Effect of Boundary Conditions on Residual Stress and Distortion in T-joint Welds', *Journal of Constructional Steel Research, 102, pp.121-135.*

[15] Gabler, M. (2015) 'Time for Action', *Bridge: Design & Engineering* (80), pp. 88-89.

[16] IIW (2008) *Recommendations for Fatigue Design of Welded Joints and Components.*IIW Commission XIII and XV edn. France: International Institute of Welding.

[17] Lee,C., Chiew,S. and Jiang,J. (2012) 'Residual Stress Study of Welded High Strength Steel Thin-Walled Plate-to-plate Joints, Part 1: Experimental Study', *Thin-Walled Structures,* 56, pp. 103-112.

[18] Maddox, S.J. (2011) 'Fatigue Design Rules for Welded Structures' in K.A. Macdonald (ed.) *Fracture and Fatigue of Welded Joints and Structures.* Woodhead Publishing Limited, pp. 168-206.

[19] Miner,M.A. (1945) 'Cumulative Damage in Fatigue', *Journal of Applied Mechanics,* 12 (3), pp. 159-164.

[20] Muhamad Khairussaleh, N.A. (2016) *Fatigue of Cable Anchorages on a Cable-Stayed Bridge.* Unpublished PhD. University of Surrey.

[21] Miki, C. (2004) 'Bridge Engineering Learned From Failure - Fatigue and Fracture Control' Proceeding of the 2nd International Conference on Bridge Maintenance, Safety, Management and Cost. Kyoto, Japan, 18 - 22 October 2004. CRC Press Taylor & Francis Group.

[22] Niemi, E. (1995) *Stress Determination for Fatigue Analysis of Welded Components* .The International Institute of Welding edn. Abington Publishing.

[23] Niemi, E., Fricke, W. and Maddox, S.J. (2006) *Fatigue Analysis of Welded Components: Designers's Guide to the Structural Hot-Spot Stress Approach* . The International Institute of Welding.

[24] Nussbaumer, A., Borges, L. and Davaine, L. (2011) Fatigue Design of Steel and Composite Structu*re.* 1st edn. Portugal: European Convention for Constructional Steelwork (ECCS).

[25] Russell, H. (2016) 'Rapid Respond', *Bridge: Design & Engineering* (82), pp. 82-84.

[26] Tabatabai, H., Ciolko, A. & Dickson, T. (1995) 'Implications of Test Results From Full-Scale Fatigue Tests of Stay Cables Composed of Seven-Wire Prestressing Strand', Proceeding of the Fourth International Bridge Engineering Conference - Transportation Research Board Conference. San Francisco, California, 28 - 30 August 1995. Washington, D.C.: National Academy Press, 1995.

[27] Wang,Y., Li,G. and Chen,S. (2012) 'The Assessment of Residual Stresses in Welded High Strength Steel Box Sections', Journal of Constructional Steel Research, 76, pp. 93-99.

[28] Zhao,M., Chiew,S. and Lee,C. (2016) 'Post - Weld Heat Treatment for High Strength Steel Welded Connections', Journal of Constructional Steel Research, 122, pp. 167-177.

IOP Conf. Series: Materials Science and Engineering **419** (2018) 012028 doi:10.1088/1757-899X/419/1/012028

Dynamic behaviour of the steel footbridges with spatial pipe truss girders

Marek Pańtak[1], Bogusław Jarek[1], Kinga Marecik[1]

[1]Cracow University of Technology, Institute of Building Materials and Engineering Structures, Warszawska 24, Cracow 31-155, Poland

E-mail: mpantak@pk.edu.pl

Abstract. The excessive vibrations of the steel footbridges can be an important problem during the verification of the requirements of the serviceability limit state of these structures. In the paper the issues of dynamic characteristics and dynamic behaviour of the steel footbridges on the basis of the results of the field tests and numerical analyses of the four single span steel footbridges with spatial pipe truss girders and span length from 30 to 50 m were presented. The dynamic properties of the structures (stiffness, mass, damping) were characterized and the dynamic response of the structures was studied. Results of the analyses and researches showing, in most cases, a very high dynamic susceptibility of the structures subjected to dynamic loads from slow running (jogging) users, mainly due to the very low damping value, which is important parameter affecting the vibration amplitudes.

1. Introduction

At the outset the information concerning primary rules of dynamics of the single degree of freedom system will be quoted to remind them for further analysis. Single degree of freedom system (SDOF), in the form of a mass suspended on a spring with added damping is described by three basic parameters *M, K* and *C* (mass, stiffness and damping respectively). It is well know that the natural vibration frequency of the SDOF system *f* depends on its stiffness and mass. This dependence is described by the equation (1):

$$f = \frac{1}{2\pi}\sqrt{\frac{K}{M}} \tag{1}$$

Changing of the natural vibration frequency of the SDOF system is possible by changing of its stiffness *K* [N/m] or mass *M* [kg]. The natural vibration frequency of the SDOF system is directly proportional to the stiffness of the system *K* and inversely proportional to its mass *M*. Modifications of the *M* and *K* parameters changes the natural frequency of the system. It is worth noting that natural vibration frequency of a SDOF system does not depend on damping value *C*. The equation (1) does not describe the effect of changes in *K* and *M* parameters on the amplitude of vibrations.

The influence of the *M, K, C* parameters on the amplitude of forced vibrations of the system can be described by the dependencies (6), (8) and (10) resulting from the solution of the equation of motion of the SDOF system under harmonic excitation (2).

$$M\,\ddot{x}(t) + C\,\dot{x}(t) + K\,x(t) = P\sin(2\pi \cdot f_P \cdot t) \tag{2}$$

The solution of equation (2) can be written in the form (3) [1]:

$$x(t) = A\sin(2\pi \cdot f_P \cdot t - \varphi) \tag{3}$$

where:

$$A = \frac{P}{\sqrt{\left(K - 4\pi^2 f_P^2 M\right)^2 + 4\pi^2 f_P^2 C^2}} \tag{4}$$

$$tg\varphi = \frac{2\pi\, f_P\, C}{K - 4\pi^2 f_P^2\, M} \tag{5}$$

M, K, C – parameters of SDOF system (mass, stiffness, damping),
P – amplitude of the harmonic excitation,
f_P – frequency of the harmonic excitation,
A – amplitude of forced vibration,
φ – phase shift.

Using the equation (4), it is possible to analyse the influence of parameters *M, K, C* on the amplitude of forced vibrations *A* in three cases the relations between the frequency of the excitation f_P and the natural vibration frequency of the system *f*.

Case No. 1: $f_P << f$ – frequency of the excitation is significantly lower than the natural vibration frequency of the system. Assuming a small values of f_P and a small damping *C* (this assumption is correct for the majority of footbridges [2]) one obtains:

$$A = \frac{P}{\sqrt{\left(K - 4\pi^2 f_P^2 M\right)^2 + 4\pi^2 f_P^2 C^2}} \approx \frac{P}{\sqrt{K^2}} = \frac{P}{K} \tag{6}$$

Case No. 2: $f_P >> f$ – frequency of the excitation is significantly greater than the natural vibration frequency of the system. Assuming a small values of *f* and a small damping *C* for:

$$K = 4\pi^2 f^2 M \tag{7}$$

one obtains:

$$A = \frac{P}{\sqrt{\left(K - 4\pi^2 f_P^2 M\right)^2 + 4\pi^2 f_P^2 C^2}} = \frac{P}{\sqrt{\left(4\pi^2 f^2 M - 4\pi^2 f_P^2 M\right)^2 + 4\pi^2 f_P^2 C^2}} \approx \frac{P}{4\pi^2 f_P^2 M} \tag{8}$$

Case No. 3: $f_P = f$ – frequency of the excitation is equal to the natural vibration frequency of the system (resonance effect) for:

$$K = 4\pi^2 f^2 M = 4\pi^2 f_P^2 M \tag{9}$$

one obtains:

$$A = \frac{P}{\sqrt{\left(K - 4\pi^2 f_P^2 M\right)^2 + 4\pi^2 f_P^2 C^2}} = \frac{P}{\sqrt{\left(4\pi^2 f_P^2 M - 4\pi^2 f_P^2 M\right)^2 + 4\pi^2 f_P^2 C^2}} = \frac{P}{2\pi\, f_P C} \tag{10}$$

It can be seen that the first two cases are approximate solutions achieved with the adoption of simplified assumptions. Whereas, the case no. 3 is a strict solution applicable in the case of the resonance effect.

Equation (10) indicate that amplitude of forced vibration of the system in case of resonance is inversely proportional to the damping *C*. Decreasing of the vibration amplitude in case of resonance

IOP Conf. Series: Materials Science and Engineering **419** (2018) 012028 doi:10.1088/1757-899X/419/1/012028

requires increasing the damping of the system. Changes in mass M or stiffness K of the system do not reduce the vibration amplitudes in case of resonant vibrations. According to equation (1) the correction of M or K parameters of the system can help to avoid the resonance effect by changing the vibrations frequency of the system. However, if the resonance effect can occur in a wide frequency range (e.g. in frequency range 1.40-3.40 Hz – vertical vibration or 0.70-1.20 Hz – horizontal vibration as it is in the case of vibration of footbridges caused by their users), changes in mass M or stiffness K of the structure may not lead to the desired effect. As can be seen from (10) changes in M or K do not helping to reduce vibration amplitudes in the case of resonant vibrations. Moreover, improper correction of M and K may result in increased dynamic susceptibility of the structure to other dynamic loads, in particular dynamic loads from wind action.

2. Characteristic of the footbridges

In order to examine the basic dynamic parameters and dynamic behaviour of steel footbridges with spatial pipe truss girders four exemplary structures presented in Fig. 1 were tested and analysed. The footbridges are single span structures with spans: 30.00 m, 44.77 m, 47.00 m, 50.05 m.

Figure 1. General views of footbridges a) footbridge in Kraków, span 30.00 m, b) footbridge in Sławięcice, span 44.77 m, c) footbridge in Biecz, span 47.00 m, d) footbridge in Osjaków, span 50.05 m

Footbridge in Krakow, span 30.0 m, is designed as a spatial truss with truss nodes spaced in intervals of 3.0 m and with orthotropic deck constructed with steel plate (thickness 8.0 mm) stiffened by longitudinal open ribs (thickness 10 mm) and supported on stiff transverse beams spaced in intervals of 1.50 m (beam web thickness 14.0 mm). Truss girder consist of 10 pyramidal parts forming the spatial truss. The details of the cross-section of the footbridge are presented in Fig. 2a.

Footbridge in Sławięcice, span 44.77 m, is designed as a spatial truss with truss nodes spaced in intervals of 4.07 m. Truss girder consist of 11 pyramidal parts forming the spatial truss. The deck of the footbridge is constructed using a system known as the "battle-deck-floor" ("b-d-f") consisting of longitudinal wide flange beams (HEB 100) supported on transverse beams (HEB 140, arranged in intervals of 2.035 m) and covered by steel plate thickness 8.0 mm. The basic dimensions and construction details of the footbridge cross-section are presented in Fig 2b.

Footbridge in Biecz, span 47.0 m, is designed as a spatial truss with truss nodes spaced in intervals of 3.00 m in central part of a span (L = 42.0 m) and 2.50 m in initial part of the girder at both ends. Truss girder consist of 16 pyramidal parts forming the spatial truss. The upper chord of the truss girder is curvilinear with radius of 330.0 m. The distance between axis of the bottom chord and the axis of upper chord of the truss girder in the middle of the span equals 1.50 m and decreasing to 0.63 m at both ends of the girder near to the supports. The deck of the footbridge is constructed using a "battle-deck-floor" system. The cross-section of the footbridge containing the basic dimensions is shown in Fig 2c.

Footbridge in Osjaków, span 50.05 m, is designed as a spatial truss with truss nodes spaced in intervals of 3.85 m. Truss girder consist of 13 pyramidal parts forming the spatial truss. The structural solutions used in a footbridge and its dimensions are similar to solutions applied in footbridge in Sławięcice. The details of the cross-section of the footbridge are presented in Fig. 2d.

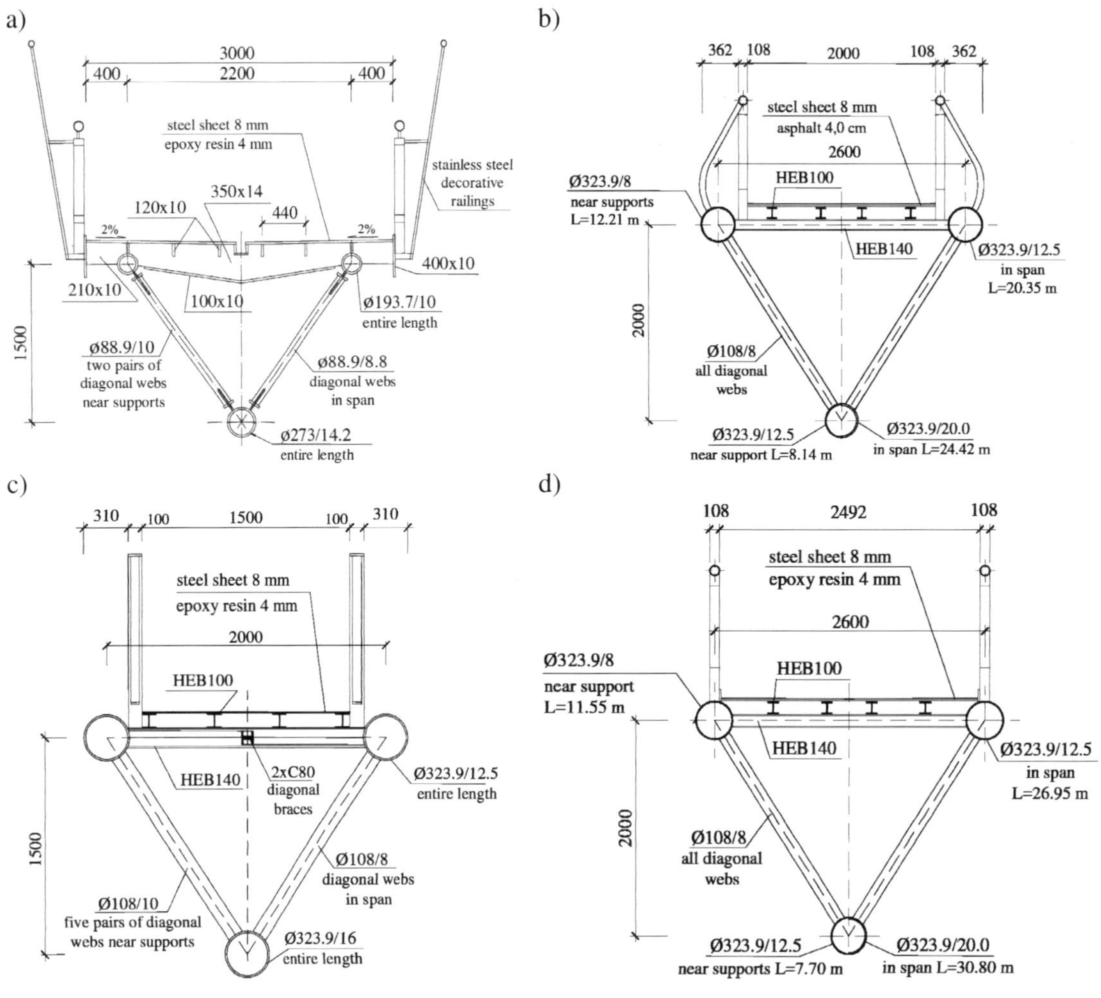

Figure 2. Cross-sections of the footbridges a) footbridge in Kraków, b) footbridge in Sławięcice [3], c) footbridge in Biecz, d) footbridge in Osjaków [4]

3. Dynamic characteristics of the footbridges

The dynamic characteristics of the footbridges were diagnosed during a series of field tests. In the time of the tests, the natural vibration frequency of the structures were identified. Acquired measurement signals allowed to determine the value of the maximum vibration acceleration generated during resonant excitation of vibrations by running and jumping or squatting people with a frequency adjusted to the frequency of the fundamental mode shape of the footbridge as well as to determine the vibration damping parameters in the form of the logarithmic decrement δ and corresponding value of the damping ratio ζ. The results of the field tests are shown in Tab. 1, Fig. 3 and Tab. 2. In the case of vibrations excited by jumping/squatting people the values of vibration accelerations are presented after 10 s and 20 s of vibration excitation.

Table 1. Natural vibration frequencies of the footbridges (the results of the field tests)

Mode shape	Kraków - 30.0 m		Sławięcice - 44.77 m		Biecz - 47.0 m		Osjaków - 50.0 m	
	Frequency f [Hz]							
1	1st vertical	4.21	1st vertical	2.50	1st vertical	2.38	1st vertical	2.39
2	1st horizontal	6.23	1st horizontal	3.23	1st horizontal	4.71	1st horizontal	2.72
3	1st torsional	8.57	1st torsional	6.26	2nd vertical	6.59	1st torsional	6.56
4	2nd vertical	12.85	2nd vertical	7.38	1st torsional	6.79	2nd vertical	7.45

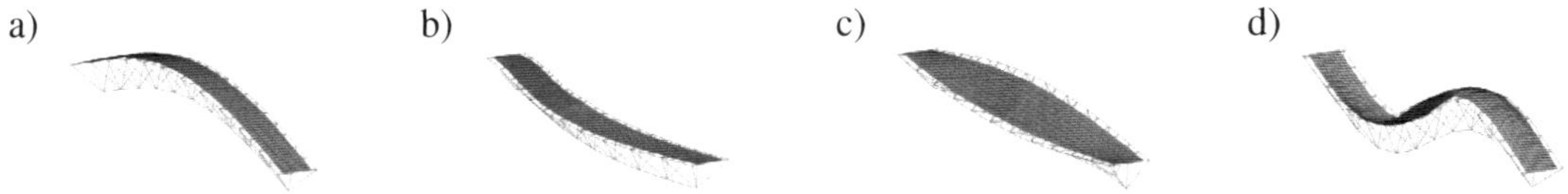

Figure 3. Visualizations of the four mode shapes of the footbridges: a) 1st vertical mode shape, b) 1st lateral mode shape, c) 1st torsional mode shape, c) 2nd vertical mode shape

Table 2. Vibration damping and vibration accelerations for 1st vertical mode shape f_1 of the analyzed footbridges (the results of the field tests)

Footbridge	Frequency f_1 [Hz]	Damping (mean value for 1st mode shape) Logarithmic decrement δ_{test}	Damping ratio $\zeta_{test} = \delta_{test}/2\pi$	Maximum accelerations [m/s^2] One running person	One jumping person t_j = 10 s	t_j = 20 s
Kraków	4.21	0.0562	0.0089	0.53*	0.83*	0.97*
Sławięcice	2.50	0.0224	0.0036	2.67	2.37	3.44
Biecz	2.38	0.0433	0.0069	2.85	2.12	3.32
Osjaków	2.39	0.0084	0.0013	3.27	2.35	4.59

* excitation frequency $f_{0.5} = ½ f_1$ = 2.10 Hz
t_j – duration of jumps

Analysis of the presented results allow to notice a relatively high fundamental vibration frequency of the footbridge in Kraków f_1 = 4.21 Hz > 3.0 Hz and much lower frequencies of the other footbridges. The 3.0 Hz value of vibration frequency is defined in some standards [5, 6] as a lowest allowable fundamental frequency of the footbridges. The structures with fundamental frequency in vertical mode less than 3.0 Hz (and 1.3 Hz in lateral mode) can experience excitations of resonance vibrations by pedestrians. If the fundamental frequency cannot satisfy these limitations, or if the second harmonic is a concern, an evaluation of the dynamic performance shall be made [5].

The fundamental frequencies of the footbridges in Sławięcice, Biecz and Osjaków are lower than 3.0 Hz, the dynamic analyses of the footbridges should be performed. However, it is worth noting that natural vibration frequencies of these footbridges are out of the range of the frequency of pacing during walking (1.4–2.4 Hz) and are in the range of frequency of running or jumping/squatting (1.90 – 3.40 Hz). This can be considered as a beneficial factor. Forces generated during running and jumping/squatting can be considered as a dynamic load cases which occur rarely on footbridges and for this reason resonance excitations are less likely. However, in the case of footbridges located along an important walking routes, connecting attractive and popular areas or footbridges located near important transport hubs (buss, railway or underground stations), footbridges located within sports and recreational areas and footbridges with long span/spans the loads generated by running people can be an important dynamic load case.

The case of a jumping/squatting people should be treated as an act of vandalism (vandal load [7, 8, 9]) which should be considered as an exceptional load case. This type of dynamic load usually leads to large vibration amplitudes. Analysis and prevention of vibrations generated by jumping or squatting people should be considered individually if it is important from point of view of comfort and safety of use of the structure.

In general all analyzed footbridges characterize a low damping (damping ratio $\zeta \approx 0.10–1.00\%$), especially footbridge in Osjaków with $\zeta \approx 0.13\%$. This low values of damping are insufficient to damp the vibrations forced by users of the footbridges. With regard to vibration damping, it can be generally stated that uncomplicated and rigid structures with small number of flexible connections have low damping. Various looseness, gaps and imperfections of connections increase the damping value. Steel structures with spatial truss girders made of pipes can be classified to first group of uncomplicated and rigid structures characterized by low damping. In the case of the footbridge in Kraków and Biecz the damping ratios have the relatively high values (generally two and half to six and half times higher than in the Sławięcice and Osjaków footbridges respectively). It may be a consequence of a more complex structural form of the footbridges in Kraków and Biecz and large number of welded joints applied in these footbridges (welded connections between deck, cross-beams and longitudinal ribs, additional decorative railings in Kraków; welded connections between longitudinal beams HEB100 and diagonal braces C80 placed under the deck in Biecz). This structural complexity may cause a slightly more efficient dissipation of energy from the vibrating structure.

The values of vibrations accelerations obtained during dynamic tests of the footbridges (during resonant vibrations caused by running and jumping/squatting people) indicate that the damping values in all analysed footbridges are insufficient to limit the amplitudes of forced vibrations of the footbridges (vibration acceleration) to an acceptable level. In almost all cases of resonant excitations the value of vibration acceleration (Tab. 2) significantly exceed the values of $a = 0.50$ m/s^2 which was defined in [10] as the level of maximum comfort. The vibration accelerations also exceed values of $a = 0.70$-1.00 m/s^2 which are generally acceptable but these vibrations can be slightly felt by walking people [11]. In almost all cases the vibration accelerations of the analysed footbridges are in the range of vibration accelerations causing significant walking disturbances i.e. vibration accelerations $a_{max} > 1.20$ m/s^2 for the vibration frequency range 2.0 – 3.0 Hz [12, 13] (Fig. 4).

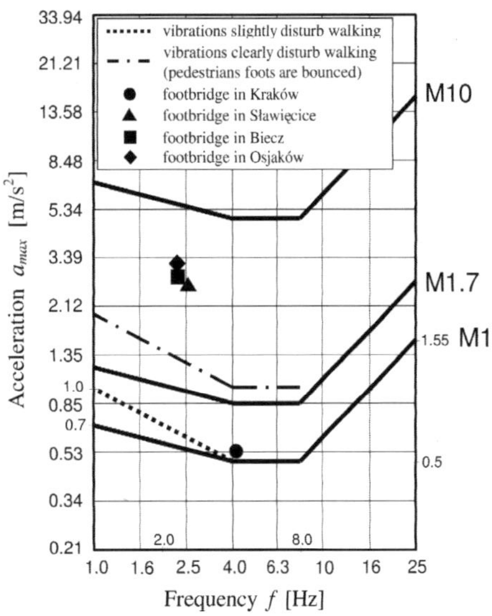

Figure 4. Vibration comfort criteria for footbridges in case of vertical vibrations [12, 13] and evaluation of comfort of use of the footbridges in Kraków, Sławięcice, Biecz and Osjaków during vibrations induced by running users:

M1 – basic comfort curve, frequently occurring vibrations, vibrations are slightly felt or perceptible and do not disturb walking; *M1.7* – rarely occurring vibrations, vibrations are clearly felt (fully perceptible), vibrations disturb walking; *M10* – vandal intentional actions, comfort is strongly disturbed (free walking is impossible, standing or running is difficult and strongly disturbed). The *M10* vibrations limits have been proposed to protect pedestrians against body injury (mainly legs) that can be caused by high-level vibration acceleration.

Only in the case of the footbridge in Kraków vibration accelerations induced by running or jumping/squatting people moving with frequency $f_{0.5} = \frac{1}{2} f_1 = 2.10$ Hz (movement with a frequency of 4.21 Hz is very difficult to implement by people) are relatively small. The value of vibration acceleration $a = 0.97$ m/s^2 in case of vibrations intentionally induced by vandals can be considered as a very low value of vibration. It can be regarded that the fundamental vibration frequency of the footbridge in Kraków $f_1 = 4.21$ Hz (resulting from the relatively high stiffness of the structure – see Tab. 3) effectively reduce the risk of pure resonant excitation. Nonetheless, during the several tests of vibration excitation by one jumping/squatting person with frequency $f_1 = 4.21$ Hz the vibration acceleration of the footbridge deck reaches the value of 2.5 m/s^2 after 10 s of vibration excitation.

4. Analyses of the dynamic parameters of the footbridges

In order to investigate the possibility of increasing the fundamental frequencies of the footbridges (to avoid the resonant vibrations of the structures) and the possibility of reduction of vibration amplitudes (vibration acceleration), the analyses of influence of the M and K changes on the fundamental frequency of the footbridges as well as influence of the changes of C on the vibration amplitudes were carried out. Three footbridges with natural frequencies below 3.0 Hz were analyzed using the equivalent SDOF systems determined for the footbridges. In Tab. 3 the parameters of the equivalent SDOF systems are presented. SDOF for footbridge in Krakow was presented for comparative purposes.

The changes in mass M of the structures were introduced by changing the construction system of the footbridge deck from "battle-deck-floor" into the orthotropic deck by replacing the longitudinal HEB100 beams by longitudinal ribs 100x10 mm, removing all diagonal braces and additionally changing the deck surface from asphalt to epoxy resin in the case of the footbridge in Sławięcice.

The changes in stiffness K was introduced by changing the structural height of the spatial truss girder by increasing the initial height of the spatial truss girders H (Fig. 2) by 10 cm, 20 cm and 50 cm.

Table 3. Equivalent parameters of the SDOF systems of analyzed footbridges

Footbridge	Weight of the structure [kg/m]	Equivalent parameters of the SDOF systems for 1st mode shape of natural vibration		
		Equivalent mass M_e [t]	Equivalent stiffness K_e [kN/mm]	Equivalent damping C_e [kg/s]
Kraków	696.40	10.189	7.129	2156.17
Sławięcice	1101.50	19.677	4.855	2225.42
Biecz	717.40	13.979	3.126	2884.77
Osjaków	833.80	17.312	3.904	675.93

Moreover, the analyses of the forced vibration of the footbridges were performed to investigate the impact of the changes of damping on the amplitudes of vibrations. During the analyses the dynamic load in form of squats performed by one person (body weight – G = 850 N) with frequency equal to the fundamental vibration frequency of the structure was used. The SDOF system was loaded by single harmonic equivalent force determined by the time-dependent force function presented in [8, 9] applicable to frequency range $f_{sq} > 1.60$ Hz. The results of the analyses are presented in Tab. 4 and Tab. 5. It can be seen that changes in mass of the structures slightly changes the fundamental natural vibration frequency of the footbridges, the resonant excitations are still possible. The changes in the height of the truss girder can help to increase the fundamental natural vibration frequency to the value of $f \geq 3.0$ Hz only in two cases of the footbridges in Sławięcice and Osjaków with constant height of the truss girder. In the case of the footbridge in Biecz with variable height of the truss girder (variable stiffness), in spite of increasing the truss girder height, the fundamental vibration frequency of the structure remained less than 3.0 Hz, further increase of the height of the girder is required.

Table 4. Dynamic parameters of the modified structures

Footbridge	M_{or} [t] K_{or} [kN/mm]	f_{or} [Hz]	Equivalent stiffness K_i [kN/mm] / fundamental frequency f_i [Hz]											
			H+10 cm				H+20 cm				H+50 cm			
			"b-d-f"		orthotropic		"b-d-f"		orthotropic		"b-d-f"		orthotropic	
			$K_{bd,10}$	$f_{bd,10}$	$K_{or,10}$	$f_{or,10}$	$K_{bd,20}$	$f_{bd,20}$	$K_{or,20}$	$f_{or,20}$	$K_{bd,50}$	$f_{bd,50}$	$K_{or,50}$	$f_{or,50}$
Sławięcice	15.601 4.760	2.78	5.226	2.61	5.153	2.89	5.652	2.71	5.574	**3.01**	6.964	**3.01**	6.872	**3.34**
Biecz	12.416 2.800	2.39	3.355	2.47	3.020	2.48	3.604	2.56	3.262	2.48	4.460	2.84	4.100	2.88
Osjaków	15.720 3.817	2.48	4.248	2.49	4.155	2.59	4.600	2.59	4.500	2.69	5.685	2.88	5.567	**3.00**

M_{or}, K_{or}, f_{or} – parameters of the structures with orthotropic deck,
K_{bd}, f_{bd} – parameters of the structures with "battle-deck-floor" system ("b-d-f").

IOP Conf. Series: Materials Science and Engineering **419** (2018) 012028 doi:10.1088/1757-899X/419/1/012028

Table 5. Vibration acceleration of the footbridges induced by one squatting person in a function of damping (results of numerical analyses)

Footbridge	Vibration acceleration a_{max} [m/s^2] (duration of the squats t_{sq} = 10 s)				
	ζ_{test} (see Tab. 2)	ζ = 0.01	ζ = 0.03	ζ = 0.05	ζ = 0.1
Sławięcice	2.42	1.57	0.67	0.41	0.20
Biecz	2.74	2.26	0.96	0.58	0.29
Osjaków	3.27	1.82	0.77	0.47	0.24

The changes in the damping value allows to reduce the maximum vibration acceleration to the acceptable levels without changing the height of the girder. Attaining the damping ratio value of $\zeta \geq 0.05$ is required.

5. Summary

The issues presented in the paper relate to the subject of dynamic susceptibility of footbridges with steel spatial truss girders. It can be seen that analyzed structures have the proper dynamic characteristics in the range of the span length up to about 35.0 m. For longer spans the fundamental vibration frequency of the structures is less than 3.0 Hz and structures can experience resonant vibrations excited by users. The vibration acceleration can reach a large values (Tab. 2). Proper modifications of structural system of the truss girder, consisting in increasing the girder height, allow to increase the fundamental vibration frequency of the structures ($f \geq 3.0$ Hz) and minimize the risk of the resonant excitations. Increasing of the vibration damping level e.g. by installing TMD system, allows to effectively reduce the vibration acceleration of the structure to acceptable levels without changing the height of the girder if the damping ratio $\zeta \geq 0.05$ is achieved.

References

[1] Lewandowski R 2006 Dynamic of building structures (Poznan: Publishing House of Poznan University of Technology)

[2] Salamak M 2003 Experimental methods for determining the level of vibration damping in footbridges. (Silesian University of Technology, Faculty of Civil Engineering)

[3] Biliszczuk J Rabiega J 1995 Pedestrian bridge in urban park in Sławięcice *Civil Engineering and Building* **7-8/95** pp 357-360

[4] Biliszczuk J Barcik W Machelski C *et al.* 2004 *Design of steel footbridges* (Wrocław: DWE)

[5] AASHTO 2009 *LRFD Guide Specifications for the Design of Pedestrian Bridges* (Washington: American Association of State Highway and Transportation Officials)

[6] KPRM 2000 *Regulation of the Ministry of Transport and Maritime Economy on the technical conditions of road engineering objects and their locations (Journal of Laws No. 63 Item 735)*

[7] Caetano E Cunha Á Moutinho C 2011 Vandal loads and induced vibrations on a footbridge ASCE *Journal of Bridge Engineering* **6/11** pp 375-382

[8] Pańtak M 2017 Partial squats – the dynamic load of the footbridges *Proc. Int. Conf. Footbridge 2017 (Berlin)* ed M Schlaich, A Goldack (Berlin: Technische Universität Berlin)

[9] Pańtak M 2017 Ground reaction forces generated during rhythmical squats as a dynamic loads of the structure *IOP Conf. Series: Materials Science and Engineering* ***245*** (IOP Publishing)

[10] SÉTRA 2006 *Footbridges: Assessment of vibrational behaviour of footbridges under Pedestrian loading - technical guide* (Sourdun, Paris: SÉTRA and AFGC)

[11] Bachmann H Ammann W Deischl F *et al.* 1995 *Vibration Problems in Structures: Practical Guidelines*, (Basel, Boston, Berlin: Birka□user Verlag)

[12] Flaga A Pańtak M 2008 Vibration comfort criteria for pedestrians on footbridges, *Proc. 3rd Int. Conf. Footbridge 2008 (Porto)* ed E Caetano (Porto: FEUP) pp 263-268

[13] Pańtak M 2007 *Analysis of vibration comfort criteria for steel footbridges susceptible to dynamic actions* (Cracow University of Technology, Faculty of Civil Engineering)

IOP Conf. Series: Materials Science and Engineering **419** (2018) 012029 doi:10.1088/1757-899X/419/1/012029

Vibration damping in steel footbridges

Marek Pańtak[1], Bogusław Jarek[1], Kinga Marecik[1]

[1]Cracow University of Technology, Institute of Building Materials and Engineering Structures, Warszawska 24, Cracow 31-155, Poland

E-mail: mpantak@pk.edu.pl

Abstract. The subject of the paper concerns the problem of vibrations and vibration damping values in steel footbridges with various structural systems (truss, beam and slab, cable-stayed, suspension). In the paper values of vibration damping in steel footbridges determined on the basis of the results of numerous dynamic field tests of the footbridges are presented. The vibration damping values are given as a logarithmic decrement δ. Presented information may constitute a database useful at the stage of designing a new or analyzing existing structures.

1. Introduction

In the analyses of the forced vibration of the structures the value of the vibration damping plays an important role. Proper estimation of the vibration damping is important to determine the correct value of the dynamic response of the structure. In general, the steel footbridges are the structures characterized by low damping [1]. This can lead to excessive vibration amplitudes and significant reduction of the comfort of use of the structures.

It is well know that in the case of resonant vibrations the amplitudes of vibrations strongly depends on the value of damping (Fig. 1a).

The situation in which the excitation frequency is equal to the natural vibration frequency of the system (resonance effect) the amplitudes of vibrations are inversely proportional to the damping C (eq. 1) [1]. In other words, in the case of resonant vibrations decreasing of the vibration amplitudes A requires increasing the damping of the system.

$$A = \frac{P}{2\pi f_P C} \tag{1}$$

where:

A – amplitude of forced vibration,
P – amplitude of the harmonic excitation force,
f_P – frequency of the harmonic excitation force,
C – damping coefficient.

Damping in a vibrating structures can be defined as the mechanism of dissipation of the mechanical energy of the system i.e. irreversible absorption of a part of the work of external forces. This energy-loss mechanism due to its complexity is not fully explained. For this reason the damping value in most structural systems is evaluated by experimental methods using time or frequency domain analysis and for simplicity the damping mechanism is modeled as viscous damping.

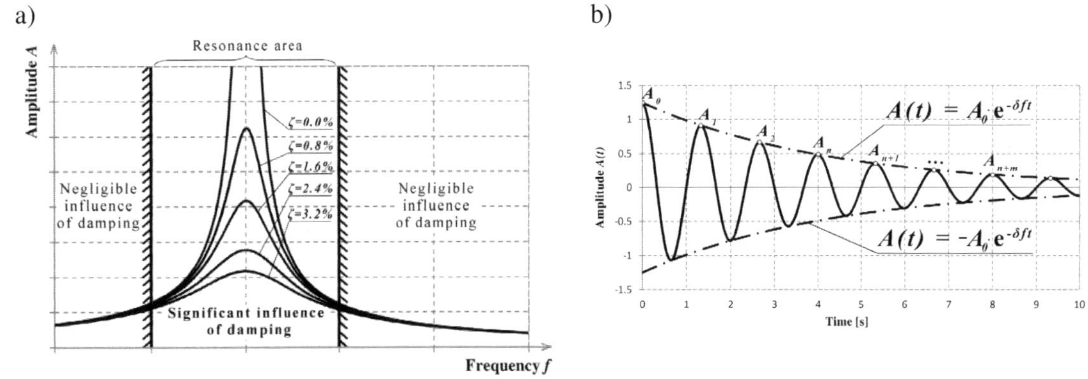

Figure 1. Damping a) effect of damping on the amplitude of resonant vibrations, b) free vibration decay method (ζ – damping ratio, δ – logarithmic decrement)

One of the simplest and most frequently used method for estimating of the value of logarithmic decrement δ (or damping ratio ζ) is a free vibration decay method performed in the time domain (Fig 1b).

To estimate the logarithmic decrement value by using the free vibration decay method the vibrations can be initiated by any convenient method and only the amplitudes of vibrations (displacement or acceleration) must be measured [3]. However a "convenient method" of vibration excitation can not change the dynamic parameters of the structure. It is necessary to perform the dynamic tests in appropriate conditions and to eliminate all factors that can affect the free vibration decay (e.g. unnecessary accessories, machine, devices, people staying on the structure as well as dynamic impact of environment including strong wind and machines or cars working in the vicinity of tested structure or passing under the structure).

In the free vibration decay method the logarithmic decrement δ can be estimated by means of equation (2):

$$\delta = \ln \frac{A_n}{A_{n+1}} = \frac{1}{m} \ln \frac{A_n}{A_{n+m}} \tag{2}$$

where:
A_n – amplitude of *n-th* cycle of vibrations, $n = 0, 1, 2,...$,
A_{n+m} – amplitude of *n+m-th* cycle of vibrations,
m – number of full vibration cycles (periods) counting from the n cycle, $m = 1, 2, 3,...$

Determined value of damping can be verified by fitting exponential curve to decaying vibration amplitudes (Fig 1b) using equation (3):

$$A(t) = A_0 \cdot e^{-\delta \cdot f \cdot t} \tag{3}$$

where:
A_0 – initial amplitude of the analysed part of the free vibration decay,
δ – logarithmic decrement,
f – vibration frequency,
t – time steep.

The damping ratio value can be determined using equation (4):

$$\zeta = \frac{\delta}{\sqrt{4\pi^2 + \delta^2}} \approx \frac{\delta}{2\pi} \tag{4}$$

Further part of the paper contains tabular summary of the vibration damping values determined for numerous steel footbridges with various structural systems based on the results of the dynamic field tests of the footbridges.

2. Vibration damping in steel footbridges

Generally the footbridges characterised by low damping. The logarithmic decrement δ for footbridges is in the range of low damping ($\delta < 0.10$). As set forth in [2] on the basis of the dynamic field tests of 35 footbridges: 50% of tested footbridges had $\delta = 0.05 - 0.10$; 30% of footbridges had $\delta = 0.03 - 0.05$ and only 11% footbridges had $\delta = 0.10 - 0.15$. Characteristic for the footbridges is also lack of the structures with a large damping $\delta > 0.20$. These results do not contain informations about static scheme, structural scheme, frequency, mode shape, span of the structures and should be considered as illustrative. Nonetheless, these results confirm the general thesis of greater dynamic susceptibility of footbridges compared with other types of bridge structures.

a) b)

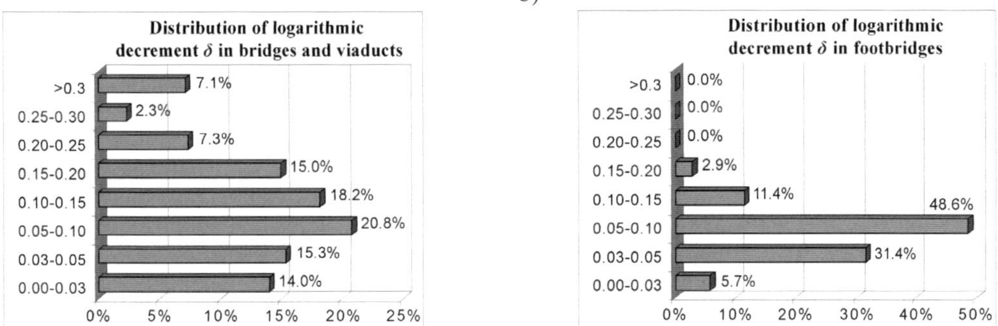

Figure 2. Distribution of logarithmic decrement δ a) for bridges and viaducts, b) for footbridges [2]

Recommendations presented in [4, 5, 6, 7, 8, 9] regarding the vibration damping value for steel footbridges with welded joints define the logarithmic decrement value $\delta = 0.01 - 0.03$ for vibrations of small and medium amplitudes in a frequency range 1.0 – 3.5 Hz. In the case of vibrations of large amplitudes $\delta = 0.04 - 0.06$ is recommended.

In Tab. 1 the logarithmic decrement values δ established on the basis of the dynamic field tests of numerous steel footbridges with various structural systems carried out by authors during several years of research are presented. The logarithmic decrement values δ were determined using a data (vibration acceleration signals) acquired during dynamic field tests of the footbridges during vibrations excited by one walking or running person. The person inducing the vibration was passing the footbridge and leaving the structure (the person was not remaining on the footbridge deck).The logarithmic decrement values δ were determined using free vibration decay method taking into account 10-40 vibration pariods. Examples of determination δ using free vibration decay are presented in Fig. 3.

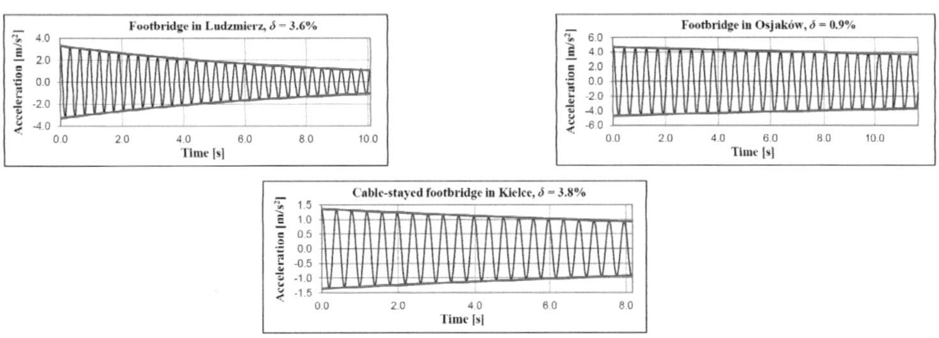

Figure 3. Examples of determination of the logarithmic decrement δ by free vibration decay method

IOP Conf. Series: Materials Science and Engineering **419** (2018) 012029 doi:10.1088/1757-899X/419/1/012029

Table 1. Natural vibration frequencies and logarithmic decrements δ for 32 various footbridges

No.	Footbridge location	General view of the footbridge	General informations	δ [%]
1.	Brno, Czechia		three span L = 8.5+27.5+8.5 m, lightweight orthotropic deck, anchored in abutments f = 5.83 Hz - vertical.	4.6-5.4
2.	Beroun, Czechia		three span L = 21+42+21 m, rectangular box girders, lightweight openwork deck, f = 2.88 Hz - vertical.	4.0-4.8
3.	Kraków, Poland		two span L = 35+32.5 m, box-girder deck, thin epoxy resin pavement, f = 2.68 Hz - vertical.	9.0-12.0
4.	Ludźmierz, Poland		three span L = 16+36+16 m, heavy asphalt pavement f = 3.19 Hz - vertical.	2.9-4.8
5.	Eberswald, Germany		three span L = 13.5+30+13.5 m lightweight orthotropic deck, f = 4.98 Hz - vertical.	5.7-6.6
6.	Kassel, Germany		steel-UHPC concrete composite footbridge, six spans L = 19.2+24+21+36+21+12 m f = 3.47 Hz - vertical.	2.2-3.2
7.	Kraków, Poland		single span L = 35.m, lightweight orthotropic deck, thin epoxy resin pavement f = 2.71 Hz - vertical.	6.8-9.2
8.	Trstena, Slovakia		single span L = 35.m, light deck (thin steel plate without pavement) f = 2.98 Hz - vertical.	3.6-4.1
9.	Svatava, Czechia		single span L = 26.m, wooden deck f = 5.22 Hz - vertical.	24.4-26.3
10.	Roztoky, Czechia		single span L = 30 m, lightweight steel openwork deck f = 4.20 Hz - vertical.	14.2-15.1

IOP Conf. Series: Materials Science and Engineering **419** (2018) 012029 doi:10.1088/1757-899X/419/1/012029

11.	Frydek Mistek, Czechia		single span *L* = 45 m, heavy asphalt pavement *f* =3.30 Hz - vertical.	5.1-6.2
12.	Svatava, Czechia		single span *L* = 34 m, wooden deck, structure anchored in abutments *f* = 4.88 Hz - vertical.	9.6-11.0
13.	Rožnov pod Radhoštěm, Czechia		single span *L* = 42 m, concrete deck, *f* =3.76 Hz - vertical.	2.6-3.3
14.	Brno-Obřany, Czechia		single span *L* = 33 m, concrete deck, *f* =4.30 Hz - vertical.	5.7-6.6
15.	Biecz, Poland		single span *L* = 47 m, heavy orthotropic deck, thin epoxy resin pavement, *f* = 2.38 Hz - vertical.	3.9-4.8
16.	Sławięcice, Poland		single span *L* = 44.8 m, heavy orthotropic deck, heavy asphalt pavement, *f* = 2.50 Hz - vertical.	2.1-2.6
17.	Osjaków, Poland		single span *L* = 50 m, heavy orthotropic deck, thin epoxy resin pavement, *f* = 2.39 Hz - vertical.	0.72-1.33
18.	Kraków, Poland		single span *L* = 30 m, lightweight orthotropic deck, thin epoxy resin pavement, *f* = 4.21 Hz - vertical.	4.9-6.3
19.	Jaroměř, Czechia		single span *L* = 60.4 m, prestressed steel structure, lightweight openwork deck, *f* = 0.98 Hz - horisontal, *f* = 2.48 Hz - torsional.	1.9-2.3 0.25-0.32
20.	Węgierska Górka, Poland		single span *L* = 51.3 m, wooden deck, *f* = 1.44 Hz - vertical.	2.6-3.3
21.	Wola Wieruszycka, Poland		single span L = 63 m, concrete prefabricated deck, *f* = 2.68 Hz - vertical.	3.6-4.8

22.	Korwinów, Poland		single span L = 32.0 m, lightweight orthotropic deck, thin epoxy resin pavement, f = 3.11 Hz - vertical.	7.1-10.5
23.	Beroun, Czechia		single span L = 37.5 m, wooden deck, crossing inclined hangers, f = 4.88 Hz - vertical.	17.6-23.3
24.	Włocławek, Poland		single span L = 41.8 m, prestressed steel structure, wooden deck, f = 2.44 Hz - vertical.	17.4-19.8
25.	Cieszyn - Český Těšín (Polish - Czechia border)		four spans L = 17+45+18+13 m, box-girder deck, f = 2.85 Hz - vertical.	2.4-3.5
26.	Łapanów, Poland		single span L = 45 m, wooden deck, f = 2.14 Hz - vertical.	4.4-5.8
27.	Plzeň, Czechia		single span L = 46 m, lightweight orthotropic deck, thin asphalt pavement, f = 3.98 Hz – vertical.	4.8-6.4
28.	Řež, Czechia		three spans L = 37.3+62.3+37.3 m, wooden deck, f = 3.32 Hz - vertical.	1.7-2.4
29.	Tvrdošín, Slovakia		single span L = 65 m, wooden deck, f = 1.60 Hz - vertical.	1.3-1.6
30.	Kielce, Poland		two spans L = 12.7+29 m, heavy orthotropic deck, thin epoxy resin pavement, f = 2.52 Hz - vertical.	3.7-4.5
31.	Piwniczna Zdrój, Poland		single span L = 102 m, suspension footbridge, lightweight orthotropic deck f = 2.18 - vertical (three extremes).	2.5-3.2
32.	Beroun, Czechia		three spans L = 31.2+61.4+31.2 m, suspension footbridge, wooden deck, wooden railings, f = 1.90 Hz - torsional.	3.0-5.8

The results presented in Tab. 1 refer mainly to vertical vibrations in the frequency range 1.0-6.0 Hz. Additionally, in Fig. 4 the mean values of the logarithmic decrements δ from Tab. 1 are presented.

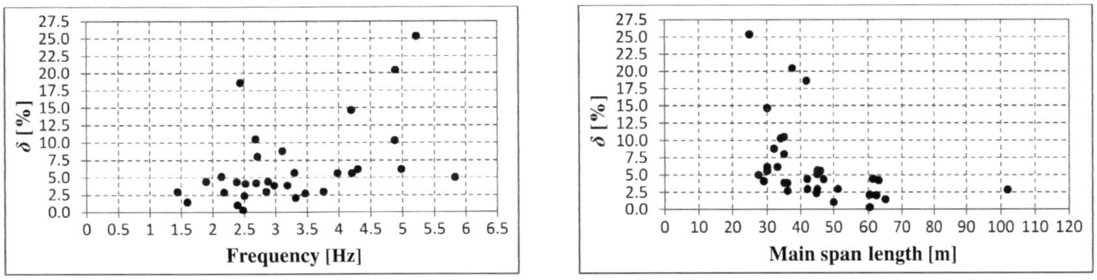

Figure 4. Mean values of the logarithmic decrements δ of the footbridges presented in Tab. 1 in relation to fundamental vibration frequency and main span length

The analysis of the results shows that 80% of the examined footbridgess have $\delta \leq 10\%$, 55% of the footbridges have $\delta \leq 5\%$. Moreover, natural vibration frequency of 70% tested footbridges is in the frequency range of vertical forces generated during walking and running (1.4 – 3.4 Hz). This increases the risk of excitation of large vibration amplitudes during the daily use of these footbridges.

3. Summary

The results presented in the paper indicate the low vibration damping ability of footbridges compared with other types of bridge structures. Relatively low damping can result in high vibrations caused by dynamic action of users (walking or running person or crowd) or by dynamic environmental influences (e.g. wind action).

Knowledge of the damping value is necessary to perform forced vibration analysis and to determine the dynamic response of the structure. Data presented in the paper allows to estimate the damping value in constructions with structural systems similar to constructions presented in Tab. 1.

It should be remembered that in situations of resonant (or near-resonant) dynamic action the value of damping has a significant effect on the vibration amplitude. When designing steel footbridges the possibility of increasing the damping value by installing the damping devices (e.g. passive, semi-active or active TMD device) should be taken into account at the design stage.

References

[1] Lewandowski R 2006 Dynamic of building structures (Poznan: Publishing House of Poznan University of Technology)

[2] Salamak M 2003 Experimental methods for determining the level of vibration damping in footbridges. (Silesian University of Technology, Faculty of Civil Engineering)

[3] Clough R Penzien J 1993 *Dynamic of Structures* (New York: McGraw-Hill)

[4] Heinemeyer Ch Butz Ch et al. 2009 *Design of lightweight footbridges for human induced vibrations*. JRC Scientific and Technical Reports (Luxemburg: JRC-ECCS)

[5] Bachman H Amman W Deischl F et al. 1995 *Vibration problems in structures – practical guidelines* (Basel: Springer Birkhäuser Verlag AG)

[6] Tilly G P Cullington D W Eyre R 1984 Dynamic behaviour of footbridges, IABSE Surveys **S-26/84** pp 13-24

[7] Butz C Feldmann M Sedlacek G et al. 2008 *Advanced load models for synchronous pedestrian excitation and optimised design guidelines for steel foot bridges* (Luxembourg: European Commission CORDIS)

[8] Grundmann H Kreuzinger H Schneider M 1993 Schwingungsuntersuchungen für Fußgängerbrücken, *Bauingenieur* **68** pp 215-225

[9] Schlaich M Brownlie K Conzett J et al. 2005 *fib Bulletin 32: Guidelines for the design of footbridges* (Stuttgart: Sprint - Digital – Druck and *fib*)

IOP Conf. Series: Materials Science and Engineering **419** (2018) 012030 doi:10.1088/1757-899X/419/1/012030

Comparative analysis of dynamic behaviour of two cable-stayed footbridges made entirely of steel and GFRP composite

Piotr Górski[1], Beata Stankiewicz[1] and Marcin Tatara[1]

[1]Opole University of Technology, Faculty of Civil Engineering and Architecture, Department of Roads and Bridges, Katowicka Street 48, Opole 45-061, Poland

E-mail: m.tatara.po@gmail.com

Abstract. This paper is focused on the comparative numerical analysis of the dynamic behaviour of two cable-stayed footbridges made entirely of steel and Glass Fiber Reinforced Polymer (GFRP) composite, with similar geometrical parameters and bearing capacity. The analysis was executed using the Autodesk Robot Structural Analysis Professional software. Firstly, the determination of modal parameters of both footbridges was done. For the further finite element analysis the damping ratio of the all-GFRP footbridge was measured *in situ,* whereas for the all-steel footbridge was taken from data published in the literature. Then, the dynamic responses of both structures were determined based on the Footfall harmonic analysis. In this case, the numerical models of analyzed footbridges were subjected to dynamic effects of human gait with various moving frequency. The maximal values of vertical acceleration of structural vibrations were obtained. The vibration comfort criteria for both footbridges were verified. The numerical results were compared and differences in dynamic responses of analyzed footbridges were examined. The obtained results delivered some important practical data regarding the dynamic response as well as vibration comfort criteria of cable-stayed footbridges which could be used at a design stage to make a decision on the construction material solution.

1. Introduction

Always at the beginning of a design process, bridge designers must decide on the choice of construction materials from which bridge structures will be made. The materials-selection decisions making by designers require consideration of important aspects in terms of material performance requirements which can be divided into five categories, i.e. functional requirements, resistance to service conditions, reliability, processability requirements, and costs, including the initial costs of materials and maintenance costs over the lifetime of a bridge structure. Practical guidelines concerning of these aspects are well discussed in the publication [1]. Considering materials selection problem of newly-designing footbridges a very important aspect is also their dynamic susceptibility to dynamic effects of the pedestrian traffic.

Over the past few decades, the number of newly-designed light-weight and slender footbridges significantly increased. The community also demands more architecturally attractive footbridges. Unfortunately, this leads to more likely serviceability problems. Contemporary slender footbridges are often susceptible to human induced vibrations. In some cases, these vibrations can attain high levels, especially when the walking pace of pedestrians approaches the natural frequency of a footbridge. Such a vibration state could result in a situation where the pedestrian feel uncomfortable or even unsafe [2]. Although codes of practice were developed enabling a designer to evaluate the vibration serviceability of the structure based on simplified models of crowd load [3], the performance requirements for construction materials are higher than in the past. For this reason, besides the traditional materials, like steel, aluminum, concrete or wood, some innovative materials are introduced in contemporary footbridge structures. It should noticed that hybrid solutions combining elements made of different materials are also willingly used in bridges.

One of the most innovative materials being slowly introduced in contemporary bridge structures is Glass Fiber Reinforced Polymer (GFRP) composite material. There are several advantages of the GFRP material such as a high tensile strength, low self-weight, corrosion resistance, electromagnetic transparency, easy and quick assembly and creation of innovative shapes of pultruded profiles as well as whole structures. Besides these advantages GFRP has also some limitations like high initial costs, relatively low modulus of elasticity (Young's modulus), low alkaline resistance, low fire resistance and low long-term strength due to stress rupture [4]. Among others, the relatively low Young's modulus requires the use of relatively large cross-sections of structural elements in order to minimize serviceability problems of all-GFRP footbridges. Compared to GFRP composite, steel has the advantage of lower initial costs and several times higher Young's modulus which leads to thin-walled cross-sections with smaller outer dimensions.

This paper is focused on the comparative numerical analysis of the comprehensive dynamic behaviour in the vertical direction of two medium span cable-stayed footbridges made entirely of steel and GFRP composite material, with similar geometrical features and bearing capacity. A three dimensional (3D) Finite Element (FE) model of the all-GFRP footbridge was created based on the geometrical and material data of the existing Kolding Footbridge, acquired from its available technical documentation. Then, the FE model of the all-steel footbridge was made taking into account new structural steel elements providing a similar bearing capacity of both footbridges. The first part of the paper presents the determination of modal parameters of analyzed footbridges, i.e. their natural frequencies and corresponding mode shapes. For the further finite element analysis the damping ratio of the all-GFRP footbridge was measured *in situ,* whereas for the all-steel footbridge was taken from data published in the literature. In the second part, the dynamic responses of both structures were identified based on the Footfall harmonic analysis. In this case, the numerical models of analyzed footbridges were subjected to dynamic effects of human gait with various moving frequency. The maximal values of vertical acceleration of structural vibrations were obtained. The vibration comfort criteria for both footbridges were verified according to Eurocode [5] and Sétra [6] guidelines. The numerical results were compared and differences in dynamic responses of analyzed footbridges were examined. The obtained results delivered some important practical data regarding the dynamic behaviour of cable-stayed footbridges which could be used at a design stage to make a decision on the construction material solution.

2. Description of analyzed footbridges

Two construction solutions of medium span cable-stayed footbridges were analyzed. The first one was the existing Kolding Footbridge made entirely of GFRP composite material. The geometrical and material data of the all-GFRP footbridge were acquired from its available technical documentation. The second one was the newly-designed all-steel footbridge with similar geometrical features and new structural steel elements providing a similar bearing capacity as in the case of the all-GFRP footbridge.

2.1. All-GFRP composite footbridge

The Kolding Footbridge is located in Kolding city, Denmark, providing pedestrian and cycle traffic, and crossing the main railway line leading to the Kolding city center. The whole bridge construction, including bridge deck, pylon and cables, is made of GFRP composite using 12 different pultruded profiles. The 40.30 m long and 3.21 m width footbridge comprises two continuous spans with lengths of 26.35 and 12.40 m, supported by a single A-shaped 17.61 m tall pylon. Four pairs of cables with a length of 17.85, 17.66, 13.58 and 13.36 m were secured to the top of the pylon. The outer dimensions of the pylon are varying from 5.22x1.29 m at the bottom to 1.39x0.49 m in the upper part.

The structural system of the deck comprises two I-shaped pultruded profiles with a height of 1.5 m, thickness of 12 mm and length of 3.1 m used as main girders. Only bolts for assembly and abutments at the foundations were made of steel. The upper belt of the girders was designed from C-profiles with dimensions of 200x60x10 mm and 75x75x8 mm, being the handrail of the balustrade. Additional protection against buckling was made of transverse and longitudinal ribs constructed from angle beams with dimensions of 100x100x10 mm and 75x75x8 mm, respectively with a distance of 1.55 m and 0.73 m. The platform deck was made of composite platform gratings with a thickness of 50 mm. The usable width of the platform is 3.02 m. The supporting structure of the deck comprises cross-bars with C-shaped profiles and dimensions of 240x72x12 mm, and side-bars with I-shaped profiles and dimensions of 200x100x10 mm. All cables have square cross-sections with dimensions of 100x100x8 mm. The pylon structure was made of C-profiles with dimensions of 300x150x12 mm, while the side bracing beams were constructed from angle beams with dimensions of 75x75x8 mm. The longitudinal and cross-sections of the all-GFRP Kolding Footbridge are shown in Fig. 1(a-b).

2.2. *All-steel footbridge*

The structural system of the steel deck was designed from typical I-shaped profiles, i.e. comprises two HEB 300 profiles as the main girders, IPE 240 profiles as the cross-bars with a distance of 1.55 m, and IPE 200 as the side-bars. The platform deck was made of sheet metal with a thickness of 12 mm and with an epoxy resin surface of 5 mm. The usable width of the platform is 3.15 m. The balustrade was constructed from tie rods with a diameter of 3.0 cm, as security elements, and from tie rods with a diameter of 6.0 cm, as handrail elements connected by vertical banisters with I-shaped profiles with a distance of 1.55 m. The Macalloy 520 cable system was used to suspend the steel footbridge structure. The pylon structure was made of C-profiles with dimensions of 300x150x12 mm, while the side bracing beams were constructed from angle beams with dimensions of 75x75x8 mm. The cross-section of the all-steel footbridge is depicted in Fig. 1(c).

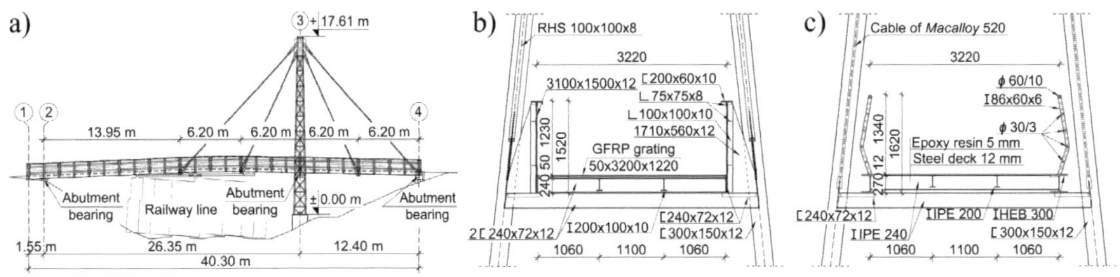

Figure 1. (a) Longitudinal and (b) cross-sections of the all-GFRP Kolding Footbridge, and (c) cross-section of the all-steel footbridge.

Both analyzed structural solutions, i.e. the all-GFRP and all-steel structures fulfill the condition of the geometric similarity as well as a similar load capacity of the main structural elements. To assess the differences in the bending stress σ and degree of carrying load capacity of structural elements, the static analysis were carried out including the steady and service loads acting on both footbridges. The analysis was executed using the Autodesk Robot Structural Analysis Professional software. The limit stress value for GFRP composite was assumed in accordance with Fiberline guidelines [9] as $f_{b,0°,d}$=185 MPa, while the assumed elastic limit of steel was f_y=235 MPa [10]. Fig. 2 presents the bending stress σ and degree of carrying load capacity of main structural elements of the all-GFRP and all-steel footbridges identified numerically taking into account the steady and service loads according to the Eurocode [7] and Polish Standard [8]. Based on the results presented in this figure, it can be concluded that the values of the capacity utilization factor of structural elements of both footbridges are very close. Thus, the structural similarity of the all-GFRP and all-steel footbridge was confirmed.

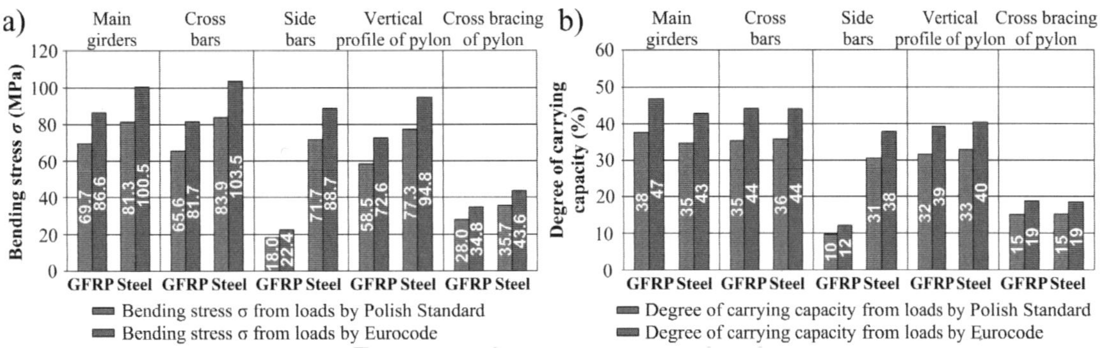

Figure 2. Comparison of (a) the bending stress σ and (b) degree of carrying load capacity of main structural elements of the all-GFRP and all-steel footbridges identified numerically according to the Eurocode [7] and Polish Standard [8].

In addition, the maximum static deflections q of both footbridges due to the steady and service loads according to the Eurocode [7] were numerically predicted in the mid-span of the main span and compared with the limit deflection values q_{dop} [9,10]. The results of calculations are presented in Table 1. It should be noticed that the maximum static deflection of the all-GFRP footbridge is about twice higher than comparably value of the all-steel footbridge.

Table 1. Predicted the maximum static deflections q of the all-GFRP and all-steel footbridges compared with the limit deflection values q_{dop} according to [9,10].

Construction solution of footbridge	Deflection due to the steady load (cm)	Deflection due to the service load [7] (cm)
all-GFRP	2.5	$5.0 < q_{dop} = 8.8$ [9]
all-steel	1.4	$2.7 < q_{dop} = 10.5$ [10]

3. FE models and modal analysis

In order to identify the dynamic characteristics, i.e. the natural frequencies and corresponding mode shapes of both considered footbridges, a 3D FE models were modeled using Autodesk Robot Structural Analysis Professional 2018 software. The FE model of the all-GFRP footbridge was created using the geometrical and material data of the existing Kolding Footbridge, acquired from the available technical documentation (see subsection 2.1). The FE model of the all-steel footbridge was made, including newly-designed structural steel elements describing in subsection 2.2. The Young's modulus of the GFRP material was adopted based on the test carried out in 2012 as E=23.4 GPa [11], while for the steel was assumed as E=210 GPa. The density of the GFRP profiles was 18.0 kN/m^3 while for the steel was assumed as 78.5 kN/m^3. The moment of inertia of all profiles was estimated from the geometry of its cross-sections. In order to predict a more realistic structural behaviour of footbridges all structural elements and accessory parts were taken into account in modeling process. The concrete foundations, considerably much stiffer than other structural elements, were not modeled.

Two kinds of finite elements were used in FE models of the all-GFRP and all-steel footbridges, i.e. 4429 and 5525 one-dimensional frame elements, and 11010 and 11843 two-dimensional shell elements, respectively, while the total number of active degrees of freedom was 59884 and 65626, respectively. The final FE models of both footbridges are shown in Fig. 3.

Sets of primary five natural frequencies and corresponding mode shapes of both footbridge models in the vertical direction are given in Fig. 4 and are summarized in Table 2. Based on the predicted fundamental natural frequencies, it can be stated that the all-steel footbridge is slightly stiffer than the all-GFRP footbridge, despite relatively smaller cross-sections of structural elements. Moreover, the same mode shapes sequences are observed for both footbridges.

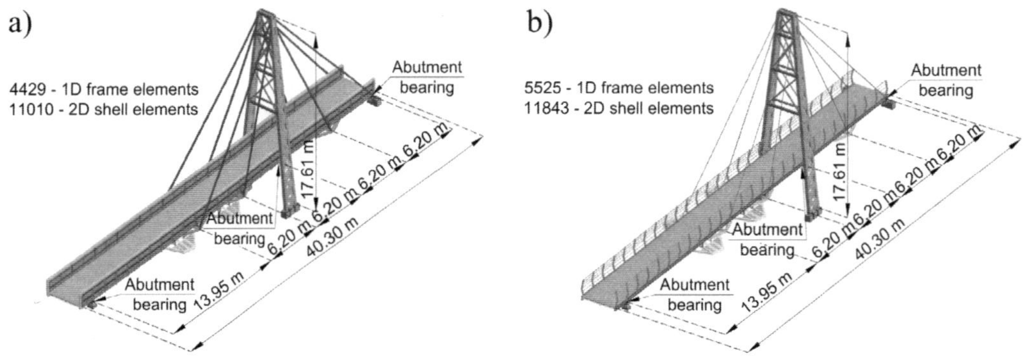

Figure 3. 3D FE models of (a) the all-GFRP and (b) all-steel footbridges.

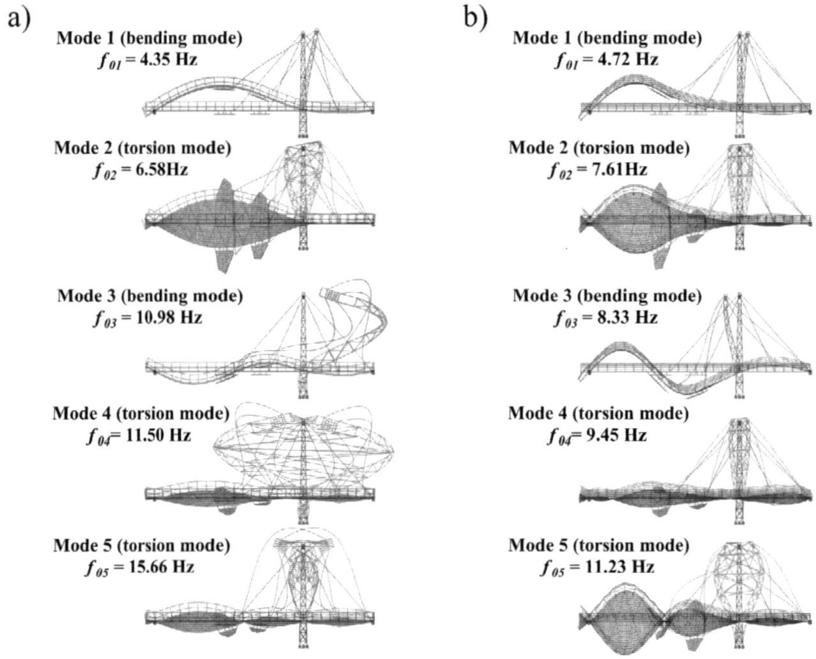

Figure 4. Primary five natural frequencies and mode shapes in the vertical direction of (a) the all-GFRP and (b) all-steel footbridges.

Table 2. Comparison of five natural frequencies predicted for the all-GFRP and all-steel footbridge models using FE analysis.

Mode No.	Mode type	Natural frequency (Hz)		Difference (%)
		all-GFRP	all-steel	
1	Bending	4.35	4.72	8.5
2	Torsion	6.58	7.61	15.6
3	Bending	10.98	8.33	24.1
4	Torsion	11.50	9.45	17.8
5	Torsion	15.66	11.23	28.3

4. Numerical Footfall harmonic analysis

Since human footfalls are the main source of footbridge vibration and it could cause discomfort to the structure users when the vibration level exceeds the recommended level, the serviceability performance of both considered footbridge was numerically predicted. Footbridge structures are subjected mainly to the loads of pedestrians walking or running on them. Thus, in the analysis two loading cases were considered, i.e. one walking or running person acting on the footbridge.

To estimate the maximum acceleration of vertical vibrations due to human excitations the Footfall harmonic analysis option available in Autodesk Robot Structural Analysis Professional 2018 software was used. The Footfall analysis is an appropriate tool to predict the dynamic response of a structure to footfall-induced loading. The pedestrian activities are modeled by an application of harmonic load model applied to selected structural nodes with the assumed frequency range [12].

The calculations of both considered footbridges were carried out on the assumption that the harmonic load was applied to all nodes of the footbridge deck. This approach enables to locate deck regions of highest acceleration response. According to recommendations given by Bachmann [13], the pedestrian walking was modeled at frequencies between 1.4 and 2.4 Hz, while the pedestrian running at frequencies between 1.9 and 3.3 Hz. The pedestrian mass assumed in the analysis was 90 kg. As the appropriate case study the SCI P354 analysis was chosen [12]. The damping ratio of the all-GFRP Kolding Footbridge was measured *in situ* as 2.87%, and this value was included in the Footfall analysis. The detailed description of the methodology and results of dynamic characteristic investigations of the existing Kolding Footbridge are presented in the paper [14]. For the all-steel footbridge the damping ratio was selected as 0.5% from data published in the literature for similar light-weight steel footbridges [14]. In the analysis, the natural mode shapes and corresponding frequencies of both footbridges up to 20 Hz were considered.

Fig. 5 shows the Footfall analysis results as the contour plots of acceleration levels on the deck of both analyzed footbridges for two considered load cases, i.e. one walking and running person. In this figure, the nodes with highest acceleration response were also marked. Fig. 6 presents the representative results of acceleration of the deck of both footbridges in the node with highest acceleration response depending on the frequency of the harmonic forces.

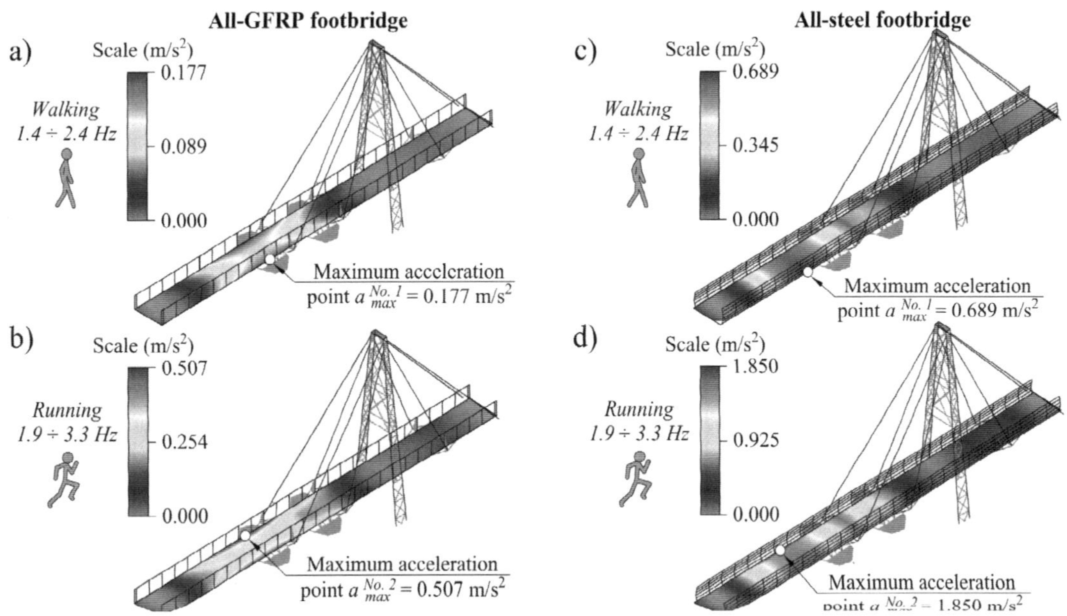

Figure 5. Footfall vertical acceleration response contours predicted for (a-b) the all-GFRP and (c-d) all-steel footbridges excited by one walking or running person, respectively.

5. Evaluation of comfort criteria

The critical acceleration values obtained from the Footfall analysis were compared with the limits provided by Eurocode [5] and Sétra guidelines [6]. According to the guidelines presented in the

Eurocode standard, the maximum vertical accelerations of spans of newly-designed footbridges should not exceed of 0.7 m/s^2, however, a dynamic analysis is required if the natural frequency of structure in the vertical direction is below 5 Hz. Much more advanced recommendations are provided by the Sétra technical guide. In these recommendations four classes of comfort criteria were formulated for footbridges regarding the measured critical accelerations. The risk of resonance could appear within the frequency range of 1÷5 Hz. Evaluation of vibration comfort criteria for the all-GFRP and all-steel footbridges for the vertical vibrations induced by one walking and running person is shown in Fig. 7.

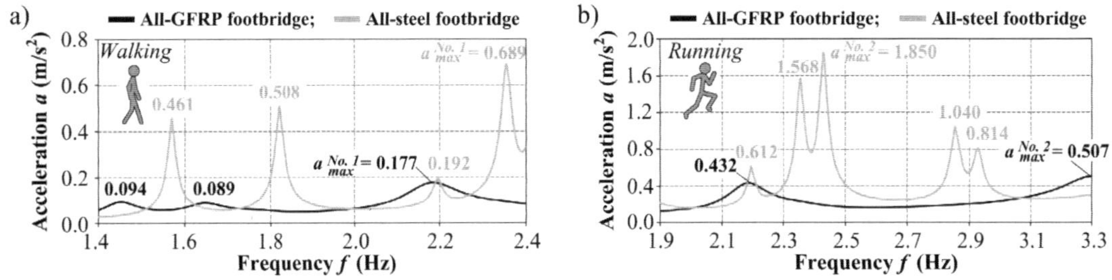

Figure 6. Footfall analysis results of acceleration of the deck of considered footbridges in the node with highest acceleration response depending on the frequency of the harmonic forces for: (a) one walking person (1.4 and 2.4 Hz), (b) one running person (1.9 and 3.3 Hz).

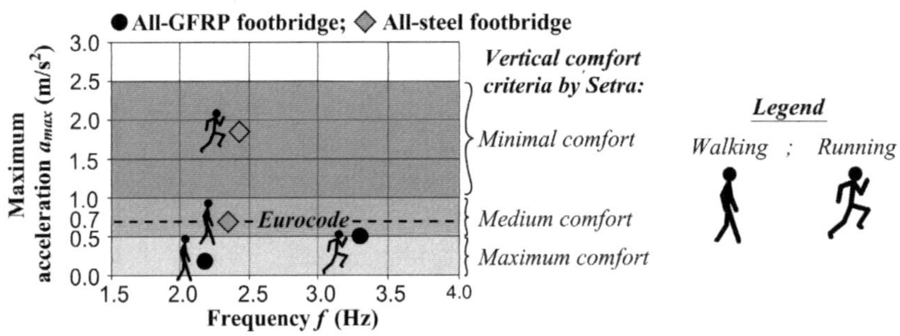

Figure 7. Evaluation of vertical comfort criteria of the all-GFRP and all-steel footbridges by Eurocode [5] and Sétra guidelines [6].

6. Conclusions

This comparative study analyzed numerically the comprehensive dynamic behaviour in the vertical direction of two medium span cable-stayed footbridges made entirely of steel and GFRP composite material. The first considered structure was the existing all-GFRP Kolding Footbridge, whereas the second one was newly-designed all-steel footbridge with similar geometrical features and new structural steel elements providing a similar bearing capacity as in the case of the all-GFRP footbridge. Considering the steady and service loads according to the Eurocode [7] and Polish Standard [8], it was proved that the degree of carrying load capacity of main structural elements of the all-GFRP and all-steel footbridges was very close (see Fig. 2b), e.g. for the main girders of both footbridges it was between 35 and 47%, while for the main pylon structure between 32 and 40%. The predicted maximum static deflections of the main span of both footbridges were not exceeded the limit deflection values. However, the maximum static deflection of the all-GFRP footbridge was about twice higher than the calculated value for the all-steel footbridge.

Based on the results of the FE modal analysis sets of primary five the natural frequencies and corresponding mode shapes of both footbridges in the vertical direction were compared (see Fig. 4 and Table 2). As a result of nine times higher Young's modulus and four times higher density of structural steel elements, the predicted fundamental natural frequency of the all-steel footbridge was 8.5% higher than for the all-GFRP footbridge. Thereby, it can be concluded that the structural stiffness of the all-steel footbridge is slightly higher than in the case of the all-GFRP footbridge, despite relatively larger cross-sections of GFRP structural elements. Both footbridge models have similar mode shapes and the same mode shapes sequences were observed for both structures. The damping ratio of the all-GFRP footbridge identified experimentally as 2.87% is much higher than the value of 0.5% (based on the literature data) for similar all-steel footbridge. This advantageous dynamic property of the all-GFRP footbridge has a crucial influence on the reduction of vibrations caused by pedestrian excitation.

The fundamental frequencies of both footbridges in the vertical direction were found within the critical ranges given by Eurocode [5] and Sétra guidelines [6]. In this case further analysis with respect to the maximum acceleration threshold was required. Based on the Footfall analysis results it was found that the predicted maximum accelerations of vertical vibrations due to human excitations of both considered footbridges are different. The maximum dynamic response in the middle of main span of the all-steel footbridge induced by one walking pedestrian, at frequencies between 1.4 and 2.4 Hz, was about 3.9 times higher compared to the all-GFRP footbridge, and about 3.6 times higher induced by one running pedestrian at frequencies between 1.9 and 3.3 Hz (see Figs 5 and 6).

Comparing the maximum acceleration values obtained from the Footfall analysis for both footbridges to the limit values provided by Eurocode [5] and Sétra guidelines [6] it can be concluded that the all-GFRP footbridge satisfied the vibration serviceability requirements given by Eurocode, and the comfort level was classified as maximum according to Sétra guidelines under both considered load cases. However, the all-steel footbridge was not fulfilled the limit value suggested by Eurocode, and according to Sétra guidelines the comfort level was classified as medium and minimal under one walking and running person, respectively.

References

[1] Kutz M 2002 *Handbook of Materials Selection* (New York: Wiley-Interscience)

[2] Meinhardt C 2009 *Footbridge Vibration Design* (London: CRC Press)

[3] Van Nimmen K, Lombaert G, De Roeck G and Van den Broeck P 2014 Vibration serviceability of footbridges: Evaluation of the current codes of practice *Engineering Structures* **59** pp 448-461

[4] Stankiewicz B 2017 Durability of glass fiber-reinforced polymer bridge panel based on differential thermal analysis, dynamic mechanical analysis, and differential scanning calorimetry analysis *Journal of Composite Materials* **51**(16) pp 2301-2313.

[5] European Committee for Standardization (CEN) 2004 Basis of structural design; *Eurocode, Annex A2 (EN 1990:2004/A1:2008)*

[6] Sétra 2006 *Assessment of vibrational behaviour of footbridges under pedestrian loading* (French Association of Civil Engineers Sétra/AFGC)

[7] European Committee for Standardization (CEN) 2003 Actions on Structures; Traffic loads on bridges *Eurocode 1, Part 2 (EN 1991-2)*

[8] Polish Committee for Standardization (PKN) 1985 The bridge structures; Loads *Polish Standard (PN-85/S-10030)*

[9] Fiberline Composites A/S 2003 *Fiberline Design Manual* (Kolding: Fiberline Composites A/S)

[10] European Committee for Standardization (CEN) 2005 Design of steel structures; General rules and rules for building *Eurocode 3, Part 1-1 (EN 1993-1-1)*

[11] Technical Report 2013 *Generaleftersyn samt undersøgelse af egenskaber for glasfibermateriale,* ref. no. 1100002180\LF00001-2-NTO

[12] Autodesk Robot Structural Analysis Professional 2018 *User's Guide*

[13] Bachmann H 2002 „Lively" Footbridges – a real challenge *Proc. Int. Conf. on Design and Dynamic Behavior of Footbridges (Paris)* pp 18-30

[14] Górski P, Stankiewicz B and Tatara M 2017 Modal parameter identification of all-GFRP composite cable-stayed footbridge in Denmark *Proc. Conf. on Dynamics of Civil Engineering and Transport Structures and Wind Engineering (Slovakia)* ed J Melcer, K Kotrasova (Slovakia: MATEC Web of Conferences) **107** pp 26-33

IOP Conf. Series: Materials Science and Engineering **419** (2018) 012031 doi:10.1088/1757-899X/419/1/012031

Advanced layout of a steel bridge - Ullevaalskrysset footbridge

Andri Gunnarsson[1] and Magnús Arason[1]

[1]EFLA Consulting Engineers, Höfðabakki 9, Reykjavík 110, Iceland

E-mail: andri.gunnarsson@efla.is

Abstract. The Ullevaalskrysset Footbridge replaces two temporary and ageing truss girder bridges across a busy road intersection with a single, modern steel structure, shaped to suit the requirements of pedestrians and cyclists. Different layouts and bridge types were examined by EFLA in the conceptual phase, resulting in the client choosing a steel box girder bridge with cantilever ribs stretching out from a continuous spine. The bridge is constructed of four ramps with fairly constant construction height and span lengths. The ramps connect to the main spans of the bridge across the two roads at an intersection. The height of the cross section varies across the main spans. A pair of V-shaped steel columns support the bridge. The four ramps join up in an integration zone with increased width to allow for safe intersection of the flow of cyclists and pedestrians in all four directions. The steel bridge has a total length of 290 m in 18 spans and the total width of the structure is 7,2 m. The longest span is 37,5 m, over the main highway (Ring 3), and 27,5 m, over a side road. Tuned mass dampers are installed in both spans to secure acceptable comfort for pedestrians. Practical aspects of steel as a structural material, are widely integrated in the design, resulting in light structure with seamless flowing aesthetic.

1. Introduction

The bridge site is in Ullevaal, a northeastern district of Oslo, Norway, at the busy road intersection of Ring 3 and Sognsveien. The Ullevaal district is a trafficked area, heavily used by motorist, pedestrians, and cyclists. Various infrastructure landmarks such as the University hospital, University of Oslo, several colleges, and the national football stadium, Ullevaal, are located in the vicinity of the site [1].

The principal aim of the project is to replace the existing bridges at the site, shown in Figure 1, and thereby improve the accessibility, safety and capacity for the pedestrians and cyclists in the area. The project has also been aimed at providing a bridge structure with aesthetic qualities. The existing bridges no longer meet the requirements of pedestrians and cyclist with ramp slope of 1:10 as well as being only 2,5 m wide. The bridges also tend to oscillate when exposed to high traffic load.

The Ullevaalskrysset Footbridge, shown on Figure 2, replaces the ageing truss girders with a single, modern steel structure, shaped to suit the requirements of pedestrians and cyclists.

A preliminary design phase for new bridges was carried out in conjunction with areal planning in 2008. This focused on a post-tensioned concrete bridge. However, the design work was put on hold and was not resumed until January 2015. At that point, EFLA Consulting Engineers, in cooperation with BEaM architects, were engaged by the Norwegian Road Administration to compare steel and concrete based bridge solutions, devise new alignments for the bridges and subsequently to prepare the tender design of the favoured alternative.

The bridge is a steel box girder bridge with cantilever ribs stretching out from a continues spine. The steel bridge has a total length of 290 m in 18 spans and the total width of the structure is 7,2 m (free width 6,0 m). The bridge is constructed of four ramps with fairly constant construction height and span lengths. The ramps connect to the main spans of the bridge across the two roads at an intersection. The longest span is 37,5 m, over the main highway (Ring 3), and 27,5 m, over the Sognsveien secondary road. The height of the cross section varies across these main spans and tuned mass dampers are installed in both spans to secure acceptable comfort for pedestrians. Pairs of V-shaped steel columns support the bridge. In turn, the columns are supported on pile caps founded on composite piles, i.e. steel tubes filled with concrete, extending 20-45 m down to bedrock.

Figure 1. The ageing truss girder bridges of today.

Figure 2. The new Ullevaalskrysset footbridge (rendering by BEaM).

The tender design was finalized in the spring of 2017 and a tender was opened in the summer of 2017, resulting in the appointment of a contractor in October 2017. The construction work started in November 2017 and planned completion of the construction project is in late 2019.

2. Ullevaalskrysset area and background of the project

The bridge, currently situated at the site, were constructed as preliminary structures in 1992 in conjunction with large road and tunnel projects in the area. Work on the revision of the areal planning started in 2007 and completed in 2012, defining the footprint for a new bridge [2].

The existing bridges are heavily used by pedestrians and cyclists, commuting to and from the large infrastructure landmarks in the area and the bus stops at both sides of Ring 3. Furthermore, the cycling routes along Ring 3 and Sognsveien are part of the main cycling network in Oslo [2].

The bridges are narrow, only 2.5 m wide, which results in low capacity and danger of collisions between pedestrians and cyclists. Their ramps slope 1:10 without landings, making them unsuitable for the disabled. The bridges also tend to oscillate when exposed to high traffic associated with events at Ullevaal stadium, making users feel unsafe.

The cycling route across Sognsveien is currently at a zebra crossing, very close to the Ring 3 slip road. This causes danger of collisions between motorists and cyclists crossing the road due to limited visibility and high speed of both parties. In addition, using the existing bridge to cross Ring 3 on a bicycle is currently very tricky, leaving a gap in the defined cycling route along Sognsveien at this point.

3. Design criteria

The main design criteria were to increase the traffic safety, capacity and accessibility for pedestrians and cyclists at the Ring 3 – Sognsveien intersection.

To this end, the bridge has a free width of 6 m width between railings, defined as a 3.5 m wide cycling lane and a 2.5 m wide footway separated by a 20 mm high curb. Maximum longitudinal slope is 7% and the vertical clearance above the road surfaces is to be 4.9 m at minimum.

The design load criteria included the following:

- Pedestrian load according to NS-EN 1991-2.
- Vibrations from pedestrians should be accounted for according to SÉTRA Guidelines
- Wind load according to NS-EN 1991-1-4. Base value of v_b = 22 m/s.
- Temperature load according to NS-EN 1991-1-5
- Collision load according to NS-EN 1991-1-7

The new bridge crosses a busy highway, Ring 3 (60.000 vehicles/day) and Sognsveien (12.000 vehicles/day) in Oslo. High emphasis has therefore been put on defining a structural solution which minimizes the effect on the motorists in the construction phase. This also highlights the need for a structure with a low maintenance demand, especially where the bridge crosses the roads.

There are also various other constraints that affect the choice of solution. Distance to bedrock is 20-45 m and the soil mainly consists of sensitive clay which makes for a challenging foundation design. There is a lot of various technical installations in the ground, including drainage, electrical and telecommunications, leading to a requirement for relocation of cables and/or pipes at all the 19 bridge foundations. The areal planning of the site also entails various challenges as it was developed assuming a narrower bridge of smaller footprint compared to the final design. A time-consuming re-evaluation of the planning process, with new width criteria for the bridge, was not an option within the project framework.

4. Conceptual phase and choice of solution

The conceptual phase took place in the winter of 2015 with the main objectives of defining a solution in terms of plan and profile and bridge type. The main criteria for the choice of solution where cost, feasibility, functionality, traffic safety for the pedestrians and cyclists, effect on infrastructure in the ground, operation and maintenance, environmental impact, aesthetics and effect on the motorists.

Several alternatives were considered for the bridge alignment in plan, resulting in the selected alternative which is shown on Figure 3. The chosen layout has four adjacent ramps which connect in a large integration zone at the centre of the bridge. This is where pedestrians and cyclists coming from all four directions will intersect each other in a safe manner. There are also two stairs which connect two of the main bus stops in the area to the bridge. The chosen plan layout defines the basic geometry for the bridge, i.e. a total width of 7.2 m and a total length of 290 m which is divided to 18 spans. The spans on the ramp sections vary from 14 to 18 m in length while the two main spans above Ring 3 and Sognsveien are 37.5 m and 27.5 m long respectively.

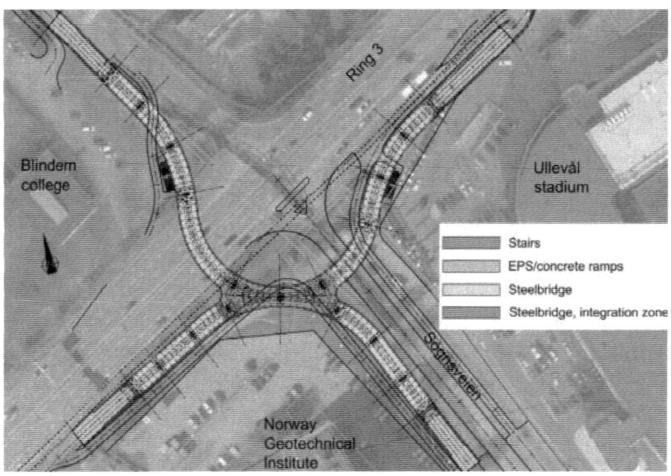

Figure 3. Plan view for the bridge.

A pre-design phase was carried out in 2008, resulting in a proposal for a post-tensioned concrete bridge. However, when the design recommenced in 2015, the client also wanted to explore the alternative of a steel bridge. Both alternatives where considered for the same layout.

For the concrete bridge, a post-tensioned beam with an elliptically shaped underside was considered as shown on Figure 4, while the three following sub-alternatives where considered for the steel bridge:

- A cable stayed bridge where the two main spans are supported by the main cables and the stay cables are attached to two of the ramps. The proposed cross section is shown on Figure 5. A pylon was then planned at the centre of the bridge which would also support the intersection zone of the bridge.
- A girder bridge with a triangular structural edge beams.
- A box girder bridge with a box section forming a narrow spine below the deck with cantilevered ribs extending from it to support the full width of the bridge.

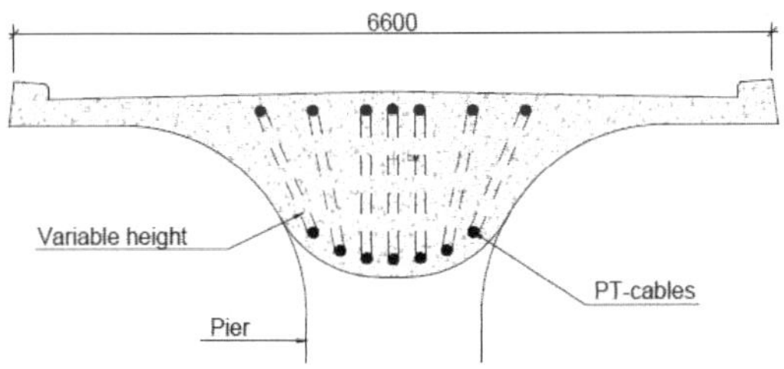

Figure 4. Cross section for the proposed post-tensioned concrete bridge solution.

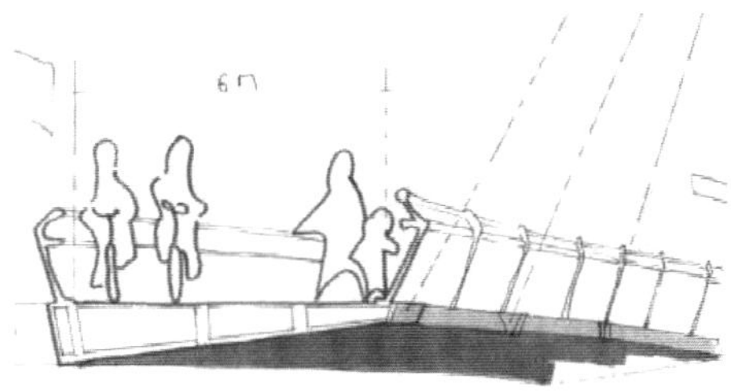

Figure 5. Cross section for the proposed cable stayed bridge solution (Sketch by BEaM).

The client eventually chose the steel box girder bridge mainly with respect to its feasibility and traffic safety during construction, along with the perceived flowing, lightweight and simple aesthetic of this bridge. Figure 6 and Figure 7 show parts of the rendering work done in the conceptual phase, visualizing the planned appearance of the bridge.

Figure 6. The new Ullevaalskrysset footbridge viewed from Ring 3 (Rendering by BEaM).

Figure 7. The new Ullevaalskrysset footbridge view from the bridge deck (Rendering by BEaM).

5. Structural system and design phase

The structural system is a box girder section running along the centreline of the bridge below deck level, see Figure 8. It has a height of 0.6 m at the ramps but varying height above the main spans, 0.9-1.2 m above Ring 3 and 0.7-0.9 m above Sognsveien. The width of the box section is approximately 1/3 of the total width of the bridge and it forms a seamless flowing spine below the bridge deck as shown in Figure 6.

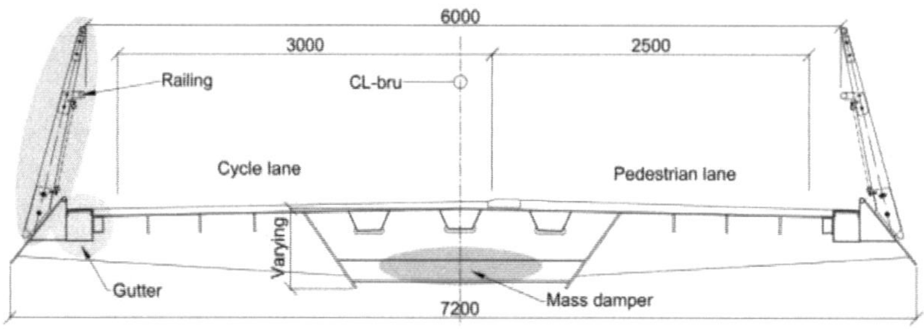

Figure 8. Principal cross section of the bridge deck.

The stiffened box section forms the main bearing element of the bridge, and the use of a stiffened box section is a widely known solution for the span lengths of this project.

Cantilevered V-shaped beams extend out from the spine at 2.3-2.5 m intervals supporting the rest of the bridge deck. At the intersection zone, the beams extend further out and connect with the beams from the adjoining bridge part. These cross beams support the large transition zone between the spines. The weight of the steel deck is varying between 1.4-2.4 tonne/m. The total steel weight is 530 tonnes. At the ends, lightweight ramps of concrete and EPS are constructed to provide connection between the bridge deck and ground level. Longitudinal sections are shown in Figure 9 and Figure 10 below.

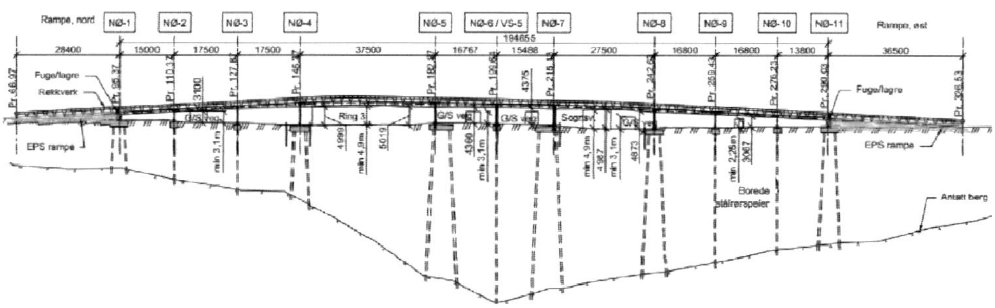

Figure 9. Longitudinal section for the bridge, stretching from the north end to the east end.

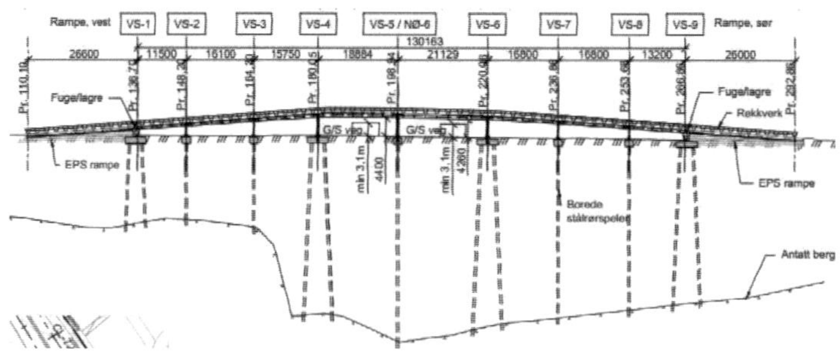

Figure 10. Longitudinal section for the bridge, stretching from the west end to the south end.

The bridge is supported by V-Shaped pairs of columns with circular cross sections. Their deck connections are pinned along the ramps to allow for thermal expansions, see Figure 11. Five pairs of columns supporting the transition zone are welded to the underside of the bridge deck to provide rigidity, see Figure 12. The bridge is supported on bearings at the ramp ends. The bearings sit on concrete abutments. Expansion joints are also integrated in all of the four bridge ends.

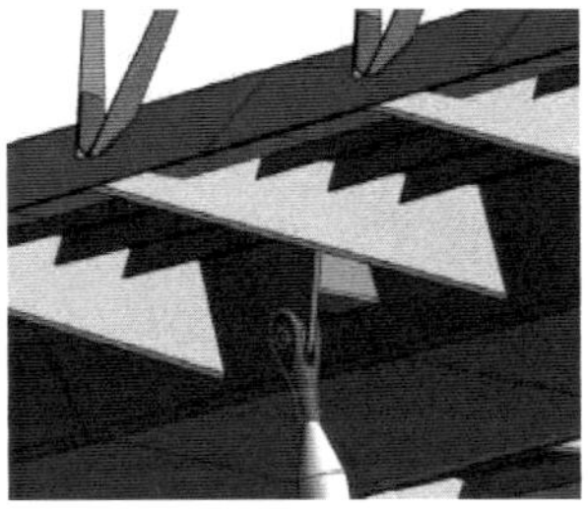

Figure 11. Pinned connection at the ramp sections.

Figure 12. Fixed connection at the transition zone.

The columns are supported on concrete foundations (pile caps) which rest on Ø508-610 mm drilled composite piles (steel tubes filled with reinforced concrete) which extend to bedrock.

Some of the lowest natural frequencies of the bridge are quite close to the frequencies for pedestrians and joggers. The first horizontal mode along with the two first vertical modes for the Ring 3 span (37.5 m) and the first vertical mode for the Sognsveien span (27.5 m) are considered vulnerable for oscillation due to vibration from pedestrians and are considered in the dynamic analysis of the bridge. Models for crowd loading are used to estimate acceleration and comfort criteria. A finite element model for the bridge, constructed of linear elements in SAP 2000 (V17.2.0) and defined with the relevant foundation stiffness as calculated by the advanced pile group program (Geosuite Pilegroup), is used to carry out the dynamic analysis.

The calculated acceleration in the vertical direction, both for the Ring 3 and the Sognsveien spans, exceeds limits. The main reason for this vulnerability regarding the response of the structure in the vertical direction is the relatively slender cross section for the main spans along with the relatively low modal mass, especially on the Sognsveien span. Vulnerability to dynamic loading was not recorded for other spans of the bridge, ranging from 14 to 18 m in length.

Tuned mass dampers will be installed in both the Ring 3 and the Sognsveien spans, 2000 kg and 1000 kg respectively. The effect of the dampers for the Ring 3 span is shown in Figure 13, and this clearly demonstrates how effective they will be in enhancing the comfort of pedestrian and securing a dynamic behaviour which is up to standard.

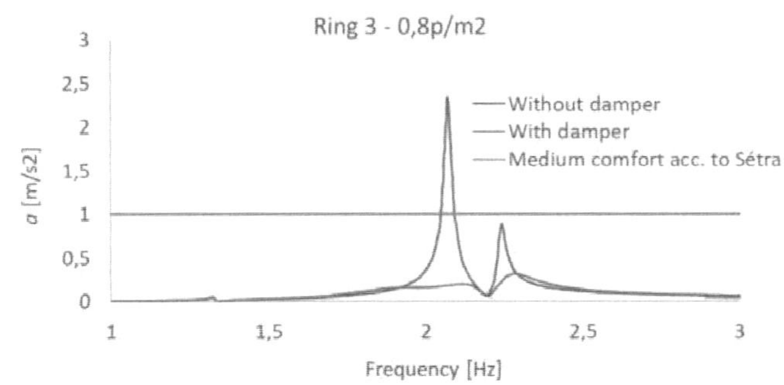

Figure 13. Calculated vertical response of the Ring 3 span of the new Ullevaalskrysset footbridge when loaded with crowds of density 0.8 pers. / m^2, with and without damper.

A gutter is embedded into the edge beam along both sides of the bridge deck. It will be made of stainless steel to minimise the need for maintenance. The railing is also made of stainless steel, designed with V-shaped posts extending from the V-shaped transverse beams. A stainless steel meshing spans between the posts. An LED lighting is integrated into the handrail to provide adequate illumination of the bridge deck.

6. Construction phase

The construction phase started in November 2017 and planned completion of the construction is in late 2019.

The main contractor is HAG Anlegg AS of Norway but the subcontractor responsible for the fabrication of the steel bridge is Western Construction of Lithuania.

Space is limited at the construction site. It is split in half by the Ring 3 motorway, and Sognsveien also splits the southern site in half. This makes for a challenging planning of the construction works, especially regarding work with piles and mounting of the steel bridge.

The work schedule for the bridge consists of four main aspects, installation of piles, casting of foundations, fabrication of steel and erection of the steel bridge. Work with installation of piles is carried out separately for the three main quarters of the site followed by casting of the foundations. Work with installation of piles, shown on Figure 14, started in February 2018 and planned completion of the concrete foundation and abutments is in August 2018.

Figure 14. Work on pile-drilling in the southwestern quarter of the construction site.

Figure 15. Work on relocating electrical cables in the ground.

Fabrication work for the steel bridge is ongoing during the work with piles and foundations. The steel bridge will be transported to Oslo from Lithuania in relatively small parts. It will then be welded together to larger parts, approximately one span, on site and each span will then be lifted to its final position. The plan is to begin by erecting the transition zone of the bridge to create a rigid base, and then to construct the four ramp sections in all directions. The biggest challenge of the erection phase is the main span over Ring 3. The road must be closed for as little time as possible (approximately 2 days) making it very important to carry out the task in a quick and efficient way. Planned beginning of the erection works is in the fall of 2018 and planned finish is in the spring of 2019.

In addition to the bridge works, a lot of electrical cables, se Figure 15, and drainage and communications infrastructure is relocated or restructured. The Sognsveien road in its entirety will also be relocated and pedestrian- and cycle lanes constructed at ground level. These additional works are integrated into the work schedule for the bridge construction.

7. Discussion

A new footbridge providing separated cycle- and pedestrian lanes, high capacity, a safe and innovative layout and a refined structure has been presented. The bridge will improve considerably the accessibility and traffic safety for the pedestrians and cyclists in the area.

The design team has fulfilled the main objectives defined by the client and provided the means for creating an important link in the infrastructure in Oslo.

The complexity of the project and the construction site makes steel the favourable material selection for the bridge because:

IOP Conf. Series: Materials Science and Engineering **419** (2018) 012031 doi:10.1088/1757-899X/419/1/012031

- Steel is appropriate when there is a demand for a lightweight and advanced geometry.
- A steel solution minimizes the effect on the underlying highway due to short construction time of the crossing, i.e. the whole span, crossing Ring 3, can be lifted in over 2-3 days. This is a big advantage compared with a concrete bridge which requires extensive scaffolding and closure of the underlying road during the curing period of the concrete.
- It is possible to transport the bridge to the site in relatively small elements and erect it simultaneously making the most of a tight construction site.
- It shortens the construction time to be able to fabricate the steel parts simultaneously to foundation works.

References

[1] Ernst C, Arason M 2016 Cycle Superhighways: Ullevaal Footbridge, Oslo, Norway *Challenges in Design and Construction of an Innovative and Sustainable Built Environment, 19th IABSE Congress (Stockholm)* ed L Elfgren, J Jonsson *et al* (Zurich: ETH Hönggerberg) pp 2662-2666

[2] Cowi AS 2008 *Rv 150 (Ring 3) og Sognsveien – Kryssing ved Ullevål Stadion for gående og syklende – Detaljplan – Planhefte 1* (Oslo: Norwegian Public Road Administrastion)

IOP Conf. Series: Materials Science and Engineering **419** (2018) 012032 doi:10.1088/1757-899X/419/1/012032

Railway Road Bridge in Novi Sad – Steel tied network arches across the Danube

A Bojović[1], A Mora[2], Z Marković[3], M Pavlović[4], N Novaković[5], M Spremić[6]

[1] Lead bridge structure designer, DEL ING DOO, Belgrade, Serbia
[2] Project manager, Azvi S.A., Seville, Spain
[3] Designer, University of Belgrade Faculty of Civil Engineering, Serbia
[4] Associate designer, Delft University of Technology, Netherland
[5] Associate designer, DEL ING DOO, Belgrade, Serbia
[6] Associate designer, University of Belgrade Faculty of Civil Engineering, Serbia

aleksandar.bojovic@deling.rs
amora@azvi.es
zlatko@grf.bg.ac.rs
M.Pavlovic@tudelft.nl
novak.novakovic@deling.rs
spremic@imk.grf.bg.ac.rs

Abstract. The Railway road bridge in Novi Sad is situated on the international railroad Belgrade-Budapest. The bridge is designed for two railway tracks (with design speed of 160 km/h), two road lanes and two footpaths. The bridge structure consists of four structures: two approach composite bridges at the banks and two steel tied network arch bridges over the river and transition structure between two arches. The spans are 27.0+177.0+3.0+219.0+48.0 m, totally 474.0 m in length. The rises of arches are 34.0 m and 42.0 m respectively. The width of the bridge is 31.440 m. The arches and ties, as well as the girders of the approach spans, are steel box girders. The decks of all bridge structures are the composite reinforced concrete slabs with thickness of 300 mm, locally 400 mm. The launching was very complex, in both analysis and construction. The arch bridges were fully assembled on the banks and launched by skids over the bank and by pontoons over the river, to the final position on piers. The bridge was designed fully according to European standards and additionally – according to requirements of German codes DIN-Fachbericht 101 to 104 Edition 2009 and Deutsche Bahn Richtlinie 804, Edition 2003. The bridge is, despite of heavy loads and structural complexity, very rational in steel volumes and construction costs as well. The bridge is also, with the arch span of $L = 219$ m, the bridge with the longest span in the world in the category of tied network arch bridges with two rail tracks. The bridge is open to rail traffic at 7th April 2018.

1. Introduction

The New Railroad Bridge across the Danube River in Novi Sad is situated on the location of the old concrete road railway bridge – destroyed in air strikes in 1999, on the major international electrified railroad line No 2 Belgrade – Budapest, over the Danube, *figure 1*.

The contractor for the bridge construction is Spanish-Italian consortium AZVI-Taddei-Horta Coslada. The list of participants is given in section 6.

Figure 1. Location of the Bridge in Novi Sad.

Terms of Reference, composed by Serbian Railways includes the following basic references:

- Bridge location: location of the old bridge (1961-1999);
- Traffic across the bridge: 2 tracks + 2 traffic lanes + 2 foot-cycling paths;
- Structural system, as urban planning condition: steel arch structures over the river;
- Maximum line (train) speed: passenger trains – 160 km/h, freight trains – 120 km/h;
- Maximal vertical acceleration: $a_v \leq 1.3$ m/s^2;
- Road and rail: rail tracks axes distance = 4.20 m, road lanes = 2 x (3.50+0.35) m;
- Bridge accessories: water pipes deck drainage system, accessories and equipment for electrical power, telecommunications, illumination (public, intern and decorative), navigation signals (for ships and airplanes), traffic signalization of the road and railway, road safety barriers;
- Foundation: supporting of the new bridge onto the existing fundaments as much as possible;
- Design codes: key code Richtlinie 804:2003 [2] and related to that code series DIN-Fachbericht 101 to 104 Edition 2009 [3] to [6], i.e. appropriate codes series EN 1991, EN 1992, EN 1993, EN 1994 and EN 1998 in actual editions. Alpha factor for railway traffic load is 1.0.

2. Bridge structures

2.1. General

Essential general data about the Railway Road Bridge in Novi Sad from design [1], *figure 2*:

- Track alignment: according to the traffic solution of traffic line management – satisfying the requirement for the as short as possible connection with the existing traffic line and securing of navigation clearance under the bridge;
- Structural system: Four independent structures as simple beams, left bank approach bridge 25.30 m, main (river) bridges 177.00 and 219.00 m; right bank approach bridge 45.30 span;
- Position of the piers: Position of the axis *3* (axis of the middle pier in the Danube), is done according to the requirement that the pier should be positioned centrically onto the existing foundation of the old bridge. Positioning of the axes *2* and *4* of the designed bridge is done bearing in mind the positions of the existing foundations, position of the piers required by

the conditions of proper institution and structural raster of the bridge structure (as *n* x 3000 mm). The piers distances *1-2-3-4-5* are: 27.00 + 178.50 + 220.50 + 48.00 m;

- Main bridge dimensions: total length is 474.00 m; bridge width between the outer edges of the masks is 31.440 m; spacing of the ties axes is 23.50 m, *figure 3*.

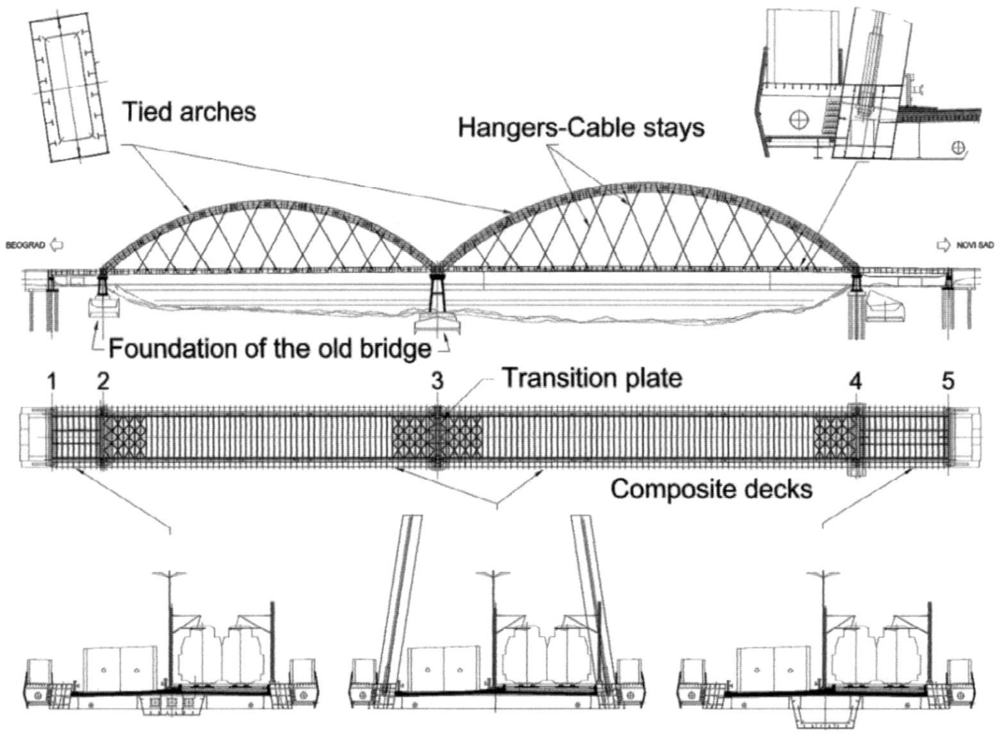

Figure 2. Layout of the bridge structure.

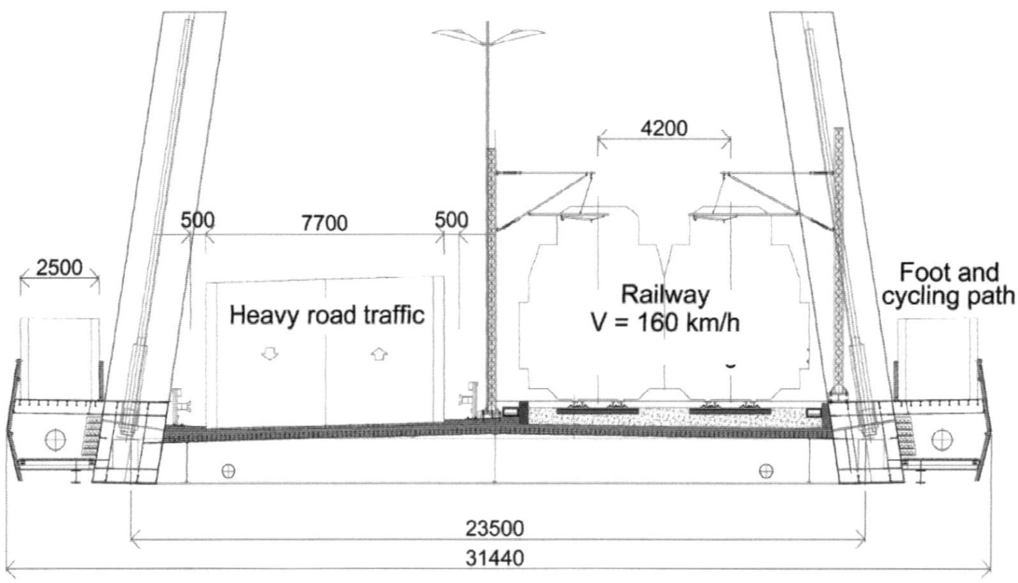

Figure 3. Cross-section of the deck of the arch bridge structures.

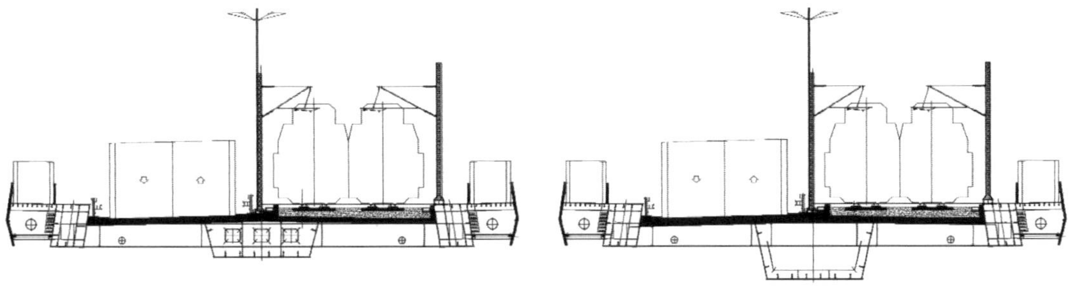

Figure 4. Cross-sections of the approach bridge structures *1-2* and *4-5*.

The choices and decisions in the design process were influenced by: elements of the Terms of Reference, design codes requirements, the requirements of rational erection and the need to achieve the optimal quantity of structure material.

Alignment of railway, and hence the location of the bridge structure was adapted according to the navigation clearance of the Danube River under the bridge. Also technical requirements for the design of railway and connection to the existing railway tracks on both banks were considered. The ballast depth is min 350 mm.

The transversal slopes of bridge deck are 2.5% and 1.5% which corresponds to conditions of drainage, both road and rail respectively, *figures 3 and 4.*

Structural materials:

- Structural steel grades are S355 mainly and in very limited areas of the nodes arch/tie is S460 applied, with thickness constraints according to the terms of Richtlinie 804:2003 [2] and DIN-Fachbericht 103:2009 [5]; the maximal plate thickness is 50 mm;
- The concrete deck slab is casted in situ, with concrete class C35/45 according to EN 1992-1-1:2004 and DIN-Fachbericht 102:2009 [4]:
- Headed studs are Ø25 x 200 mm according to EN ISO 13918:1998 and EN ISO 14555:1998;
- Reinforcement bars are from steel B500C according to EN 10080:2005;
- The hangers of arch bridges are parallel strand cables with the ultimate (nominal tensile) strength f_u = 1860 N/mm^2.

The structure and hangers are painted in signal white (RAL 9003) which reduces influence of thermal actions on the bridge structures.

2.2. Main bridges

The main bridge structures are tied arches with hangers in network arrangement.

Basic characteristics of arch structures:

- Spans and rises: bridge *2-3*: *L*/*H* =177.0/34.0m=5.21; bridge *3-4*: *L*/*H* =219.0/42.0m =5.21;
- Main structural system acts as a spatial structure, arches + transversal arch connecting beams + ties + hangers + cross beams + concrete deck plate;
- Box cross-sections dimensions of arches: bridge *2-3*: 4000x2000 mm; bridge *3-4*: 4900x2000 mm;
- Cross-sections dimensions of ties: box 2550x2000 mm, footpath cantilever length *b*=2.90m;
- Hangers: cable stays as 55 parallel strands; each cable is with the magnetic sensor on the reference strand (randomly selected strand); the cable sensors for registering of actual cable stays axial forces are the part of monitoring system of the bridge structure; the number of hangers at the arch bridges – 2x18 and 2x22;
- Deck slab: *t* =30cm and *t* =40cm at the ends – in bracings zone; three layers of reinforcement bars in longitudinal direction and two layers in transversal direction,(*figure 8*);

- Supports: 2 x 3 bearings at system axes. Longitudinal and transversal displacements are restrained at central bearings in axis *3*, (*figure 6*).

About the main structural designing decisions and most important structural analysis results see the details in the references [7], [8] and [9].

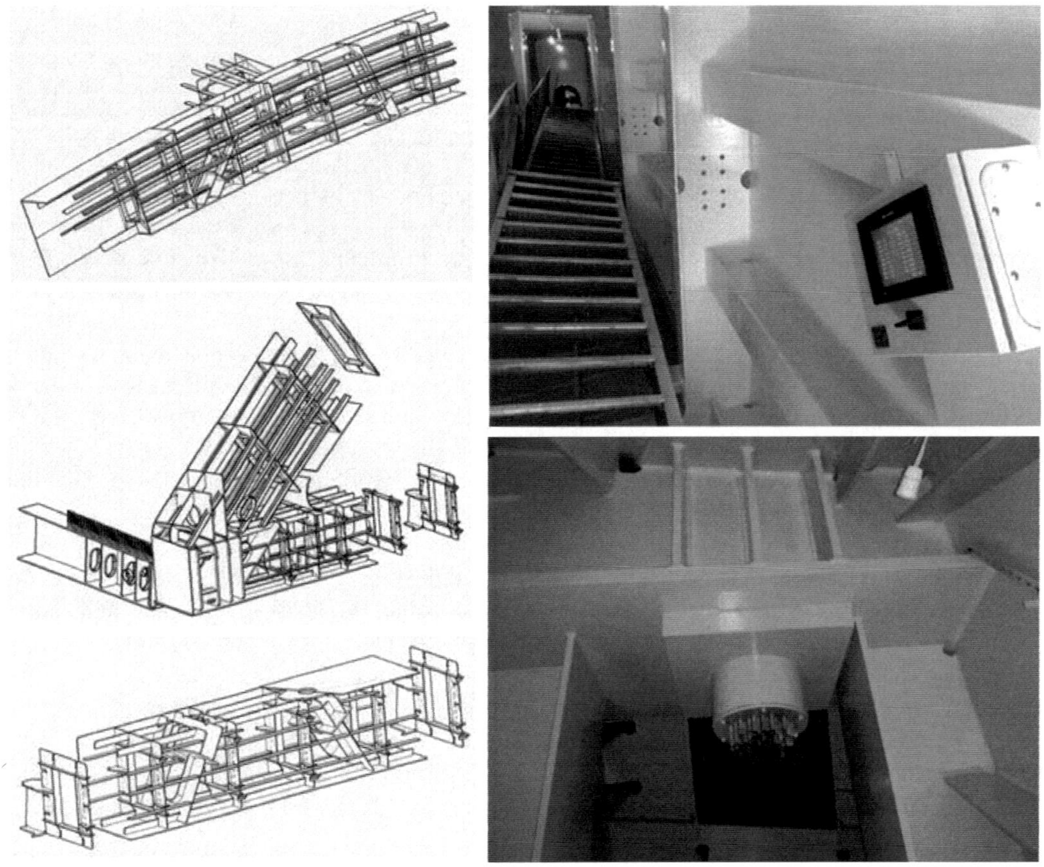

Figure 5. Fragments from 3D-Workshop drawings and interior of arch and tie.

The approaching structures *1-2* and *4-5* consists of three girders (*figure 4*), connected transversally by cross beams. The side girders are steel structures, shaped as ties of arch bridges for sake of visual compliance. Side girders are also horizontally connected with the deck slab by the headed studs. The middle girder is a composite beam.

The mutual work of three girders in the grillage system is well balanced considering the eccentricities of traffic loads.

3. Some specific details of the bridge structure and construction of the bridge

3.1. Horizontal supports on Pier 3

Central pier – pier in axis *3*, *figure 2* – is placed over residual caisson from the previous bridge (1961-1999). The dimensions of the old caisson are significant, 39.5x24.5 m in the layout and therefore capable to resist significant horizontal forces and corresponded moments from the bridge arch structures. The maximal magnitude of horizontal forces is 37000 kN per horizontal support in axes *3B*, (*figure 6*). This fact was used to design in [1] horizontal fixed supports *3A/M* and *3B/M*, longitudinally and transversally, on pier 3. Horizontal support structures are the bearings designed to transfer the horizontal forces, but not the vertical, (*figure 6*). Horizontal supports are connected to end cross beams of both arch bridge structures, transported with the bridge structures during launching and welded connected in situ after the final space tuning on the pier. The structural solution is presented in *figure 7*.

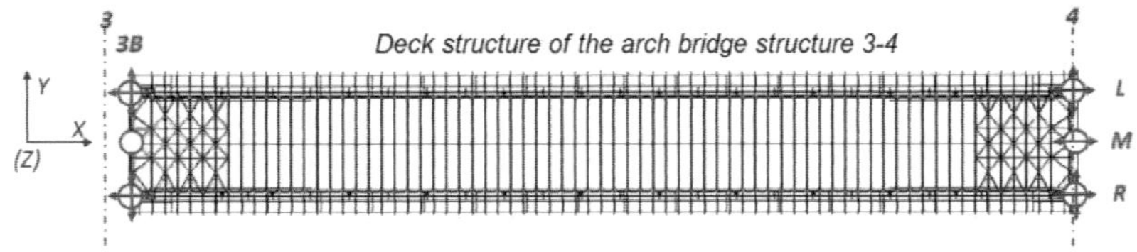

	Bearings L& R			Bearings M		
Axis	X	Y	Z	X	Y	Z
1	-	-	+	-	+	+
2A	-	-	+	+	+	+
2	-	-	+	-	+	+
3A	-	-	+	+	+	-
3B	-	-	+	+	+	-
4	-	-	+	-	+	+
4B	-	-	+	+	+	+
5	-	-	+	-	+	+

Figure 6. Bearings arrangement of the arch bridge *3B-4* and support reactions scope.

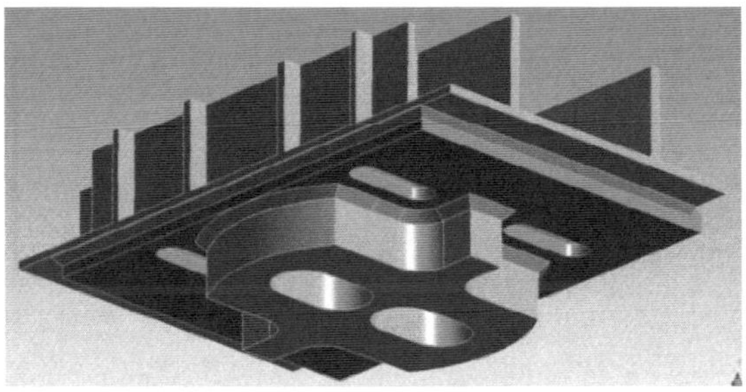

a

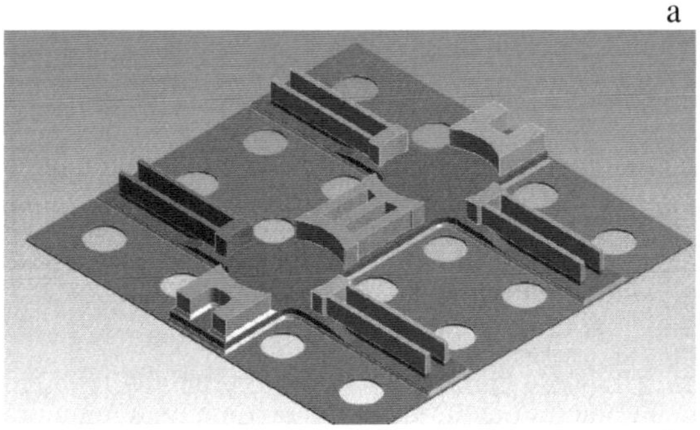

b

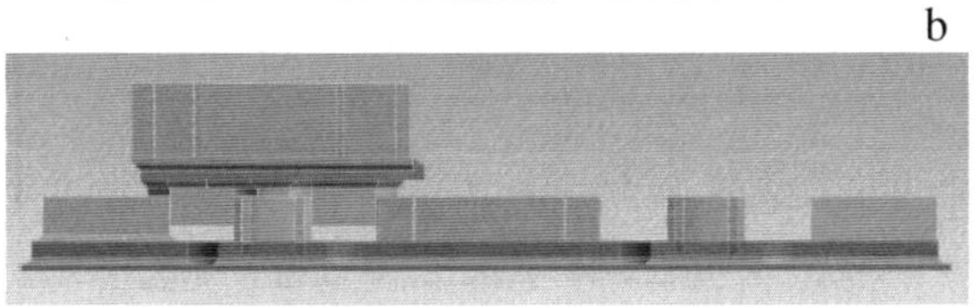

c

Horizontal support structure:
a) Superior part – Pin, under the end cross beam,
b) Inferior part – Pot, over the concrete pier,
c) Complete structure as Pin + Pot.
Horizontal forces tranfere from End cross beam, over the Horizontal support to the steel tube pins in concrete top beam of the central pier.

Figure 7. Horizontal support structure in the design and on the site.

3.2. Deck slab

The deck structure of all four bridge structures is the composite one, with cross beams and reinforced concrete slab. Deck slab is joined with the cross beams (transversal composite action) and with ties of arch bridges and side girders at approach spans for purpose of longitudinal and lateral composite action.

The deck structures of arch bridges are also exposed to tension forces from main structural system. Limitation of crack widths is realised thru the appropriate structural detailing of reinforcement bars, strictly according to very strong requirement of DIN-Fachbericht 104:2009 [6] for reinforced concrete deck plates of arch bridges, see *figure 8*.

Additionally, the deck plate is double protected: superior surface – water insulation methyl methacrylate resin layer; inferior surface – permanent steel formwork.

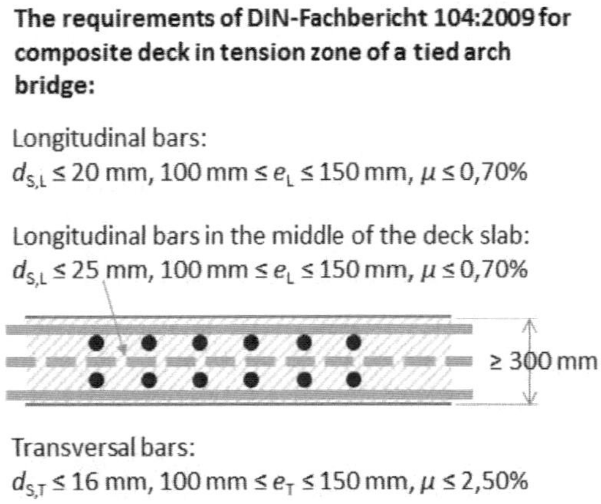

Deck reinforcement installation. The temporary columns for launching are still on place.

Figure 8. Reinforcement of concrete deck.

3.3. Launching

The launching of arch bridge structures (*figure 9*) has been realized through the movement on the banks – with skids and over the water – on pontoons. The bridge structures were reinforced with the temporary columns for launching, securing the structure to overstressing.

Figure 9. Launching of arch bridges over the Danube.

Special attention has been paid to calculation of forces and deformations of entire structure during cable stay installation as well as to erection analysis in all phases of the erection, *figure 9*. The hanger cable stays forces vary from tension to "compression" from step to step, according to changes of structural system during launching. The analysis was performed using Third Other Theory (analysis with large cable stays sags), avoiding the long iterative calculations with strait members and effective modulus of elasticity of cable stays.

3.4. Test load
The test load of the bridge (*figure 10*) is an obligation according to Serbian standard SRPS U.M1.046:1984. The tests were practised as static tests in seven phases and dynamic tests in four stages. The used vehicles were: three axle trucks of 23.5 t mass, four axle hopper-dozator wagons of 72 t and locomotives of 110 t.

In final phase of static test, the bridge structures were loaded on both road lanes and both tracks, across the whole span. For example, on the arch bridge of 219 m span, the total load was 32000 kN (3200-ton mass) of fully loaded vehicles.

Dynamic tests were performed with three coupled locomotives, with following velocities: On track No 1, $V = 10$; 55 and 80 km/h and on track No 2, (downstream side, *figure 3*) $V = 30$ km/h.

The measured deformations (displacements and support rotations), stresses and cable stays forces, vibration frequencies and accelerations (in vertical and horizontal plane) as well, were in very good correspondence to calculated values. For example – vertical displacements of four bridge structures (*figure 2*) with the spans of 25.3 m, 177.0 m, 219.0 m and 46.3 m, in the final static test phase, the values Measured/Calculated were respectively: 4.0/6.6 mm, 64.4/71,9 mm, 86.5/89.3 mm, 21.0/21.2 mm.

Figure 10. Test load of the Bridge, static and dynamic.

4. Structural materials quantities and construction costs
Quantities of structural materials for the bridge structures:

- Structural steel (for *1-2*, *2-3*, *3-4* and *4-5*): 300+3900+5300+700 = 10200 t
- Secondary steel structures: 400 t
- Cables, 55 parallel strands: 1000+1500 = 2500 m
- Concrete for decks: 3730 m^3
- Reinforcement: 1250 t.

The construction cost is about 55 Million EUR, (state April 2018).
The bridge is open to rail traffic at 7th April 2018.

5. Conclusion
The Railway Road Bridge in Novi Sad is excellent example of high efficiency of the network arch systems for railway bridges, especially regarding very strict serviceability limit states requirements. The specific consumption of structural steel (kg/m^2 or kg/m^1 of rail track) is below the expected, statistically defined, mean values according to literature, (e.g. [10]).

The bridge itself (*figures 11 and 12*), with the arch span of L = 219 m is a bridge with the longest span in the world in in the category of tied network arch bridges with two rail tracks.

6. Participants

Structure: Railway Road Bridge across the Danube in Novi Sad.
Employers: Delegation of the European to the Republic of Serbia, Enterprise Serbian Railways.
Financing: Delegation of the European to the Republic of Serbia
Autonomy Province of Vojvodina
Municipality of Novi Sad
Contractor: JV AZVI S.A., Taddei S.p.A., Horta Coslada S.L.
Subcontractors for designing:
Bridge structure: DEL ING DOO, Belgrade, Serbia.
Foundation and piers: ENCODE DOO, Belgrade, Serbia.
Subcontractors for construction:
Assembling of the bridge structure: Mostogradnja AD, Belgrade, Serbia.
Cable stays: VSL Ltd., Poland.
Launching: ALE Heavylift Iberica, Madrid, Spain.
Bearings and expansion joints: FIP Industriale, Servazzano, Italy.
Technical control of the Detailed Design: "Kirilo Savić" Institute AD, Belgrade, Serbia.
Engineer: JV DB International and Egis International.

a b

Figure 11. Railway Road Bridge in Novi Sad
a – Footpath with parapets and masks inside; b – The cargo train on the Bridge

Figure 12. Railway Road Bridge in Novi Sad.

References

[1] *Detailed design of the bridge structure of the Railway Road Bridge across the Danube in Novi Sad* 2011. (DEL ING DOO, Belgrade as Subcontractor for designing of JV Azvi-Taddei-Horta Coslada).
[2] DB Richtlinie 804: *Eisenbahnbruecken (und sonstige Ingenieurbauwerke). Planen, Bauen und Instand halten*. (Deutsche Bahn AG, München, 2003).
[3] DIN-Fachbericht 101: *Einwirkungen auf Bruecken*. (*Actions on bridges*). (Deutsches Institut fuer Normung, Beuth Verlag GmbH, Berlin, 2009).
[4] DIN-Fachbericht 102: *Betonbruecken*. (*Concrete bridges*). (Deutsches Institut fuer Normung, Beuth Verlag GmbH, Berlin, 2009).
[5] DIN-Fachbericht 103: *Stahlbruecken*. (*Steel bridges*). (Deutsches Institut fuer Normung, Beuth Verlag GmbH, Berlin, 2009).
[6] DIN-Fachbericht 104: *Verbundbruecken*. (*Composite bridges*). (Deutsches Institut fuer Normung, Beuth Verlag GmbH, Berlin, 2009).
[7] Bojović A., Marković Z, Mora A, Blom J, Aleksić D, Pavlović M, Spremić M, Novaković N, Janjušević B 2013: *Railway Road Bridge across the Danube in Novi Sad – Design and construction*. (The Eight International Conference „Bridges in Danube Basin". Timisoara – Belgrade, 4-5 October 2013. Procedings, p.139-150. Springer Vieweg, Springer Fachmedien Wiesbaden 2013).
[8] Bojović A, Mora A., Marković Z, Aleksić D, Pavlović M, Spremić M 2014: *Railway Road Bridge in Novi Sad – Design and Erection*. (Engineering for progress, nature and people. Proceedings, p.2163-2170. 37th IABSE International Symposium Madrid, Spain, 3-5 September 2014)
[9] Bojović A, Mora Munoz A, Marković Z, Novaković N 2018: *Network arches over the Danube – Railway Road Bridge in Novi Sad*. (Bauingenieur 93 (2018), Heft 3, p. 110-115).
[10] Herzog M 1975: *Stahlgewichte moderner Eisenbahn- und Strassenbrücken*. (Der Stahlbau, 44 (1975), Heft 9, p. 280-282).

IOP Conf. Series: Materials Science and Engineering **419** (2018) 012033 doi:10.1088/1757-899X/419/1/012033

Comparison of acoustic phenomena in the vicinity of two types of steel plate girder bridges

Lucjan Janas

Rzeszow University of Technology, al. Powstanców Warszawy 12, 35-959 Rzeszów, Poland

E-mail: ljanas@prz.edu.pl

Abstract. This paper discusses the results of own research of acoustic phenomena in the surrounding of two railway bridges of different type of decks. One of the objects has an open deck with cross girders and stringers, next has an orthotropic deck and track laid on ballast. The biggest threat for the environment is posed by objects with open deck. The extent of increase in noise reaches as much as 20 dB. Objects with ballast may also affect worsening of acoustic climate around the bridge, in particular under the bridge.
Reasons for an increase of noise when a train crosses the bridge in general split in-to two groups: - vibrations coming from vehicle wheels, transmitted through rails to a bridge structure make its components vibrate; large surface of structure may act as membranes of a loudspeaker and emit unwanted sounds; - for bridges with open platform the lack of attenuation by the deck (ballast) make the sounds at point-to-point contact of wheels with a rails freely disseminate.

1. Introduction

Railway bridges may affect the environment to much greater extent than railway lines on sections between bridges. This impact may be manifested as the excessive emission of noise which is significantly important within agglomeration areas.

The problem of railway bridges generating noise is discussed for example in [1, 2]. In Eurocode 3 [3], in a chapter relating to serviceability limit states, in general provisions, states that one should limit the frequencies of own vibrations, among others to limit the fatigue damages and the excessive noise emission. In point 7.7 regarding the criteria of use of railway bridges it was written, that any requirements regarding the noise emission may be specified in design objectives. Some recommendations for noise reduction are given in [4].

2. Method of research

In order to establish an impact of bridge objects on noise, simultaneous measurements from a distance of 7.5 m from the axis of track at the bridge were conducted, and at the same distance from the track on a track section beyond the bridge (usually 50 to 75 meters before or after the bridge). In both cases, microphones were placed 1.5 m above the head of rail. At the same time, noise was measured additionally under the bridge, 1.5 m above the ground. The research used Bruel&Kjaer microphone set and Pulse Reflex software. The acoustic phenomenon was registered during the passage of passenger trains, including long-distance trains, regional trains and railbuses, as well as cargo trains. Measurements were conducted at a temperature of 15 – 25 °C and relative humidity of 50 - 70 % and wind speed not exceeding 5 m/sec. Microphones were covered with anti-wind covers. Prior to tests, the measurements tracks were checked using the reference sound source.

In order to determine the magnitude of the noise exposure, measurements were taken at the following points (figure 1): near the track beyond the bridge (measuring point [m.p.] #1), near the bridge (m.p. #2) and under the bridge (m.p. #3).

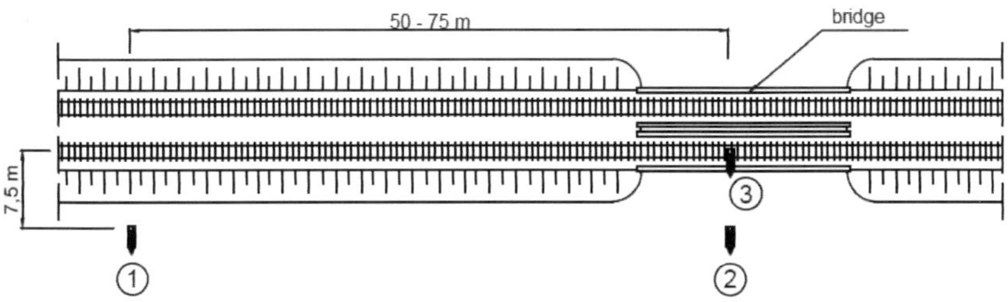

Figure 1. Distribution of noise measurement points

The recorded signals were analysed in one-third octave bands. The analysis included a comparison of the levels of acoustic pressure and determination of the summary levels. The values represent the total sound level (noise) with the characteristic (A) taken into account. The narrow frequency band analysis included a detailed identification of the frequency composition of the generated sounds are presented in [5].

3. A plate girder bridge with open deck

First object under analysis was one-span plate girder bridge, with free supported beam and stringer-traverse grid, i.e. the so-called open platform - figure 2. The span length is 24.15 m. Rails were affixed to wooden bridge sleepers using PM-60 washers and PKW-type spacers. Wooden sleepers were supported on stringers. The object technical condition during the test was good.

Figure 2. A plate girder bridge with an open deck: a) front view, b) bottom view.

The acoustic phenomenon was registered during passages of cargo and passenger trains - selected results of measurements are presented in table 1.

The noise beside the bridge was higher by 3.3 to 10.1 dB than the noise beside the track beyond the bridge. The noise under the bridge was even higher by approx. 20 dB than the noise beside the track. In few instances the noise under the bridge exceeded 100 dB. Figure 2 shows the typical characteristics of sound pressure levels as a function of time recorded simultaneously in three points (showed in figure 1) during the passage of the cargo and passenger trains. Comparison of acoustic pressure levels in one-third octave frequency bands is shown in figure 4.

Table 1. Sound levels in the surrounding of plate girder bridge with an open deck.

No.	Type of train	Speed of passage [km/h]	A sound level in [dB (A)]: beside the track (1)	beside the bridge (2)	under the bridge (3)
1	passenger train (traction unit)	50	80.3	83.8	96.2
2	passenger train (long-distance)	60	83.8	87.4	99.9
3	passenger train (light suburban)	50	74.8	78.1	90.5
4	passenger train (long-distance)	70	84.7	89.2	102.0
5	passenger train (long-distance)	50	77.3	87.4	100.5
6	cargo train	50	84.3	88.9	100.7
7	cargo train	40	79.6	88.6	101.3

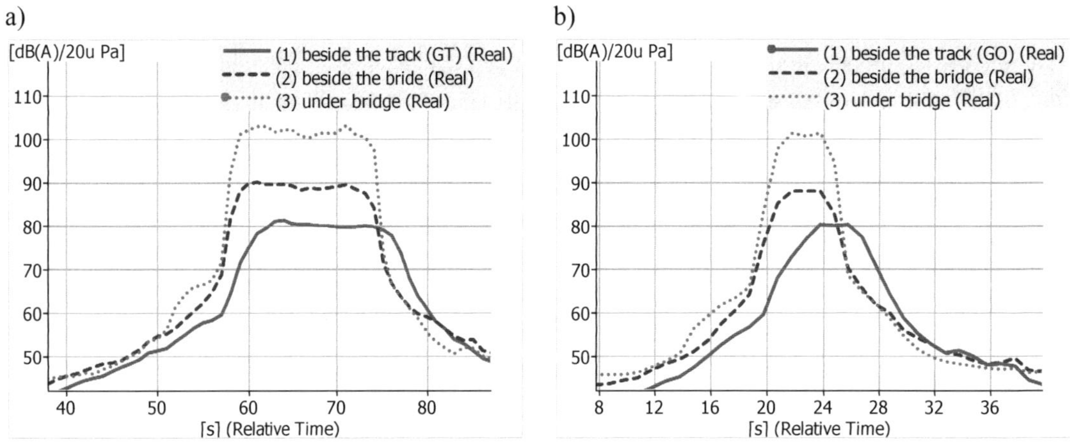

Figure 3. Sound levels as a function of time in the vicinity of a bridge with an open deck during passage: a) cargo train – 50 km/h, b) passenger train – 60 km/h.

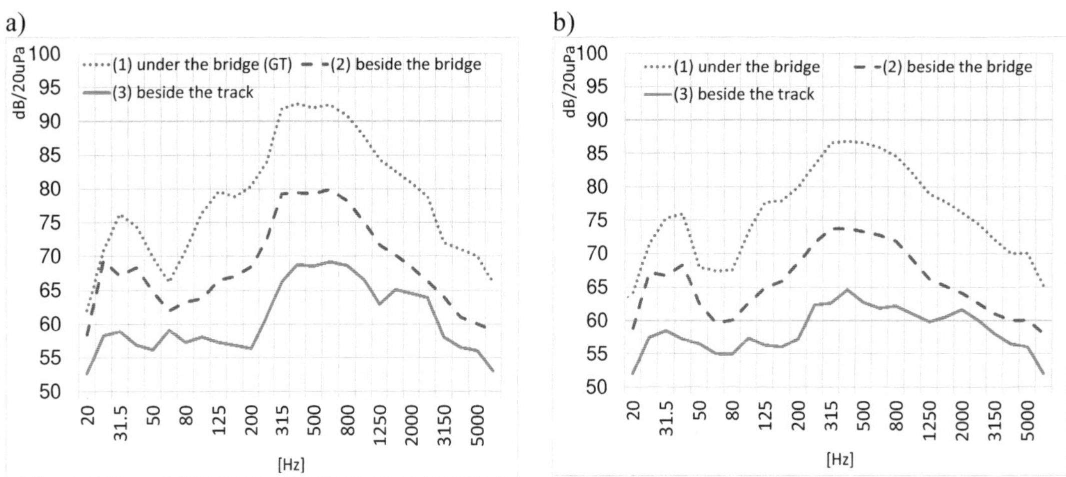

Figure 4. Level of acoustic pressure versus one-third octave frequency bands in the vicinity of a bridge with an open deck during passage: a) cargo train – 50 km/h, b) passenger train – 60 km/h.

The main reason of such big noise was the type of structure - the open platform is not a barrier for sounds generated at point-to-point contact of wheels with a rails. In addition, vibrations of large surface plate girder's web emit sounds within the range of low and medium frequencies - this question was discussed in details in [5].

The increase in the sound pressure level is observed throughout the entire frequency range – figure 4. The highest acoustic pressure level was observed in low frequencies (around 31.5 Hz) and between 200 and 2000 Hz.

4. A plate girder bridge with a track laid on ballast

The example of currently designed objects is a plate girder bridge with steel orthotropic platform and a track laid on ballast (figure 5). The length of span of the bridge is 31.68 m, height of girders 2.47 m. The ballast was laid in a deck made of ribbed metal plate. The thickness of the ballast layer under the concrete sleepers was at least 0.35 m. There was no vibroisolation under the ballast.

The bridge was in a very good technical condition and was opened to public just few months before the test.

a) b)

Figure 5. A plate girder bridge with a steel orthotropic deck and a track laid on ballast: a) front view, b) bottom view.

The table 2 presents the results of measurements of sound level made during the passage of passenger and cargo trains. Figure 6 shows sound pressure levels as a function of time recorded simultaneously during the passage of the two trains.

Table 2. Sound levels in the surrounding of plate girder bridge with a track laid on ballast

No.	Type of train	Speed of passage [km/h]	A sound level in [dB (A)] beside the track (1)	beside the bridge (2)	under the bridge (3)
1	passenger train (regional)	80	76.9	75.9	84.7
2	passenger train (regional)	60	80.8	76.7	85.6
3	passenger train (suburban)	80	76.8	75.5	80.8
4	cargo train	35	74.1	73.0	79.8
5	cargo train	50	81.7	81.3	87.4
6	cargo train	40	73.6	74.3	80.9

The noise beside the bridge does not differ significantly from the noise beside the track beyond the bridge. In most instances it is even smaller, which results from the fact that sounds generated at point-to-point contact of wheels with a rails are damped by plate girders. The noise under the bridge was higher from 4 to 7.8 dB than the noise beside the track. Since the object analysed was low over the ground, sounds were damped by the ground and did not disseminate. However, it must be noted that this type of structure used on populated areas e.g. in fly-over on high supports may be inconvenient for the environment. Exemplary comparison of acoustic pressure levels in one-third octave frequency bands is shown in figure 7.

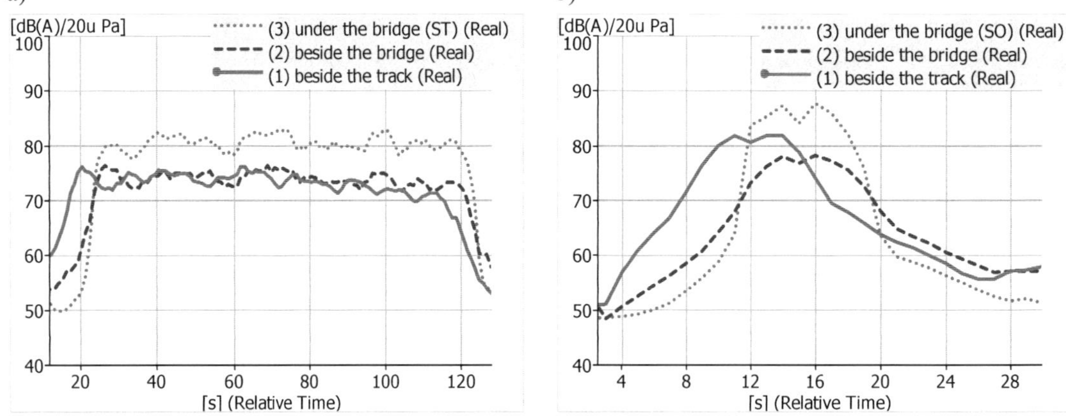

Figure 6. Sound levels as a function of time in the vicinity of a plate girder bridge with track laid on ballast during passage: a) cargo train - 50 km/h, b) passenger train – 60 km/h.

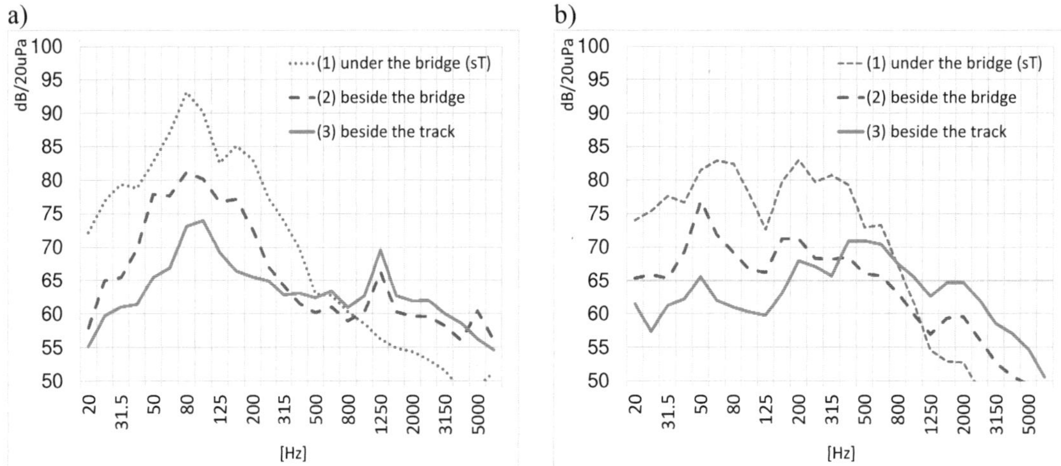

Figure 7. Level of acoustic pressure versus one-third octave frequency bands in the vicinity of a bridge with track laid on ballast during passage: a) cargo train – 50 km/h, b) passenger train – 60km/h.

The increase in the acoustic pressure level under the bridge was observed in frequencies range up to approx. 800 Hz. The main reason of noise increase was excessive vibrations of steel plate of the deck. Above 800 Hz the acoustic pressure under the bridge was lower than the acoustic pressure beside the track – sounds in this range were dumped mainly by ballast.

The acoustic pressure beside the bridge was lower than the acoustic pressure beside the track up to approx. 315 Hz. In this case, plate girder beams ware a type of acoustic screens.

5. Conclusion

Steel bridges may pose a threat to environment. They may contribute to the dissemination of noise generated at point-to-point contact of wheels with a rails or independently emit the unwanted sounds.

Objects with open deck belong to type of structures in which noise is mainly due to free dissemination of sounds generated at contact of wheels with rails and as a result of vibrations of large-surface elements. These types of bridges and viaducts should not be used on urbanized areas. If they are present there, they should be replaced if possible with others, with better acoustic properties. The partial reduction of noise be obtained by separating the track from the bridge structure using vibroisolation and application of horizontal sound barriers.

If one wishes to design the silent bridges, a track should be laid on ballast, but this type of structures may also emit noise. If the frequency of excitation of vibrations matches the frequency of own vibrations of large-surface elements of structures, then the emission of air sounds may occur. Since these sounds will appear most often under the structure, objects with a track laid on ballast may present an inconvenience if they are situated high above the ground on urbanized areas. Acoustic properties may be improved by using vibroisolation pads or matts.

The analysis of railway bridges for the emission of noise should be the important element in the process of preparation of investment projects. This recommendation applies not only to newly constructed objects but also to reconstructed and modernised bridges and viaducts.

References

[1] Li X, Yang D, Chen G, Li Y, Zhang X 2016 Review of recent progress in studies on noise emanating from rail transit bridges. *Journal of Modern Transportation* **24**, Issue 4, pp 237–250.

[2] Janas L 2017 The acoustic specificity of steel railway bridges, *Przegląd komunikacyjny* **9**, pp 22–25, (in Polish).

[3] European Committee for Standardization (CEN) 2003 Design of steel structures: Steel bridges *Eurocode 3, Part 2 (EN 1993-2)*

[4] International Union of Railways (UIC) 2010 Recommendations for the design of bridges to satisfy track requirements and reduce noise emissions. *UIC Code 717R 2nd edition*

[5] Janas L 2015 Vibroacoustic tests of the plate girder bridge. *Zeszyty Naukowo-Techniczne Stowarzyszenia Inżynierów i Techników Komunikacji w Krakowie. Seria: Materiały Konferencyjne,* ***2(106)***, pp 47-60, (in Polish).

IOP Conf. Series: Materials Science and Engineering **419** (2018) 012034 doi:10.1088/1757-899X/419/1/012034

Experimental results of steel truss nodes strengthen by high strength concrete encasement

Stoyan Ivanov[1], Roman Geier[2], Carlos Rebelo[3] and Bruno Pedrosa [3]

[1]Schimetta Consult Engineering Ltd, Trepetlika Str. No5, Sofia 1407, Bulgaria
[2]Schimetta Consult Ziviltechniker Gmbh, Arndtstraße 89, Vienna A-1120, Austria
[3]ISISE, Department of Civil Engineering, University of Coimbra, P-3004 516 Coimbra, Portugal

E-mail: stoyan.ivanov@schimetta.at

Abstract. In the old steel truss bridges one particular point of interest are the nodes. While for strengthening/refurbishment of the elements of the truss engineers can rely on more conventional methods, usually, for the nodes unique solutions have to be adopted. One integral solution, which could be implemented in a variety of steel truss nodes, is strengthening/refurbishment by totally encasing the node in high performance fibres reinforced concrete. In order to validate this idea laboratory testing of steel truss nodes strengthened by encasement with different concrete classes was initiated. The connection between the concrete and the structural steel is ensured by application of epoxy resin, no mechanical elements are added. The results of the strengthened specimens are compared with one reference bare steel node test. The strengthened specimens show reduced values of the stresses in the gusset plate as well as increase in the load bearing capacity of the connection of the elements entering the node. These results are observed even for encasement in normal class concrete of C40/50.

1. Introduction

The idea of strengthening steel bridges (structures) with the help of concrete, in order to take advantage of the joint work of the two materials in one united composite section is not new [1]. Perhaps it is not so wildly spread because of the additional self-weight of the concrete, its low tensile strength and the possible durability and maintenance problems. In the recent decades with the development of the new materials like ultra-high performance concrete the above mentioned restrictions have been overcome. Examples for strengthening steel bridges with ultra-high performance concrete are recorded under [2], [3] and [4].

Most of the cited examples are related to orthotropic steel bridge decks. Until now no evidence was found by the authors for the implementation of this technic to old steel trusses and more specifically to their nodes, which are of particular interest to the structural engineers. While for strengthening the elements of the truss, design engineers can rely on more conventional methods, usually, for the nodes unique solutions have to be adopted [5].

The main problems found in the truss nodes are related with:

- Gusset plates - too thin or corroded gusset plates, where local buckling or shear tearing might govern the design;

- Connecting rivets – strengthening the connection of the elements entering the node might be restricted by the possibilities to add additional bolts or rivets, because of the small space available between the existing connectors and the optimised geometry of the gusset plates.

One integral solution to the above problems, which could be implemented in a variety of steel truss nodes, is strengthening by totally encasing the node in high performance concrete (HPC). The authors believe that this solution will not only lead to higher load bearing capacity of the node, but will also improve its fatigue behavior by reducing the stresses in the gusset plate and connecting elements from live loads. Other benefits as higher durability and reduced maintenance costs are also expected.

In order to prove the efficiency of the proposed strengthening method an experimental program has been developed under the research project PROLIFE. Five full scale steel truss nodes were tested: one bare steel reference node, one strengthened by normal class concrete and three strengthened by High-Strength Fiber Reinforced Concrete (HSFRC). In order to ease the encasement process no additional conventional reinforcement was envisaged and the connection between structural steel and the concrete is realized by epoxy resin only without any shear connectors.

2. Description of the specimens

The truss nodes chosen for the experiments are full scale having the geometry shown in Figure 1. In order to make it possible to test the nodes in the laboratory i.e. apply load and add supports, it was decided to incorporate them into truss girder.

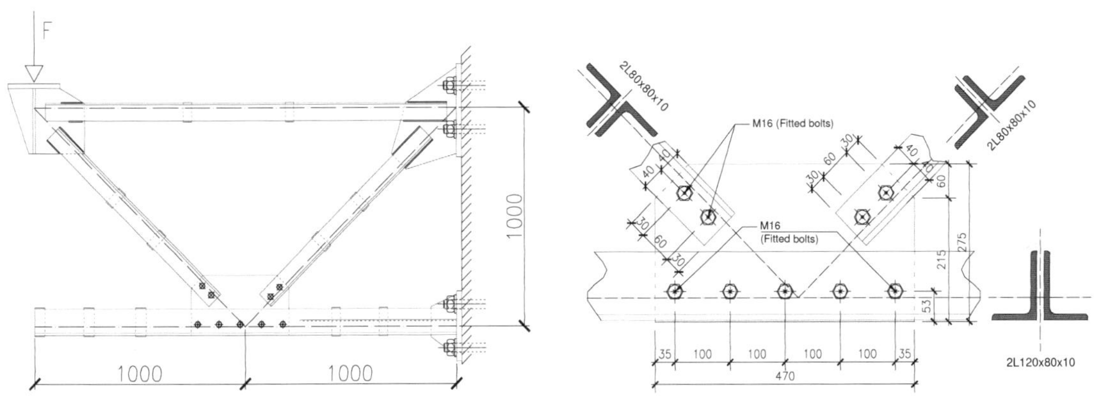

Figure 1. Geometry of the test specimen (left) and the steel node (right).

Unlike the original riveted truss nodes, the connector used in the specimens are fitted bolts M16 (placed in holes Φ16.5mm). This was accepted in order to ease the production as rivets are not widely produced nowadays. All other nodes of the truss are designed as welded having higher load bearing capacity than the one under consideration.

Except the reference bare steel test four more tests are performed, where the node under consideration is encased in concrete (Figure 2).

The chosen parameters to be varied for the strengthened specimens are: concrete class, concrete cover and encasement length of the diagonal truss elements. Two of the specimens have concrete cover of 30mm and the other two of 50mm. The encasement length varies between 140mm and 210mm and the different concrete strengths are listed in Table 1.

According to the initial design of the specimens, the tuned failure mode is shear of the two fitted bolts connecting the diagonal members. This failure mode is later proven by the tests.

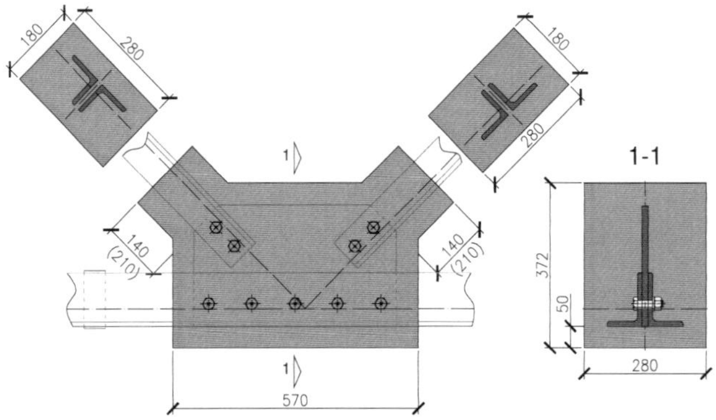

Figure 2. Geometry of the encased nodes with concrete cover c=50mm.

Table 1. Variable parameters in the specimens.

Specimen	Concrete cover	Encasement Length	Concrete			
	c [mm]	L [mm]	$f_{cm,cube}$ [MPa]	f_{ck} [MPa]	f_{ctm} [MPa]	Class
No1 - RN	-	-	-	-	-	-
No2 - C50L140	50	140	57,6	40	3,4	C40/50
No3 - C50L210	50	210	71,4	55	3,8	C55/67
No4 - C30L140	30	140	102,1	80	11,0	C80/95
No5 - C30L210	30	140	102,1	80	11,0	C80/95

All specimens are loaded with static monotonic loading with a load-displacement protocol of 0.01mm/s except the last one, in which cyclic loading to verify the bonding between the two materials has been applied.

3. Measuring Equipment

During each test, the displacements are recorded with minimum of 7 displacement transducers (LVDT's), additional transducers were installed in two specimens to measure the slip between the two materials (Figure 3).

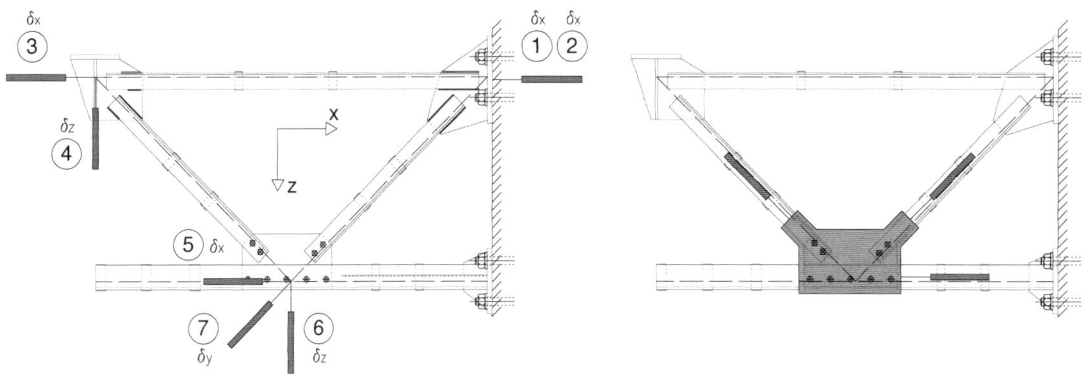

Figure 3. Position of the displacement transducers (LVDT's).

Linear strain gauges (TML FLA-6-11) are placed in the mid-cross section (outside the encased node) of each truss element in order to catch the overall stress state of the truss during loading.

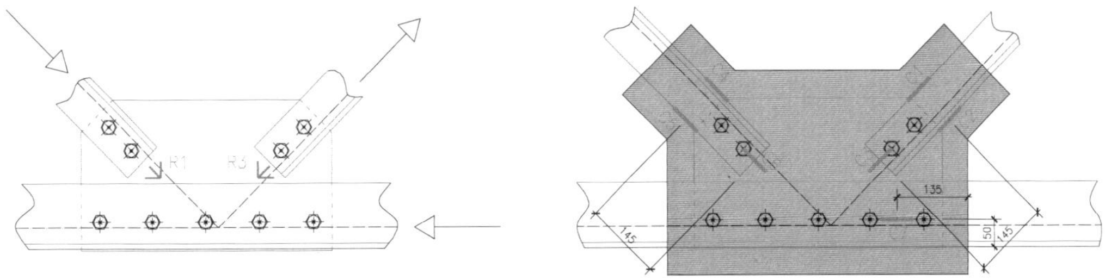

Figure 4. Position of the rosettes and strain gauges at the node.

In order to measure the stresses into the steel gusset plate two rosettes are placed at the end point of the diagonals entering the node. Their position was determined after linear elastic analysis performed over initial FEM of the bare steel specimen. Strain gauges were also installed over the concrete after the encasement (Figure 4).

In Figure 5 the reference specimen as well as one encased (strengthened) specimen are shown prior to start the testing. The out of plane restrains could also be seen on the pictures (with green color).

Figure 5. Specimen prior to testing, reference (left) and strengthened (right) one.

4. Materials properties

The nominal class of the structural steel for the profiles of the specimens is S275. Three coupon tests were taken from unstressed pieces in order to find the real mechanical properties of the steel. The mean yielding strength obtained by the tests is 329 MPa, which is around 19% higher than the nominal value. The tests were performed in accordance with EN ISO 6892-1 using a Multipurpose Servo hydraulic Universal Testing Machine rated to 200kN.

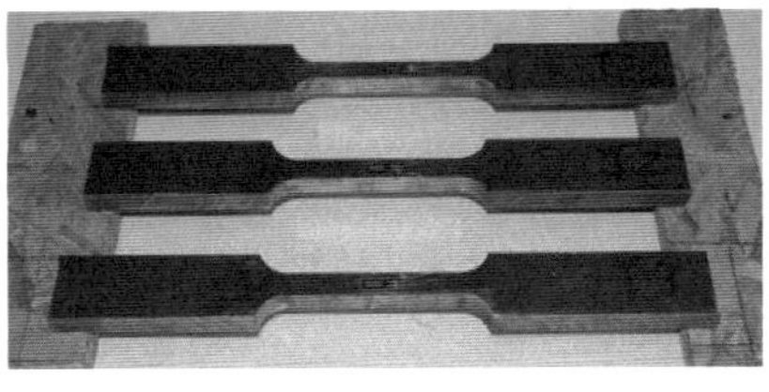

Figure 6. Steel coupons test specimens.

The concrete properties were tested with cube specimens sized 150x150x150mm after each casting. The compressive tests were performed according to EN 12390 with 8 concrete cubes in total, two are tested at age of 3 days, three are tested at age of 14 days and the remaining three at age of 28 days.

The tensile strength was obtained in an indirect way through the application of a compressive load on a small area along the length of cube until it splits into two halves along a vertical plane due to indirect tensile stress generated by Poisson's effect (Figure 7). Thereby, 3 concrete cubes with 150 mm size were tested at the age of 28 days.

The mean values of the concrete tensile and compressive strengths, for different concrete castings, are presented in Table 1.

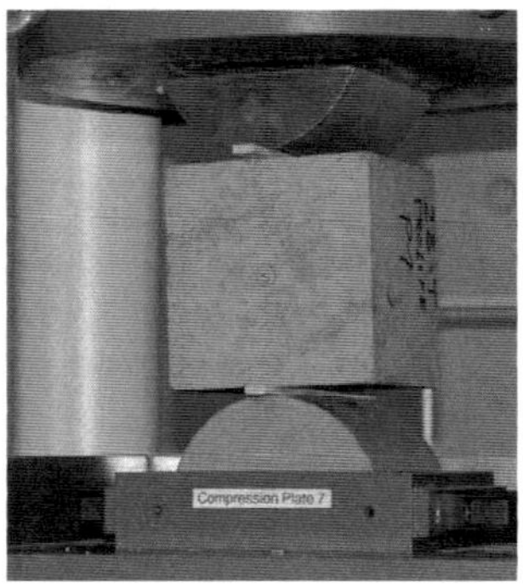

Figure 7. Indirect test to obtain tensile concrete strength.

5. Results from the experiments

Graphical representations of some of the obtained results are presented under this section. The results from the last test specimen are not presented as they were not fully processed to the date of the submission of the article. Face 1 and 2 on the presented graphs are used to denote the two strain gauges situated symmetrically over z axis of one and the same cross section.

In Figure 8 the vertical displacement, measured at the loaded node, in relation to the applied load are plotted.

In Figure 9 the normal stresses in the tension diagonal of the truss are plotted against the applied load.

In Figure 10 the stresses in the gusset plate at 45° angle, found from rosettes R1 (compression side diagonal) and R3 (tension side diagonal) are plotted against the applied load.

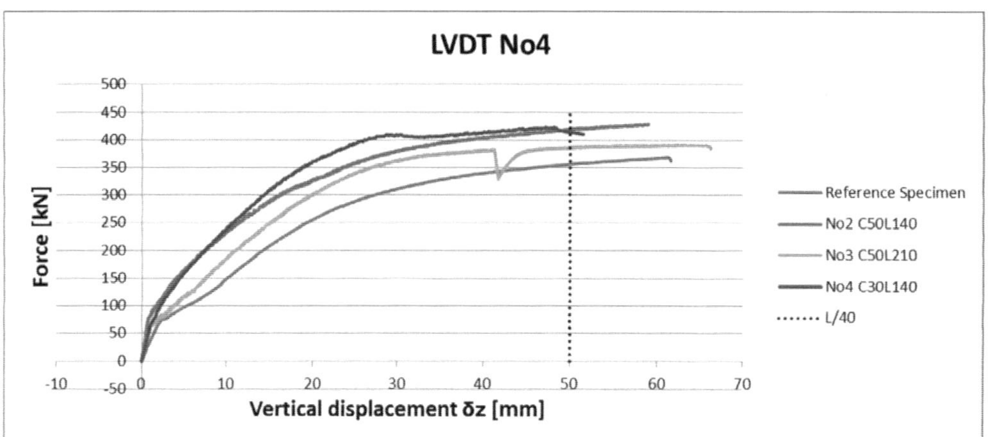

Figure 8. Applied load – Vertical Displacement graphs for LVDT 4.

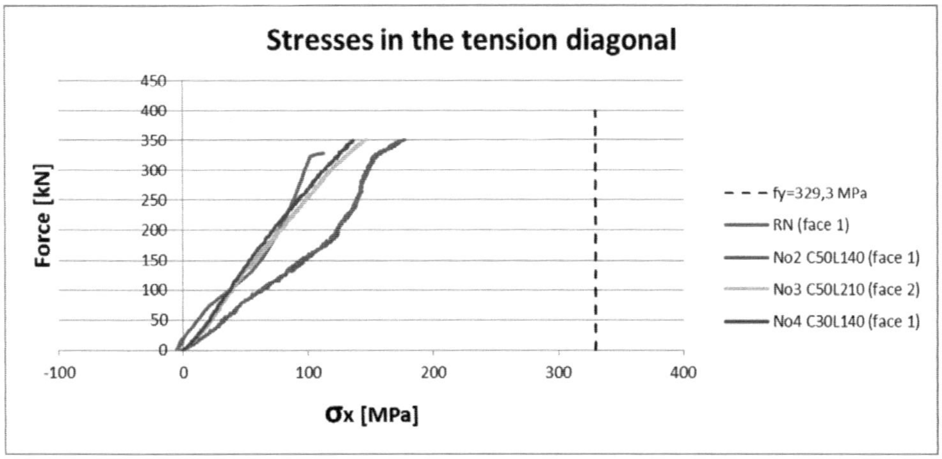

Figure 9. Applied load – Stress in the tension diagonal.

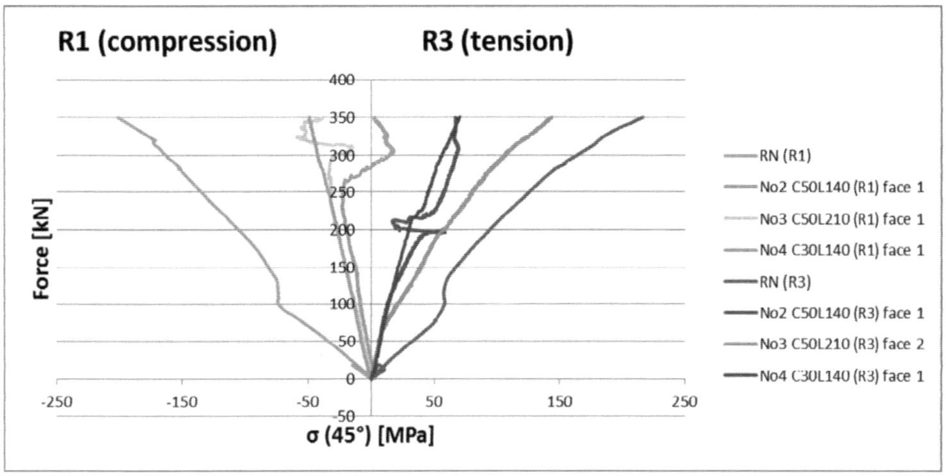

Figure 10. Applied load – Stress in the gusset plate (Rosettes R1 and R3).

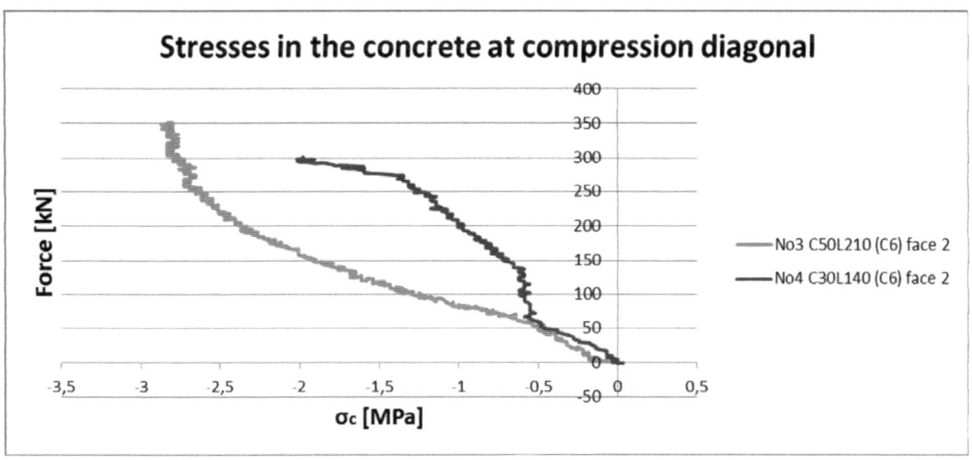

Figure 11. Applied load – Stress in the concrete (Strain gauge C6).

In Figure 11 the stresses in the concrete from strain gauge C6 (compression side) are plotted against the applied load.

6. Summary and conclusions

Failure mode: All specimens failed by shear of the two bolts connecting the tension diagonal to the node, as for the encased nodes it is accompanied with prior cracking of the concrete. This fully matches with the initially expected failure mode from the preliminary calculations;

Figure 12. Failure of the reference specimen.

Figure 13. Failure of the encased node.

Debonding: The connection between the concrete and steel is ensured only by the application of epoxy resin. The maximum measured slip between the concrete and the tension diagonal, for both specimens, in which the slip has been recorded, is around 0,50mm. It could be stated that debonding has not occur due to the preceding cracking of the concrete in the tension side of the node;

Breaking load: If we accept failure criteria of 50mm vertical displacement at loaded point, the respective breaking load for the specimens is 356kN, 418kN, 386kN and 423kN.

Stresses in the gusset plate: the stresses obtained from rosettes R1 and R3 located right after the compression and tension diagonals are listed in Figure 10. The average reduction of the stresses at 45° angle, for loading magnitude equal to the theoretical breaking load of the bolts in the reference specimen (~250kN), is 77% for R1 (compression side) and 52% for R3 (tension side). This reduction is confirmed also by the results presented on Figure 11, where the stresses into the concrete are plotted.

In summary it could be stated that the obtained experimental results are quite promising even for concrete classes C40/50 and C55/67. The encasing appears more effective for the gusset plate itself by reducing the stresses and preventing local buckling, rather than the loadbearing capacity of the connecting elements. Nevertheless increase of the breaking load and the overall stiffness is also

IOP Conf. Series: Materials Science and Engineering **419** (2018) 012034 doi:10.1088/1757-899X/419/1/012034

observed. Positive effects, which are expected to be even more, pronounce for higher concrete classes and especially for encasing in UHPFRC.

Acknowledgements
The research leading to these results has received funding from the European Union's Research Fund for Coal & Steel (RFCS) research programme under grant agreement n° RFSR-CT-2015-00025.

References

[1] Hayashi K, Ono S and Nakamura S 1997 Experimental studies on Retrofit by Partially Encased Concrete to the Steel I-Girder subjected to Buckling Deformation *Public Works Research Institute of Japan* https://www.pwri.go.jp/eng/ujnr/tc/g/pdf/19/7-3ono.pdf;
[2] Buitelaar P, Braam C R and Kaptijn N 2004 Reinforced high performance concrete overlay system for steel bridges *Proc. of 5th International CROW-workshop on Fundamental modelling of the design and performance of concrete pavements (Istanbul)*;
[3] Hordijk D A, Kaptijn N and Van der Vossen S 2009 HSB-BETONOVERLAGINGEN OP STALEN BRUGGEN (in Dutch) *Rapport 6742-1-5 (Rijswijk);*
[4] Hajar Z, Novarin M, Servant C, Genereux G, Przybyla D and Bitar D 2013 Innovative solution for strengthening orthotropic decks using UHPFRC: The Illzach Bridge *Proc. of RILEM-fib-AFGC International Symposium on Ultra-High Performance Fibre-Reinforced Concrete, (Marseille)*;
[5] Tarrell J 2017 Hagwilget Bridge, Canada: Replacement of entire truss nodes under live traffic *Proc. 39th IABSE Symposium Engineering the Future (Vancouver)* pp 562-569;
[6] European Committee for Standardization (CEN) 2003 Actions on Structures; Traffic loads on bridges *Eurocode 1, Part 2 (EN 1991-2)* – this is an example of a reference to a standard

IOP Conf. Series: Materials Science and Engineering **419** (2018) 012035 doi:10.1088/1757-899X/419/1/012035

Application of Research into Deck Bridges with Encased Filler-Beams in Design Practice

Viktória Kožlejová[1], Ľubomír Kožlej[2] and Vincent Kvočák[1]

[1]Faculty of Civil Engineering, Technical University of Košice, Vysokoškolská 4, 042 00 Košice, Slovakia
[2]Amberg Engineering Slovakia, s.r.o. Somolického 1/B, 811 06 Bratislava - Palisády

E-mail: viktoria.kozlejova@tuke.sk

Abstract. Research into deck bridges has been conducted at the Faculty of Civil Engineering of the Technical University in Košice, Slovakia. Encased steel filler-beams in a wide variety of shapes and dimensions were designed for deck bridges, whereby reducing steel consumption by a significant 40 per cent, and maintaining the same resistance and stiffness of the bridges. Our theoretical assumptions were verified by experiments. Specimens were tested for static, dynamic, and long-term loads. Consequently, with the assumptions proved to be correct, a design protocol was elaborated to serve as a template or code of practice for designers and engineers.

Using this design protocol, a designer working for *Amberg Engineering Slovakia* designed a load-bearing structure for the reconstruction of a bridge in the village of Prituľany. The proposed technical solution took into account all local restrictions given by the horizontal and vertical alignment of the road structure on the bridge as well as its low building height. The plan of the load-bearing structure as viewed from above (i.e. the angle between the axis of the bridge and the axis of the obstacle being bridged) is oblique.

1. Introduction

Deck bridges with encased steel beams present a structural system that is commonly applied in small and middle-span constructions. Initially, in first designs, rails were employed as rigid reinforcement. Later, welded or rolled steel sections came into use. The currently valid and applied Eurocode 4 standard contains structural requirements and design procedures for such bridges; however, it also prescribes the application of I-sections only. At the same time, it allows for the plastic design, while the position of the plastic neutral axis is to be situated in the steel beam web. The upper flange of a beam thus appears to be near the neutral axis, and its contribution to the bending resistance of the beam is minimal. The fulfilment of this condition is required to ensure composite action as the structural system of this kind does not make use of any sheer connectors [12-19].

Research into deck bridges with encased steel filler-beams has been carried out at the Civil Engineering Faculty of the Technical University in Košice for several years. Our goal is to make the design of such bridges more efficient by lowering the consumption of steel while maintaining the same load-carrying capacity and stiffness of the structure. T-sections were suggested for deck bridges, and steel savings in the upper flange of the section were achieved as a result. The theoretically calculated resistance and stiffness did not change For the components to act together, several types of strip connectors were designed, and their design further refined and verified during the research at the Faculty in the past. The dimensions of the sections remained identical for all beams. We have been researching the deck bridges for over 10 years. In the following, we give only basic information [7-11].

2. Research results

2.1. Test specimens

Complying with the standard specifications, a single type of test specimens were designed for experimental research that served comparative purposes (Figure 1).

The width of the concrete section: 670 mm
The height of the concrete section: 270 mm
IPE200 rolled steel section
Theoretically calculated resistance moment using the measured material properties: 107.93 kNm
Theoretically calculated bending stiffness of the section: 11 959 kPa.m^4 [1-5]

Another type of test specimens contained a steel T-section, manufactured by the longitudinal flame-cutting of a rolled steel I-section in a comb-shaped manner (as can be seen from Figure 2), creating a profiled upper edge of the steel section.

The width of the concrete section: 670 mm
The height of the concrete section: 270 mm
IPE200/2 rolled steel section
Theoretically calculated resistance moment using the measured material properties: 99.91 kNm
Theoretically calculated bending stiffness of the section: 13 036 kPa.m^4 [1-5]

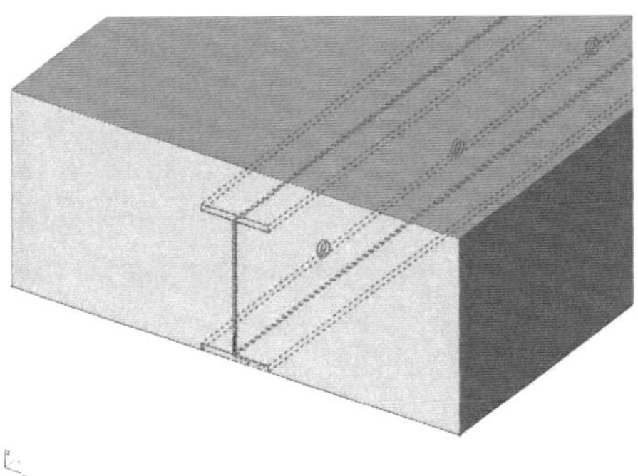

Figure 1: Specimens with an I-section

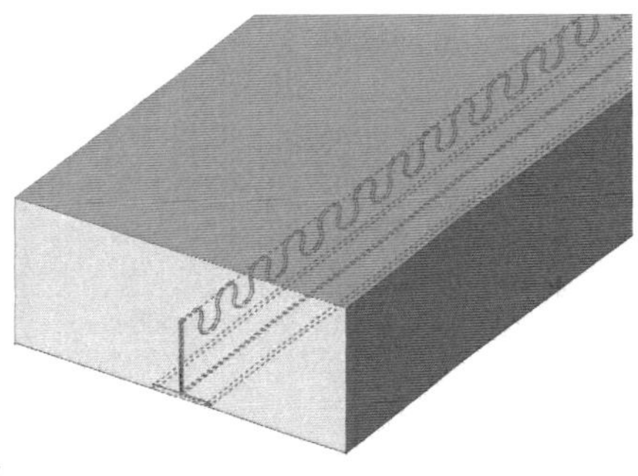

Figure 2: Specimens with a T-section

2.2. Test results

Deflections were monitored during experimental measurements. All changes depending on the amount of load applied are shown in Figure 3.

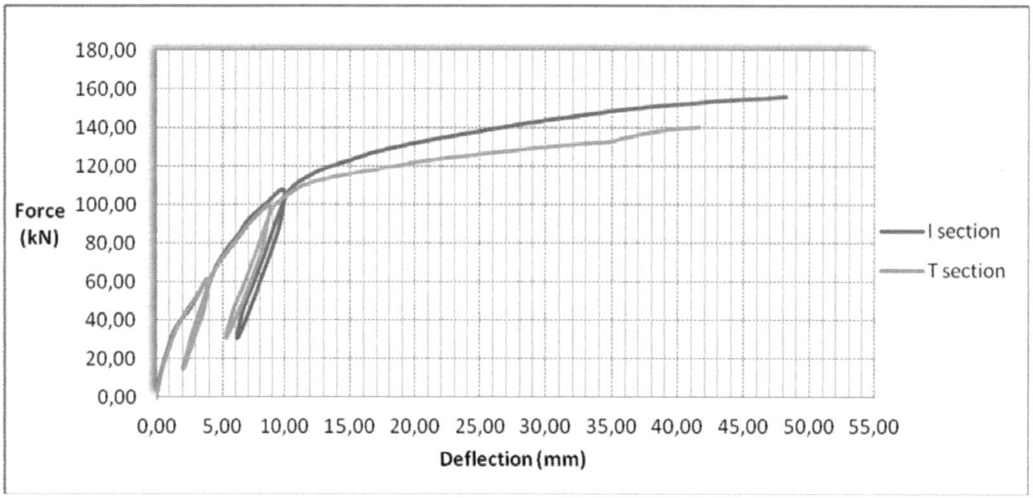

Figure 3: The load-deflection curves: Change in the development of deflections depending on the amount of force applied to the beams

Testing was completed when the deflections in the specimens increased excessively, signalling the plasticization of the sections. At this stage of the experiment, plastic resistance moments were detected as follows:

Specimens with an I-section: 124.8 kNm, meaning an increase of 15.6 per cent
Specimens with a T-section: 117.0 kNm, meaning an increase of 17.1 per cent

The specimens with T-sections were further verified by dynamic and long-therm load.

The results of the experiments proved that the designed solution is suitable for deck bridges. The assumption that resistance and stiffness are the same has been confirmed. However, limitation on the use of such a design is for bridges supported during their construction, not having a self-carrying function during the concreting process.

2.3. Application of the results

Based on the experimental results, a design protocol was elaborated to serve as a template or code of practice for designers and engineers. It contained calculation procedures for the verifications of bending resistance of deck bridges with encased T-section filler-beams, their serviceability, and the verification of composite action. Moreover, constructional requirements for this type of bridge were provided as well.

3. Reconstruction of 3822-002 Pritul'any Bridge

Parts of the village of Pritul'any are divided by the Pritulianka stream. Crossing the stream is possible over a composite bridge made of steel and concrete. Significant degradation of the concrete in the load-bearing structure of the bridge has taken place over the years, and increased permanent deformations and lateral buckling of the steel sections occurred. The bridge *per se* is in terrible condition. Neglecting the current status quo may lead to severe damage to the bridge structure.

Its reconstruction is to consist of the renewal of the original composite bridge structure, construction of a new bridge, alteration of some road sections before and after the bridge, and the remodelling of the stream bed under the bridge.

3.1. The technical solution for a new bridge

Because of serious space restrictions in the village area, a method of construction, a shallow building height of the bridge structure due to the maintenance of the stream flow rate, the best appropriate solution seemed to be a composite deck bridge with encased filler beams.

The load-bearing structure was designed as a reinforced concrete slab with an average width of 430 mm, made of C35/45 concrete, encased steel sections (structural steel grade S355) and soft secondary reinforcement (steel graded B500 B). In a transverse direction, the load-bearing structure consists of 10 welded T-shaped beams with a centre-to-centre spacing of 700 mm. The lower flange is 280 mm in width and 20 mm in thickness, while the web of the beam is the vertical flange, 260 mm high and 10 mm thick (Figure 4).

The width of the load-bearing structure is 7 100 mm, uniformly along the whole bridge. The length of the bridge is 11 000 mm (oblique), and the length of the load-bearing structure is 12 770 mm (oblique).

The load-bearing structure is placed on a reconstructed reinforced bearing block of concrete graded C35/45 and a flexible 20mm-thick strip. Bond at the interface between the new piece of the bearing block and the original part is made by a glued board.

The geometry of the new structure is determined by the horizontal and vertical alignment of the road SO 101-00. It is designed with a 0.66 per cent longitudinal and 2.5 per cent transversal roof-like inclines of the roadway [1-7].

Used materials

Concrete/Load-bearing structure: C35/45 – XC4, XD1, XF1 (SK) – Cl 0,4 – Dmax16 – S3
Structural steel: S355 X0
Reinforcing steel: B500B

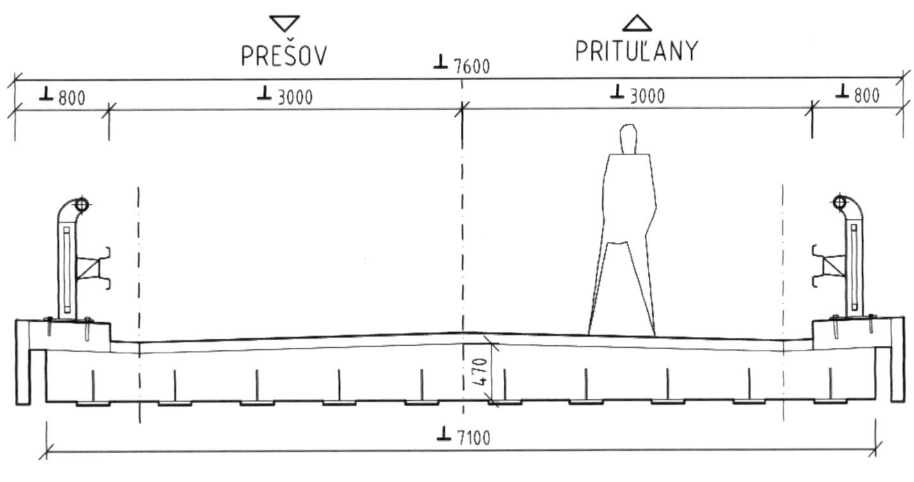

Figure 4: The cross-section of the bridge

The substructure of the bridge is supported by two existent supports – OP1, OP2. Both supports are massive, founded on flat bases. After the partial demolition of the original supports, a bearing block is built on each of them. The new load-bearing structure is supported using a transverse beam with elastic rubber base. All parts of the substructure below ground level, permanently in direct contact with earth, are protected by a damp-proof membrane.

New slab cornices in the form of prefabricated components were designed for the bridge. Both left and right cornices are 800 mm wide, protruding 250 mm from the edge of the load-bearing structure. The height of the end of the cornice/fascia is 750 mm, its lateral slope 4.0 per cent, leaning towards the roadway.

3.2. Model of a load-bearing structure

Midas Civil, a software programme, was used to design a grid-type model of the load-bearing structure of the bridge (Figure 5). Apart from the sequence of the bearing structure construction, the rheological properties of the concrete were considered in the model (Figure 6). Temporary point supports of steel T-sections were designed in the model during both the assembly of the steel construction and casting concrete. For the sequence of construction, both the loading states of placing fresh concrete in situ and loading by workers on the load-bearing structure acting compositely with still the green concrete were taken into account.

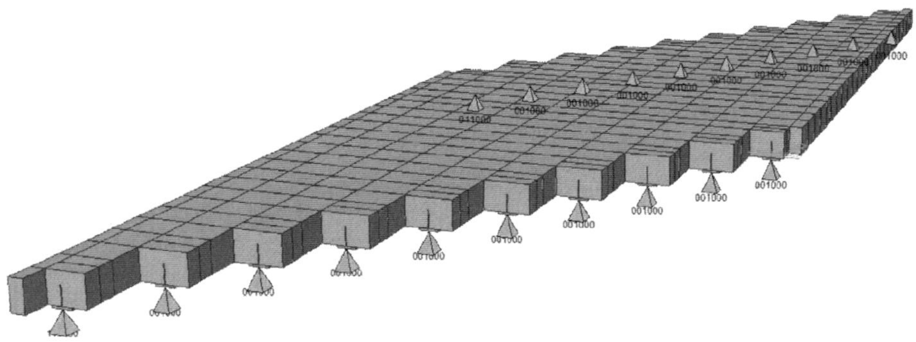

Figure 5: A new bridge model using the Midas software

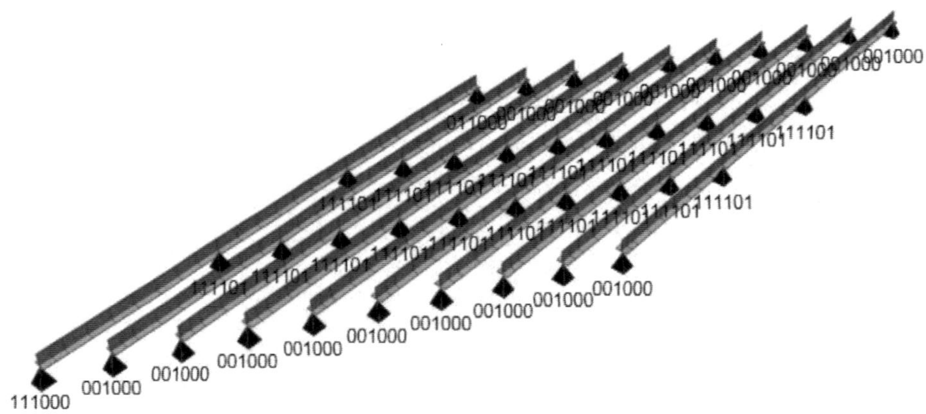

Figure 6: Point supports as considered during the sequence of construction according to Midas

The temporary supports were removed in the programme after the green concrete had set, hardened and reached the required strength. The load of the superstructure (such as the cornices, insulation, etc.) and transport is borne by the composite structure.

Stresses in the T-section were observed at each stage of construction, whereas those in the concrete started to be monitored at the time when it was possible to consider loading the green concrete by construction workers. The section in the Midas Civil programme was designed as a composite-general section. The bridge was verified for seismic effects as well.

The grid-type model used for modelling this load-bearing structure describes the actual behaviour pattern of this type of structure correctly enough. In its layout, it is an oblique bridge, and the grid-type model was constructed according to the recommendations provided by the designers of the Midas Civil programme for grid-type oblique bridge models. Long-term observation of bridge deformations will take place in the future. The results are display in the axes of steel beams (Fig. 7).

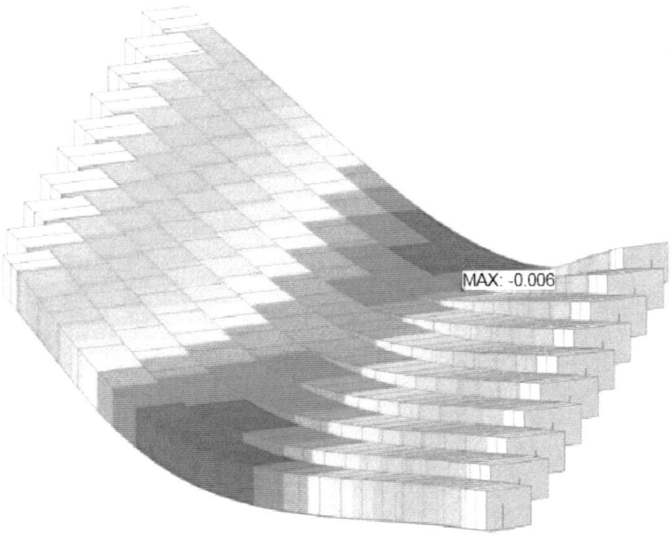

Figure 7: Deflection in model of bridge

4. Conclusion

The load-bearing structure of the bridge was verified under the stringent Eurocode standards currently in force. Verification of the stress condition of the bridge structure at each stage of its construction was based on an analytical model. The bridge, as well as all its structural details, was designed according to *Deck Bridges with Encased Filler-Beams of Modified Sections [5],* a publication written at the Civil Engineering faculty of the Technical University in Košice during the research in progress.

These days, the bridge is under construction. Measurements are being carried out during the construction stage, and a loading test on the bridge is being prepared to follow successful completion of the construction work.

Acknowledgements

This work was supported by the Slovak Research and Development Agency under the contract No. APVV-15-0486 and by the Ministry of Education, Science, Research and Sport of the Slovak Republic under the contract VEGA 1/0108/16.

References

[1] STN EN 1994 – 1 – 1: Eurocode 4 Design of composite steel and concrete structures. Part 1 – 1: General rules and rules for buildings

[2] STN EN 1994 – 2: Eurocode 4 Design of composite steel and concrete structures. Part 2: General rules and rules for bridges

[3] STN EN 1992 – 1 – 1: Eurocode 2 Design of concrete structures. Part 1 – 1: General rules and rules for buildings

[4] STN EN 1993 – 1 – 1: Eurocode 3 Design of steel structures. Part 1 – 1: General rules and rules for buildings

[5] Kvocak, V. et al.: Deck Bridges with Encased Filler-Beams of Modified Sections (Technical University of Košice)

[6] Reconstruction of the Bridge 3822-002 Prituľany on the Road III-3822-HS, Documentation on Construction / Documentation on Offer, Amberg Engineering Slovakia, s.r.o. Bratislava, Kosice, 2017

[7] Kvocak, V. - Kozlejova, V. 2011 Research into Filler-Beam Deck Bridges With Encased Beams of Various Sections *Technical Gazette*.Vol. 18, no. 3, p. 385-392.

[8] Kvocak, V. - Kozlejova, V. 2013 First Load Phase of Long-Term on Composite Beams *Concrete and Concrete Structures 2013* (Elsevier) 423-427

[9] Kvocak, V. - Kozlejova, V. - Dubecky D. 2015 Software Modelling of Composite Steel and Concrete Structures *Applied Mechanics and Materials* (Switzerland: Trans Tech Publications) 320-325

[10] Kvocak, V. - Kozlejova, V. - Dubecky D. 2016 Another Stage in the Research into Deck Bridges with Encased Filler-Beams - Dynamic Load Tests *Bridges in Danube Basin 2016* (Elsevier) 207-211

[11] Kvocak, V. - Dubecky D. - Kozlejova, V. 2018 Experimental Verification of Design Models in a Static and Dynamic Loading Test *Applied Mechanics and Materials* (New Zealand: Trans Tech Publications) 394-399

[12] Machacek, J. - Cudejko, M. 2011 Composite steel and concrete bridge trusses *Engineering Structures* (Elsevier) 3136-3142

[13] Studnička, J. - Macháček, J. - Peleška, K. 1996 Nové prvky spražení pro ocelobetónové konstrukce *Stavební obzor* 42-45

[14] M. Rovňák, M. - Ďuricová, A. - Kundrát, K. - Naď, Ľ. (2006) Spriahnuté oceľo-betónové mosty, (Košice: Elfa)

[15] Bujňák, J. - Hubinská, L. 2007 The Real Behaviour of the Bearing Plate Structures with Encased Beams *Metal, Composite and Wooden Structures and Bridges, XXXIII. Workshop of Steel Structures Specialists* 49 – 54.

[16] Vičan, J. - Odrobiňák, J. - Gocál, J. - Hlinka, R. (2006) Railway Bridges with Encased Steel Beams in Operation - Diagnostics and Calculation *Application of Steel, Composite and Wooden Structures and Bridges for Construction, XXXII. Workshop of Steel Structures Specialists* 147 – 152

[17] Samec, J. – Studnička, J. (2003) Analysis of the behaviour of the connection strip *Steel structures and bridges 2003, 20th czech – slovak conference* (Prague) 379 – 384

[18] Schmitt, V. - Seidl, G. - Hever, M. - Zapfe, C. (2004) Verbundbrücke Pöcking – Innovative VFT-Träger mit Betondübeln *Stahlbau* (Wiley) 1437-1049

[19] W. - Kołakowski, T. (2014) Modern composite bridges by VFT-WIB method in Poland realized at new express road S7 at Olsztynek-Nidzica sector *Economical Bridge Solutions based on innovative composite dowels and integrated abutments* (Springer) 93-110

Using fracture mechanics principles in steel bridge renovation projects

Maarten Rikken[1], Daan Tjepkema[1] and David Gration[2]

[1]Arup, Naritaweg 118, 1043 CA Amsterdam, The Netherlands
[2]Arup, 13 Fitzroy Street, W1T 4BQ London, United Kingdom

E-mail: maarten.rikken@arup.com

Steel bridges built in the post war period were not built for the current traffic intensities and are therefore prone to fatigue damage. The combination of high traffic intensities, low quality steel and suboptimal weld detailing can generate fatigue problems in existing bridges that could compromise the structural integrity of the bridge. Fracture mechanics principles can be used in steel bridge renovation projects, when S-N based calculation procedures do not predict sufficient structural capacity, as an alternative to potentially unnecessary strengthening of the considered detail. These principles have been used in the renovation of a 231 m long, plate-girder bridge in The Netherlands. Inspection intervals were derived for a non-load carrying, double fillet welded connection between transverse stiffeners and an I-shaped main girder. Such an analysis requires a detailed insight of the material's condition and therefore should be completed in close cooperation with the inspection operator to verify calculation assumptions. Using fracture mechanics allows the bridge owner to choose between inspecting a detail at a certain frequency or completing preventative strengthening.

1. Introduction

The Managing Contractor (MC), a combination of Royal HaskoningDHV, Arup, and Greisch was appointed by Rijkswaterstaat, The Dutch national bridge authority, in 2009, to refurbish 8 long span steel bridges in The Netherlands. The scope of the MC project is to recalculate the capacity of the existing bridges and to design strengthening schemes to increase the capacity of the bridges for both fatigue and static strength. The MC has also a leading role in the procurement and construction management.

The bridges in the contract were built between 1969 and 1990, during a period of rapid infrastructure development. They were not designed to withstand the current traffic intensities and are therefore prone to fatigue damage. The span length of the bridges varies between 100 and 300 m and they are all located in strategically important locations in the Dutch highway network. All 8 bridges have steel orthotropic decks, with trapezoidal longitudinal stiffeners, but have different types of superstructure e.g. tied-arch, beam-girder and cable-stay. Cracks in the deck plates have been observed since 1997. Additionally, certain of the bridges have other fatigue problems in the deck and superstructure.

To assess the fatigue capacity of the bridges, the MC used fatigue load models that were derived during traffic measurements on Dutch highway bridges, discussed in [1] and implemented in NEN8701 [2]. These loads, in combination with a detailed finite element model of the deck and the experimental results

IOP Conf. Series: Materials Science and Engineering **419** (2018) 012036 doi:10.1088/1757-899X/419/1/012036

of [3] were used to calculate the predicted fatigue damage in the orthotropic deck. The bridge decks were strengthened with a High Strength Concrete (HSC) overlay initially developed as part of [4]. The asphalt is replaced during the renovation with this new, stiff overlay, reducing the stress in the deck plate and increasing the fatigue life of the orthotropic deck.

The MC used the same fatigue load model, in combination with a finite element model of the entire bridge, to determine the fatigue capacity of the welded joints in the superstructure of the bridges. The derived stress ranges were then used in a S-N based approach, using detail categories from [5]. Where insufficient fatigue capacity was predicted, measures to guarantee the structural safety of the superstructure were developed. The corresponding structural upgrading can cause excessive hindrance for the traffic using the bridge and the surrounding infrastructure, i.e. roads, railway or waterways, during the execution of the works and could also be expensive.

An alternative solution is presented in this paper which is based on fracture mechanics and inspections to demonstrate adequate structural capacity of the bridge without any structural upgrading of the bridge.

2. Fracture mechanics principles

Fracture mechanics assessors the propagation of cracks within structures. These cracks, or flaws, can be attributed to various processors including: the welding practice, voids produced during the production process of steel or propagating fatigue cracks due to cyclic loading conditions. In most cases a Linear Elastic Fracture Mechanics (LEFM) based approach is used to assess these existing cracks. This method is valid as long as the plastic zone around the crack tip is relatively small. This approach uses the Stress Intensity Factor (SIF) which defines the crack tip conditions and relates internally or externally applied stress states to the region of the structure containing the flaw. The SIF is an important parameter in LEFM and extensively discussed by Anderson [6].

Fracture mechanics can be used to determine the fatigue crack growth (da) per stress induced cycle (dN) using the SIF approach for cyclically loaded structures, such as bridges. The most simple, and frequently applied crack growth law is the one derived by Paris and Erdogan [7], shown in Equation 1. This equation uses material parameters (C and m) which can be found in guidelines, such as BS7910 [8]. This guideline also includes parametric equations for the SIF (ΔK), as a function of the load configuration, crack geometry and flaw size.

$$\frac{da}{dN} = C\Delta K^m \tag{1}$$

A second application of LEFM is to determine the critical flaw size before failure is predicted under given static loading conditions. This procedure often uses a Failure Assessment Diagram (FAD), which relates the SIF to the fracture toughness of the material (the fracture ratio) and the static stress level to the yield stress (the load ratio) to determine the acceptability of a flaw. The flaw is deemed to be unacceptable whenever the combination of the fracture and load ratio exceeds the FAD assessment line.

A key step in this approach is the Probability of Detection (PoD) of the applied Non Destructive Testing (NDT) technique. The PoD is defined as the capability of a particular NDT technique to find a crack like flaw, of a particular size, during inspections. The PoD curves can are based on a significant amount of testing and are included in certain fitness-for-service guidelines, e.g. BS7910 [8].

Fracture mechanics can be used to determine the fatigue design life of a structural member containing a flaw until failure (i.e. fast fracture or plastic collapse) is predicted. The fracture resistance assessment determines after each crack increment (da) if the flaw size is acceptable or not using the FAD. Based on the outcome of this procedure, a certain inspection frequency can be calculated which allows for sufficient time for taking measures if a crack is found on site. This approach can be utilised even when a S-N based approach results in predicted fatigue damage exceeding one.

3. Case study: Suurhoff bridge

A fracture mechanics analysis was carried out for the Suurhoff bridge, which is one of the 8 bridges within the MC project. The Suurhoff bridge is a 231 m long beam-girder bridge with an orthotropic, steel deck, spanning the Hartel canal in the Port of Rotterdam area. This bridge was open for traffic in 1972 and carries 4 lanes of traffic of the A15 motorway.

The original contract requirement for the Suurhoff bridge was to strengthen the steel deck and superstructure to allow for a remaining service life of 30 years. The HSC overlay is however less effective for this bridge as the longitudinal, trapezium shaped stiffeners are discontinuous and welded in between the cross beams. Fatigue cracks are therefore expected in welds for which the stress range is not significantly reduced by the HSC overlay. Moreover, the increase in self-weight due to the application of HSC introduces a static capacity problem in the superstructure. Therefore the cost to strengthen the bridge for the full 30 year life extension is prohibitively expensive and would cause significant traffic hindrance.

Rijkswaterstaat has decided to replace the Suurhoff bridge in the future, but in the meantime, the functionality and structural safety of the bridge needs to be maintained. The necessary structural upgrading should be limited, because the bridge will be replaced. All connections with insufficient fatigue capacity have been identified and the MC has looked for solutions that are cost effective and do not cause excessive hindrance for traffic or the underlying canal.

The Suurhoff bridge contains two I-shaped main girders with cross beams welded to the main girder, typically spaced at 4.54 m. Transverse stiffeners are welded by means of two fillet welds to the web and bottom flange of the main girder at each cross beam location, shown in Figure 1. The transverse stiffeners increase the capacity against lateral torsional and local plate buckling of the main girder and are positioned either side of the main girder web at the location of interest. A fatigue crack initiating at the toe of the weld and growing into the main girder web or bottom flange could compromise the structural integrity of the bridge. Fatigue damage exceeded one was predicted at this detail using a S-N based approach, implying insufficient fatigue capacity.

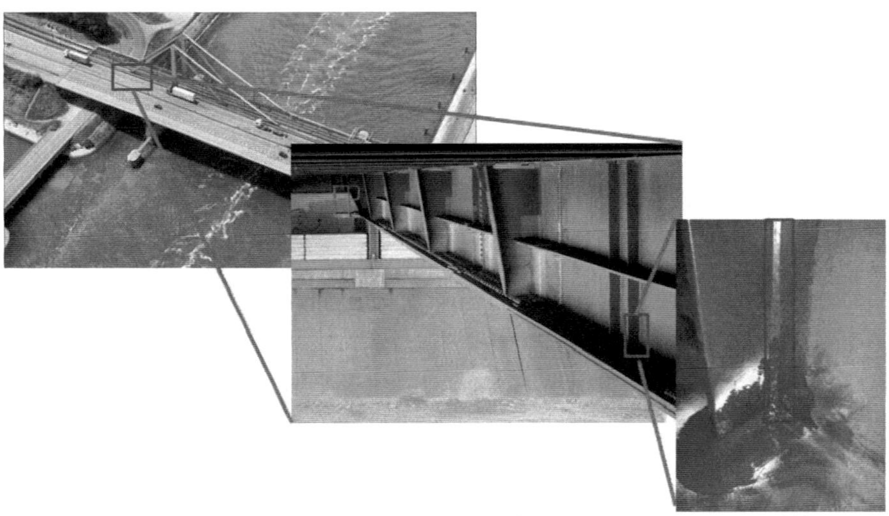

Figure 1. Encircled double fillet welded connection between transverse stiffeners and the I-shaped main girder, zoomed in to illustrate the location of the welded details in the bridge.

It was decided to inspect these connections for the remaining service life of the bridge until it is replaced. The inspection frequency was derived using fracture mechanics; a combination of fatigue crack growth and fracture as outlined in Section 2. This approach corresponds to a damage tolerant design philosophy, which requires a partial safety factor of 1.15 for the bridge superstructure [5]. This partial safety factor is normally applied on the resistance side, i.e. to the detail category, but has been applied to the stress ranges in the fracture mechanics approach. The calculation procedure was carried out for a connection between a transverse stiffener and the web of the main girder, referred to as 'detail I2' in the remainder of this paper, located in the bridge as shown in Figure 1.

Two inspection techniques were chosen to inspect the fillet welded connections between the transverse stiffeners and main girder; Time of Flight Diffraction (TOFD) and Magnetic Particle Inspection (MPI). TOFD uses ultrasonic principles to detect a flaw and is capable of providing crack length and depth information of planar flaws in a structure. MPI can detect surface breaking flaws in ferromagnetic materials, measuring the crack length of flaws only. MPI was chosen to verify whether indications picked up by TOFD, were surface breaking fatigue cracks, initiating at the weld toe.

The minimum flaw size that can be reliable detected using TOFD, when inspecting from the opposite side of the main girder web or bottom flange, is 2 mm deep by 10 mm long [8]. The position of the TOFD probes for a fatigue crack growing into the main girder web is shown in Figure 2. This detection threshold is determined using a probabilistic approach with PoD curves. This threshold indicates that a fatigue crack slightly smaller than 2 mm x 10 mm could remain undetected by an operator during a TOFD inspection. This therefore represents the initial flaw size, or starting flaw size, for the fracture mechanics assessment. The flaw size length that can be reliably picked up using MPI is 10-20 mm for as-welded joints [8], dependent on the weld quality and surface roughness.

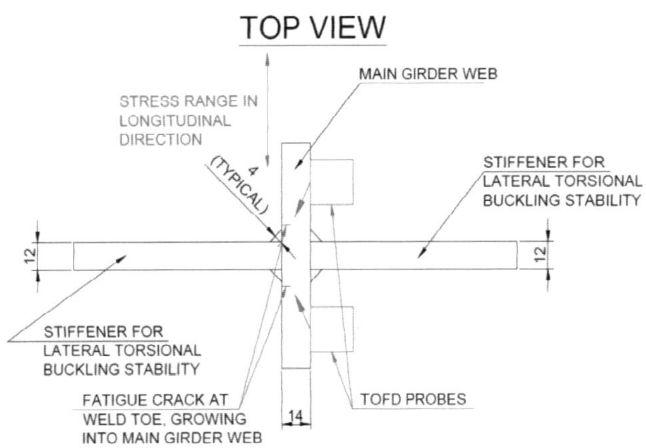

Figure 2. Fatigue crack initiation location for detail I2 and indicating the position of the TOFD probes during inspections

A two-staged crack growth law has been used for the fatigue crack growth calculation. This crack growth law accounts for decelerated growth of cracks in the near-threshold regime and uses two different sets of crack growth parameters (C and m, Equation 1). The usual practice for fatigue assessments of welded joints is to neglect the mean stress effect and to assume high, through-thickness, tensile residual stresses are present, generated by the welding process. This results in the simplification that the entire stress range ($\Delta\sigma=\sigma_{max}-\sigma_{min}$) is effective in terms of fatigue crack growth, not a function of the mean stress and the stress ratio ($R=\sigma_{min}/\sigma_{max}$). The crack growth parameters, corresponding to $R\geq0.5$ and mean plus 2 times the standard deviation were taken from BS7910 and are shown in Table 1. A fatigue crack growth threshold, relating to $R\geq0.5$ and equal to 63 MPa$\sqrt{mm}$ was also applied [8].

Table 1. Crack growth parameters for a two staged crack growth law

Crack growth parameter	Stage A	Stage B
C	2.10E-17	1.29E-12
m	5.10	2.88

The fatigue crack initiates at the weld toe for detail I2 and grows into the main girder web (da/dN) and along the weld toe profile (d2c/dN). This surface breaking flaw is often simplified as a semi-ellipsoid, shown in Figure 2. Annex M, Section M.4 of BS7910 [8] contains parametric equations for surface breaking flaws, applicable for membrane and local bending stress conditions. Local bending over the plate is considered negligible for this detail. The membrane stress contains the axial and bending stress components over the strong axis of the I-section main girder, these induce a stress range perpendicular to the crack (Mode 1 loading). Shear stresses have been included for detail I2, as recommended in [5] because the stiffener ends in the web of the main girder. The shear stress components have been added to the total membrane stress in which mixed mode loading has been disregarded. Parametric equations for the Mk factor, that represents the stress raising effect caused by the weld profile, were taken from Section M.11.1.3 of [8].

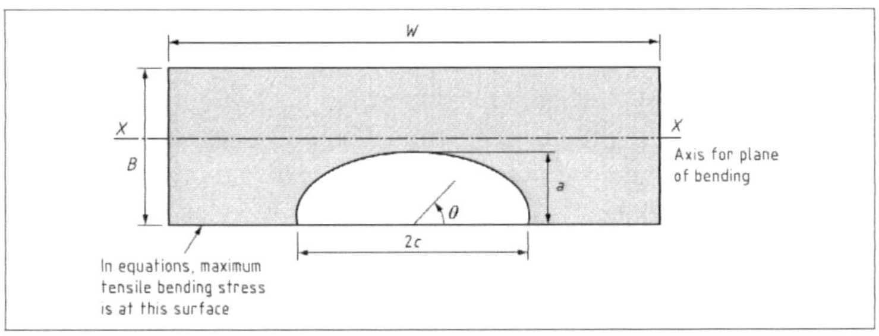

Figure 3. Semi-elliptical surface breaking flaw *[8]*

The Palmgren-Miner linear damage summation rule, used in S-N based approaches, disregards load sequence effects. A certain stress range results in a predicted damage, irrespective of the current crack length. Also the cut-off limit is not dependent on the current crack size. This simplification is not valid in fracture mechanics based analyses. Fatigue crack growth is only observed when the SIF, which is a function of the current crack size, exceeds the fatigue threshold. The fatigue load model of NEN8701 [2] includes various time-dependent factors, including: vehicle configuration, increase of heavy traffic numbers per year and an estimated axle load increase in the future. The stress ranges were therefore randomly ordered per year and applied in the fatigue crack growth assessment. Figure 4 and 5 show the stress ranges for detail I2 in 2017 and the Finite Element Model (FEM) used to derive the stress history.

Figure 4. Stress history for detail I2 in 2017

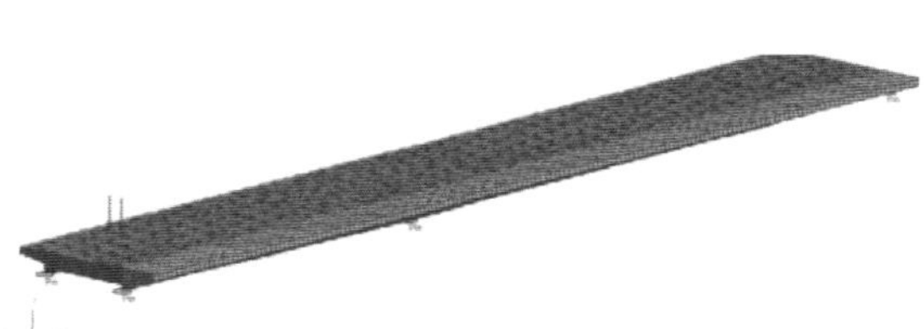

Figure 5. FEM of the bridge

The fracture assessments are carried out at the lowest possible service temperature of the bridge, assumed to be -20 °C in The Netherlands. This is because steel becomes more brittle at low temperatures, resulting in a lower fracture toughness. An accidental load combination [9], to combine (quasi-)static loads, may be used for fracture assessments due to the low probability of the combined occurrence of a low service temperature, large critically positioned flaws, poor material properties and onerous combination of actions [10]. The static stress level, using the accidental load combination for detail I2, is equal to 129 MPa.

A steel grade of Fe 510 D3 was used during the original construction of the bridge, this corresponds well with a modern S355 steel grade. The FAD assessment line has been created assuming a yield discontinuity in the stress-strain curve, a yield and tensile strength of 355 and 510 MPa respectively and a modulus of elasticity of 210000 MPa. The fracture calculation procedure verifies, for every crack increment (da) in the crack growth procedure, whether or not the total flaw size exceeds the assessment line which could correspond to failure of the remaining ligament.

NEN-EN 10025-2 [11] contains technical delivery conditions of structural steels, but these were not applicable during the original construction of the bridge. These standards require a minimum Charpy impact value of 27J at -20°C for low carbon steels. Other delivery standards were applicable in the late 1960's, and required comparable Charpy impact values, this value was therefore assumed in this assessment. The fracture toughness was calculated based on this minimum Charpy impact value using the master curve presented in Annex J of BS7910 [8]. A tensile residual stress distribution of yield magnitude (355 MPa) was also assumed for calculating the fracture ratio. The SIF equations for a semi-elliptical surface breaking flaw were used with the static stress level as input and the Mk factors based on 3D FEM calculations [8]. The load ratio relates the static stress level to the yield stress of the steel grade, accounting for the reduction in cross sectional area due to the propagating crack, using the reference stress equations in Annex P of BS 7910 [8].

An initial surface breaking flaw of 2 mm x 10 mm in the main girder web breaks through thickness for detail I2 in less than 5 years as shown in Figure 6. Brittle fracture is however predicted at smaller crack sizes using the fracture assessment, shown in Figure 7 and Table 2. The fracture and load ratio are a function of the crack size and exceed the FAD assessment line at a flaw size of 4.7mm by 21.8mm when fracture is considered in the through-thickness (a-)direction for a fracture toughness based on 27J. This crack size is predicted to be reached after 41 months, by using Figure 6. A less stringent critical flaw size is predicted for the (2c-)direction along the weld profile.

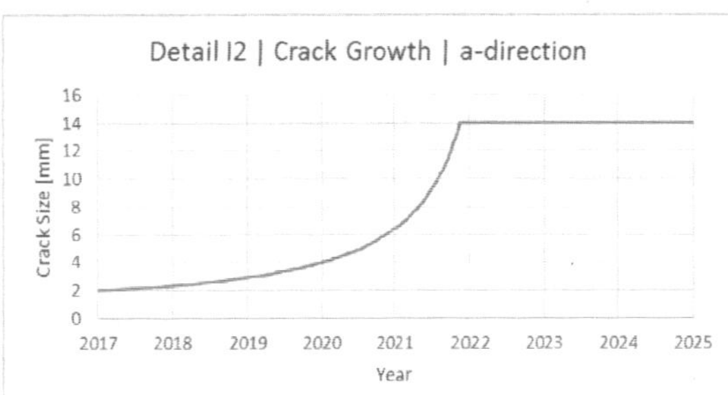

Figure 6. Through-thickness fatigue crack growth rate for a surface breaking flaw

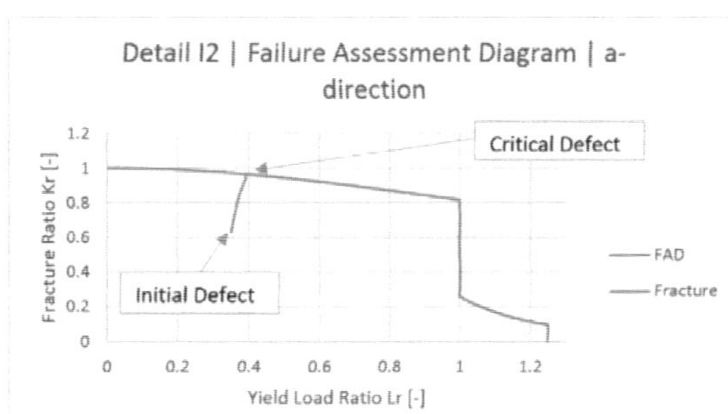

Figure 7. FAD for a surface breaking flaw fatigue crack growing through thickness

Table 2. Remaining fatigue design life of a surface breaking flaw for detail I2

	Charpy impact value [in J]	Fatigue Life [In Months]	(a) at brittle failure [In mm]	(2c) at brittle failure [In mm]
Brittle Fracture a-direction	27	40.9	4.741	21.772
Brittle Fracture 2c-direction	27	46.5	5.970	25.675

Based on the fracture mechanics analyses it is recommended that the details are inspected every two years with a back-surface TOFD procedure, combined with MPI. This provides Rijkswaterstaat sufficient time to react in case a fatigue crack is found during the inspections and to prepare necessary precautions. The inspection frequency also accounts for the fact that 1. the remaining fatigue life decays if for example 2023 is used as a starting point, due to the expected heavy traffic number and axle load increase and 2. the 27J Charpy impact value has not been proven by testing on heat affected base material. If cracks are found during inspection suitable measures could include repair welding the fatigue crack, by-passing the fatigue crack with a bolted splice connection or using burr grinding as post weld treatment to remove any historic damage and to improve the detail classification.

The assumptions taken for this analyses could be refined to predict a less stringent inspection frequency. Sensitivity results are given in Table 3 for various Charpy impact values and residual stress magnitudes. The fracture toughness can be measured by taking samples of the bridge. Residual stresses can be measured using various experimental techniques or studied numerically using non-linear finite element analysis. Arup is currently developing numerical welding simulations that can, amongst other things, predict residual stresses to provide benefit in asset management of existing, steel structures.

Table 3. Sensitivity results to determine the remaining fatigue life for detail I2

Residual Stress [in MPa]	Charpy impact value [In Joules]	Fatigue life [In Months]	(a) At brittle fracture [in mm]	(2c) at brittle fracture [in mm]
355	27	40.9	4.741	21.772
355	50	47.4	6.272	26.620
275	27	48.4	6.509	27.382
200	27	53.5	8.823	34.948
200	50	55.5	10.403	40.608

NEN-EN 1993-1-10 [10] and its background document [12] also provides guidance for deriving inspection intervals for joints that have expired their fatigue service life. The crucial assumption in the derivation of 'safe service periods' using this design standard is that the virgin, as-welded, situation may be assumed once the inspection has been carried with no fatigue cracks found. This approach does not assume that a crack may be left behind in the structure due to detection capabilities of the applicable inspection technique. This assumption may result in too optimistic inspection intervals and could potentially compromise the structural safety of the bridge.

4. Conclusion

The design life of existing steel bridges exposed to cyclic loading conditions induced by traffic is often assessed using a S-N based approach provided by various guidelines and standards. The outcome of such an analysis is a theoretical, predicted damage number which only indicates whether or not the fatigue capacity of the joint is according to standards used. This doesn't yield information to determine the inspection regime. Predictions of insufficient fatigue capacity can result in structural upgrading which is often expensive, could cause hindrance for traffic and/or underlying infrastructure and could be unnecessary due to model assumptions and conservatisms.

An alternative approach is to derive bespoke inspection intervals using fracture mechanics and to inspect the joints predicted to have insufficient capacity. Sufficient time should be included in the inspection frequency for the bridge owner to prepare actions in case of detected fatigue cracks. Furthermore the measures to be taken when a crack is detected should be known on forehand. This alternative approach for joints with insufficient fatigue capacity is less robust than implementing a comprehensive strengthening, but could be considered a more cost effective solution for the bridge which reduces traffic hindrance. This option is especially relevant for bridges with a relatively short remaining service life.

References

[1] Frank van Dooren, Gerland Nagtegaal, David Gration, David Ashurst, Sander den Blanken, Piet Kunst, "Orthotropic deck fatigue: renovation of 8 bridges in The Netherlands," 2010.

[2] "NEN8701 Assessment of existing structures in case of reconstruction and disapproval - Actions," december 2011.

[3] M.H. Kolstein, "Fatigue classification of welded joints in orthotropic steel bridge decks," TU Delft, 2007.

[4] F.B.P. de Jong, "Renovation techniques for fatigue cracked orthotropic steel bridge decks," TU Delft, 2007.

[5] "NEN-EN 1993-1-9 Eurocode 3: Design of steel structures - Part 1-9: Fatigue," januari 2006.

[6] T.L. Anderson, Fracture Mechanics; Fundamentals and Applications; Third Edition, Boca Raton, United States of America: Taylor and Francis Group, 2005.

[7] P.C. Paris and F. Erdogan, "A critical analysis of crack propagation laws," *Journal of Basic Engineering,* vol. 85, pp. 528-533, 1963.

[8] "BS 7910:2013+A1:2015," Guide to methods for assessing the acceptability of flaws in metallic structures.

[9] "NEN-EN 1990+A1+A1/C2 Eurocode: Basis of structural design," december 2011.

[10] "NEN-EN 1993-1-10+C2: Eurocode 3: Design of Steel Structures - Part 1-10: Material toughnes and through-thickness properties," december 2011.

[11] "NEN-EN 10025-2 Hot rolled products of structural steels - Part 2: technical delivery conditions for non-alloy structural steels," november 2004.

[12] G. Sadlacek, M. Feldman, B. Kuhn, D, Tshcikardt, S. Hohler, C. Muller, W. Hensen, N. Stranghoner, W. Dahl, P. Langenberg, S. Munstermann, J. Brozetti, J. Raoul, R. Pope, F. Bijlaard, "Commenatry and worked examples to EN 1993-1-10 "Material toughness and through thickness properties" and other toughness oriented rules in EN 1993," European Commission, Joint research Centre, 2008.

The strengthening of a steel bridge with prestressed CFRP strips

Tomasz W Siwowski[1], Paulina Siwowska[2] and Agnieszka Wiater[1]

[1]Rzeszow University of Technology, Poznanska 2, Rzeszow 35-959, Poland
[2]Promost Consulting, J.Niemierskiego 4, Rzeszow 35-307, Poland

E-mail: siwowski@prz.edu.pl

Abstract. The use of CFRP (carbon fibre reinforced polymer) strips in steel strengthening works has gone through a big development during the past 15 years. However, only in a few studies the effect of prestressing on the behaviour of steel beams strengthened with the CFRP strips has been examined. The paper briefly describes the research on the new CFRP strips prestressing system with special steel anchorages and its first on-site application for a steel bridge strengthening. The bridge proof tests before and after strengthening have been carried out. The results of testing proved the strengthening efficiency, however the final effect was moderate. As a result of strengthening the following effects has been gained: max. 18.3 % reduction of mid-span deflection and max. 13.9 % reduction of stresses in steel beams. The CFRP strengthening slightly improved the overall dynamic performance of the bridge span as well. The tests results showed moderate strengthening effects, mainly due to beams geometrical constraints not allowing to apply more than one 60 x 14 mm strip per beam. Regardless the moderate efficiency of strengthening, the research results clearly revealed that the new strengthening technology could be efficiently applied for steel bridges.

1. Introduction

The CFRP materials, as viable alternative to structural strengthening, offer several advantages such as high strength to weight ratio, light weight, ease of transport and installation, excellent fatigue characteristics and non-corrosiveness (durability). A few different approaches have been investigated to assess the effectiveness of the CFRP materials for strengthening of steel structures. Since the beginning of the new millennium, more than one hundred research papers on strengthening of steel structures with the CFRP materials have been published in scientific journals and conference proceedings. The most important research findings were reviewed and summarized in the following state-of-the-art reports [1-4]. These reviews showed that the load-carrying capacity of the CFRP-strengthened beam can be increased regarding stiffness and strength. In the case of the degraded beams, it was found out that the load-carrying capacity could be partly or fully restored by the strengthening system. Field applications have also been carried out on bridge girders, see [5-7] for example, which resulted in the stresses in the original beams being reduced. On the basis of these studies, it was concluded that using bonded CFRP plates as a strengthening system can be effective.

By prestressing the CFRP plate, which combines benefits of passive bonded plate with the advantages due to external prestressing, the ultra-high tensile strength of carbon fibres can be utilized and much higher effectiveness of the strengthening technique is achieved. A greater CFRP's tensile capacity is employed and it contributes to the load carrying capacity of the strengthened steel structure.

IOP Conf. Series: Materials Science and Engineering **419** (2018) 012037 doi:10.1088/1757-899X/419/1/012037

under both service and ultimate conditions. However, only in a few studies the effect of prestressing on the behaviour of steel beams strengthened with the CFRP plates has been examined [8-13]. The anchor zones in plate end locations has been the main problem of steel beam strengthening with the CFRP prestressed plates. Without mechanical anchorages, the peeling failures at the plate ends were observed. Premature debonding of the CFRP plates is one of the most common failure modes, and it could occur before the stress in the CFRP material reaches its tensile strength. Nevertheless, more detailed investigations are needed to obtain a good understanding of the way the forces acting inside the strengthening system and the way different parameters of the CFRP plate affect the behaviour of the strengthened beam.

The main goal of the R&D study was to compare the flexural behaviour of steel beams strengthened with the CFRP plates. This research evaluated two strengthening schemes, namely: (1) adhesive bonded passive plates and (2) adhesive-bonded active or prestressed plates, with different strengthening configurations and CFRP materials. The previous author's paper presents the results of an experimental investigation carried out on the behaviour of steel beams before and after strengthening using different CFRP systems [14]. In the scope of the R&D study the actual bride proof tests before and after strengthening was also carried out. The results of bridge testing proved the strengthening efficiency, however the final effect was moderate. Nevertheless the presented case study revealed the effectives of the new prestressing system, which seems to be reliable for steel bridge applications. The strengthening technology, its first application on a steel bridge and discussion of bridge proof testing results have been the main subject of the paper.

2. Novel CFRP Prestressing Technique

The new system called *Neoxe* consists of two main elements: special steel anchorages mounted on both ends of a single CFRP strip and a relevant tensioning device. There are two kinds of anchorages: an active anchorage combined with a tensioning device and a passive one. The strip with a specific length is delivered on site as ready-to-install, i.e., with two prefabricated steel anchorages mounted on both strip ends. The anchorage is made of two thin steel plates welded together along the edges to create a pocket, in which the CFRP strip end is fixed and bonded with a special epoxy-based adhesive. It is followed by gripping both materials (steel plates and CFRP strip in-between) with small mechanical fasteners: rivets or high-strength bolts (Fig. 1).

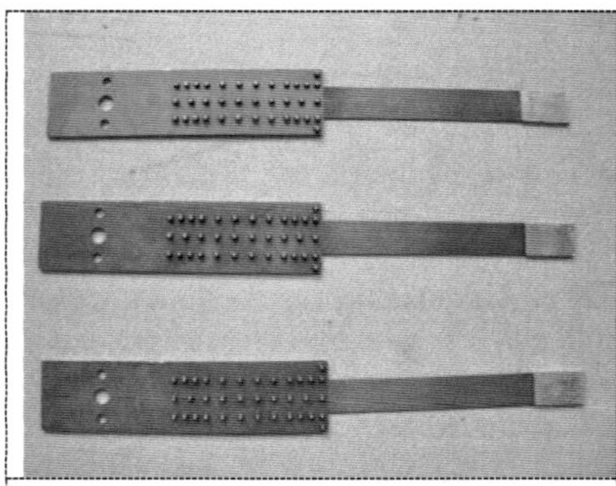

Figure 1. Steel anchorages

Figure 2. Tensioning device

However, it is often said, that galvanic corrosion may be occurred in a CFRP/steel system because of direct contact between steel and CFRP surfaces accompanied with an electrolyte like salted water. It was showed using a nonconductive fabric between two materials, an isolating epoxy layer on the steel surface, and a moisture buffer decreased significantly the rate of corrosion. Meanwhile, epoxy adhesive usually play the role of insulator [15].

The tensioning device is the second key element of the strengthening system. The tensioning device consists of three separately installed components: guide rails, carriage (bolted to the active anchorage) and hydraulic jack with resisting block (Fig. 2). The hydraulic jack is driven by a manual pump, and it can generate a subsequent tensioning force to stretch the CFRP strip. Owing to division of the device structure into three small and lightweight parts (the heaviest element weighs 37 kg), its manual application on-site is very fast and easy without using the heavy equipment.

The high-strength UHS 614 strips with a cross-section of 60 × 1.4 mm, with the ultimate tensile strength of 3200 MPa, a modulus of elasticity of 160 GPa and the strain at failure of approximately 2% are typically used in the NPS-II for strengthening. To complement the system, an epoxy-based adhesive is used for the final bonding of the strip between anchorages to the concrete substrate. The development and experimental investigation of a novel anchor and tensioning system for the flexural strengthening of steel as well as RC elements with prestressed CFRP strips has been presented in [16].

3. Bridge description

The bridge is situated along the local road over the small river. This is a simply-supported structure with the span length of 11.92 m and the total width of the deck of 7.41 m (Fig.3). The superstructure consists of five INP 500 rolled I-beams spaced at 1.40 m and laterally braced with C300 rolled channel sections spaced at 3.75 m along the bridge. This steel grid is covered with the temporary timber deck with no sidewalks, which state of repair was quite bad. The bridge span is supported by two concrete abutments founded directly on the ground. Because of deteriorated timber deck the carrying capacity of the bridge was posted to 15 metric tonnes and thus the bridge became the cumbersome bottleneck on the local road.

Figure 3. The steel bridge before strengthening.

These bridge constraints and the growing traffic on the local road caused the local road administration had made a decision to strengthen the bridge along with timber deck replacement into the new RC deck slab having wider roadway and sidewalks. The main goals of the modernization were as follows: to replace the timber deck into a durable RC slab, to increase the bridge carrying capacity up to 40 tonnes, to widen the roadway up to 6.0 m and to create at least one sidewalk. As the steel grid structure and the concrete abutments were in good condition it was decided to re-use them in the bridge modernization.

4. Bridge strengthening and modernization

Due to financial constraints the local administration divided the modernization works into two phases. In the first phase the steel grid was strengthened to carry the new RC slab deck and the increased traffic load, ultimately for 40 tonnes vehicles. The deck replacement & widening along with the approach road reconstruction were planned to be done in the second phase, one year later. However, the first phase' works enabled to increase the bridge carrying capacity up to 30 tonnes immediately after steel strengthening execution, complemented with the partial replacement of the most deteriorated timber elements.

The INP 500 rolled I-beams were strengthened with the *Neoxe* system by means of the prestressed UHS 614 *neoxeplate* CFRP strips with the average material parameters as stated above (p.2). On each beam one strip was installed, glued and anchored after prestressing. The prestressing force determined in the detailed calculation taking into account two-phase strengthening was in the range of 137÷147 kN (depending on beam location), which capitalized the ultimate strength of CFRP in the range of 50÷55% respectively. The main stages of the strengthening procedure on-site were as follows: steel and CFRP surface preparation, adhesive application, mounting the passive anchorage on steel beam (Fig.4), mounting the tensioning device on the opposite end of the beam (Fig.5), trial strip prestressing, mounting the active anchorage on steel beam, final strip prestressing, strip bonding to steel substrate along with removing the adhesive excess. The steel anchorages were mounted on the steel beams by means of the 10.9 class HSFG bolts. No traffic limitation was required during the strengthening works.

Figure 4. Passive anchorage mounted on the steel beam

Figure 5. Mounting the tensioning device

In the second phase of bridge modernization one additional steel beam will be added to facilitate the bridge deck widening and the new RC deck slab will be executed replacing the timber deck. Before the new concrete slab is cast, the shear studs will be mounted on beams' top flanges to ensure the full composite action of the new steel/concrete beams after rehabilitation. The achievement of the composite action is as much efficient for bridge strengthening as the CFRP prestressing, implemented in the first phase.

5. Evaluation of the strengthening efficiency

5.1. Tests description

To assess the efficiency of the bridge strengthening, two proof tests were carried out during the first phase of the modernization works: before and after steel strengthening. In both tests exact the same load was placed on the bridge to compare the relevant strains and displacements and to evaluate the superstructure behaviour. Two trucks with the mass of 32 metric tonnes each were used for static tests whereas one of them was further applied for dynamic tests as well. The total proof load was limited due to timber deck capacity and its bad state-of-repair. Therefore only 40 – 45 % of the ultimate strength of steel beams was capitalized during static tests.

In both tests the vertical displacements in the mid-span of all beams were measured by means of the linear variable displacement transducers (LVDTs) with 100 mm base (points D1 – D5). Moreover, the 10 mm foil strain gauges were applied for strain measurement both in steel beams as well as in the CFRP strips. The steel strains were detected in the top and bottom flanges of each beam in the mid-span (points S1-S5) (Fig.6). The CFRP strains were measured in 8 points along the central beam's strip (C3/1-C3/7) (Fig.7). All gauges were connected via the set of QuantumX MX840A amplifier modules to the computer with Catman Easy data acquisition software. The signal of the selected LVDTs and strain gauges was also applied to determine the dynamic characteristics of the bridge span before and after strengthening.

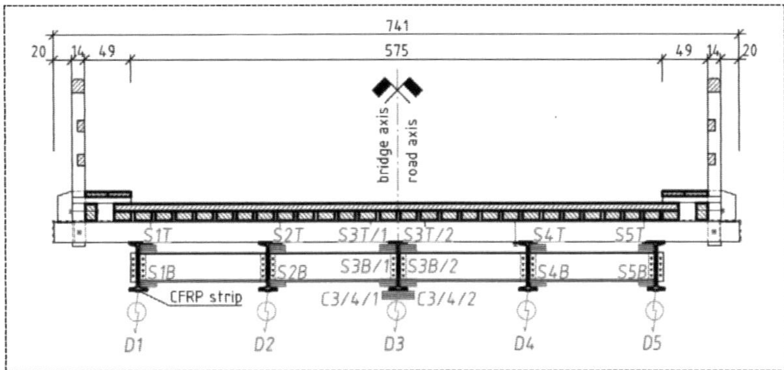

Figure 6. Displacement and strain measurement points.

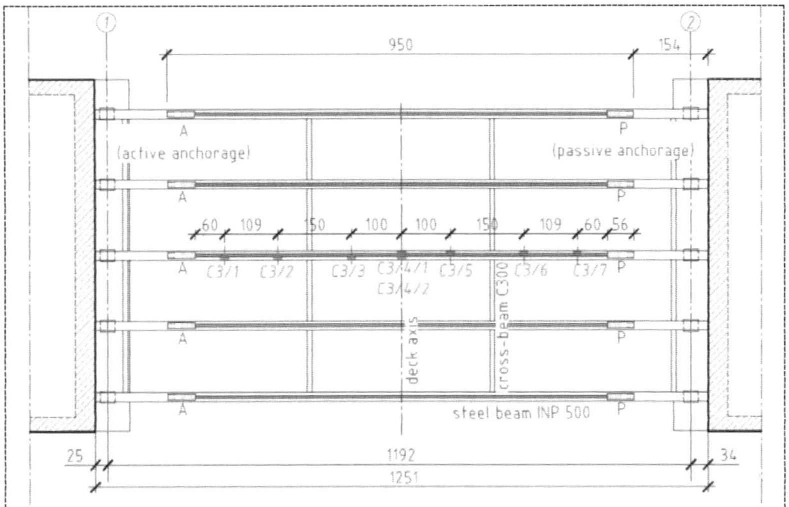

Figure 7. CFRP strain measurement points.

5.2. Static tests results

In Tables 1, 2 and 3 the both tests results (i.e. before and after strengthening) are collected to be directly compared in order to assess the strengthening efficiency. The strengthening efficiency in this particular case is defined as the percentage of the reduction of the mid-span deflection as well as the decrease of steel strains (stresses) in steel beams. To give an idea of beam's carrying capacity capitalization the experimentally obtained strain values are converted into stresses with the assumption of modulus of elasticity of 205 GPa for steel and of 160 GPa for CFRP respectively. The stresses in CFRP strips enabled to assess the utilisation of composite material under bridge full loading, having in mind the initial stresses induced by prestressing.

Table 1. Beam's mid-span deflection before and after strengthening.

Beam	Measurement point	Deflection before (mm)	Deflection after (mm)	Reduction (%)
1	D-1	18.05	16.64	7.8
2	D-2	23.69	21.92	7.5
3	D-3	30.60	24.99	18.3
4	D-4	23.40	20.90	10.7
5	D-5	18.37	16.79	8.6

The stress reduction in steel beams due to strengthening was moderate: from 6.2 to 13.9 % in the bottom flanges and from 1.1 to 8.9 % in the top flanges depending on the beam location (Table 2). The reduction in the most loaded central beam was almost equal for the both flanges, but in the adjacent, outer beams higher reduction in the bottom flanges was obtained. However, there is a scatter of strengthening effects despite the symmetry of the span structure as well as loading. It is mainly due to uneven contribution of timber deck in load carrying capacity of the steel beams, influencing the lateral load distribution. The small differences in the prestressing forces in particular beams could also influence the stress distribution after strengthening.

Table 2. Stresses in steel beams before and after strengthening.

Beam	Measurement point	Stress before (MPa)	Stress after (MPa)	Reduction (%)
1	S1T	-69.7	-65.8	3.0
1	S1B	56.0	52.5	6.2
2	S2T	-92.7	-91.6	1.1
2	S2B	88.6	83.4	7.9
3	S3T/1	-107.4	-98.4	8.4
3	S3B/1	104.9	96.6	7.6
3	S3T/2	-110.1	-100.2	8.9
3	S3B/2	102.9	93.7	9.0
4	S4T	-83.2	-77.9	6.4
4	S4B	91.2	78.5	13.9
5	S5T	-69.3	-66.0	4.7
5	S5B	67.9	63.1	6.9

The analysis of the composite strains clearly shows, that the CFRP strips contributed successfully in beam's load carrying capacity (Table 3). However, the composite strains induced by bridge loading constituted only maximum 5.17 % of the total strains of CFRP strips. The strains due to the initial prestressing (10.19 ‰) are always decisive for the capitalization of the ultimate strength of the CFRP strips. It was about 50% in this strengthening case. The strain distribution along the CFRP strip mounted on the most loaded central beam was decreasing from the mid-section towards strip ends. The strains induced in or close to steel anchorages were negligible what proved the reliability of the strengthening system in terms of fatigue. All test results confirmed the appropriate behaviour of the CFRP strengthened superstructure under loading.

Table 3. Composite strains and stresses along the CFRP strip in beam 3.

Measurement point	Strains before (‰)	Strains after (‰)	Total stresses (MPa)	Material use (%)
C3/1		0.369	1659.04	50.58
C3/2		0.472	1675.52	51.08
C3/3		0.532	1685.12	51.38
C3/4/1	10,19	0.552	1688.32	51.47
C3/4/2		0.555	1688.80	51.49
C3/5		0.514	1682.24	51.29
C3/6		0.389	1662.24	50.68
C3/7		0.278	1644.48	50.14

5.3. Dynamic tests results

The dynamic test of the bridge span before and after strengthening was carried out by means of one truck with the mass of 32 metric tonnes passing the bridge with two different velocities: 30 and 50 km/h. The dynamic response of the span was detected with LVDTs in the mid-span location for beams 3 and 4. Generally it can be said, that the CFRP strengthening slightly improved the overall dynamic performance of the bridge span. The first eigen frequencies increased in the range of 5.3 to 12.5 % after strengthening and simultaneously the dynamic factors were reduced by 15% in average. The strengthening enhanced also damping characteristics of the span increasing the logarithmic decrement by more than 20% in average. The measured dynamic characteristics had a slight scatter and did not clearly depend on the vehicle velocity and beam location.

5.4. Analysis of results

The main goals of a bridge strengthening are always at least to decrease deflections and/or to relieve a structure, which means to reduce the stresses in subsequent bridge elements, mostly main beams or girders. Taking into account these goals it can be said, that the bridge strengthening with CFRP prestressing let to achieve only moderate effects. The quantitative estimation of these effects are as follows: max. 18.3 % reduction of mid-span deflection and max. 13.9 % reduction of stresses in steel beams. Moreover, the CFRP strips were almost not affected by the proof load, because the only maximum 5.2% increase of composite strains was detected.

There are several reasons the strengthening efficiency is so moderate. First of all the proof load was considerably limited due to deteriorated and only partially repaired timber deck. Under the applied loading only about 40% of the load carrying capacity of the bridge was capitalized in terms of steel strength. The larger steel capitalization under service load, the higher strengthening efficiency. The efficiency limitation was caused by the fact that only one CFRP strip per beam might be applied due to its geometrical constraints. On the other side the higher prestressing force might jeopardize the local stability of the beam's top flange, not supported by a compositely acting deck slab. In more likely cases when beams are connected to the RC deck slab the presstressing could reach the allowable level, i.e. about 70% of the ultimate strength of the CFRP strip. In our case however the strengthening CFRP strips take part of the self-weight of the future RC slab, thus increasing the utilization (and efficiency) of composite material.

6. Conclusions

The first phase of the steel bridge modernization including strengthening, redecking and widening has been presented in the paper. In this phase the strengthening of steel beams with the prestressed CFRP strips was the main intervention. Two bridge proof tests: before and after strengthening, enabled to check and prove the strengthening efficiency. The strengthening was applied to typical rolled steel beams covered with the timber deck not structurally connected with beams. The comparison of two tests results showed moderate strengthening effects, mainly due to beam's geometrical constraints not allowing to apply more than one strip per beam. The limited proof load used in proof tests also affected the final results and thus the obtained efficiency. However, the CFRP strips contributed successfully in load carrying capacity of the bridge. The moderate quantitative results of the strengthening will be considerably enhanced in the second phase of bridge modernization.

Regardless the moderate efficiency of strengthening, the research clearly revealed that the new strengthening technology using the prestressed CFRP strips could be efficiently applied for steel beam bridges and enabled to strengthen them up to the required carrying capacity. The strengthening level depends on particular geometrical and technological constraints of the existing bridge and works to be undertaken for its modernization. The strengthening system can be quite quickly implemented on-site without traffic limitations and significant structural intervention. Moreover, due to long-lasting material used for strengthening the final effect seems to be much more durable than in case when another methods of steel bridge strengthening would be applied.

Acknowledgements
The authors wish to acknowledge the financial support provided within the framework of the research project named *Innovative system for strengthening of structures with prestressed strips of carbon fibre reinforced composites*, Project No. POIG.01.03.01-18-010/12. The project was co-funded by the European Union under the Innovative Economy Operational Programme.

References
[1] Hollaway L C and Cadei J 2002 Progress in the technique of upgrading metallic structures with advanced polymer composites *Progress in Structural Engineering and Materials* **4** pp 131–148
[2] Shaat A, Schnerch D, Fam A and Rizkalla S 2004 Retrofit of steel structures using fiber-reinforced polymers (FRP): state-of-the-art *Proc. 83rd Annual Meeting of the TRB (Washington)* on CD pp 1-18
[3] Zhao X L and Zhang L 2007 State of the art review on FRP strengthened steel structures *Engineering Structures* **29** pp 1808–1823
[4] Teng J G, Yu T and Fernando D 2012 Strengthening of steel structures with fiber-reinforced polymer composites *Journal of Constructional Steel Research* **78** pp 131–143
[5] Luke S 2001 The use of carbon fibre plates for the strengthening of two metallic bridges of an historic nature in the UK *Proc. Int. Conf. on FRP Composites in Civil Engineering (Hong Kong)* ed J G Teng (Amsterdam: Elsevier) pp 975-983
[6] Wipf T J, Phares B M, Klaiber F W, Lee Y S, Al-Saidy A H, Abu-Hawash A and Monk C 2004 Use of CFRP plates in the strengthening of steel girder bridges *Proc. 2nd Int. Conf. on Bridge Maintenance, Safety and Management (Kyoto)* ed E Watanabe, D M Frangopol and T Utsunomiya (Boca Raton: CRC Press) pp 18- 22
[7] Miller T C, Chajes T J, Mertz D R and Hastings J N 2001 Strengthening of a steel bridge girder using CFRP plates *Journal of Bridge Engineering* **6** pp. 514-522
[8] Al-Emrani M and Kliger R 2006 Analysis of interfacial shear stresses in beams strengthened with bonded prestressed laminates *Composites Part B: Engineering* **37** pp 265–272
[9] Benachour A, Benyoucef S, Tounsi A an Adda Bedia E A 2008 Interfacial stress analysis of steel beams reinforced with bonded prestressed FRP plate *Engineering Structures* **30** pp 3305–3315
[10] Soudki K, Schumacher A and Motavalli M 2009 Flexural strengthening of a steel beam with prestressed strips-preliminary investigation *Proc. 9th Int. Symp. on Fiber Reinforced Polymer Reinforcement for Concrete Structures (Sydney)* ed D J Oehlers, M C Griffith and R Seracino (Adelaide: University of Adelaide) pp 1-4
[11] Ghafoori E and Motavalli M 2013 Flexural and interfacial behaviour of metallic beams strengthened by prestressed bonded plates *Composite Structures* **101** pp 22–34
[12] Siwowski T and Pasko P 2014 Research on the new CFRP prestressing system for metal bridge strengthening *Proc. 15th European Bridge Conf. at Structural Faults & Repair (London)* on CD pp 1-8
[13] Ghafoori E and Motavalli M 2015 Innovative CFRP - prestressing system for strengthening metallic structures *Journal of Composites for Construction* **19** pp 04015006-1-14.
[14] Siwowski T and Siwowska P 2018 Experimental study on CFRP-strengthened steel beams *Composites Part B: Engineering* **149** pp 12-12
[15] Tavakolizadeh M and Saadatmanesh H 2001 Galvanic corrosion of carbon and steel in aggressive environment *Journal of Composites for Construction* **5** pp 200–210
[16] Piatek B, Siwowski T, Michalowski J and Blazewicz S 2018 The flexural strengthening of RC beams with prestressed CFRP strips: Development of a novel anchor & tensioning system *Journal of Composites for Construction* (in press)

IOP Conf. Series: Materials Science and Engineering **419** (2018) 012038 doi:10.1088/1757-899X/419/1/012038

Investigation on stiffened panels subjected to biaxial compression

Martin Mensinger[1], Ulrike Kuhlmann[2], Joseph Ndogmo[1], Vahid Pourostad[2] and Nadine Maier[1]

[1]Technical University of Munich, Faculty of Civil, Geo and Environmental Engineering, Chair of Metal Structures, Munich, Germany
[2]University of Stuttgart, Institute of Structural Design, Chair of Steel, Timber and Composite Construction, Stuttgart, Germany

E-mail: nadine.maier@tum.de

Abstract. This paper intends to provide an overview on large-scale buckling tests, which were carried out at the Technical University of Munich in January 2018. In the experiments, the load introduction was simulated as it usually occurs during incremental launching of a steel bridge when a biaxial stress state from direct stresses from the bending moment due to cantilever action and local transverse compression stresses due to load introduction at the bearings occurs. The topics and background of the research project will be presented, explaining the test program as well as the structure of the whole test setup and first numerical calculations. Initial results from the experiments are presented and compared with the numerically validated model, which is carried out at the University of Stuttgart. For this purpose, two of the specimens are compared and differences of their buckling behavior are analyzed.

1. General

Large bridge constructions are frequently erected by incremental launching. During launching, the cross-section is stressed by bending due to cantilever action and at the same time, local compression in the vertical direction from the load introduction at the launching bearings occurs. In this case a biaxial compressive stress state arises in the web panels in the area of the bearings. At least in German bridge construction the use of stiffeners for this type of large steel bridges is common. One of the existing verification methods is, among others, the reduced stress method according to EN 1993-1-5 (10.5) [1].

$$\eta = \left(\frac{\sigma_{x,Ed}}{\rho_{cx} \cdot \frac{f_{y,k}}{\gamma_{M1}}}\right)^2 + \left(\frac{\sigma_{z,Ed}}{\rho_{cz} \cdot \frac{f_{y,k}}{\gamma_{M1}}}\right)^2 - V \cdot \left(\frac{\sigma_{x,Ed}}{\rho_{cx} \cdot \frac{f_{y,k}}{\gamma_{M1}}}\right)^2 \cdot \left(\frac{\sigma_{z,Ed}}{\rho_{cz} \cdot \frac{f_{y,k}}{\gamma_{M1}}}\right)^2 + 3 \cdot \left(\frac{\tau_{Ed}}{\chi_w \cdot \frac{f_{y,k}}{\gamma_{M1}}}\right) \leq 1 \quad (1)$$

The reduced stress method allows the utilization of individual components only up to the yield limit, so that the weakest cross-sectional part may only be exploited up to yielding. As already shown in [2] and [3], the reduced stress method in EN 1993-1-5 [1] section 10 seems not always lead to safe results for panels under biaxial compression. Therefore, for panels under biaxial compression in [2], a modification factor V for the current verification format equation (10.5) according to EN 1993-1-5 [1] was proposed and also introduced as an amendment to EN 1993-1-5 by TC250/SC3.

In contrast to unstiffened plates, longitudinally stiffened buckling plates have a pronounced anisotropy and thus a pronounced column-like buckling behavior. For the biaxial stress state they are not systematically investigated so far and the application of the reduced stress method has been verified by some numerical calculations but not by real testing. For this reason, EN 1993-1-5 does not contain any specific rules for the verification of the global buckling when taking into account the effects of column-like behavior in a longitudinal stiffened plate under biaxial compressive stresses. Due to the lack of rules, in [4] and [5] additional design checks of the flexural buckling of the longitudinal stiffeners according to Second Order Theory by taking into account deviation forces from transverse compression, are suggested. Since the existing verification method [1] does not explicitly take account of these deviation forces of transverse compression in longitudinally stiffened plates, the here presented tests were carried out in order to get a better knowledge about the behavior of biaxially loaded panels with longitudinal stiffeners.

2. Experimental Setup

As part of the planning of the replacement of the Thulba viaduct, a composite bridge with a box cross-section in northern Bavaria, which will be erected by incremental launching in 2019, six large-scale buckling tests were carried out at the Technical University of Munich.

The test setup was designed as a modular U-frame with external dimensions of 7.64 meters x 11.08 meters. Test specimens with a size of 3.0 meters x 4.0 meters were used. The test setup is laid horizontally and supported by in total six HEB 600 profiles. The bending moment and the normal force F_1 are applied through the legs of the U-frame by a hydraulic jack and two tension rods. Furthermore two pressure-rated hydraulic jacks initiate the lateral loads F_2 in the middle of the specimen. This force is short-circuited by means of a traverse, which is clamped to the U-frame with the help of four tension rods. This allows a maximum bending moment of 24 MNm combined with an external normal force of 4.3 MN (F_1) and a maximum lateral load of 8.6 MN ($2*F_2$) on the test specimen. Figure 1 shows the test setup and the loading schematically. The specimen is highlighted by the grey area.

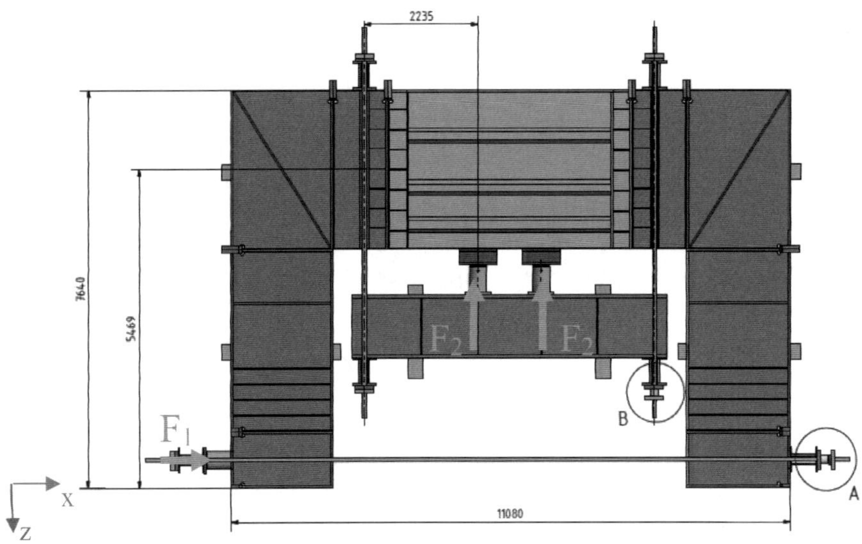

Figure 1. Test setup, load transmission

The forces F_1 and F_2 can be controlled independently. As a first step, the F_1 is brought to a certain level, which results from numerical pre-calculations realized at the University of Stuttgart. Subsequently the transverse loads F_2 are applied and increased until failure of the specimen. During the test, the loading, the strains and the deformations are continuously measured. In addition, the deformation is measured by a 3D scanner before the beginning of tests as initial imperfections, when reaching the transverse loads obtained from numerical pre-calculations and after the ultimate load. The measurement results could be used to investigate the behavior of panel and for the comparison with numerical models.

In the six tests, the stiffness and the position of the stiffeners and the stress ratio σ_z/σ_x are varied. In the following the tests V3 and V4 are presented, in which only the stress ratio σ_z/σ_x varies.

3. Results from tests V3 and V4

3.1. General

The geometry of the specimen is the same for test V3 and V4. It can be taken from Table 1. Table 1 also shows two different planned stress ratios σ_z/σ_x. In this case, the stress ratio σ_z/σ_x is 1.0 at V3 and 0.5 at V4. So the influence of the stress ratios could be studied since V3 has the same geometry and boundary condition like V4 but with significantly higher transverse stresses than V4. Figure 2 shows the geometry of the specimens in general and the acting stresses.

Table 1. Main dimensions and varied parameters of the cross section

	h_w (mm)	t_w (mm)	a (mm)	b_f (mm)	t_f (mm)	B_{st} (mm)	t_{st} (mm)	b_1 (mm)	b_2 (mm)	b_3 (mm)	b_4 (mm)	γ_{st}	Stress ratio σ_z/σ_x
V3	3000	8	4000	500	30	300	6	300	400	700	700	27	1.0
V4	3000	8	4000	500	30	300	6	300	400	700	700	27	0.5

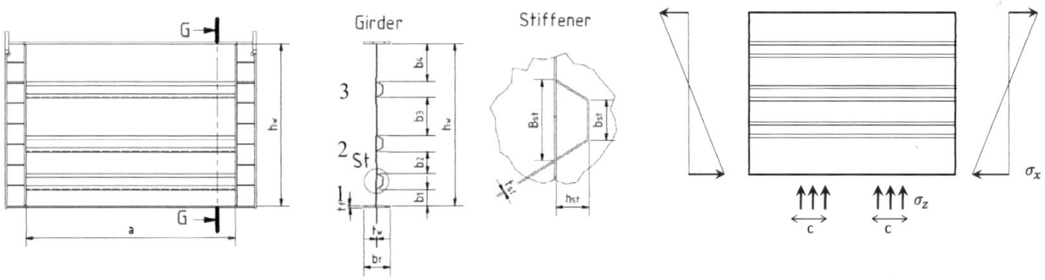

Figure 2. Cross-section of specimen V3 and V4 and acting stresses

3.2. Imperfections

Figure 3 and Figure 4 show the initial imperfections of specimen V3 and V4. Red areas indicate positive, blue indicates negative pre-deformation. The positive direction is defined in the direction of the stiffeners. It can be seen from both figures that the position of the maximum deflection is in the 3rd sub-panel. However, the imperfection of specimen V3 is significantly higher than that of specimen V4. It can be seen from the position of the first stiffener that the specimen of test V3 has a larger negative imperfection than the one of V4.

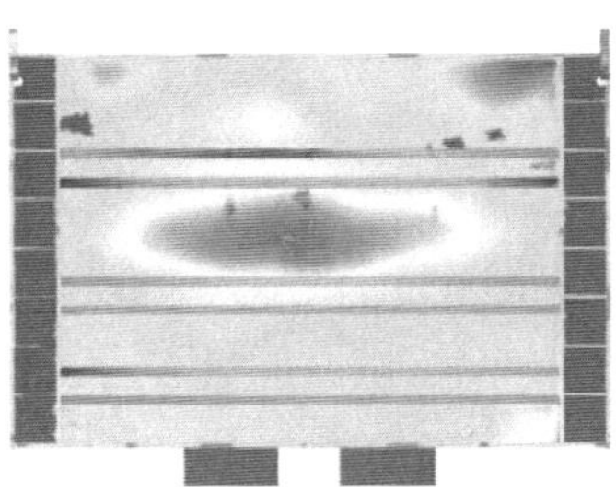

Figure 3. Imperfections of specimen V3

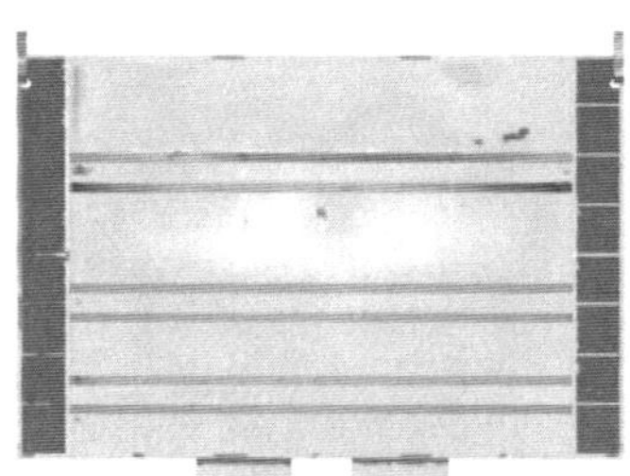

Figure 4. Imperfections of specimen V4

3.3. Deformations

For the measurement of the deformations transducers were mounted under the specimen directed vertically on the specimen and thus measuring the deformations out of the plane. In Figure 5 and Figure 6 the deformations of panel subjected to longitudinal stresses σ_x due to F_1 are shown.

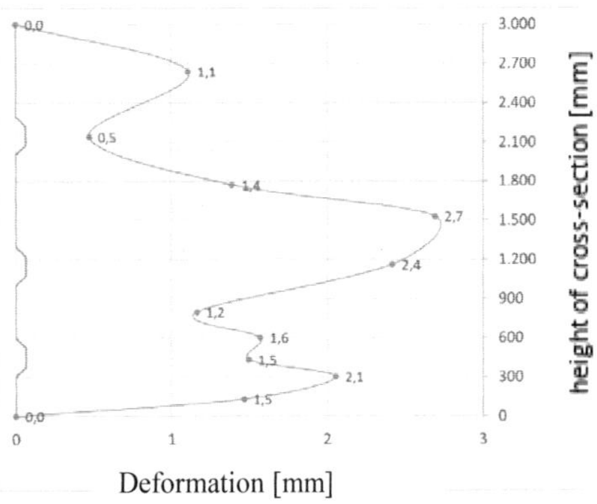

Figure 5. Deformations out-of-plane under longitudinal stresses (V3)

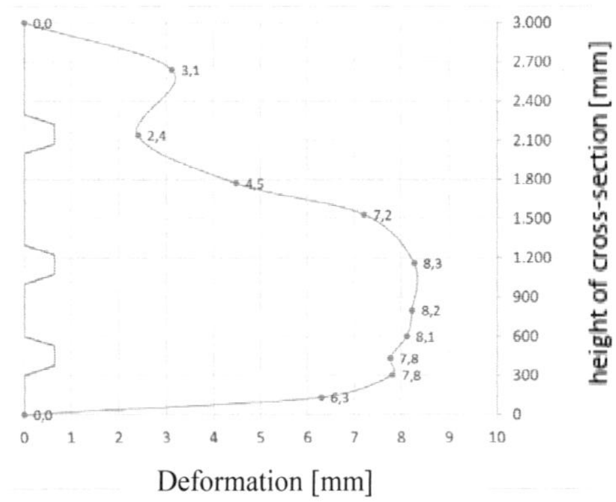

Figure 6. Deformations out-of-plane under longitudinal stresses (V4)

At this state of the tests, no transverse loads F_2 were applied. The deformations were all in the positive range. This resulted from the fact that the load was applied in the center of the web panel. Due to the longitudinal stiffeners, the overall center of gravity of the panel was slightly above the web. The resulting eccentricity between load introduction and the center of gravity of the longitudinal stiffened panel led to hogging bending and caused the deformation in positive direction (see Figure 7). The deformations observed during test V4 were higher because the longitudinal loading and therefore also the hogging bending moment were 1.6 times bigger than in test V3. In addition, Due to the different initial imperfections of the position of the flanges of the first stiffener, a rotation of the first stiffener in the counterclockwise direction was observed.

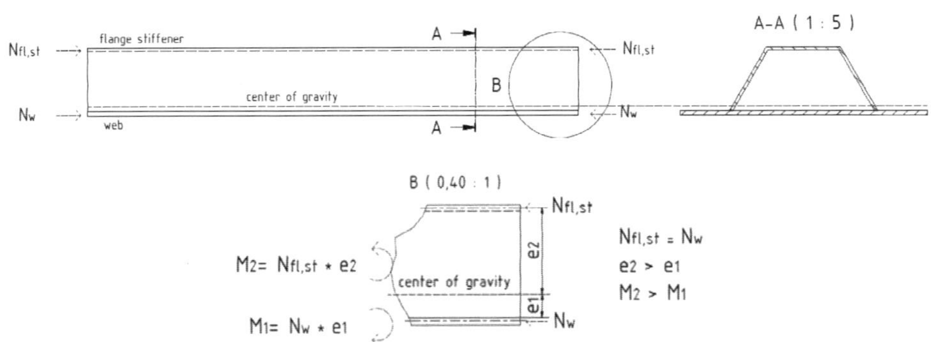

Figure 7. Internal forces due to direct compressive stresses

In Figure 8 and Figure 9 the deformations under biaxial compression for test V3 and V4 are shown. They show the deformation at maximum transverse compression at failure. The ratio of the longitudinal stresses between V3 and V4 remained 1.6 as before. The ratio of the transverse loads was 1.0. Therefore, in case of V3 a different ratio $\sigma_z/\ \sigma_x$ between longitudinal and transverse stresses was realized compared to V4. The influence of the transverse load is higher in test V3 than in V4, since the applied longitudinal load is smaller in case of V3. In the shape of the deformations at failure, a clear difference between the two tests can be seen. While in test V3, the deformations in the region of the first stiffener switched to the negative direction, the deformations in test V4 showed a deformation until failure in the positive direction. On the one hand, this behavior may also be due to initial imperfections, which are more pronounced in test V3 than in test V4. On the other hand, this effect can also result from the influence of the transverse load and the resulting internal forces. Similarly, in test V3, a change in the direction of rotation of the first stiffness can be observed. This effect does not occur in test V4 and may be attributed to the influence of transverse load because the imperfections occurred exactly opposite.

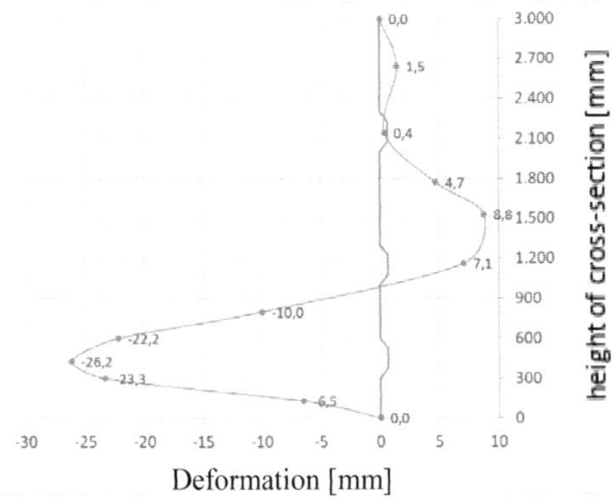

Figure 8. Deformations out-of-plane under biaxial stresses (V3)

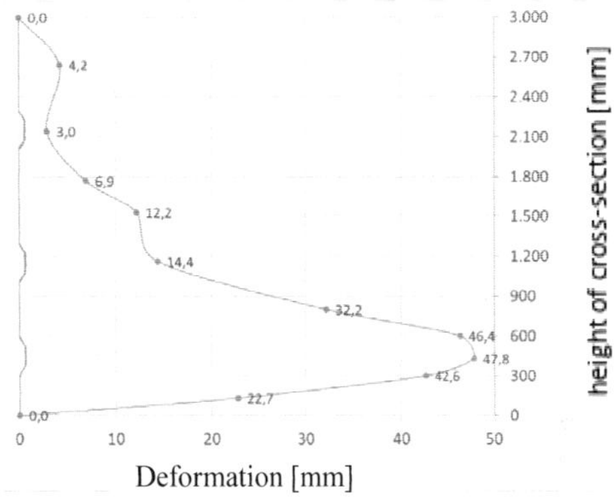

Figure 9. Deformations out-of-plane under biaxial stresses (V4)

The mechanical background of how a change in the direction of rotation of the stiffener can result is explained in Figure 10.

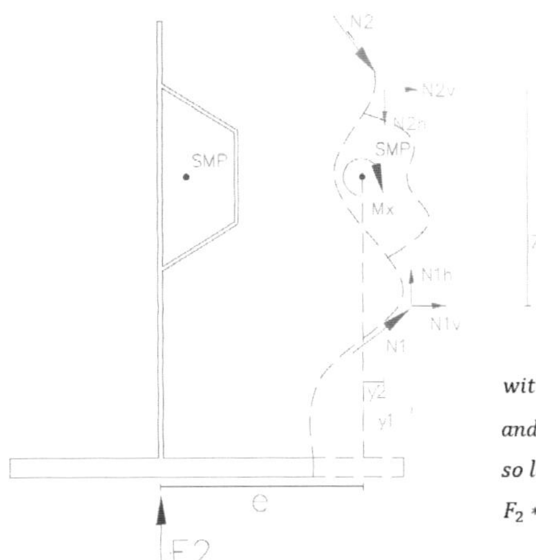

$with: M_x = F_2 * e - N_{1,H} * z_1 - N_{1,V} * y_1 + N_{2,H} * z_2 + N_{2,V} * y_2$

$and\ N_1 > N_2$

$so\ long\ as:$

$F_2 * e + N_{2,H} * z_2 + N_{2,V} * y_2 > N_{1,H} * z_1 + N_{1,V} * y_1$

Figure 10. Influence on the rotation of the stiffener

4. Numerical investigations

4.1. Imperfections

In slender structures, imperfections are unavoidable. The imperfections can influence the ultimate load, the buckling behavior as well as the shape of failure of structures. Generally, in buckling structural and geometrical imperfections are distinguished. They can be triggered by production or processing.

Residual stresses are mainly caused by the welding process. *Sinur* stated based on his investigations [7] that the influence of residual stresses in slender plate girders is rather small and may be neglected. Therefore, the measurements and simulations presented here mainly concentrated on the effects of geometrical imperfections.

Out-of-plane deflections were measured with laser scanner as well as Linear Variable Displacement Transducer (LVDT). For the recalculation of the tests, the imperfection shape has been determined and simplified taking only the maximum measured geometrical imperfections of each sub-panels and stiffener (13 different points) into account. Figure 11 shows the applied imperfection shape in the numerical model of V3 and V4.

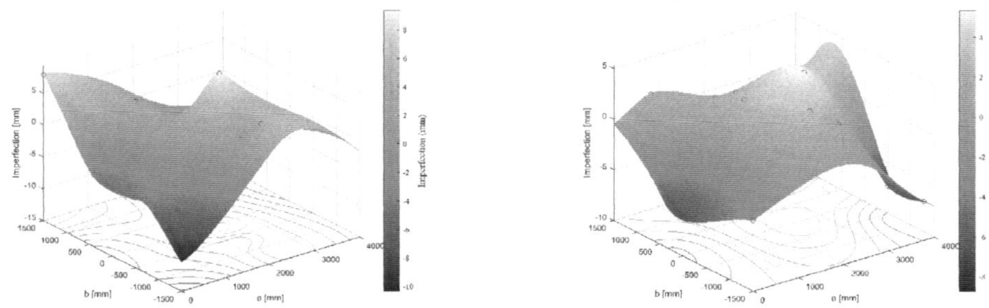

V3 V4

Figure 11. Out-of-plane of tests

4.2. Material

The steel was modeled as an elastic-plastic material. Swedish Steel Standard BSK07 [8] provides a parameterized material curve, see Figure 12. It can represent very simple the material behavior and consider the strain hardening. For the numerical simulation of the tests, the material parameters of each panel were obtained from the tensile tests. The recalculation of the tests e.g. in [7] with this assumption showed good results. Therefore, it is used in the numerical investigation. However, the elastic modulus E, f_y and f_u were obtained from the tensile material tests and the Poisson's ratio assumed to $\nu = 0.3$. These engineering material curves were, for finite element simulation, modified in the form of the true Cauchy stresses on logarithmic strains acc. to Annex C of EN 1993-1-5.

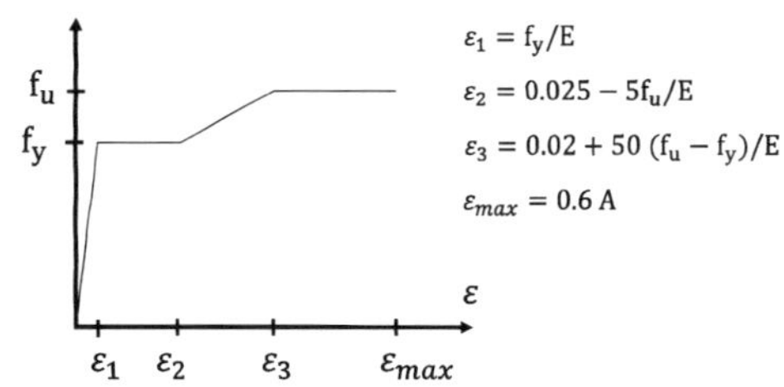

Figure 12. Material curve acc. to BSK07 [7]

5. Numerical Model

5.1. General

A nonlinear numerical model based on the Finite Element Method was developed with ABAQUS [9] applying the same boundary conditions as in the tests, see Figure 13. The specimens and test setup were modeled to simulate exactly the boundary condition of the test panel. However, the traverse girder that formed a support for loading jacks in the transverse direction, was not modeled. The transverse loads were applied to the specimens by means of two jacks. The load contact areas on the pressure flanges were coupled with a reference point in the center of the surface so that the load introduction in ABAQUS was modelled relatively close to reality.

Since the ratio of width to thickness of the steel panel is small, the shell element called "S4R" [9] was used which is a suitable element for the geometrical and materially non-linear calculation of thin-walled structures. The application of loads has been simulated in two steps like in the tests. At the first step, only F_1 is applied then GMNIA analysis is conducted, in the second step F_2 is applied and the ultimate load is obtained using arc length method.

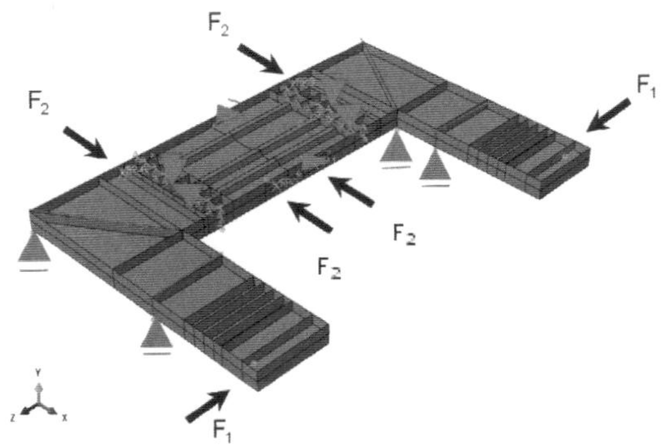

Figure 13. Numerical model

5.2. Load-carrying capacity

In Table 2, the maximum force obtained from the numerical simulations is compared with the test results. It can be seen that with increasing the transverse load the difference between the numerical results and test results slightly increases. The out-of-plane deformations of the tested panel in Figure 14 and Figure 15 are compared to results from the numerical simulations.

Table 2. Comparison of the ultimate load of tests and FE.

Test	$F_{1,FE}/F_{1,Test}$	$F_{2,FE}/F_{2,Test}$
V3	1	1.15
V4	1	1.02

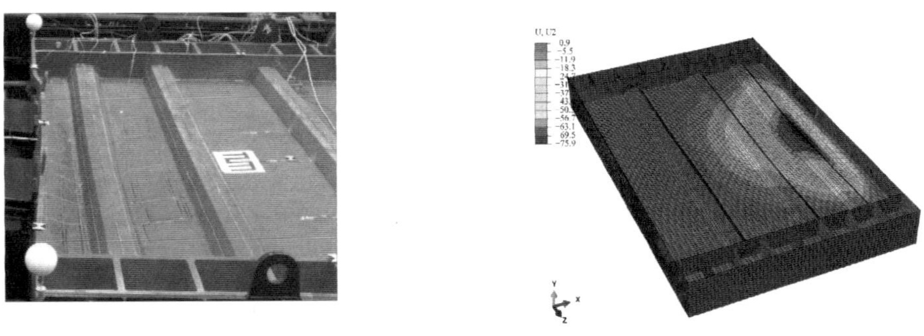

Figure 14. Out-of-plane deformations of test V3 (left) and numerical model (right)

It can be seen that the numerical results have a similar out-of-plane deformed shape as the test results. Global buckling had been already pre-calculated as decisive buckling shape and was observed for both tests. The direction of buckling is the only difference which was found for the test V4. This difference may occur due to the simplification of the imperfection shape or small eccentricity in load introduction in the transverse direction. This will be investigated as the next step based on more refined measurements. Generally, a good agreement between numerical models and tests can be concluded, especially in view of the complexity of the system and acting forces.

Figure 15. Out-of-plane deformations of test V4 (left) and numerical model (right)

6. Summary

Experimental investigations were conducted on six welded girders. Here, the results of two tests are presented. The tests focused explicitly on the influence of the interaction of biaxial stresses induced from a bending moment and transverse loading. In the tests the stress ratio σ_z / σ_x was varied. In both cases a global buckling failure was observed including the longitudinal stiffeners. In the case of a higher transverse compression, a failure in the negative deformation direction (on the other side of the longitudinal stiffener) in the area of the first stiffener was observed. It may be as a result of the rotation of the lowest stiffener and should be investigated. The tests are numerically re-calculated. The numerical models were developed using the material parameters from the tensile tests and measured geometrical imperfection shapes. The results show a good agreement in view of loading and shape of failure.

It can be seen, that an increasing σ_z / σ_x ratio increases the deviation of the ultimate load. A sensitivity under higher transverse loads and the concomitant torsional stress of the lowest stiffener seem decisive. Since effects of imperfections, as well as influences on the torsional resistance of the stiffener, such as local plasticization and/or distortion of the cross-section, may play a significant role, these influences have to be considered in further investigations on the numerically validated model, in order to analyse the behavior of longitudinally stiffened plates under biaxial buckling and improve the verification.

References

[1] EN 1993-1-5, Eurocode 3: Design of steel structures - Part 1-5: Plated Structural Elements, (October 2006).

[2] Braun, B., "Stability of steel plates under combined loading", Thesis, University of Stuttgart, Mitteilungen des Instituts für Konstruktion und Entwurf No. 2010-3, 2010.

[3] Braun, B.; U. Kuhlmann, "Reduced stress deisgn of plates under biaxial compression", Steel Construction 5 (2012), No. 1 pp. 33- 40.

[4] Frickel J., Pourostad V., Kuhlmann U., Schmidt-Rasche C., "Untersuchungen zum Beulnachweis nach DIN EN 1993-1-5", BASt. FE.89.0313/2015, Research report 2017, http://bast.opus.hbz-nrw.de/volltexte/2017/1858/pdf/B140_Internet_PDF.pdf

[5] Kuhlmann, Ulrike; Mensinger, Martin; Frickel, Jörg; Pourostad, Vahid; Ndogmo, Joseph: „Beulfelder unter mehraxialer Beanspruchung", (Panels under multiaxial stress states). In: Stahlbau 86 (2017), Issue 10, pp. 917–931

[6] Kuhlmann, U., Feldmann M., Galéa Y., Johansson B., Degée H., J., Raoul J., Davaine L. et.al. "Competitive Steel and Composite Bridges by Improved Steel Plated Structures (COMBRI), Final Report, RFS-CR-03018, European Commission, Reseach Fund for Coal and Steel. 2007.

[7] Sinur, F., Behaviour of longitudinally stiffened plate girders subjected to bending-shear interaction. Dissertation. University of Ljubljana, 2011.

[8] BSK. Boverkets Handbok om Stålkonstruktioner, BSK 07, November 2007.

[9] ABAQUS 6.11. Documentation. Dassault Systèmes Simulia Corp, 2011.

IOP Conf. Series: Materials Science and Engineering **419** (2018) 012039 doi:10.1088/1757-899X/419/1/012039

Automation of evaluation of data measured on steel bridges, category detail of welded lamellas

Stanislav Vejvoda

Rebesovice, Dily 241, 664 61 Rajhrad, Czech Republic

E-mail: stanislav.vejvoda@seznam.cz

Abstract. The paper introduces an automatic data evaluation system on the server or on the connected PC to it via SW ALERT and POKOF. The system prepares measurement results for the graphical representation on the PC screen not only of the daily files but also of the selected period. The results of the fatigue assessment are stored in a permanently supplemented text file. Design of detail category of welded lamellas on the basis of fatigue tests.

1. Automation of evaluation of data measured on steel bridges

The automatic evaluation of the measured data was developed at the time of the construction of the steel-concrete bridge over the Lochkov valley. Measurement using strain gauges and thermocouples placed on the carrier welds of the lamellas was started out from the beginning of the assembly of the bridge to its gradual assignment of welded parts of the bridge, the concrete road and the first few years of the operation of the bridge. The written records of the assembly run are possible to assign the measured data to them. Five EMS DV 803 fully-equipped measuring units were used. The built-in system for the automatic evaluation of the measured data allowed to omit the evaluation process and thereby reduce the cost of processing. There was no detail category available for evaluating the damage to the welded joints by the fatigue process. Thanks to support from the state budget, his suggestion was based on fatigue tests of large test sticks made from welded joint at a scale of 1: 1.

The strain gauges, accelerometers, displacement meters, deflectors, and temperature sensors are commonly used to determine the response of bridges' steel structures to operational loads, including climatic loads and possible seismic effects. For normal measurements, sampling of measured data is usually 25 Hz or 12.5 Hz. In order to evaluate the response to rapidly changing events, measured e.g. accelerometers, it is necessary to specimen the measured data, e.g. by the frequency of approx. 400 Hz. The large amount of data is measured, if the bridge is equipped with up to 60 sensors. The sampling of measured 400 Hz data is usually performed for up to 15 minutes. The measuring system must allow the sampling of the measured data to be switched at a distance.

It is expedient to transfer the measured data, for example, by a GSM system to the server of the organization owning the bridge, or to carry out a measurement evaluation. The data files should be closed every day after midnight and ready for immediate evaluation. Measured data are assessed by the automatic system on the server or on the connected PC. Using of the SW ALERT and POKOF allow specialists not to be burdened by routine activities, but they can engage in other highly specialized activities. The system prepares measurement results for the graphical representation on the PC screen not only of the daily files but also of the selected period. The results of the fatigue assessment are stored in a permanently supplemented text file.

The measured data transported to the server are stored in one folder named eg MOSTD1. Suppose we have two daily files, labeled: DRRMMDD_001.EMS and DRRMMDD_002.EMS. The letters RR denote the year, MM denotes the month and DD denote the day. The digits for underscores indicate the serial number of the measuring device.

On the bridge, for example, sensors are located in two areas, far apart. To shorten the length of the interconnection wires between the sensors and the measuring device, each of the two measuring units is placed as close as possible to the sensors. Then the same types of sensors can be found in both measuring panels. However, we have the interest to separately evaluate: 1) separate strain gauges G together with strain gauges placed in the T shape, including the measured temperatures; 2) separately R - rosettes of strain gauges, including the measured temperatures; 3) separately displacement meters P, including the measured temperatures; 4) separately accelerometers A, including the measured temperatures.

To speed up the evaluation of the measured data, we decide that each of the four groups will be processed on separate subfolders. We will call subfolders MOSTD1GT; MOSTD1R; MOSTD1P; MOSTD1A. This division also has the advantage if the transmission of the measured data from the measuring devices to the server is interrupted for several days, and their evaluation would have to perform retrospectively. Therefore, it is necessary to have a time reserve, daily evaluation must take place in a much shorter time than 24 hours. Each case of measurement on a operated bridge may be different. Therefore, for each measurement case, the control program will be modified and appropriately labeled. In this case, this is a modification called ALERTD1.

SW ALERTD1 in the present case, when the daily data files DRRMMDD_001.EMS and DRRMMDD_002.EMS are closed, creates the four text files. In doing so, it will keep the order of the sensors according to the input it has available to solve the task. Lastly, it usually stores compensatory strain gauges, when they are used, and temperature sensors, again in the specified environment. In this example, the following files are created: DRRMMDDG.TXT, which is sent to the MOSTD1GT subfolder; DRRMMDDR.TXT, which is sent to the MOSTD1R subfolder; DRRMMDDP.TXT, which is sent to the MOSTD1P subfolder; DRRMMDDA.TXT, which is sent to the MOSTD1A subfolder.

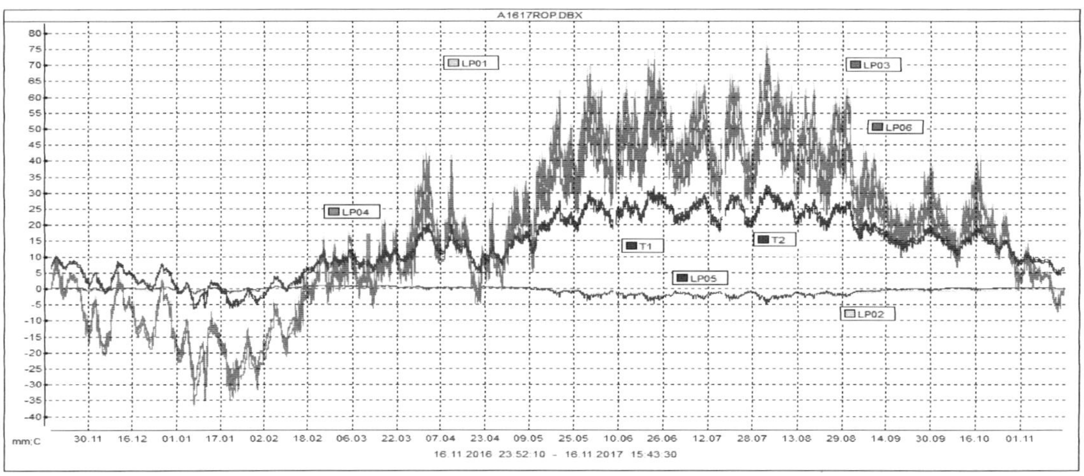

Figure 1. The displacements of the measured bridge LP01 and LP03 in the longitudinal direction and LP02 in the transverse direction relative to the support on the side of Plzen town and the displacement LP04 and LP06 in the longitudinal direction and LP05 in the transverse direction relative to the support on the side of Brno town. They were measured throughout the year from 17.11.2016 to 16.11.2017, [3].

After that, the SW ALERTD1 control runs on each subfolder the POKOF evaluation software, currently it is a version 8 to 10. The SW POKOF for its operation needs the auxiliary files that are installed on each of the sub-tabs, such as the header of output text files with the symbol DBX after the dot. In some cases, a large number of calculated variables can be assessed, and SW POKOF can be entered to base the results on several DBX output text files. To distinguish them, the first letter "D" in the output files is replaced with the letters "A", "B", "C", eg ARRMMDDR.DBX. In all cases, these are daily closed text files. The first column in all TXT and DBX text files represents the time and is taken from the base file with the EMS end, again according to the prescribed protocol. If a fatigue assessment is entered, the accumulation results of fatigue damage for a given day for the rated site are stored on the appropriate text file. For example, text files for example would be labeled VMOSTD1G.TXT and VMOSTD1R.TXT.

The SW MONITOR supplied to the EMS DV 803 measuring devices allows to combine the result sets into the file DBX for a period for which the results are available. For example, for a year or three years it is possible to plot the shifts of the bridge over the supports, Figure.1. The ALERT control program can also be complemented by other features, such as sending SMS messages or emails about unusual situations, such as unexpected interruptions in measured data transmission or unexpected stresses, shifting, and so on.

2. Experimental determination of the category detail of welded lamellas.

The welded lamellas joint of two lamellar flanges, is not listed as the "category of detail" in EN 1993-1-9 [2] or CSN 27 7008 [3]. The paper presents the results of fatigue tests of test specimens made from the 1: 1 welded lamellar flanges, which were stressed both in the area of high-cycle fatigue (HCF) and in the area of low-cycle fatigue (LCF). The results of the fatigue tests were compared with the fatigue assessment using fatigue strength curves experimentally measured on standard test specimens [5].

The 1:1 welded lamella flanges model was made of S355NL + N and S355K2 + N steels. Both are classified as quality steel, used for steel structures, even at low temperatures. Two lamella flanges of 60x1200x2000 mm were made of S355NL + N steel. The lamellae flange measuring 45x1260x1950 mm was made of S355K2 + N steel. It is the material used for the same weld joint of the lamella flanges on a steel-concrete bridge, as built over the Lochkovské Valley on the southwest highway bypass of Prague. The model 1:1 of this welded joint was produced by Bögl - Krysl, k.s., which also carried out assembly welds on the bridge across the Lochkov Valley.

The sheets used for the production of the test model were rolled in the longitudinal direction, the carrier weld of the lamella flanges being perpendicular to the rolling direction. The longitudinal axis of test specimens made from the 1:1 model coincides with the direction of rolling. After measuring residual stresses by the drilling method, the model of 3900 mm length was shortened to 800 mm (400 mm on each side of the longitudinal axis of the weld), ie to the required length of the test specimens, Figure 2.

The test specimens were equipped with strain gauges to determine the distribution of the cyclically variable loading force in both lamellar flanges and to identify the number of cycles at which the fatigue crack initiates. A fatigue crack occurred at the seam weld root (Figures 4-9) of the lamella flanges with a wall thickness of 45 mm. Following this finding, the four strain gauges (G5 to G8) were transferred to the place of the fatigue crack on other test specimens (Figure 2). The number of load cycles N_p, in which the plastic deformations are the first formed in the region of the sealing weld root, as well as the number of cycles at which the fatigue crack of the N_{in}, so they were determined as accurately as possible. According to standards, eg [2] and [3], the lifetime of the steel structure is calculated into the initiation of the fatigue macro crack of about 0.5 mm, with a probability of 5% (PP05).

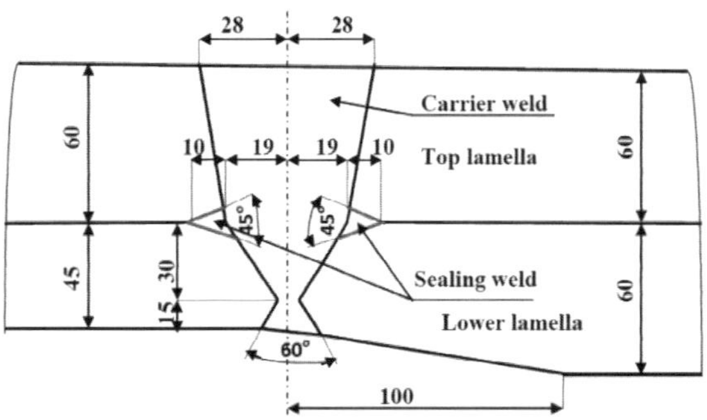

Figure 2. Dimensions of the carrier and sealing welds.

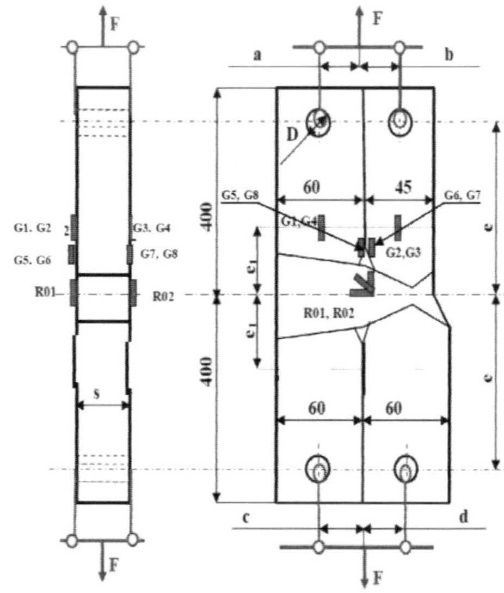

Figure 3. Place the strain gauges on the test specimen after the fatigue crack was detected. G5 to G8 were used to determine the number of cycles at the crack initiation.

Figure 4. Specimen No.17 suspended in the test frame before the fatigue test in the HCF

Figure 5. Specimen No. 6 after the fatigue test in the LCF area, terminated by broken.

Figure 6. Specimen No. 6, view of the fracture area.

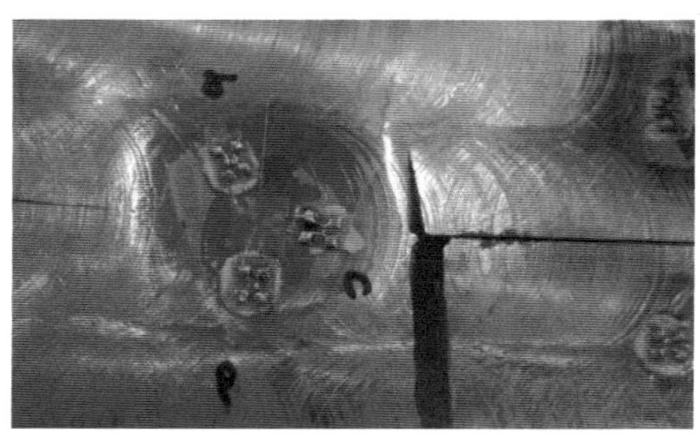

Figure 7. Specimen No. 8 after the end of the fatigue test. Crack formation was initiated at the sealing weld

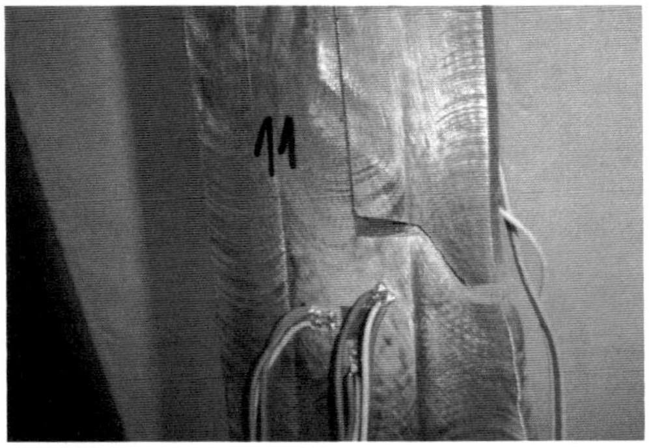

Figure 8. Crack detail in specimen No. 11. Crack formation was initiated at sealing weld. These trajectory was the only one case.

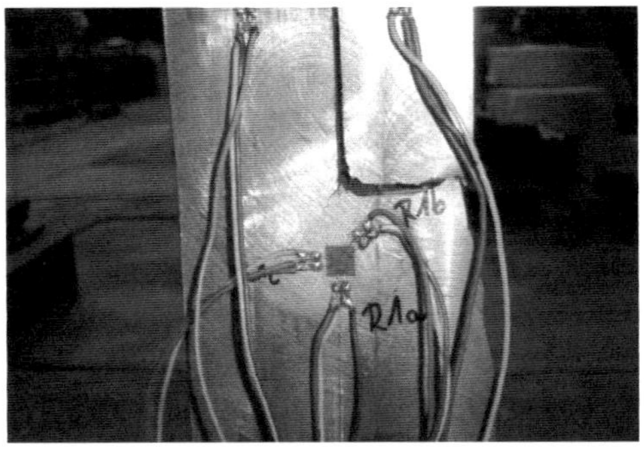

Figure 9. Detail of crack in specimen No.26 Crack formation was initiated at sealing weld. The crack in the 60 mm lamella began to emerge later.

In the standard EN 1993-1-9 [1] the dependence of $\Delta\sigma = f(N)$ is plotted in the range of 10^4 to 10^9 cycles. The specimens were tested from the beginning of this range to 2.10^7 cycles, both in the LCF and in the HCF area. The results of the cyclic loading of the test rods in the LCF and HCF areas are graphically illustrated in Figure 11. The potted points represent the results in which a fatigue macro crack of about 0.5 mm in size occurred in the area of the timed strength to the number of cycles N_D = 5000000. Results are also plotted when the crack did not occur, or cyclic loading was completed at 2000000 cycles. These range of stresses were even greater than measured on a real bridge across the Lochkov Valley [4] and [5].

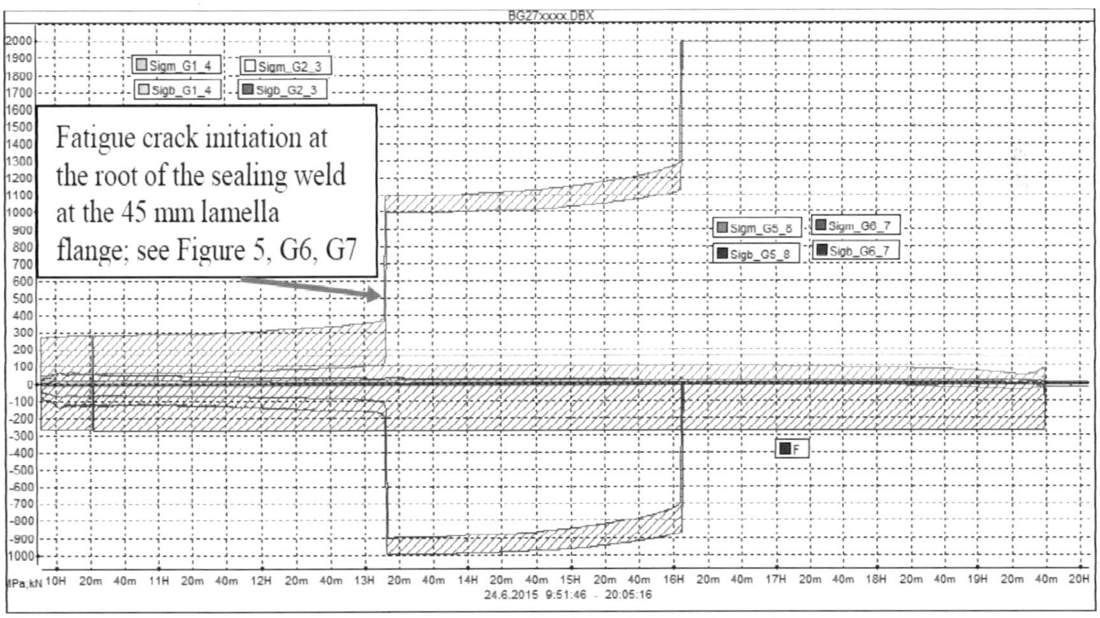

Figure 10. Specimen No. 27, the course of average and bending stresses $\sigma_{m,1\text{-}4}$ and $\sigma_{b,1\text{-}4}$ to $\sigma_{m,5\text{-}8}$ and $\sigma_{b,5\text{-}8}$ [MPa] of the measured G1 to G8 strain gauges. Load force F [1 kN = 1 unit]. The resulting crack in the 45 mm lamellar flange was passed through the G6 and G7 strain gauges, see Figure 3. In the second 60 mm lamellar flange, the fatigue crack did not occur, the G5 and G8 strain gauges therefore did not record their occurrence.

The 8 specimens were tested in the LCF area and the 13 specimens in the HCF area, the total of 21 specimens. However, in the timed strength area to $\Delta\sigma_D$, the fatigue crack occurred in only 12 specimens of 21. Measured fatigue strength curves evaluated for both PP 50% and PP 05% are plotted in Fig. 11.

Dependence the range of nominal stresses $\Delta\sigma_{nom}$ on the number of cycles N_o, until the beginning of fatigue crack initiation of the size a = 0.5 mm, can be written in the form:

1) The fatigue strength curve with a probability of failure PP50% y = - 0.3007.x + 3.6848 can be expressed in the range of 10^4 to 5.10^6 cycles in the form:

$$N_O = N_C\,(\Delta\sigma_C/\Delta\sigma_n)^{3,007}, \qquad (1)$$

in which the number of cycles $N_C = 2.10^6$ corresponds to the range of stresses $\Delta\sigma_C$ = 61.67 MPa.

2) The fatigue strength curve y = -0.3159.x + 3.6899 in the range of 10^4 to 5.10^6 cycles determined by the least squares method was interpolated by experimentally measured points reduced to PP05%. It can be expressed in the form:

$$N_O = N_C\,(\Delta\sigma_C/\Delta\sigma_n)^{3,166}, \qquad (2)$$

in which the number of cycles $N_C = 2.10^6$ corresponds to the range of stresses $\Delta\sigma_C$ = 50.05 MPa. Exponent 3.166 is close to exponent 3, which uses standards [1] and [2] in the range of cycles 10^4 to 5.10^6.

3) Fatigue strength curve PP05% y = - 0.2096.x + 2.9776 in $N_D = 5.10^6$ cycles up to 1.10^8 cycles, represent the linear curve (2) for $N_D = 5.10^6$ cycles with the test terminated near 10^8 cycles. This can be expressed in the form:

$$N_O = N_D\,(\Delta\sigma_D/\Delta\sigma_n)^{4,771}, \qquad (3)$$

in which the number of cycles N_D = 5.106 corresponds to the range of stresses $\Delta\sigma_D$ = 37.47 MPa. Exponent 4,771 is close to exponent 5, which uses standards [1] and [2] in the range of cycles 5.10^6 to 1.10^8.

4) Permanent strength limit of range stresses $\Delta\sigma_L$ = 20.26 MPa for the number of cycles 1.10^8 and above can be calculated from y = 1.3066.

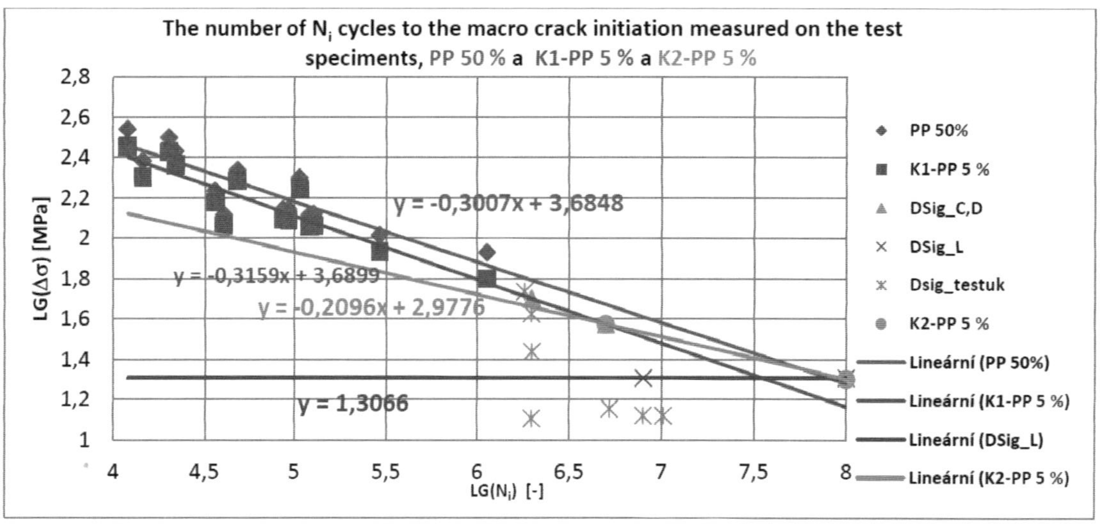

Figure 11. The red line applies in the region of $N_i = 10^4$ to 5.10^6 cycles; the orange line is in the $N_i = 5.10^6$ to 1.10^8 region and the purple line in the area of the permanent strength above $N_i = 1.10^8$ cycles.

Fatigue tests have shown that the fatigue crack originated first at the sealing weld root of the lamellar flange with a wall thickness of 45 mm and then started in a 60 mm thick lamellar flange at the same root of the sealing weld. On the other side of the carrier weld, where both lamellar flanges had a wall thickness of 60 mm, no fatigue crack was found on any specimen's test even when cycling loading continued to rapture of the specimens.

3. Conclusions

Using of the SW ALERT and POKOF allow specialists not to be burdened by routine activities, but they can engage in other highly specialized activities.

Conclusions from the fatigue tests can be generalized for design of welds of lamellar flanges with a wall thickness of up to 60 mm. According to EN 1993-1-9, the detail category of $\Delta\sigma_C = 50$ MPa can be taken for it. The fatigue strength curves for the wall thicknesses of 25 mm are plotted in EN 1993-1-9. According to the experiment, the detail category $\Delta\sigma_C$ = 50,05/(25/60) 0,2 = 59,63 MPa would correspond to a thickness of 25 mm, falling within range of detail category 56 MPa to 63 MPa.

References

[1] EN 1993-1-9: Design of steel structures – Part 1-9, Fatigue
[2] CSN 27 7008 Navrhování ocelových konstrukcí rypadel, nakladačů a zakladačů. Design of steel structures of excavators, loaders and stackers. In Czech.
[3] Vejvoda, S.: Měření na mostě přes Lochkovské údolí. Measurement on the bridge across the Lochkov Valley. Report VITKOVICE UAM a.s., archival No. V3303/17. In Czech.
[4] Vejvoda, S. – Gottvald, J. – Ryšavý, M.: Report VITKOVICE UAM a.s., archival No. V2901/11. In Czech.
[5] Vejvoda, S. – Ryšavý, P. – Titz, K.: Report VITKOVICE UAM a.s., archival No. V3086/15. In Czech.

IOP Conf. Series: Materials Science and Engineering **419** (2018) 012040 doi:10.1088/1757-899X/419/1/012040

The impact of backfill quality on soil-steel composite bridge response under seismic excitation

Tomasz Maleska[1] and Damian Beben[1]

[1]Faculty of Civil Engineering and Architecture, Opole University of Technology, Katowicka 48, 45-061 Opole, Poland

E-mail: d.beben@po.opole.pl

Abstract. The soil-steel composite bridges, typically ranging from 3 to 25 m, can be used as an effective alternative for short-span bridges and culverts. They can meet the design and safety requirements as for traditional bridges, more rapidly and at lower costs. For these reasons, soil-steel bridges are increasingly being used in road and railway projects in many parts of the world. Due to the fact, the typical static and dynamic analyses of such bridges were already conducted many times, this paper focuses on the results of the numerical analysis of soil-steel composite bridge under seismic excitation. For these purposes a bridge with span of 17.67 m and height of 6.05 m was selected. Numerical analysis was conducted using the DIANA program based on finite element method. The density and Young modulus of the backfill were varied in the numerical models. A linear model with El Centro records and Time History analysis were used. The conclusions drawn from the study can be helpful for designers and researchers dealing with the soil-steel bridges, especially for the projects situated in the seismic areas.

1. Introduction

The soil-steel bridges can be met in various parts of the world. Their span usually does not exceed 25 meters, although it is possible to build objects of a span going to up to even 40 meters. They are often a quite good alternative to traditional steel, concrete or timber bridges.

The soil-steel bridges have many advantages, amongst which the following may be included: (i) minimal use of steel, (ii) low construction and service costs, (iii) the possibility of prefabrication (partial or full prefabrication), (iv) the possibility of location the bridge in the weak ground, (v) fitting in the idea of the sustainable development. Despite numerous advantages, these structures also have some drawbacks such as: (i) a limited span, (ii) susceptibility to corrosion [1]. To some extent, it still remains unknown, how these bridges react when being impacted by seismic loads. So far, this problem has not been widely addressed in scientific works.

The main idea of the soil-steel bridges is designing a composite soil and steel structures in such a way that the loads applied to the structure do not exceed the load carrying capacity of composite bridge. Due to the phenomenon of arching in the soil and the flexibility of the steel shell, the carry of the large loads is possible. Usually, the soil-steel bridges carry the loads due to axial forces, with the minimum bending moments except of box culverts. One of the most important elements is a correct execution of the backfill surrounding the steel shell.

So far, the soil-steel bridges analyses have been carried out in the range of static loads [2,3] or dynamic loads [4,5]. Elshimi et al. [6] presented an experimental testing and numerical analysis of the culvert made from corrugated steel plates. The obtained results were also compared to the Canadian CHBDC code [7], and the maximum difference between experimental test and code results was 73%. On the other hands, Bayoglu Flener [8] analysed the influence of the backfill height on the values of internal forces. Two soil-steel structures with different spans and heights were used for the analysis. The significant differences were observed between the internal forces from the tests and the results obtained on the basis of the CHBDC code [7].

Structures subjected to seismic or paraseismic excitations are usually analysed using the Time History method, which is based on the step-by-step integration method [9,10]. For bridge structures that interact with the soil, the seismic analyses are most often conducted for reinforced concrete (RC) arch bridges or RC box culverts. Sawamura et al. [11] used the Time History method to analyse the impact of seismic excitations on the RC culverts. Maghgoub and El Naggar [12] analysed the impact of various seismic excitations on steel shell structures. The obtained values were compared to the CHBDC code [7]. It was found that compatibility of the numerical analysis with the code occurred only in the case of a short excitation, characterized by a high value of ground acceleration. In this case, the numerical analysis was performed using the FLAC program.

The main purpose of this paper is presenting the impact of the natural earthquake on the soil-steel bridge with the use of different soil parameters. Especially Young modulus and backfill density are varied. The selected bridge was subjected to numerical analysis using the DIANA FEA computer program and Time History method. The applied load is extreme El Centro benchmark excitation.

2. A short description of the soil-steel bridge

The analysed soil-steel bridge is located in Poland (Figure 1). This object is a passage for animals. The analysed object has the following parameters: the clear height of 6.05 m, the span of 17.67 m, the width of the shell in crown of 40.39 m, the width at the foundation of 53.83 m, the thickness of the steel shell of 0.007 m, the corrugation depth of 0.14 m, and the corrugation pitch of 0.38 m. The corrugated steel plate sheets were connected with one another using the bolts with a diameter of 20 mm. The bolts were tightened with a torque value of 350-400 Nm. The steel shell was fixed on RC foundations (made from concrete of C25/30) with the length of 53.83 m and the width of 4.0 m. The backfill surrounding the steel shell was made from aggregates of 10-32 mm and place with layers 0.2-0.3 m thick. In order to ensure proper interaction of the shell structure with the soil, the backfill was compacted to reach I_D = 95% according to the Normal Proctor scale. The maximum height of the backfill in the shell crown equals to 1.8 m.

Figure 1. The soil-steel bridge during construction.

3. Numerical analysis of soil-steel bridge

The DIANA FEA computer program [13] based on the finite element method (FEM) was used in the numerical analysis. The assumptions proposed by Maleska et al. [10] were used to build the numerical models:

- the corrugated steel plates were modelled as shell curved elements (Q24IF) with Young modulus of 205 GPa, Poisson ratio of 0.3, elastic-plastic model with density of 7850 kg/m^2, the yield strength of steel of 235 MPa, area moment of inertia of the unit width 21897.45 mm^4/mm, cross section area of 8.867mm^2/mm, and plate thickness of 0.007 m. The corrugated steel plate was modelled as a flat one with orthotropic characteristics. The transformation of the corrugation shell to the flat one was carried out according to the procedure contained in the works [10,14]. Based on the research carried out in [14], it was found out that the use of an orthotropic shell does not significantly affect the accuracy of the calculations.
- the backfill was modelled as solid elements (HX24L) with the Duncan-Chang nonlinear elastic hyperbolic model. The backfill parameters were: Poisson ratio of 0.2, dilation angle of 5°, angle of internal friction of 39°, cohesion of 3000 N, failure ratio $R_f = 0.7$, unloading-reloading stiffness $E_{ur} = 1000$ N/m^2, reference pressure $P_{ref} = 101350$ N/m^2, exponent for unloading reloading curve $m = 0.25$, exponent for backbone curve $n = 1.1$, minimum compressive stress 350 N/m^2, and minimum tangential stiffness of backbone curve $E_{t,min} = 1200$ N/m^3. In order to determine the impact of the backfill quality on the behaviour of the bridge under seismic excitation, five soil models were used (Table 1). Young modulus and backfill density were varied, while the parameters of model 1 were the real data for the analysed bridge,

Table 1. The value of Young modulus and backfill density for particular numerical models.

Value	Model 1 [10]	Model 2 [14]	Model 3 [15]	Model 4 [15]	Model 5
Young modulus (MPa)	100	200	114	240	150
Density (kg/m^3)	2050	2000	1600	2000	2000

- the connection between the backfill and the steel shell was modelled as automatic interface by applied function "Coulomb friction" with the angle of internal friction of 39°, dilation angle of 5°, rigidity of 100000 kN/m^3, and cohesion of 3000 N,
- the finite elements were modelled as quadratic elements with dimensions of 0.50×0.50 m for the backfill and the corrugated steel shell (Figure 2),
- the boundary conditions were modelled as non-transferable support of the shell walls and of all boundary walls of the backfill for x, y, z directions.

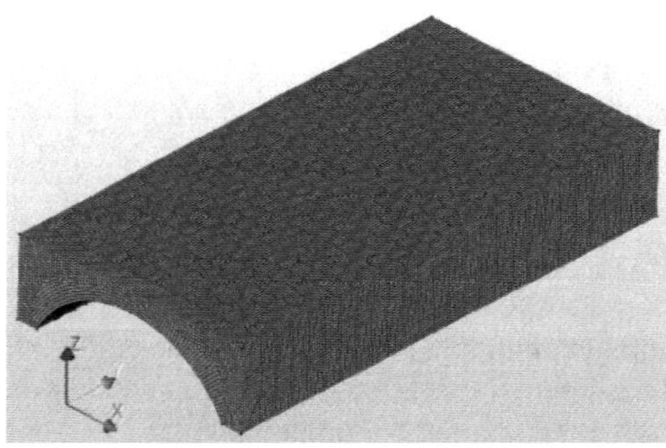

Figure 2. Numerical model of soil-steel bridge in the DIANA FEA program.

In the numerical analysis, the Time History method was applied, which allows determining the impact of seismic excitation on a soil-steel bridge. In the analysed cases, the El Centro seismic record form 1940 was used [16]. This earthquake record is considered as destructive and reaches 6.9 on the Richter scale. The ground acceleration for recording is equal to 3.44 m/s^2. The seismic tremor was applied in two directions, i.e. "X" (perpendicular to the bridge) and "Y" (parallel to the bridge). The obtained maximum values of displacements, internal forces and stresses are presented in Table 2.

Table 2. Maximum values for steel shell and backfill caused by El Centro excitation.

Value	Model 1	Model 2	Model 3	Model 4	Model 5
	Direction of seismic loads application:				
	X and Y	X and Y	X and Y	X and Y	X and Y
Shell displacements (m)	0.25	0.18	0.20	0.16	0.21
Backfill displacements (m)	0.25	0.18	0.20	0.16	0.21
Stresses in steel shell (MPa)	–108	–77.2	–84.1	–69.9	–89.2
Stresses in backfill (MPa)	–1.778	–2.241	–1.476	–2.551	–1.941
Axial thrusts in shell (kN/m)	–10900	–8113	–8623	–7405	–9229
Bending moments in shell (kNm/m)	66.16	35.05	45.44	30.00	45.51

The largest displacements of the steel shell (Figure 3) and backfill occurred for "Model 1", where the maximum value was 0.25 m. On the other hand, the smallest displacements were observed in "Model 4". In this case, they amounted to 0.16 m (Figure 4), which is 36% difference compared to the "Model 1". It is worth emphasizing that the displacements of the backfill and the steel shell were the same, which indicates the correct image of the interaction between the steel shell and the backfill. For the analysed soil models, the same displacements were obtained in both the "X" and "Y" directions. This indicates that the excitation direction is irrelevant as regards the shell and backfill displacements. The maximum displacements observed at the shell crown, in the middle part of the shell and they were directed downwards (deflection of the shell).

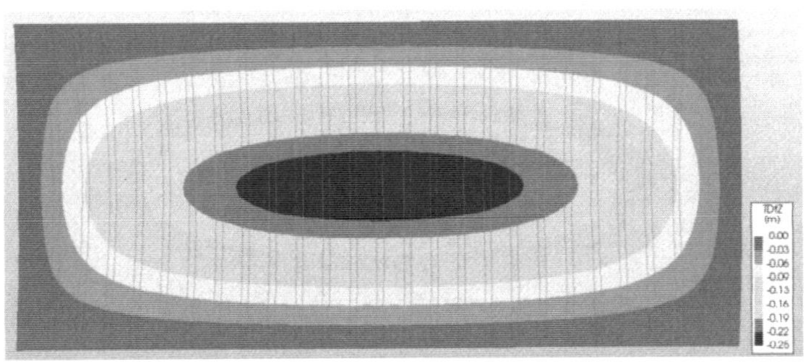

Figure 3. Top view for the maximum displacements of the shell obtained from "Model 1".

Figure 4. Distribution of backfill displacements obtained from "Model 4".

The highest stresses in the steel shell occurred in "Model 1" and their values were about –108 MPa (Figure 5). However, the smallest stresses were observed for "Model 4", where the values equalled nearly to –69.9 MPa. The difference from "Model 1" and "Model 4" was 35.3%. The maximum stresses in the particular models appeared in the vicinity of the shell supports, and they had a compressive nature. It should be noted that the location of maximum stresses is different in comparison to stresses obtained from the static or dynamic loads (usually it was at the shell crown or quarter points – depending on the shell shape).

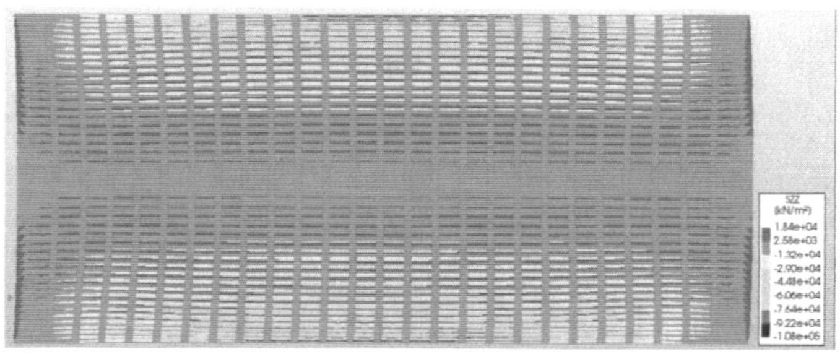

Figure 5. A view of the maximum stresses in the shell obtained from "Model 1".

Maximum stresses in the backfill were observed for "Model 4" and amounted to –2.551 MPa (Figure 6). While the smallest stresses in the backfill occurred in "Model 3" and they were smaller by 42.1% than those of "Model 4", and their values amounted to –1.476 MPa (Figure 7). The maximum stresses in the backfill occurred near the shell supports and arrived the quarter points. It was noted that the directions of seismic excitation do not affect the obtained values of stresses in the backfill.

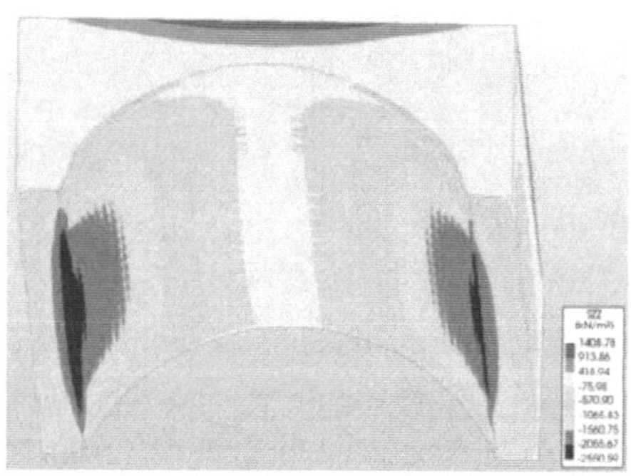

Figure 6. Distribution of stresses in the backfill obtained from "Model 4".

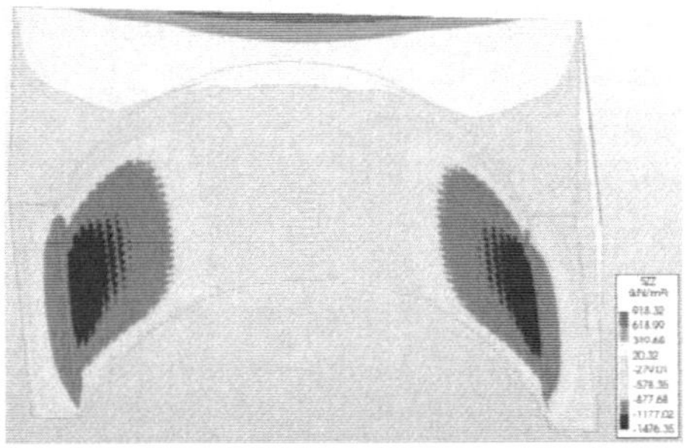

Figure 7. Distribution of stresses in the backfill obtained from "Model 3".

The maximum axial forces also occurred in "Model 1" and amounted to –10900 kN/m (Figure 8). The largest forces were the same in the "X" and "Y" direction of the seismic excitation. However, the smallest axial forces were observed for "Model 4". The axial forces were lower by about 32% compared to "Model 1" and amounted to –7405 kN/m (Figure 9). It is worth noting that the maximum axial forces were strongly compressive and occurred in the shell in the vicinity of the foundations. It was also observed that the obtained axial forces in the shell from the seismic excitation are much greater than those caused by the dynamic loads, e.g. train [5].

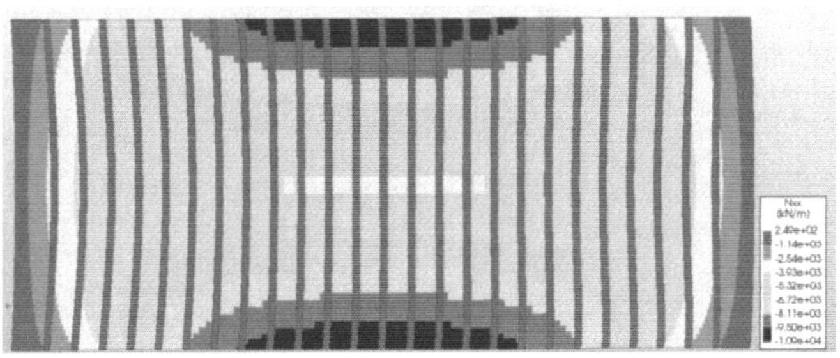

Figure 8. Distribution of maximum axial forces in the shell from "Model 1".

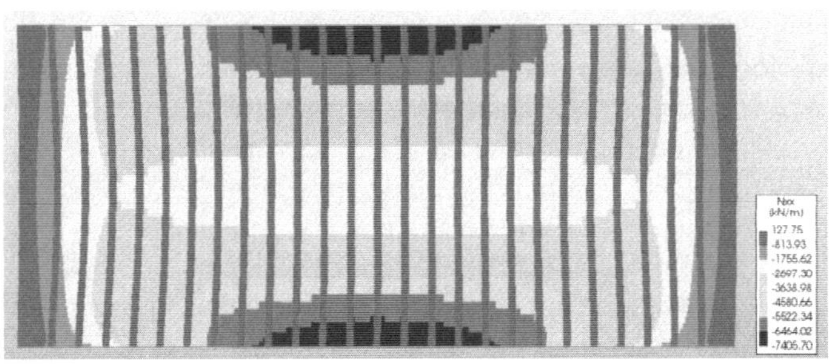

Figure 9. Distribution of maximum axial forces in the shell from "Model 4".

The highest values of bending moments were obtained for "Model 1". The maximum value was 66.16 kNm/m (Figure 10). Nevertheless, the smallest bending moment was 30.00 kNm/m and it was observed in "Model 4" (Figure 11). This is a value lower by 54.7% in comparison to the maximum bending moment obtained in "Model 1". The largest values occurred near the supports of the steel shell. They were observed in the same places as for axial forces and stresses, however, they were positive. It should be added that in the case of static or dynamic load, the maximum bending moments were usually located in the shell crown. It is also worth noting that the direction of load application is not important for the obtained values of bending moments. It should be also emphasized that the amount of bending moments obtained from the seismic excitation is much greater than in the case of static or dynamic load [5,17].

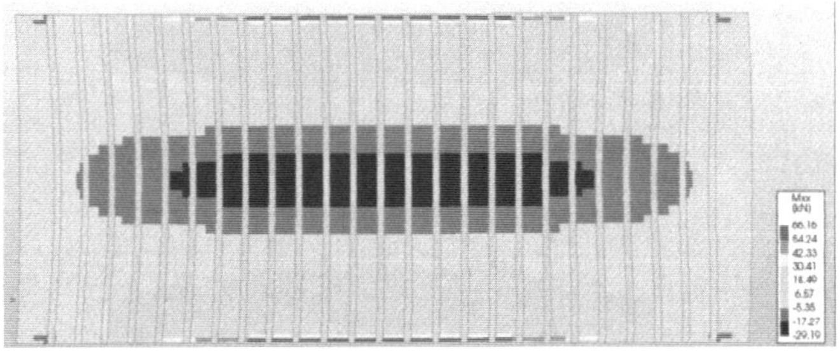

Figure 10. The distribution of the maximum bending moments in the shell in "Model 1".

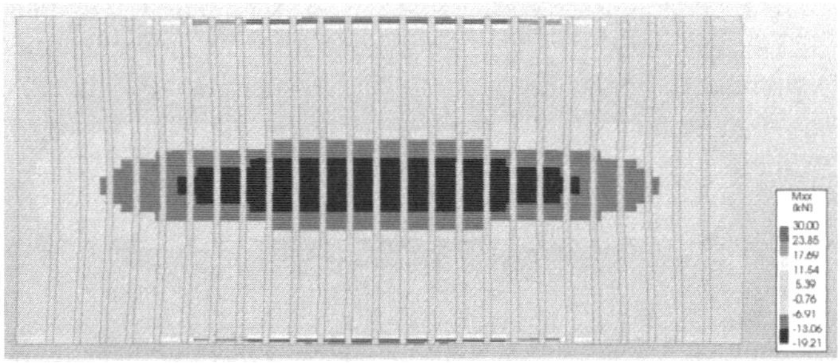

Figure 11. The distribution of the maximum bending moments in the shell in "Model 4".

4. Conclusion

As a result of the Time History analysis of the soil-steel bridge under the El Centro seismic excitation, the following conclusions can be drawn:

- the obtained results indicate that Young modulus is the most important factor. This is best seen on the basis of the results analysis from the "Model 1" and "Model 2". The difference in backfill density is very small (2.4%), while Young modulus values differ by as much as 50%. The results obtained from the "Model 2" were smaller than those of "Model 1" for: (i) displacements by 28%, (ii) stresses by 28.5%, (iii) axial forces by 25.6%, (iv) bending moments by 47%,
- the backfill density is also important for the obtained internal forces, displacements and stresses. It can be seen by comparing the results obtained from "Model 1" and "Model 3", where Young modules were similar to each other. In these models, the difference in Young modulus was about 14% between particular soil models. However, the difference in the backfill density was equal to 28%. It is worth noting that the differences in results for the steel shell between "Model 1" and "Model 3" were: (i) 20% for displacements, (ii) 22.1% for stresses, (iii) 20.1% for axial forces, (iv) 31.3% for bending moments,
- the direction of application of the seismic tremor ("X" - perpendicular and "Y" - parallel) does not affect the obtained results. In each of the considered soil models, the same results of numerical calculations were obtained (regardless of the tremor directions),
- the bending moments and axial forces in the steel shell caused by seismic excitation are much greater than those obtained during typical tests under static or dynamic loads, e.g. during the train or vehicle rides. It was also noticed that the location of bending moments and axial forces is different (near the shell support) than in the case acting on the dynamic load (the shell crown or quarter points),
- it was also observed that the place of occurrence of maximum axial forces, bending moments and stresses in the shell is the same in all the analysed numerical models. Maximum values of internal

forces and stresses occurred near the foundations of the soil-steel bridge and reached up to 1/3 of the height of the shell. In the case of displacements, the maximum values were observed near the crown of steel shell,
- the yield strength for steel has not been exceeded in any of the considered cases. The maximum stresses in the steel shell were equal to –108 MPa, while the yield strength for the analysed bridge was 235 MPa.

In order to gain the more detailed knowledge of the behaviour of the soil-steel bridges subjected to natural earthquakes, further research related to this subject is necessary.

References

[1] Janusz L and Madaj A 2009 *Engineering Structures from Corrugated Plate: Design and Construction* (Warsaw: Transport and Communication Publishers)
[2] Brachman R W I, Elshimi T M, Mak A C and Moore I D 2012 Testing and analysis of a deep-corrugated large-span box culvert prior to burial *J. Bridge Eng.* **17** pp 81-88
[3] Beben D and Stryczek A 2016 Numerical analysis of corrugated steel plate bridge with reinforced concrete relieving slab *J. Civil Eng. Man.* **22** pp 585-596
[4] Beben D 2011 Application of the interferometric radar for dynamic tests of corrugated steel plate (CSP) culvert *NDT&E International* **44** pp 405-412
[5] Beben D 2014 Corrugated steel plate (CSP) culvert response to service train loads *J. Perf. Constr. Faci.* **28** pp 376-390
[6] Elshimi T M, Mak A C, Brachman R W I and Moore I D 2011 Behaviour of a deep-corrugated large-span box culvert during backfilling *Proc. 14th Pan-American Conference on Soil Mechanics and Geotechnical Engineering* (Toronto: Canadian Geotechnical Society)
[7] CHBDC, *Canadian Highway Bridge Designe Code. CAN/CSA-S6-06* 2006 Canadian Standards Association International
[8] Bayoglu Flener E 2010 Soil-steel interaction of long-span box culverts-performance during backfilling *J. Geotech. Geoenviron. Eng.* **136** pp 823-832
[9] Bonkowski P A, Zembaty Z and Minch M Y 2018 Time history response analysis of a slender tower under translational-rocking seismic excitations *Eng. Struct.* **155** pp 387-393
[10] Maleska T, Bońkowski P A, Beben D and Zembaty Z 2017 Transverse and longitudinal seismic effects on soil-steel bridges *Eighth European Workshop on the Seismic Behaviour of Irregular and Complex Structures* Bucharest Romania
[11] Sawamura Y, Kishida K and Kimura M 2015 Experimental study on seismic resistance of a two-hinge precast arch culvert using strong earthquake response simulator *Proc. 15th Asian Regional Conf. on Soil Mechanics and Geotechnical Engineering* (Kyushu: Japanese Geotechnical Society)
[12] Maghgoub A and El Naggar H 2017 Assessment of seismic provisions of the CHBDC for CSP culverts *GeoOttawa 2017* Ottawa, Canada
[13] DIANA FEA, Available online from URL: www.dianadea.com/2018
[14] Pittino G and Gosler J 2006 Structural plate steel underpasses during backfilling – how to minimize the bending moment *Proc. 4th Int. FLAC Symposium on Numerical Modeling in Geomechanics* ed R D Hart and P M Varona (Madrid: Itasca Consulting Group)
[15] Turan A and El Naggar M H 2011 Seismic behaviour of long-span arch culverts *Proc. 14th Pan-American Conf. on Soil Mechanics and Geotechnical Engineering* (Toronto: Canadian Geotechnical Society)
[16] Yongle L, Zhu S, Cai C S, Yang C and Qiang S 2016 Dynamic response of railway vehicles running on long-span cable-stayed bridge under uniform seismic excitations *J. Str. Stab. Dyn.* **16** pp 1-25
[17] Bayoglu Flener E and Karoumi R 2009 Dynamic testing of a soil-steel composite railway bridge *Eng. Struct.* **31** 12 pp 2803-2811

IOP Conf. Series: Materials Science and Engineering **419** (2018) 012041 doi:10.1088/1757-899X/419/1/012041

Models of thermal actions for steel and composite bridges based on monitoring

J Markova[1]

[1] Senior researcher, KI CTU in Prague, Czech Republic

E-mail: jana.markova@cvut.cz

Abstract. The results of analysis of temperatures based on experimental bridge measurements, research reports and background documents to Eurocodes indicate that the partial factors of temperature components may be reduced to 1,2 for steel and composite bridges and to 1,3 for concrete bridges. It appears that the application of the unique value of the partial factor for most variable loads including climatic actions should be reassessed during the development of the second generation of Eurocodes.

1. Introduction

Thermal actions lead to the longitudinal movements of structures and in case that they are restrained, tensile or compressive stresses need to be considered in structural design. Guidance for determination of basic temperature components and their effects on construction works is given in the Eurocode EN 1991-1-5 [1] which is presently further developed in the framework of the second generation of Eurocodes. The main tasks include clarification of some provisions, simplification, more user-friendly applications and harmonization with other Eurocodes. Moreover, comparative analysis of inconsistencies in extremes of shade air temperature on the borders of neighboring countries could help to achieve much consistent characteristic values of shade air temperatures. Representative values of thermal actions on bridges are specified in EN 1991-1-5 [1] by the uniform temperature component and two temperature difference components in the vertical and horizontal directions.

For the determination of uniform temperature component, the national maps with isotherms of minimum and maximum shade air temperatures based on fifty years return period should be developed in the nationally implemented Eurocodes of each CEN Member State (MS). The characteristic values of the uniform temperatures can be based on the relationship between the extreme shade air temperatures and uniform bridge temperatures which is valid for the 10°C temperature range. This relationship need to be nationally verified taking into account national conditions and the specific ranges of daily temperatures given in the background document to EN 1991-1-5, Report 1999 [2].

For determination of the vertical temperature difference component, several models are given in EN 1991-1-5 [1]. Two alternative procedures based on the vertical linear difference temperature components or non-linear components are allowed for the national choice. The comparative analyses made in the Czech Republic in selected bridges made of different materials resulted in the national selection of the non-linear approach for the vertical temperature difference component as one of the Nationally determined parameter (NDP). For bridges with a shorter span, the linear difference approach could be also used.

It appears that the reduction of the linear temperature difference component by means of the reduction factor k_{sur} is inconsistent with respect to the reduction of the non-linear difference temperature component given in EN 1991-1-5 [1] and should be further harmonized, see (Reports 1973, 1996).

It is expected that recent analyses of long-term experimental measurements of temperatures in some bridges in the Czech Republic and in other European countries will facilitate to refine the thermal models for national conditions.

The background document [2] to Eurocode [1] provides additional information concerning experimental measurements of temperatures which formed the basis for the development of the models of thermal actions in Eurocodes. The characteristic values of temperature components are based on the fifty years return period as other climatic actions (snow, wind velocity, icing). Presently, the Eurocodes recommend a unique value of partial factor γ_Q = 1,5 for most variable actions Q with respect to the ultimate limit states. However, the reduced factor γ_T = 1,2 was applied for thermal actions T in some original national standards including the Czech Republic. It appears that the partial factors of some variable actions including thermal actions should be differentiated in Eurocodes taking into account their characteristics.

The information given in the Report [2] and the recent results of experimental monitoring of selected bridges in the Czech Republic located in Prague motorway circuit, on D1 highway between Prague and Brno, and on D8 highway across the Ohre river, are considered for the analyses of temperature components and for the assessment of partial factors for thermal actions.

2. Models of temperatures

Procedures for specification of thermal models based on experimental measurements of temperatures are described in several reports including Reports [3,4] which were worked out in the KI CTU.

The extreme value distribution should be applied for modelling of extreme values of the uniform and temperature difference components. The background report [2] provides statistical characteristics of several year temperature measurements on steel, composite and concrete bridges, and also estimated characteristics based on 10-years simulations of temperatures in different heights of the concrete bridge cross-sections using simulations and software Therm. The climatic data from the Middle Germany were considered. Moreover, recent measurements provided in several Bohemian bridges were also statistically analyzed.

For the assessment of models for thermal actions and determination of their partial factors, the statistical characteristics of temperature components need to be specified for one year and 50 years reference periods. It is shown that the Weibull distribution or three parameter Lognormal distribution may be applied in all considered cases for the assessment of the characteristic and design values of temperatures.

2.1. Shade air temperature

For specification of the uniform bridge temperature component, the characteristic values of shade air temperatures should be determined by each CEN MS in the National Annex of EN 1991-1-5 [1]. Some countries developed national maps with isotherms and they apply recommended (or calibrated) relationship between the shade air temperature and uniform (effective) temperatures considering three types of bridge superstructure (concrete, composite and steel). Some countries apply temperature ranges which are considered to be characteristic values. The Czech Republic has developed maps of minimum and maximum shade air temperatures which are based on 37 stations and 60 years of measurements, assuming the Gumbel distribution. During the new development of prEN 1991-1-5 in the framework of the second generation of Eurocodes, the consistency of the characteristic temperatures within neighboring countries should be checked. Various inconsistencies in the boarders of countries still exist, e.g. for the Czech Republic and neighboring countries, where the difference up to 12 °C in winter, and up to 6 °C in summer could be found, see Tables 1 to 4. Similar inconsistencies exists on the borders of other European countries.

Comparisons of maximum and minimum shade air temperatures in selected MS are provided in Table 1. Some countries apply directly a range of characteristic values, others have developed maps of minimum and maximum isotherms.

Table 1. Comparison of temperature differences in selected CEN MS

Country	A[1]	BE	CZ[2]	DNK	GE	NDL	SK[2]
T_{min}	-16 - *kh*	-21	-28 - (-36)	-31	-24	-25	-28 - (-30)
T_{max}	39 - *kh*	38	32 - 40	36	37	30	40

[1)] Austria recommends formulae where $k = 0{,}006$ and h = height above the sea level
[2)] Czech Republic and Slovakia have available maps with isotherms, and apply relationship in Table 5

The characteristic values of shade air temperature in the boarders of the Czech Republic with Austria and Slovakia are shown in Tables 2-3. Polish NDPs were not available. Comparison of shade air temperatures and also characteristic values of uniform temperature component for three types of bridges (1-3) is given in Table 4. The differences of shade air temperatures and uniform temperature component are more significant in case of negative temperatures.

Table 2. Characteristic values of shade air temperatures at Austrian - Czech boarder

Alt. in m	CZ - $T_{max,}$	A - T_{max}	ΔT_{max}	CZ - $T_{min,}$	A - T_{min}	ΔT_{min}
600–900	38 - 36	35,4 - 33,6	2,6 - 2,4	-32-(-34)	-19,6 - (-21,4)	-12,4 - (-12,6)
470-500	40 - 38	36,2 - 36	3,8 - 2	-30-(-32)	-18,8 - (-19)	-11,2 - (-13)
400-500	38 - 36	36,6 - 36	1,4 - 0	-28-(-30)	-18,4 - (-19)	-9,6 - (-11)
200-250	40 - 38	37,8 - 37,5	2,2 - 0,5	-28-(-30)	-17,2 - (-17,5)	-10,8 - (-12,5)

Table 3. Characteristic values of shade air temperatures at Slovak - Czech boarder

Region	CZ - $T_{max,}$	SK - T_{max}	ΔT_{max}	CZ - $T_{min,}$	A - T_{min}	ΔT_{min}
north	34 - 32	40	8 - 6	-34-(-36)	-30	-2-(-4)
central	36 - 34	40	6-4	-32-(-34)	-28	-4-(-6)
south	40 - 38	40	2-0	-30-(-32)	-28	-2-(-4)

Table 4. Comparison of temperature differences at German - Czech boarder

Type	CZ - $T_{max,}$	GE - T_{max}	ΔT_{max}	CZ - $T_{min,}$	GE - T_{min}	ΔT_{min}
Shade	37	37	0	-36	-27	-9
1	52	53	1	-36	-27	-9
2	41,5	41	0,5	-28,5	-20	-8,5
3	39	38,5	0,5	-25	-16	-9

2.2. Uniform temperatures

The Eurocode EN 1991-1-5 [1] recommends assessment of the uniform temperature component from nationally determined isotherms of minimum and maximum shade air temperatures. The relationship between the uniform temperatures and shade air temperatures is given in Table 5 which is applied by several MS. The values in Table 5 are based on the daily temperature ranges of 10 °C which is assumed to be appropriate for most European countries and which could be modified for other daily ranges following recommendations of Report [2].

The Klokner Institute in co-operation with the Czech Hydrometeorological Institute made analyses of selected national parameters. It was decided that the daily range of 10 °C can be applied for most regions of the Czech Republic based on statistically evaluated data from 151 national meteorological stations in the period from 1961 to 2000 (daily mean temperature range is 11,2 °C). The uniform bridge temperature components T_{max} and T_{min} may be specified on the basis of shade air temperatures based on 50 years return period taking into account formulae in Table 5 where for the steel truss and plate girders the maximum values given for the bridge of type 1 might be reduced by 3°C.

IOP Conf. Series: Materials Science and Engineering **419** (2018) 012041 doi:10.1088/1757-899X/419/1/012041

Table 5. Relationship between extreme shade air temperature and uniform bridge temperature

Bridge deck type	$T_{N,\max}$	$T_{N,\min}$
1	$T_{\max} + 16$	$T_{\min} - 3$
2	$T_{\max} + 4$	$T_{\max} + 4$
3	$T_{\max} + 2$	$T_{\max} + 8$

For modeling of the uniform temperature component, the extreme value probabilistic distribution might be applied. The uniform temperature component is commonly based on the three-day maxima of shade air temperature for concrete bridges and on one-day maxima for steel and composite bridges.

It should be noted that the Report [2] provides an estimated upper bound of extreme temperatures for considered Weibull distribution. For example the upper bound of temperatures for the expansion range $\Delta T_{N,\exp}$ for concrete bridges is assumed to be about 38°C and the uniform temperature $T_{N,\exp}$ for steel bridges is estimated about 43°C. The upper bound is assumed to be valid for Germany and some other neighboring countries.

However, the upper bound of temperatures depends on the national climatic conditions of each country. In some European countries less or more days of temperature measurements than three months as described in [2] might be necessary to be considered for obtaining extreme temperatures, see e.g. [3,4].

The annual statistical characteristics (mean μ_T, standard deviation σ_T and skewness α_T) of uniform temperature component (for Weibull distribution) are given in the first two rows of Table 6 for steel and composite steel concrete bridges in Germany. In the last row the results of measurements for a composite steel concrete bridge of 3 spans (slab thickness from 0,30 to 0,40 m, 5 steel girders) across the Nightingale valley situated in the Prague ring near the exit of the Lochkov tunnel are given. The composite bridge and position of sensors in concrete deck are illustrated in Figure 1.

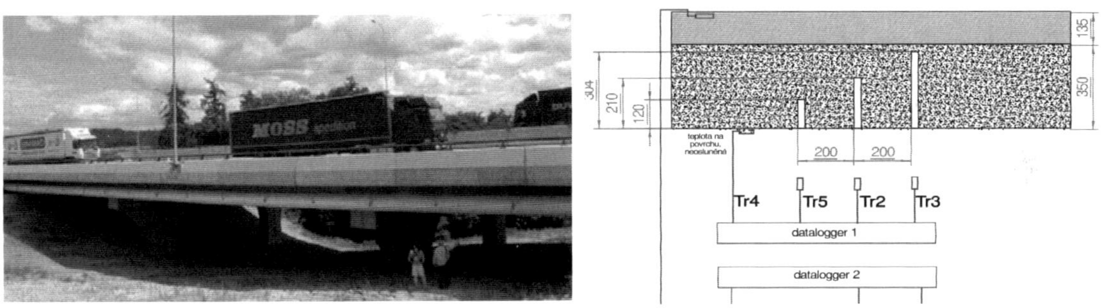

Figure 1: Temperature monitoring in the composite bridge and position of sensors in the cross-section

Table 6. Statistical characteristics of uniform temperature component T_N (in °C) and partial factor γ_T

Type of bridge	μ_T	σ_T	α_T	γ_T
Steel (GE)	40	2	0,6	1,12
Composite (GE)	35	2	0,6	1,13
Composite (CZ)	32	2	0,5	1,1

It is shown that for the uniform temperature component, the partial factor of thermal actions γ_T could be about 1,15 for steel and composite bridges.

2.3. *Difference temperatures*

For the positive vertical temperature differences (heating), the statistical characteristics and partial factors are given for steel and composite bridges in Table 7.

Table 7. Statistical characteristics of the temperature difference component $\Delta T_{My,\text{heat}}$ (in °C) and the partial factor for steel and composite bridges

Type of bridge	μ_T	σ_T	α_T	γ_T
Steel (GE)	11	1	0,65	1,16
Composite (GE)	15	1	0,7	1,17
Composite (CZ)	15	3	0,4	1,17

It is shown that for the vertical temperature difference component, the partial factor γ_T is about 1,16 for steel bridges, about 1,17 for composite bridges.

3. Analyses of partial factors

The characteristic value of a climatic action is defined in Eurocodes as the upper fractile of the probabilistic distribution for the basic time period corresponding to the 2% probability of annual exceeding. The design value of a climatic action is considered as 0,996 fractile of the probabilistic distribution. The partial factors ΔT of a temperature component can be expressed as

$$\gamma_T = \frac{x_{p,50}(\Phi^{-1}(-\alpha\beta))}{x_{p,1}(0{,}98)} \tag{1}$$

where α is the coefficient of sensitivity for the leading variable action and β_t is the target reliability index where $\beta_t = 3{,}8$ for bridges in common category CC2 as recommended in Eurocodes.

In case when the bridge needs to be classified to higher reliability class RC3 (considering the target reliability index $\beta_t = 4{,}3$), then the probability $p = 0{,}999$ should be applied in expression (5). When the bridge is to be categorised in the reliability class RC1 only ($\beta_t = 3{,}3$) as e.g. for some existing bridges with a shorter remaining life-time, then the probability $p = 0{,}99$ might be assumed.

The partial factor γ_T specified for thermal actions versus the coefficient of variation V_T for considered Gumbel and Weibull distributions (for two selected values 0,4 and 0,6 of skewness) is illustrated in Figure 2. It is shown that for the same coefficient of variation the value of partial factor significantly depends on the applied probabilistic distribution.

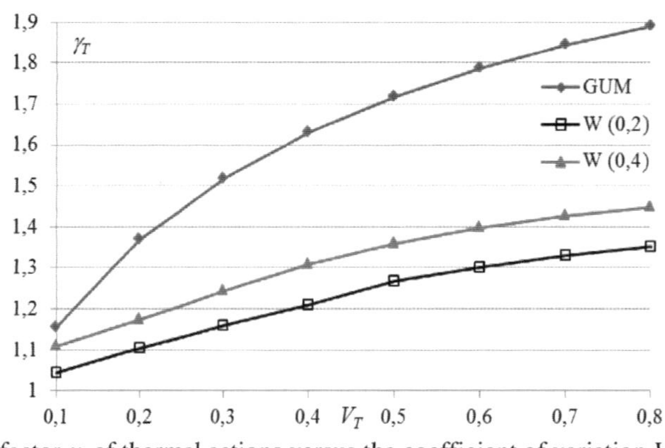

Figure 2: The partial factor γ_T of thermal actions versus the coefficient of variation V_T for considered Gumbell and Weibull distributions

The partial factor γ_T specified for thermal actions versus the coefficient of skewness α for three different reliability categories RC1 to RC3 is illustrated in Figure 3. The partial factor γ_T is increasing with increasing skewness of the probabilistic distribution of temperatures.

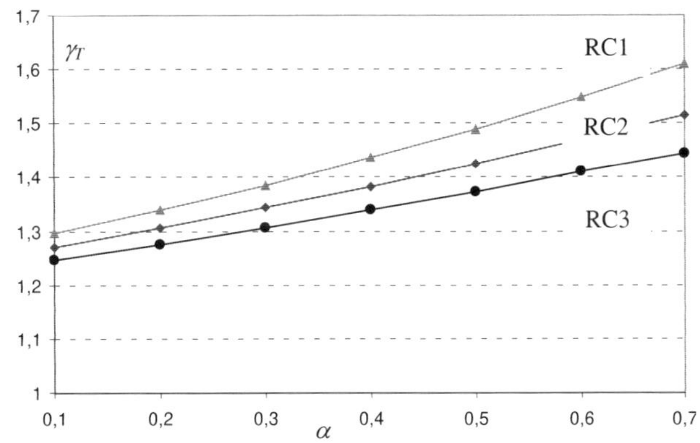

Figure 3 The partial factor γ_T of a uniform temperature component versus the skewness α (Weibull distribution) for three reliability classes RC1 to RC3 of bridges

4. Conclusions

It appears that despite the Eurocodes recommend the application of the Gumbel distribution for modeling of climatic loads, the uniform and difference temperature components could be better represented by Weibull distribution. The skewness of the statistically evaluated data of temperature measurements is in a range from 0,3 to 0,7 what is considerably less than the skewness of the Gumbel distribution. For specification of statistical characteristics for Weibull distribution, the skewness as a third parameter should be preferable applied than taking into account rough estimation of the upper limit of the distribution only as indicated in the background document to EN 1991-1-5. It appears that a three parameter Lognormal distribution might be also applied as it has no an upper bound for positive skewness.

The results of analysis of temperatures based on the experimental bridge measurements indicate that the partial factors of the both uniform and difference temperature components may be reduced to 1,2 for steel and composite steel-concrete bridges. In case that some uncertainties in the process of further development of temperatures in the next decades should also be considered, the potential enhancement factor for consideration of climate changes, if needed, could be estimated for structures with a higher working life. Application of partial factors for thermal actions calibrated for national climatic conditions will facilitate design of bridges, bearings and thermal expansions.

Acknowledgements

This study has been conducted at the Klokner Institute CTU within the framework of the research project FV20585 supported by the Ministry of Industry and Trade and under Grant LTT18003 supported by Ministry of Education, Youth and Sports of the Czech Republic.

References

[1] EN 1991-1-5 EC 1: Actions on structures - Part 1-5: General actions - Thermal Actions, 2005
[2] New European Code for Thermal Actions 1999 Report *Background document to ENV 1991-2-5*
[3] Holicky M and Markova J 2004 Research Report No. 803/120/110. *Models of thermal and accidental actions according to Eurocodes*, Czech Republic
[4] Markova J et al. 2012 Research report *Assessment of Temperatures Monitoring in Highway Bridges in the Czech Republic*. Klokner Institute CTU in Prague
[5] Markova J 2016 Analyses of climate changes for evolution of Eurocodes. In: Esrel, Portoroz, s. 2127-2133. ISBN 978-1-138-62937-0.
[6] Probabilistic Model Code 2015 Thermal actions Joint Committee on Structural Safety

Experimental investigation on the bridge segments with transversally curved bottom flange

Filip Ljubinković[1], João Pedro S C Martins[1], Helena M S Gervásio[1] and Luís A P Simões da Silva[1]

[1]ISISE, Department of Civil Engineering, University of Coimbra, P-3004 516, Coimbra, Portugal

E-mail: filj@uc.pt

Abstract. The main objective of this research project is to develop solid knowledge on the structural behavior of curved steel panels for optimized applications in steel and composite bridges. Moreover, the goal is to extend the EN 1993-1-5 design methodology for transversally curved panels subjected to various in-plane load cases. For that purpose, two identical prototype box-girder bridge segments in scale 1:3, with the transversally curved bottom flange and with two different steel grades (S460 and S690) are to be tested. Bridge segments are tested as three-point bending tests, simulating the bridge behavior near the intermediate support, where the interaction between bending moment and shear force occurs in the lower curved flange. This paper is mainly focused on the preparation process prior to the experiments and the description of the test layout. The first specimen was tested recently and thus, only the preliminary results are presented. Later on, based on these results, a numerical model will be validated and used for a parametric study, where the influence of several parameters on the behavior of these innovative bridge prototypes will be investigated.

1. Introduction

In recent years, the use of the cylindrically curved panels as a bottom flange in steel bridge applications has experienced an uprising trend. One typical example of the box-girder cable-stayed bridge with the curved bottom flange is given in figure 1.

Figure 1. Ebro River Pedestrian Bridge, Zaragoza, Spain.

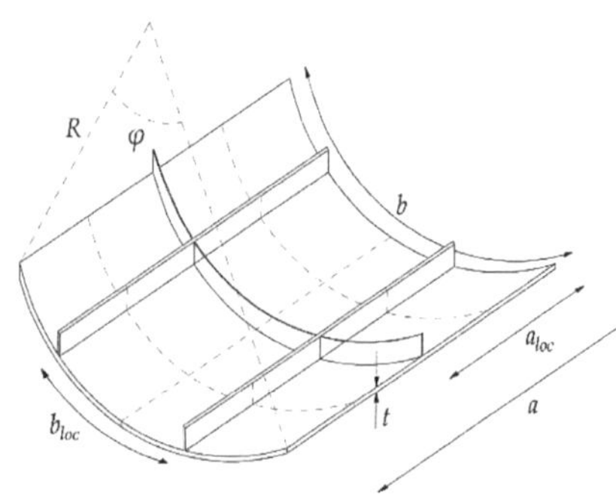

Figure 2. The geometry of a cylindrically curved panel.

The main parameters required to define the geometry of cylindrically curved panels are schematically presented in figure 2, where a and b are the length and width of the panel, a_{loc} and b_{loc} are the length and width of the corresponding subpanel, t and R are the thickness and radius of the panel, and finally, φ is the sectorial angle of the panel. The aspect ratio α is defined as the ratio between the length and width of the panel or sub-panel, whereas the curvature parameter Z is defined by equation (1)

$$Z = \frac{b^2}{R \cdot t} \tag{1}$$

An investigation carried out by Reis *et al.* [1] shows that as many as 18 bridges with curved cross-sectional parts have been successfully designed so far. The ranges of curvature parameters commonly used in these 18 applications are given in figure 3. Moreover, the global aspect ratio usually varies between 0.5 and 1, whereas the local aspect ratio varies between 3 and 6, as cited in Martins *et al.* [2].

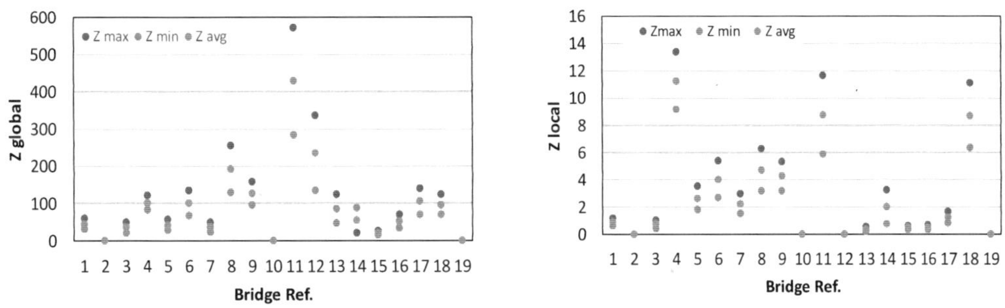

a) b)

Figure 3. Parameters range (a) Z_{global} - whole bottom flange and (b) Z_{local} - sub-panels between longitudinal stiffeners.

However, the existing design standards for plated structures (i.e. EN 1993-1-5 [3] and EN 1993-1-6 [4]) still do not provide the design rules for this kind of solutions, such as bridges with a transversally curved bottom flange. The designers are, therefore, mainly relying on the own experience and on the use of the advanced Finite Element Analysis. Therefore, a huge step forward in the understanding of the structural behavior of bridges with the curved panels is urgently required. This study aims to develop a clear design approach for curved panels, using experimental and numerical approach.

Therefore, two identical bridge prototypes, with a scale of 1:3 with respect to a real study case of a bridge with cylindrically curved bottom flange, are subjected to experimental investigation. The geometrical parameters (i.e. curvature parameter and aspect ratio) of the bridge prototype are carefully chosen to fall within the common ranges of these parameters. The exact shape and dimensions of the specimen will be described in the following chapters.

2. Predesign of the test specimen

2.1. General

The goal of this part of the study was to prepare and adapt the bridge prototype segments to fit the capacities of the laboratory at the University of Coimbra, in terms of geometry and load application. The whole process from the initial case study of the real bridge, designed by GRID International, until the final adoption of the geometry for the future test specimens is explained.

2.2. Case study

The studied structure is a continuous five-span composite steel-concrete girder bridge with standard abutments sitting on embankments. It is deemed to carry a roadway corridor with a single carriageway.

The superstructure comprises two lateral spans of 37.8 m and three central spans of 54 m, in a total of 237.6 m length (figure 4).

This cross-section shape was actually proposed as a variant of an already built bridge, designed by GRID International. The alternative solution with cylindrically curved steel panels has been developed adopting a box girder section, with a constant depth of 2.25 m. The cross-section consists of a steel girder and a concrete slab supporting the roadway. The bridge deck is 10.93 m wide, carries two traffic lanes 3.70 m wide and two pedestrian walkways 1.6 m wide in each direction.

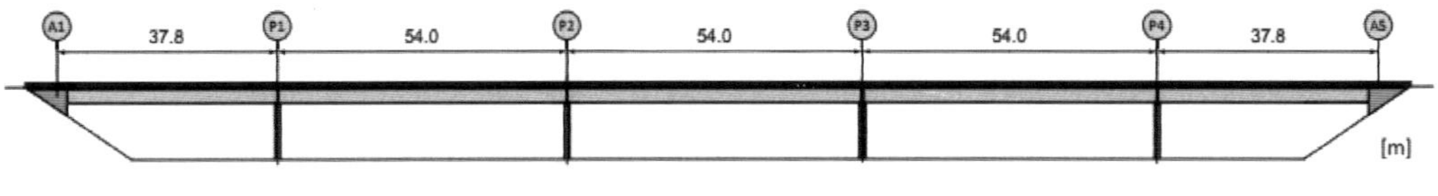

Figure 4. Elevation span layout.

The box girder is formed by: (i) a curved bottom panel, 3.4 m wide, with a radius of 2.5 m and longitudinally stiffeners (0.25x0.025 m flats each 0.95 m), (ii) webs with 46 degrees of inclination with regards to the horizontal, (iii) and standard plate top flanges, 0.8 m wide and 0.08 m thick (figure 5 and figure 6). The reinforced concrete slab is supported by cross-beams at each 2.16 m. Over the supports, a 0.03 m thick closed diaphragm was adopted. The web thickness has the same thickness and steel distribution as the cylindrical panel, and it is longitudinally unstiffened.

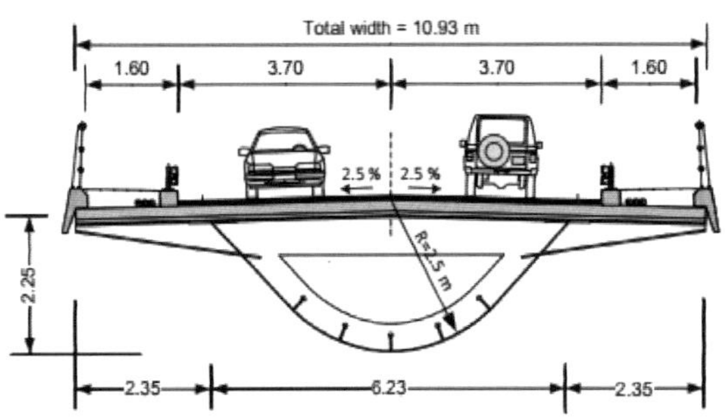

Figure 5. Span cross-section.

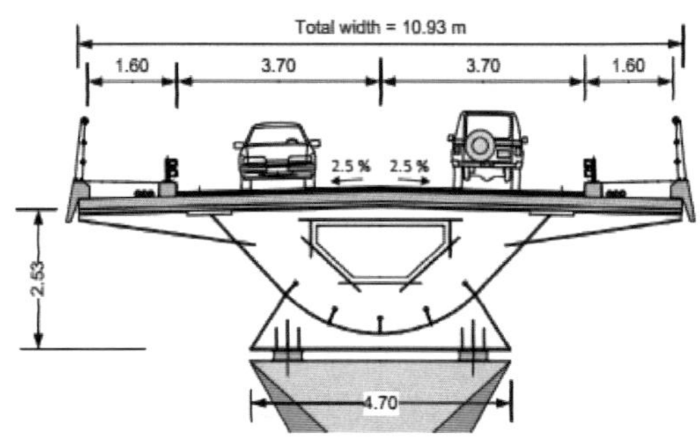

Figure 6. Pier cross-section.

The main geometrical parameters (i.e. global and local curvature parameters, aspect ratios and plate slenderness) of the proposed solution are given in table 1. It is worth noting that the global curvature, $Z_G = 235$, and the local curvature, $Z_L = 12$, are in the upper bound of existing bridge solutions (see figure 3) in order to perceive the benefits of adopting higher curvatures. Additionally, regarding the aspect ratio, sub-panels between stiffeners are classified as long panels, $\alpha_L = 2.3 > 2$, whereas the whole stiffened plate as short panel, $\alpha_G = 0.5 < 1$.

Table 1. Geometrical parameters of the cross-section

	Z_G	α_G	b_G/t	Z_L	α_L	b_L/t
Case study	235	0.5	140	12	2.3	32
Existing data (avg.)	100	0.7	165	3	4	28

2.3. Geometry of the test specimen

The experimental layout aims to imitate the behavior of the mid support of the previously defined case study, where the length of approximately 0.2xL (L ≈ 54 m) on both sides of the mid support (i.e. 0.4xL in total) corresponds to the distance between two zero-bending moments. This leads to the segment with the total length of approximately 26 m. This segment is proportionally scaled down 3 times in order to fit the laboratory capacities in terms of dimensions and loading applications. Additionally, several simplifications were adopted:

- The cross-section is considered as a homogeneous steel section, and not as a composite one, where the reinforcement steel above the mid-support is replaced by the corresponding steel plate, resulting in a top flange with the constant thickness
- The inclination of the slab of 2.5% is neglected, leading to a flat top flange
- The cross-section of the bridge which participates in load-bearing capacity is consisted only of the box, while the cantilevers that carry the load from the footways are omitted
- After the scaling down 3 times, all the dimensions are rounded to a smaller whole number in order to ease the fabrication process

Finally, the exact dimensions of all the relevant parts of the specimens are given in the in figure 7 and figure 8, whereas table 2 lists all the dimensions which may not be seen in these figures.

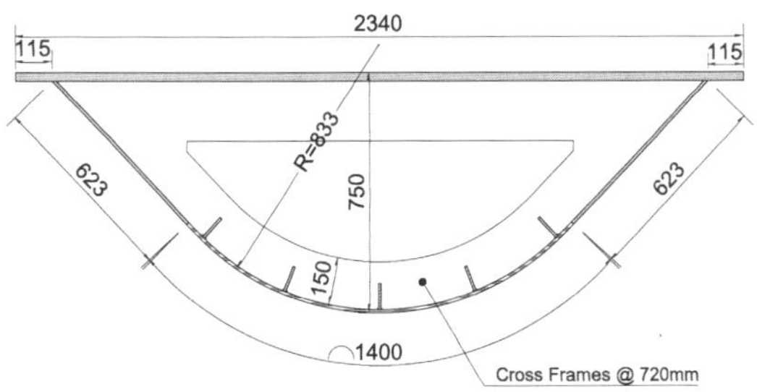

Figure 7. Adopted cross-section geometry.

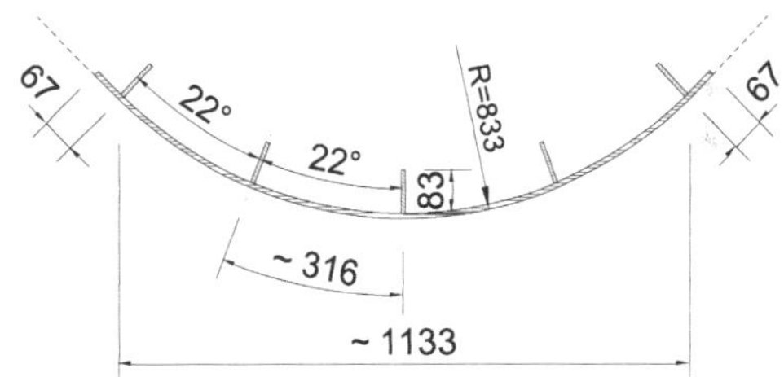

Figure 8. Geometry of the bottom flange.

Table 2. Geometry of the bridge prototype

[mm]	Symbol	R = 1:1	R = 1:3
Total length	L	25920	8640
Top flange thickness	t_f	80	25
Length (segment)	a	2160	720
Stiffener thickness	t_{st}	25	8
Cross-frame thickness	t_T	16	5
Diaphragm thickness	t_d	30	10

2.4. *Preliminary numerical study*

For the adopted geometry of the specimen from the previous step, the final step before the definition of the test setup was a preliminary numerical simulation of the test, based on the FE method. The arc-length Riks' [5] method was used to check what is the ultimate load that the bridge specimen may sustain and also, to predict the possible failure mode. This would be useful information for the instrumentation of the specimen which is described in the following section.

As for the material, the mechanical properties of the steel S460 and S690 are taken with the nominal values, increased by 15% and 10%, respectively, accounting for the hardening of the material (E = 200 GPa; v = 0.3; $f_{y,S460}$ = 530 MPa and $f_{y,S690}$ = 760 MPa). The real material properties will be tested subsequently after the test, and this law will be used for the final numerical model. Regarding the mesh, four-node shell elements with the reduced integration were used (S4R), with the mesh size of 150x150 mm^2, which is obtained after the mesh sensitivity analysis as a size that reduces CPU time and gives sufficiently correct results.

The load and boundary conditions are modeled in such a way to simulate as close as possible the future conditions of the experiment. They are schematically illustrated in figure 9. The concentrated load, which is distributed by means of a rigid loading beam, is placed on one side of the specimen, at the position of the hydraulic jack. Additionally, the self-weight of the specimen is also introduced into the model. Regarding the boundary conditions, all three translations (in x, y, and z-direction) are prevented at the middle support, whereas on the opposite side of the load, 4 DYWIDAG cables are modeled with the appropriate axial stiffened (k), with Young's modulus equal to 205 GPa.

Finally, the geometric imperfections were also accounted for with the shape affine to the first Eigenmode, obtained by Linear Bifurcation Analysis (LBA) and with the amplitude equal to the $b_{loc}/200$, which is actually suggested in Annex C of EN 1993-1-5 for the flat plates.

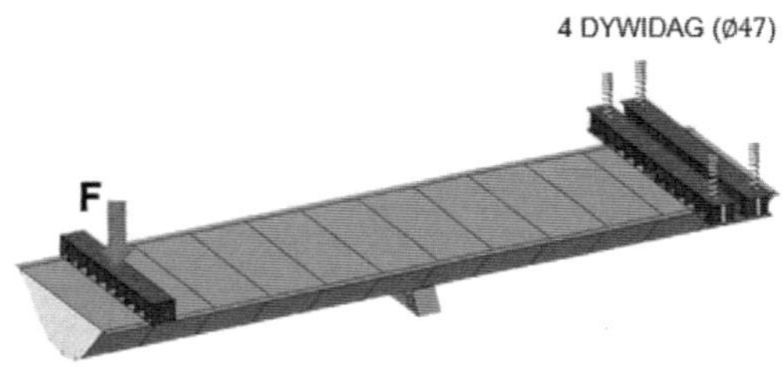

Figure 9. Load and boundary conditions.

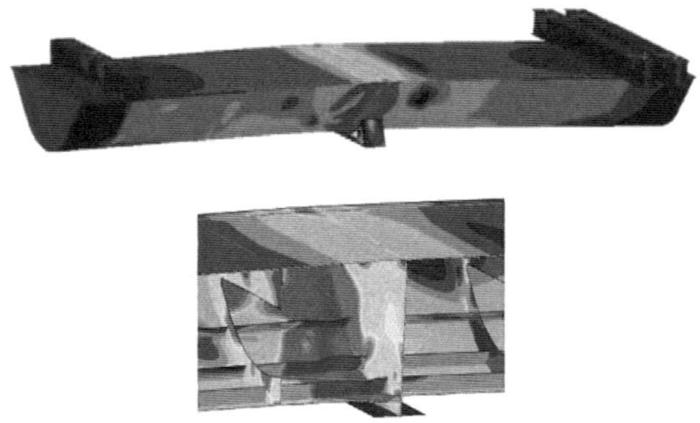

Figure 10. Expected failure mode.

The failure mode of the bridge specimen obtained numerically is presented in figure 10. As it is expected and intended, the failure occurs in the bottom (curved) flange, where both the longitudinal stiffeners and curved panel in-between buckled. The hypothetically maximum possible load of the test is equal to 2250 kN (i.e. 225 tonnes), which is obtained in case of the stronger (S690) specimen without imperfection. Based on these results, the drawings of the specimens are prepared and the specimens fabricated by MCE GmbH.

3. Test layout

3.1. General

For the experimental campaign, based on the previous results, the following material was designed and fabricated (figure 11):

- 1 bridge specimen made of steel S460 and another identical of S690
- 2 supporting feet where the mid support will be placed
- 1 loading beam (stiffened HEB300 profile)
- 2 supporting beams (back-to-back U260 profiles) to anchor 4 DYWIDAG cables
- 2 supporting beams (HEB300 profiles), to support two beams (back-to-back U260 profiles)
- 4 DYWIDAG prestressing systems, with the diameter of 47 mm, and length of 4.5 m

Except for the bridge prototypes, all the other auxiliary elements are made of steel S355. The elements were manufactured by MCE GmbH and delivered to the laboratory of the University of Coimbra, within the on-going research project OUTBURST (ref.: RFCS-2015-709782).

The bridge segment is loaded by a concentrated force on one side using the 600-*ton* hydraulic jack RCS-1002 (figure 12). The load is transferred then to the bridge over a load beam made of a stiffened HEB300 profile. The middle support of the segment is supported by two so-called support feet, designed particularly to sustain the huge axial force, which are connected between themselves by means of bolts. Finally, on the other side of the bridge segment, 4 DYWIDAG cables that are anchored to the ground, act as a counter-reaction, preventing the rotation of the bridge segment upwards. The cables are supported by two simply supported beams, each of them made of two back-to-back UPN260 profiles, with sufficient space in-between for the cables to pass. The webs of the beams are well stiffened to sustain the huge local transversal forces which appear during the test. These two beams are supported by another two beams also made of HEB300 profile, which are connected by means of welded plates in the level of top and bottom flange (figure 13).

Figure 11. Test setup

Figure 12. Hydraulic jack and loading beam.

Figure 13. End support – two 2xU260 beams (white) and two HEB300 beams (blue).

3.2. Instrumentation of the specimens

The data that are measured directly from the test are listed here:

- Externally applied force and reactions in the mid and end-support (with load cells)
- Vertical, lateral and longitudinal displacement of the bridge (with LVDTs)
- Strains distribution in the zone where failure is expected to occur (with strain gauges – SG)

3.2.1. Linear displacement traducers (LVDT). In total 10 LVDT TML are used to measure displacements in various positions along the segment. Namely, 7 of them measure vertical displacements, where two out of them (V-6 and V-7) are placed on the ends of the loading beam (see figure 12), 2 measure the longitudinal displacement (L-1 and L-2) at two ends of the specimen, and 1 measures possible horizontal (lateral) displacements (H-1) (figure 14).

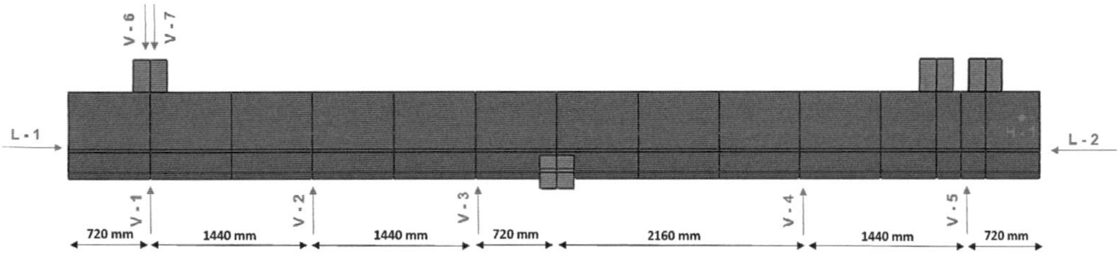

Figure 14. Position of the LVDTs.

3.2.2. Strain gauges (SG). For each of two specimens, it is planned to use 103 FLA-6-11 SG positioned in 3 different planes, namely 150 mm, 300 mm and 570 mm distant from the middle support, where the failure is expected. The SG are organized in such a way to measure both longitudinal and transversal strains, according to figure 15. To enable the installation of measuring equipment inside the section, two "windows" were left open and then closed (welded) afterward (figure 16). The cables from the inside of the bridge segment were led outside through one of the wholes (200x200 mm^2) deliberately left on the top flange for this purpose.

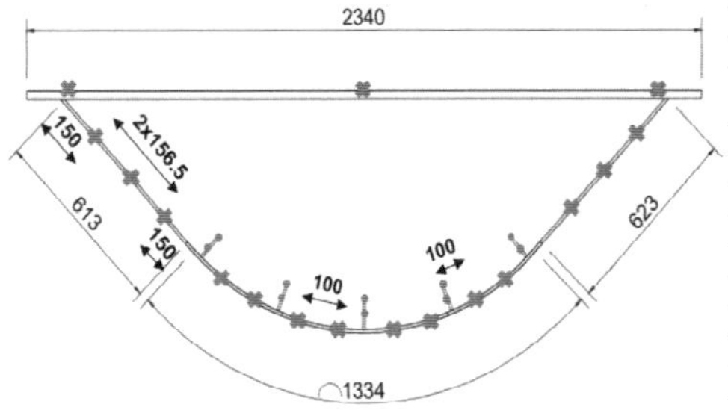

Figure 15. Position of SG.

Figure 16. SG on the welded "windows".

3.2.3. Load cells. To measure the value of the load during the test, 7 load cells of the type C6A were used. Four of them (D1, D2, D3 and D4), with the capacity of 2MN were placed at the position of 4 DYWIDAGS, whereas the other three, with the capacity of 5MN, were placed at the position of the loading point (F) and also below the middle support (R1 and R2), according to the figure 17.

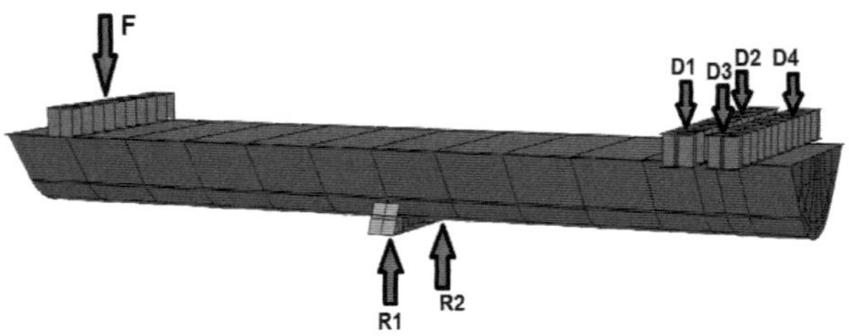

Figure 17. Position of the load cells.

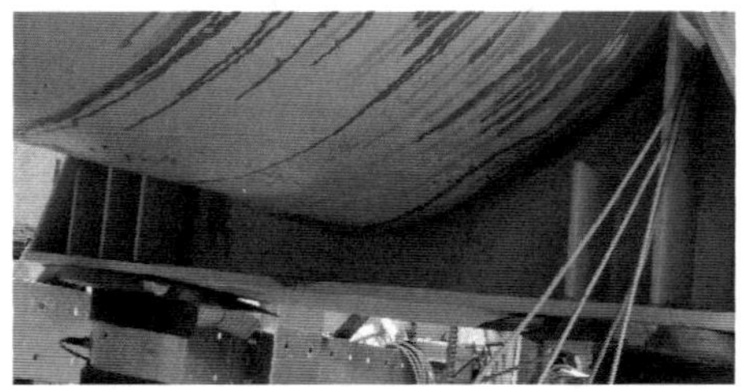

Figure 18. Failure of the curved panel.

4. Test results

The first specimen (S460) was tested only recently and thus, in this document only the preliminary results are presented. The test was displacement controlled, with the constant speed of 0.01 mm/s until the collapse. The load application was stopped when the vertical displacement at the position of the load reached the value of approximately 117 mm, which occurred in the descending part of the force-displacement curve. The location and the pattern of the failure mode completely coincide with the ones obtained in the preliminary numerical analysis (see figure 10), with the obvious buckling of the curved panel in the bottom flange close to the mid-support (figure 18).

The maximum load reached a value of 1580 kN, which is practically the same ($\Delta \approx 1.5\%$) as the one obtained in the numerical analysis (i.e. 1602 kN), as it is presented in figure 19. The data recorded by four load cells at the position of the DYWIDAG cables is presented in figure 20. The sum of these four values at the moment when the ultimate load is reached is almost equal to the applied load ($D_{tot} \approx 1400$ kN) and also almost the same values were obtained in numerical analysis ($D_{num} \approx 1350$ kN), which confirms that the bridge statically behaved in the desired way.

Based on the vertical displacements at the ultimate load measured by LVDTs, the deformed shape of the segment is schematically illustrated in figure 21. The maximum longitudinal and lateral movements reached the values of 1.5 mm and 0.7 mm, which may be considered negligible.

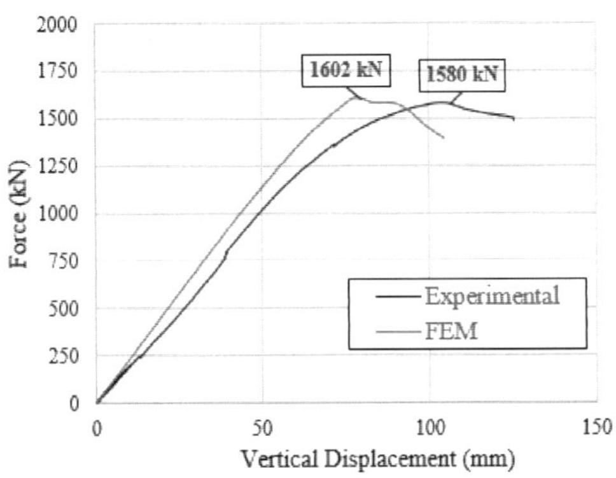

Figure 19. Force-displacement curve.

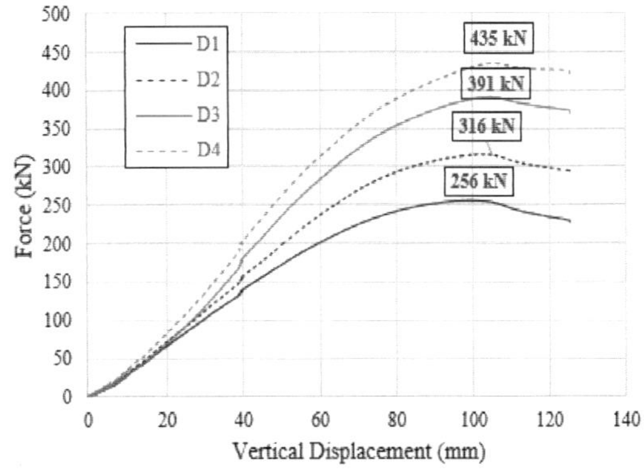

Figure 20. Forces at four end-supports.

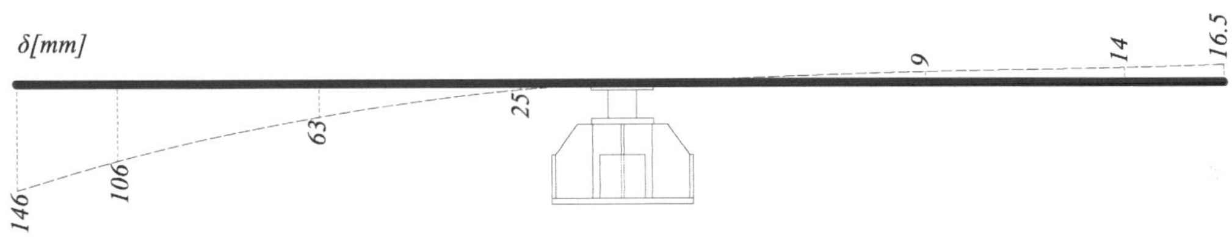

Figure 21. Vertical displacement at the ultimate load.

Concluding marks and future work

Based on these preliminary results of the first test, it was concluded that the experiment was successful and in accordance with the targeted behavior, which gives a certain confidence ahead of the second test. The minor adjustments will be subsequently performed in the numerical model, namely, the real material law will be incorporated, as well as the appropriate geometric and material imperfections (i.e. residual stresses). Also, a study on the influence of the axial stiffness of the DYWIDAG cables will be performed, since the one in the numerical model led to a slightly more rigid system, resulting in a smaller vertical displacement (e.g. 90 mm at the point of the load). Finally, it is necessary to investigate deeply the stress-strain relations in the bottom flange close to the mid-support, where the interaction between shear and compressive stresses occurs. After the second test (with the segment made of S690), it will be possible to understand the influence of various steels on the behavior, which should question the rational use of the high-strength steels in this type of bridge applications.

References

[1] Reis A, Pedro J O, Graça A B, Hendy C, Ramoli P, Simões da Silva L and Martins J P 2016 Report on the characterization of relevant parameters of curved plated bridge structures and identification of bridge cases where they can be found *RFCS Research Project OUTBURST* (Deliverable 2.1) pp 2015-709782

[2] Martins J P, Ljubinkovic F, Simões da Silva L and Gervásio H 2018 Behaviour of thin-walled curved steel plates under generalized in-plane stresses: A review *J. of Constr. Steel Research* (Elsevier) **140** 191-207

[3] European Committee for Standardization (CEN) 2006 Design of steel structures; Plated structural elements *Eurocode 3, Part 1-5 (EN 1993-1-5)*

[4] European Committee for Standardization (CEN) 2007 Design of steel structures; Strength and stability of shell structures *Eurocode 3, Part 1-6 (EN 1993-1-6)*

[5] Riks E, 1979 An incremental approach to the solution of snapping and buckling problems *Int. J. Solids Struct.* **15** 524-521

IOP Conf. Series: Materials Science and Engineering **419** (2018) 012043 doi:10.1088/1757-899X/419/1/012043

Experimental evaluation of stiffened curved plates subjected to pure compression

Sara Piculin[1], Franc Sinur[1], Primož Može[1]

[1]University of Ljubljana, Faculty of Civil and Geodetic Engineering, Jamova 2, 1000 Ljubljana, Slovenia

E-mail: sara.piculin@fgg.uni-lj.si

Abstract. The paper deals with an extensive experimental study on five stiffened curved plates with different geometry and steel grade. The objective of the study is to analyse the behaviour of curved panels and their longitudinal and transverse stiffeners subjected to pure compression. Before testing, the initial geometric imperfections of the specimens were measured. They will be further used in the finite element model as initial geometry. During tests, the displacement field was measured with photogrammetry. In this paper, some results of the experimental tests are presented together with a study of the effects of different parameters on the element's resistance to pure compression. It is shown that the displacements provided by photogrammetry are in good agreement with those measured with LVDTs in discrete points. The ultimate force of the specimen made of higher steel grade does not differ significantly from other specimens. In addition, there is almost no difference in the resistance between plates with different aspect ratios. Initial imperfections and residual stresses play an important role in the prediction of the stiffened curved plate's response when subjected to pure compression.

1. Introduction

The use of curved steel panels in bridge design of steel and steel-concrete composite decks has been increasing in the past few years due to both aesthetical and structural demands. Curved panels can represent the web of horizontally curved girder or, as discussed in the present study, transversally curved plates represent the lower flange of a box-girder. The curved lower flange is longitudinally welded to vertical or inclined web plates in order to form a closed box-section. In some cases, the curved flange can also play the role of both the web and the flange (see Figure 1). Usually, the flanges are designed to be slender and are therefore longitudinally stiffened with open or closed stiffeners that increase their load-carrying capacity and, at the same time, lower the weight of the construction.

In the past years, structural engineers are increasingly using this type of elements in bridge design. This was showed recently by Reis et al. [1] that identified in their survey almost 20 roadway, railway and pedestrian bridges in Europe, where curved bottom flange plates were adopted. The structural systems of those bridges are diverse, but mostly cable-stayed.

Although many applications including transversally curved plates are possible, the design recommendations and the knowledge in this field remain scarce. The design of curved panel segments is not covered by the Structural Eurocodes. In fact, the two codes covering shell elements, namely EN 1993-1-5 [2] and EN 1993-1-6 [3], are limited to flat or slightly curved (curvature parameter $Z < 1.0$) shell elements and shells of revolution. It has been proven by several authors, that curved panels have different characteristics from both mentioned types of shells [4, 5] and a sophisticated FEM analysis is necessary in order to properly calculate the ultimate resistance of these elements. For this reason, several authors have proposed different methods for the calculation of elastic buckling stress and ultimate resistance of curved panels [6].

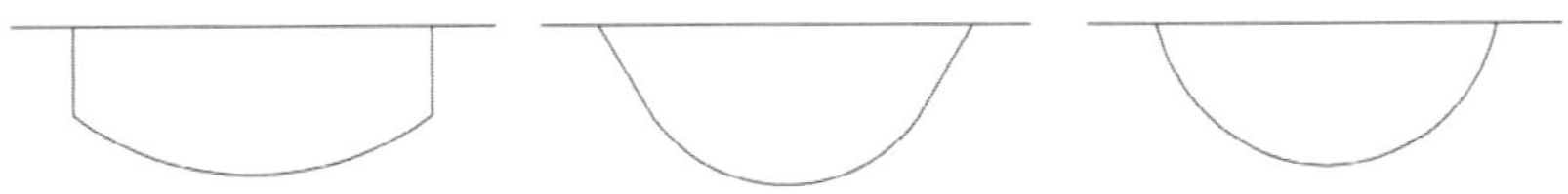

Figure 1. Cross-section shapes including curved plates.

None of the beforehand mentioned methods has any experimental background. In general, experimental tests in this field are very rare. Quite some experimental work was done on flat slander plates with stiffeners [7], but the same cannot be stated for curved panels. Ljubinkovic et al. [8] recently performed an experimental analysis of cylindrically curved unstiffened panels. One of the main conclusions was that special attention has to be given to the modelling of initial imperfections in order to achieve good correlation between experimental and numerical results. There were some experiments on stiffened curved panels carried out in the 30's (see [9]), but in all cases, the specimen's material is aluminum alloy. In a recent study, Cho et al. [10] conducted axial compression tests on six curved stiffened plates. The specimen geometry was defined based on a ship survey, also the material and shape of stiffeners are not directly applicable to bridge design. Therefore, the need of an experimental evaluation of stiffened curved bridge segments is obvious.

Since bottom flanges of box girders are subjected to high compressive forces, namely at pier sections, a series of five experimental tests on stiffened curved steel panels under in-plane compression are presented in this paper. The purpose of these experimental tests was to investigate the fundamental characteristics of the buckling and post-buckling behavior of stiffened curved panels and identify their ultimate resistance. Results of the experimental study will represent the database for the calibration of FE numerical models.

2. Experimental tests

Axial compression tests were conducted on five specimens in order to provide experimental information on the structural behavior of stiffened curved plates.

2.1. Specimen geometry and material properties

To analyze the effect of slenderness, aspect ratio and material grade on the plate's resistance, the five specimens under consideration differ in either geometry or material. Geometric parameters that are equal for all specimens are listed in the left part of Table 1, other geometric and material properties are also given in the table. Notations for the main parameters that are used to describe the geometry of a stiffened curved plate are provided in Figure 2. Additionally, the curvature of a cylindrically curved panel is denoted with the global curvature parameter Z. In case of stiffened plates, the local curvature parameter for unstiffened sub-panels is defined by the following expression:

$$Z_{loc} = b_{loc}^2 / (R \cdot t) \quad (1)$$

The local aspect ratio is the ratio between the subpanel's length and width:

$$\alpha_{loc} = a / b_{loc} \quad (2)$$

All specimens have two flat longitudinal stiffeners and all have three transverse stiffeners, except for 5C that has only one. Four specimens are made of steel grade S460 and one of grade S690. Mechanical properties of the used materials were provided with standardized tensile tests.

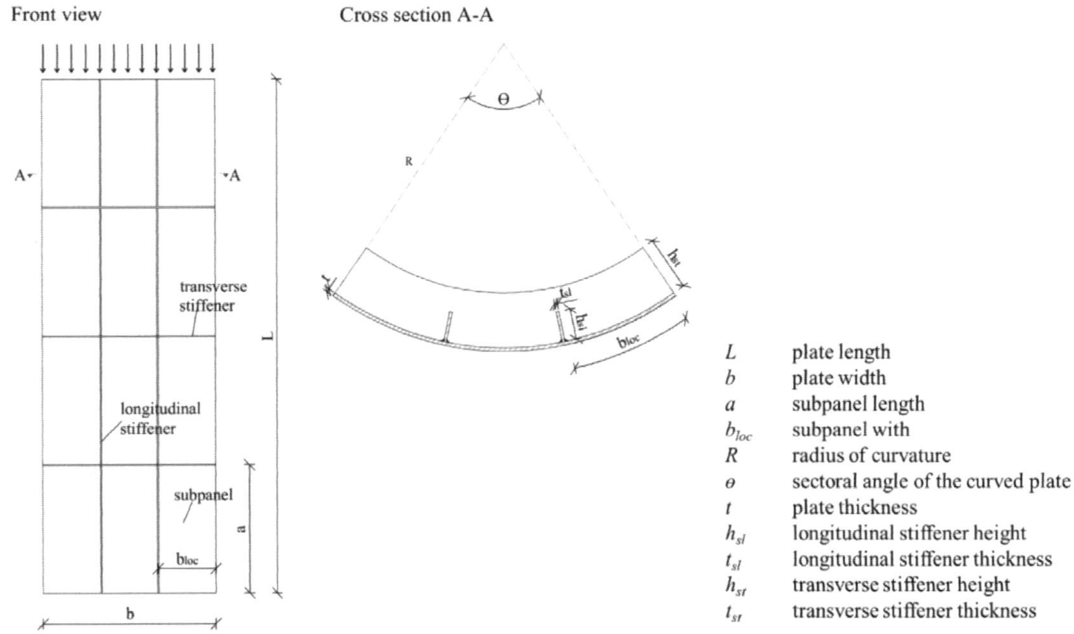

Figure 2. Geometry of test specimen

Table 1. Geometric and material parameters

Specimen	L [mm]	R [mm]	b [mm]	b_{loc} [mm]	h_{st}/t_{st} [mm]	θ [°]	t [mm]	a [mm]	α	Z_{loc}	h_{sl}/t_{sl} [mm]	Material
1C							6	432	2.3	12	50/5	S460
2C							6	432	2.3	12	50/4	S460
3C	1728	500	570	190	90/4	70	4	432	2.3	18	50/5	S460
4C							4	432	2.3	18	50/4	S690
5C							4	864	4.6	18	50/6	S460

2.2. Test layout

Experimental tests were carried out at the Laboratory of Structural and Traffic Engineering, Faculty of Civil and Geodetic Engineering, University of Ljubljana. Tests were displacement controlled with monotonously increased displacement up to collapse.

The main challenge during the design of the test layout was to imitate the desired loading and boundary conditions. Since the panels under investigation are subjected to pure compression, the configuration of the test frame had to be carefully designed in order to avoid any eccentricity during load application. For this purpose, each specimen was welded on the top and bottom edge to a 30 mm thick steel plate that served as a connection to the test frame. The center of gravity of each specimen's cross-section was positioned in the center of the plate. The top plate was further bolted on a stiff beam on which the load was applied with a hydraulic piston with maximum capacity of 3000 kN that was installed on the main testing frame. The beam was supported in the out of plane direction through a Teflon plate that allows vertical displacements. The bottom plate was also connected to a stiff beam that transferred the reaction forces to the floor through a concrete brick. Both beams were supported out-of-plane to the reaction wall. Vertical edges of the specimen were simply supported. For that purpose, special linear supports were designed in order to prevent the out of plane displacement of the vertical edge, allow the rotation around the vertical axis and the vertical and circumferential displacement. The linear supports were connected to two vertical columns that were supported laterally and out of plane. The test layout is presented in Figure 3 and Figure 4.

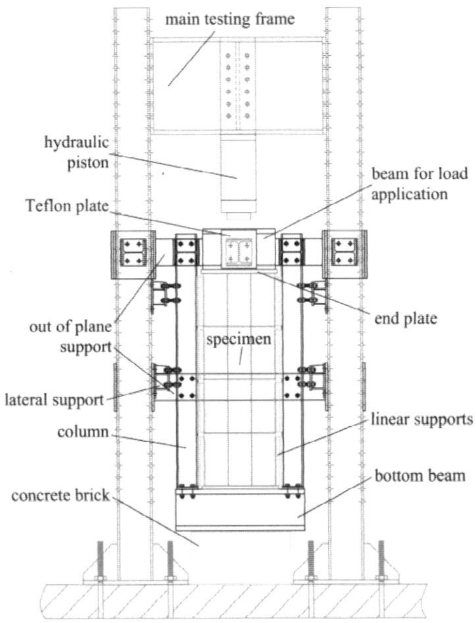

Figure 3. Schematic representation of the test layout.

Figure 4. Laboratory test setup with specimen 1C in position.

2.3. Instrumentation and measuring

With the scope of making an accurate evaluation of the response of the curved plates, several instrumentation was used in order to measure various parameters before, during and after the test. The measured quantities will be used in the calibration of the finite element model.

Before the test, a structured light portable 3D scanner was used to measure the exact initial geometric imperfections of the specimens. The output of the measurement is a very precise CAD model of the scanned specimens that can be compared to the "ideal" geometry of the specimens. This allows the definition of an exact initial geometry that will be further used in the finite element model.

The displacement field of the tested panel at different loading levels was measured by using photogrammetry. For this purpose, the panel was painted in random speckle pattern. In addition, approximately 40 coded targets were attached to the panel to track the displacements in discrete points at each step of the loading. Load application was paused according to a predefined loading protocol and during each pause 12 photos were taken from three different angles at four different heights. It is very important to provide appropriate lightning during the procedure to assure high quality of photos.

A total of 19 uni-axial strain gauges (SG) were attached to the specimen to measure the strains in the structure continuously during the loading procedure. 12 SGs were attached to the second subpanel from top and 7 to transverse and longitudinal stiffeners (see Figure 5). The displacements of the specimen and testing frame in some characteristic points were measured by using displacement transducers (LVDT) placed on the laboratory ground in the backside of the specimen. The global vertical displacement of the panel was measured with LVDT V5 that was placed on the loading beam directly beneath the piston. In addition, 2 LVDTs were used to control the vertical displacement of the loading beam, 3 for the out-of-plane displacements and 4 LVDTs to control the displacements of the loading frame (see Figure 6).

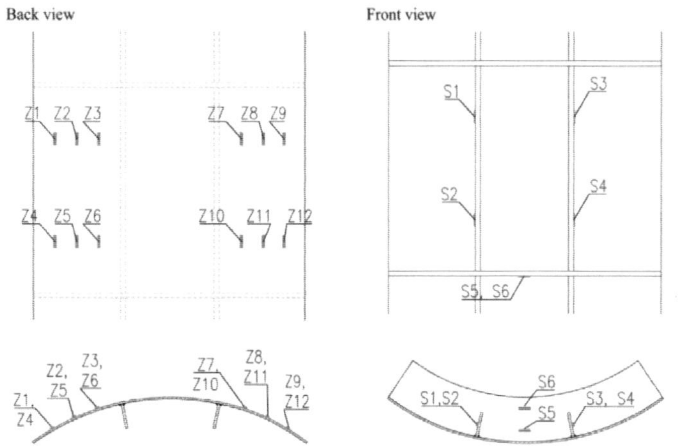

Figure 5. Position of strain gauges on the back and front side.

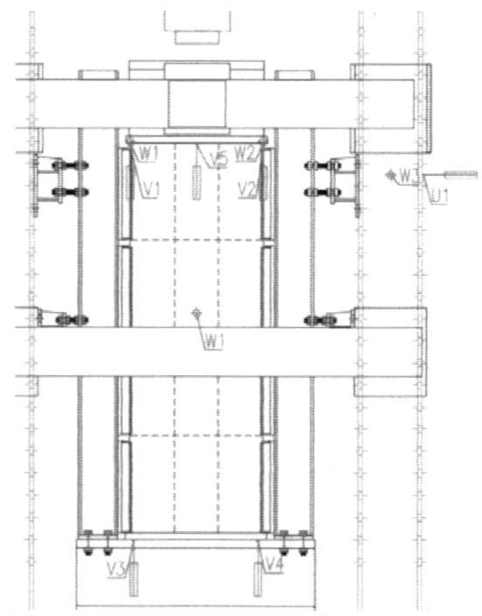

Figure 6. LVDT position.

3. Experimental results

The ultimate loads reached in the experimental tests are listed in Figure 7 together with diagrams representing the applied load versus the vertical displacement. In addition, the ultimate force of flat plates with the same dimensions was also calculated according to EN 1993-1-5 [2] although the values are not directly comparable since the tested panels are curved. In the calculation, the mechanical properties from tension tests were used. For all specimens, the resistance of curved plates achieved experimentally is 50 to 75% higher compared to the resistance of flat plates calculated according to the effective width method from the codes.

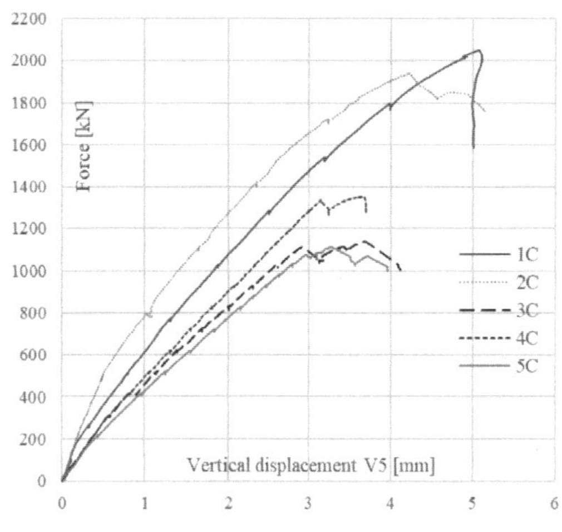

	Ultimate force [kN]		
	Experimental	EN1993-1-5	Difference
1C	2050	1252	64%
2C	1938	1198	62%
3C	1138	708	61%
4C	1353	774	75%
5C	1113	739	51%

Figure 7. Load vs. vertical displacement curves of all tests (left). Values for ultimate forces (right).

The vertical displacement was calculated as the difference between the displacement measured in V5 and the displacement of the supporting elements (V3 and V4). It can be observed, that in the case of thicker panels, the difference between the resistance of 1C and 2C is very small, around 6%. Comparing specimens with different steel grades, 4C reached a higher ultimate force compared to 3C. The difference is around 19%, while the difference in the material's yield strength is around 33%. Also for the aspect ratio, the results show that the parameter does not affect the resistance since the difference between 3C and 5C is almost negligible regardless of the doubled value of aspect ratio in 5C. The difference in the stiffness between 1C and 2C may probably be attributed to different geometric imperfections and residual stresses.

In Figure 8, an example of the comparison between the displacements measured in two discrete points with LVDTs, namely V5 and W1, and values from the photogrammetry procedure is given. Very good correlation between the results is achieved. Therefore, the displacement field can be further used for the calibration of the numerical model and comparison of deformed shapes.

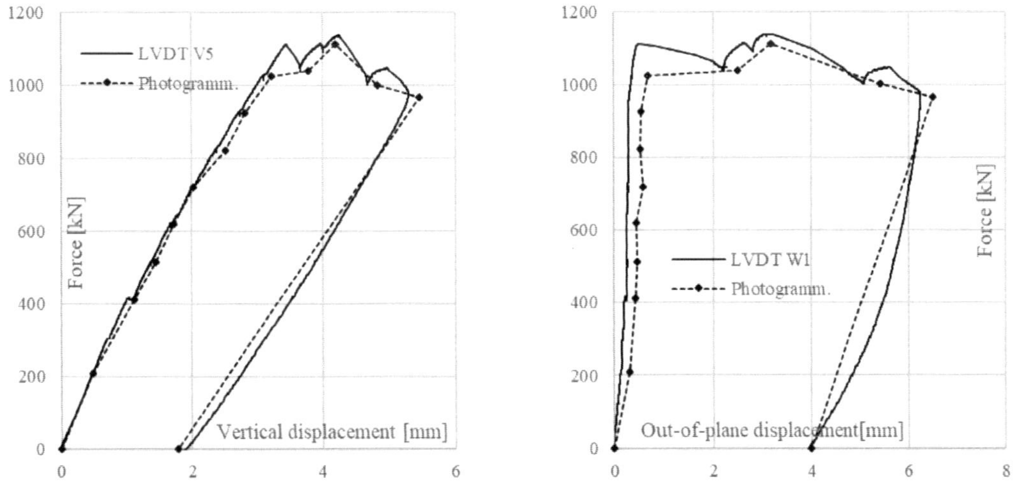

Figure 8. Comparison of displacements measured with LVDTs and photogrammetry for specimen 3C.

The collapsed shapes of models 2C, 3C, 4C and 5C are shown in Figure 9. According to the observed deformed shapes and measured strains (see Figure 10), failure of all specimens is associated to a combination of buckling and yield of material. Different buckling shapes can be attributed to the effects of initial geometric imperfections and residual stresses. Both mentioned phenomena have to be taken in account when modelling the stiffened curved plate for a finite element analysis.

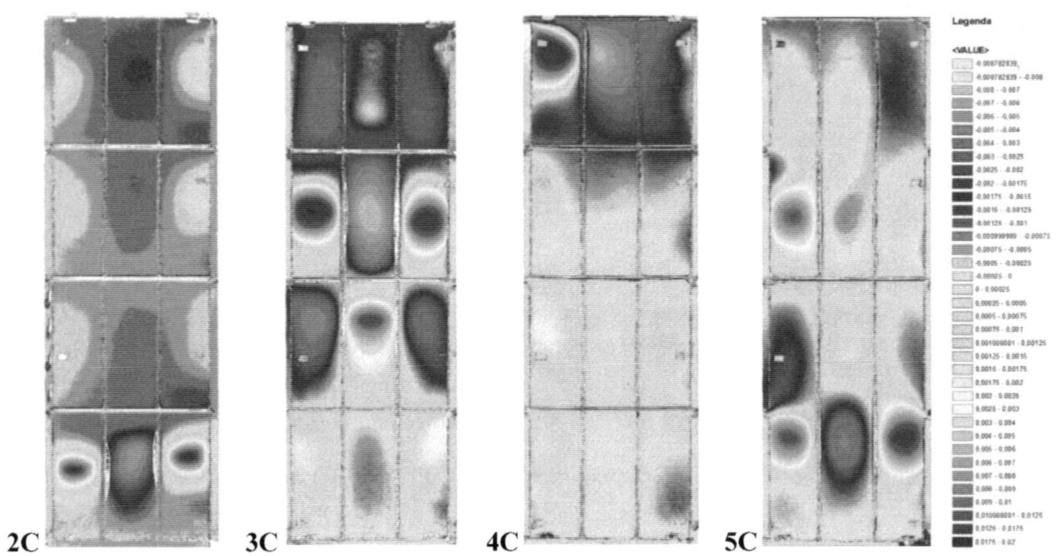

Figure 9. Deformed shape at failure for specimens 2C to 5C acquired with photogrammetry.

In Figure 10, the monitored strains from gauges Z1 to Z12 of specimen 3C are plotted. The cross-section is subjected to pure compressive stresses. In the first phase, the measured strains are elastic and almost constant through the whole cross-section. The first half-wave due to buckling happens in the left subpanel (front view) which can be seen from the strains (SGs Z7 to Z12), that instantaneously change from compression to tension. Later, the same happens with the other half of the measured strains (SGs Z1 to Z6), when the second half-wave appears.

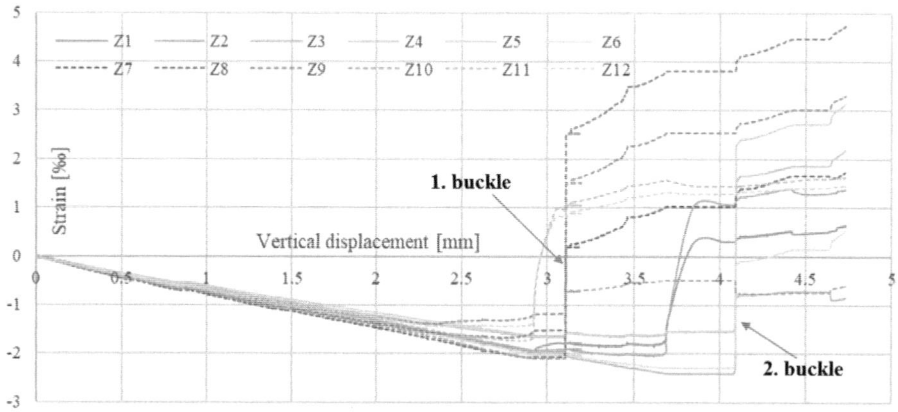

Figure 10. Measured strains on the panel for specimen 3C.

The objective of a stiffener is to increase the resistance of the whole plate. Generally, stiffeners may be distinguished between weak that buckle under direct stresses together with the panel and strong stiffeners that provide a rigid support to the panel and enforce the local buckling of the subpanel. According to EN 1993-1-5 [2], when loaded axially, torsional buckling of stiffeners shall be prevented with a sufficient torsional rigidity of the stiffener's cross-section. Experimental results show that the longitudinal stiffeners in our study provide adequate support to the panel in order to avoid global buckling of the whole plate. It is shown in Figure 9 that local buckling of subpanels predominates in all cases. In all cases, longitudinal stiffeners exhibit torsional buckling.

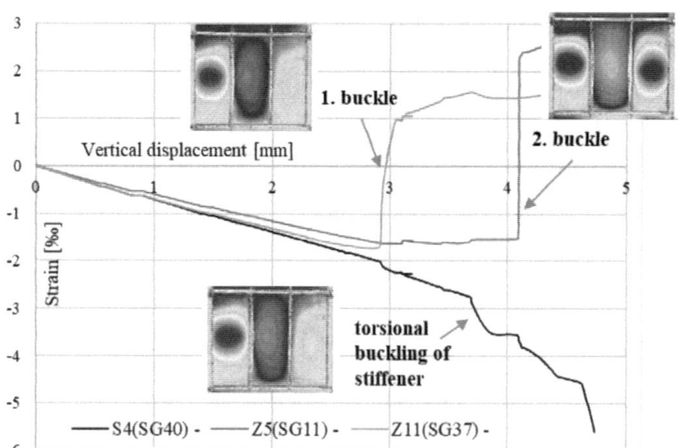

Figure 11. Comparison of strains measured in stiffener and panel for specimen 3C.

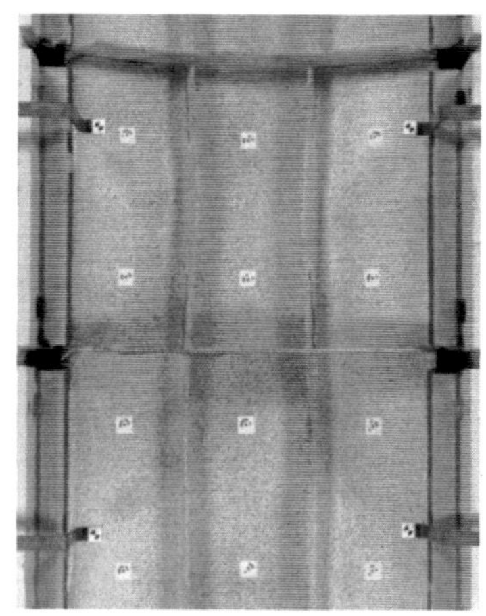

Figure 12. Deformed shape of specimen 3C at failure.

In Figure 11, a comparison of measured strains in the right longitudinal stiffener and second panel from top is presented for specimen 3C. Results for 3C are presented because in this case, buckling happened exactly in the panel where strains were measured. It may be seen from the diagram that at first, all three compressive strains are almost equal and are increasing gradually with the increased vertical displacement. The first buckle happens in the left subpanel, the strains change from compressive to tensile. Later, the compressive strain in the right longitudinal stiffener increases which is related to the torsional buckling of the stiffener. A few moments later, buckling occurs in the right subpanel. The decrease of the stiffeners stiffness is followed by the buckling of the subpanel and consequently by the decrease in the global resistance of the plate. For a more detailed study on this effect, stiffeners with different cross-sections will be studied in the numerical study.

The torsional buckling of stiffeners may be clearly seen in Figure 12, where the deformed shape of specimen 3C is presented. To avoid the problem of torsional buckling, stiffeners with a closed cross-section are often used and will therefore also be included in the numerical study.

4. Conclusions

In this paper, axial compression tests on five curved stiffened plates are reported. It was found challenging to design an appropriate experimental setup for axial compressive test avoiding eccentricity in the panel. During tests the axial compressive force, strains and displacements were measured. The displacements measured with photogrammetry showed good correlation to those measured with LVDTs. The results provided by photogrammetry will be used for the calibration of a finite element model. Test results show that failure is associated to a combination of buckling and yield of material. The deformed shapes of the specimens are highly affected by imperfections. To demonstrate the effects of imperfections on the plate's stiffness and resistance, an extensive parametric study will be carried out that will include also different plate curvatures and stiffener geometries.

Acknowledgements

This work was done under the RFCS OUTBURST project financed by European Commission contract No. 709782.

References

[1] Reis A, Pedro J O, Graça A B, Hendy C, Romoli P, Simões Da Silva L and Martins J P 2017 Report on the characterization of relevant parameters of curved plated bridge structures and identification of bridge cases where they can be found *RFCS Research Project OUTBUTST* (RFCS-2015-709782): Deliverable 2.1

[2] European Committee for Standardization (CEN) 2006 Design of steel structures, Part 1-5: Plated structural elements *Eurocode 3, Part 1-5 (EN 1993-1-5)*

[3] European Committee for Standardization (CEN) 2007 Design of steel structures, Part 1-6: Strength and stability of shell structures *Eurocode 3, Part 1-6 (EN 1993-1-6)*

[4] Tran K, Douthe C, Sab L, Dallot J and Davaine L 2014. Buckling of stiffened curved panels under uniform axial compression *Journal of Constructional Steel Research* 103 pp 140-147

[5] Piculin S, Sinur F and Može P 2017 Analysis of stiffened curved panels in compression: A preliminary numerical study for experimental tests *Eurosteel (Copenhagen)* pp 898-907

[6] Martins J P, Ljubinkovic F, Simões da Silva L and Gervasio H 2018 Behaviour of thin-walled curved steel plates under generalised in-plane stresses: A review. *Journal of Constructional Steel Research* 140 pp 191-207

[7] Gordo J M and Soares C G 2008 Compressive tests on short continuous panels *Marine Structures* 21 pp 113-137

[8] Ljubinkovic F, Martins J P, Gérvasio H, Simões da Silva L and Leitao C 2017 Experimental analysis of unstiffened cylindrically curved panels *XI Conference on Steel and Composite Construction (Coimbra, Portugal)*

[9] Becker H 1958 *Handbook on structural stability. Part IV: Strength of stiffened curved plates and shells* (Washington: National Advisory Committee for Aeronautics)

[10] Cho S, Park H, Kim H and Seo J 2007 Experimental and numerical investigations on the ultimate strength of curved stiffened plates *Proc. 10th Int. Symp. Practical design of ships and oth. Floating str. (Huston, TX)* ed R I Basu *et al* (ABS)

Advanced procedures for design of bolted connections

František Wald*, Jaromír Kabeláč[÷], Marta Kuříková*, Ondřej Perháč[÷], Pavel Ryjáček*, Oscar Minor*, Marcos Bryan Flores Pazmiño*, Lubomír Šabatka[÷], Drahoslav Kolaja[÷]

*Czech Technical University in Prague, Zikova 4, Praha 16636, Czech Republic
[÷]IDEA RS s.r.o., U vodárny 3032/2a, Brno Královo Pole, 616 00, Czech Republic

Emails: wald@fsv.cvut.cz, kabelac@idea-rs.com, marta.kurikova@fsv.cvut.cz, perhac@idea-rs.com, pavel.ryjacek@fsv.cvut.cz, oscar.minor@fsv.cvut.cz, marcos.bryan@fsv.cvut.cz, sabatka@idea-rs.com, kolaja@idea-rs.com

Abstract. Design of bolted slip resistant connection are integral part of design of steel bridges. In recent years, Component Based Finite Element Method (CBFEM) has been increasingly expanding in modelling of structural joints. The proper models of bolts play the major role in prediction of connection stiffness, resistance and deformation capacity. This contribution will describe the influence of pull and shear interaction for correct design of a group of bolts. The second part will be focussed to an advanced model of slip resistance bolts. The last chapter presents the advantages of procedure on examples of various bridge connection, which design was supported by CBFEM.

1. Introduction

Curve fitting procedures based on experimental evidence were and are still used for safe and economical design of connections. Based on analytical models of resistance of connectors, as welds, bolds, and plates, and the estimated lever arm of internal forces was and still is predicted resistance of connection. Zoetemeijer [1] was the first who equipped this model with estimation of stiffness and deformation capacity. The elastic stiffness was improved in the work of Steenhius, see [2]. Basic description of components behaviour in major structural steel connections was prepared by Jaspart for beam to column connections [3] and by Wald et al for column bases [4]. Method implemented in the current European structural standard for the steel and composite connections, see [5] and [6], is applied in majority of software for structural steel used in Europe. The idea was generalised by da Silva [7] for 3D behaviour including nonlinear parts of behaviour. Procedure starts with decomposition of a joint to components followed by their description in terms of normal/shear force deformation behaviour. After that, components are grouped to examine joint moment-rotational behaviour and classification / representation in a spring/shear model and application in global analyses. Advantage of this often called Component Method (CM) is integration of current experimental and analytical knowledge of connections components behaviour, bolts, welds and plates. This provides very accurate prediction of behaviour in elastic and ultimate level of loading. Verification of the model is possible using simplified calculation. Disadvantage of CM is that experimental evaluation of internal forces distribution is available only for limited number of the open section joint configurations. In temporary scientific papers, description of atypical components is either not present or has low validity and description of background materials. The CM´s is not developed for hand calculation but as a method for preparation of design tables or software tools. Models of hollow section connections are described in Ch. 7 of

EN1993-1-8:2006 [5] by curve fitting procedures based on mechanical and numerical experiments. Their component representation is prepared according to the curve fitting procedures available based on selection of the suitable level arms and effective widths.

The global analyses of steel structures is today carried out FEA and all the traditional procedures are not used any more (like force method, three moment equation, Cremon´s pattern, the Cross method or the method distribution moments). In current fast development of software ability connections ready to be designed by FEA and thousands experiments is available the validation process. This paper describes the model of bolts available for Component Based Finite Element Model (CBFEM) which is a multilevel FEA method to design connections of steel structures [8] and its application to steel riveted bridges. The steel plates in connection are analysed in FEA by shell elements as separate plates connected by restrains. The proper behaviour of connectors, of bolts, welds etc., is treated by introducing components representing well its behaviour in term of initial stiffness, ultimate resistance and deformation capacity.

2. Bolted connections

In CBFEM are components as bolt and rivet modelled by a dependent nonlinear springs. The deformation stiffness of the shell element, which models the plates, distributes the forces between the bolts and simulates the adequate bearing of the plate.

2.1. Bolt in tension

The spring of a bolt/rivet in tension is described by its initial deformation stiffness, design resistance, initialisation of yielding and deformation capacity. The initial stiffness is derived analytically as

$$k = \frac{EA_s}{L_b} \tag{1}$$

where E is the Young's modulus, A_s the tensile stress area of a bolt and L_b the bolt elongation length.

The model corresponds well to experimental data, see [9]. For the initialisation of yielding and the deformation capacity is assumed that the plastic deformation occurs in the threaded part of the bolt shank only. The load-deformation diagram of the bolt is shown in Figure 1 and is derived for

$$F_{t,el} = \frac{F_{t,Rd}}{c_1 \cdot c_2 - c_1 + 1} \tag{2}$$

$$k_t = c_1 \cdot k;\ c_1 = \frac{R_m - R_e}{\frac{1}{4} \cdot AE - R_e} \tag{3}$$

$$u_{el} = \frac{F_{t,el}}{k};\ u_{t,Rd} = c_2 \cdot u_{el};\ c_2 = \frac{A \cdot E}{4 \cdot R_e} \tag{4}$$

where k is the linear stiffness of bolt, k_t the stiffness of bolt at the plastic branch, $F_{t,el}$ the limit force for linear behaviour, A the percentage elongation after a fracture of bolt, $F_{t,Rd}$ the limit bolt resistance and $u_{t,Rd}$ the limit deformation of bolt. The design values according to ISO 898:2009 [10] are summarised in Table 1.

Table 1. Bolt parameters in tension, based on ISO 898:2009 [10].

Grade	R_m [MPa]	$R_e = R_{p02}$ [MPa]	A [%]	E [MPa]	c_1 [-]	c_2 [-]
4.8	420	340	14	2,1E+05	0,011	21,6
5.6	500	300	20	2,1E+05	0,020	35,0
6.8	600	480	8	2,1E+05	0,032	8,8
8.8	830	660	12	2,1E+05	0,030	9,5
10.9	1040	940	9	2,1E+05	0,026	5,0

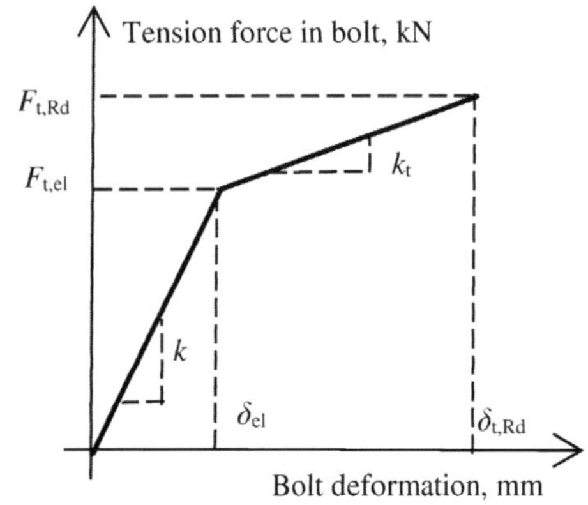

Figure 1.
Load–deformation diagram of the bolt in tension .

2.2. *Bolt in shear*

The initial stiffness and the design resistance of a bolt in shear is in CBFEM modelled according to cl. 3.6 and 6.3.2 in EN1993-1-8:2006. The spring representing the bolt in shear has bi-linear force deformation behaviour. Deformation capacity is considered according to [8] as

$$\delta_{pl} = 3\,\delta_{el} \tag{5}$$

Initialization of yielding is expected, see Figure 2, at

$$F_{v,el} = 2/3\,F_{v,Rd} \tag{6}$$

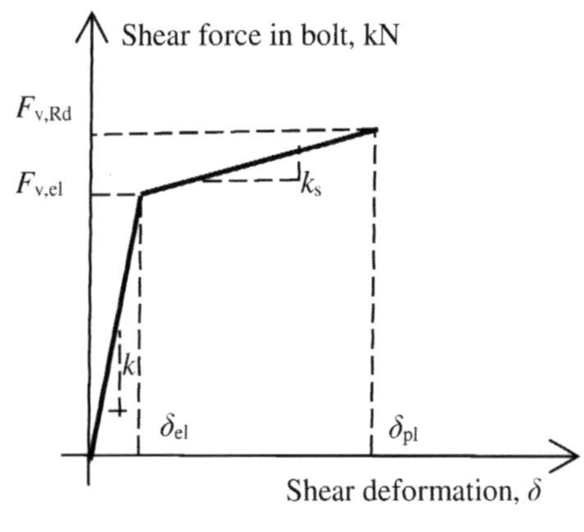

Figure 2.
Force deformation diagram of the bolt in shear.

2.3. *Bolt interaction*

A combination of a shear and a tension in a bolt is expressed in EN 1993-1-8:2006 in Tab. 3.4 by a bilinear relation and checked as

$$\max\left\{\frac{F_{t,Ed}}{F_{t,Rd}}, \frac{F_{v,Ed}}{F_{v,Rd}} + \frac{F_{t,Ed}}{1{,}4\,F_{t,Rd}}\right\} \leq 1{,}0 \tag{7}$$

where $F_{v,Ed}$ is the acting bolt shear force, $F_{t,Ed}$ is the acting bolt tensile force, $F_{v,Rd}$ is the bolt shear resistance, $F_{t,Rd}$ is the bolt tensile resistance. A condition limiting the bolt resistance is showed in Figure 3.

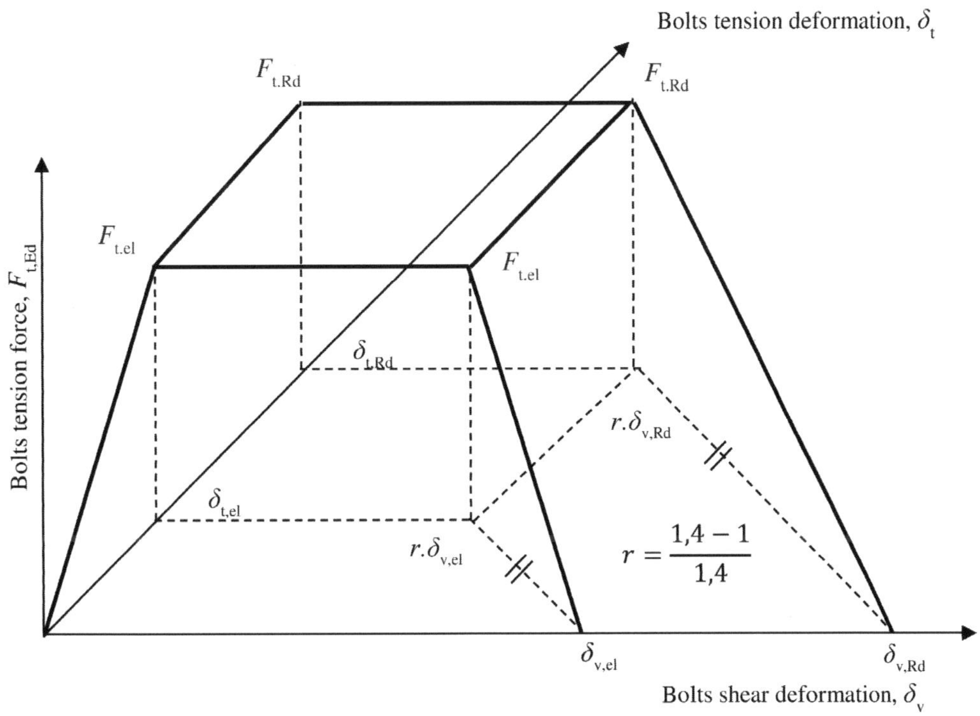

Figure 3. Bolt tension force as function of deformation in shear and tension.

2.4. *Preloaded bolts*
In connection with preloaded bolts is transferred the shear force by friction between both surfaces. Compare to regular bolts is in controlled the friction and the preloaded force. The final resistance is assured by bolt shearing and bolt and plate bearing after the slippage of bolt in the hole. In EN 1993-1-8:2006 is summarized the resistance of preloaded bolts classes 8.8 and 10.9 in Ch. 3.9. The bolts are expected to be preloaded to 70% of its strength f_{ub}. It is expected that the bolt deforms 80 % and the plates 20 %. If the external tensile force $F_{t.Ed}$ to joint in direction of the bolt is applied, the slip resistance will be reduced

$$F_{s,Rd} = k_S\,\mu \frac{0{,}7 f_{ub} A_S - 0{,}8 F_{t,Ed}}{\gamma_{M3}} \tag{8}$$

where k_s is the bolt hole size factor, μ the slip factor, f_{ub} is the bolt strength and A_s is the bolt area in tension.

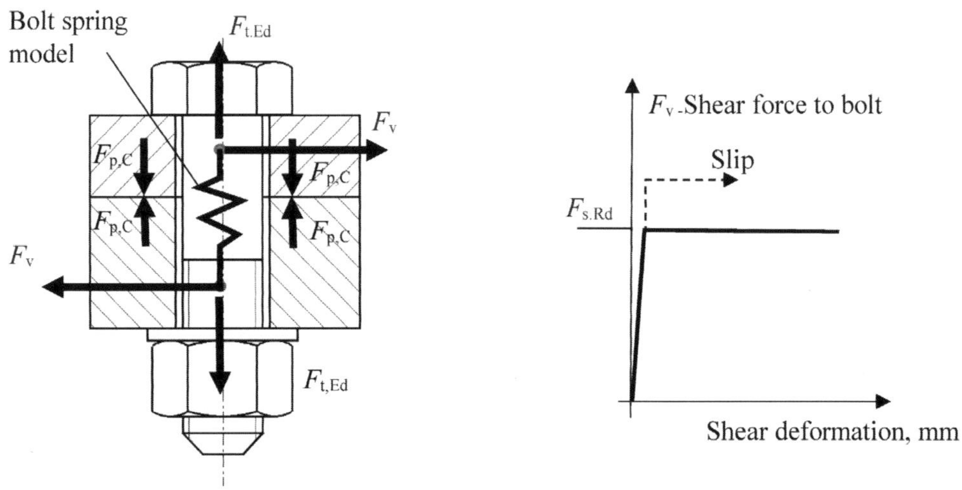

Figure 4. Shear characteristic of the preloaded bolt.

In design numerical model the component preload bolt is simulated either as nonlinear spring using the preloading force and deformation or by restrains between surfaces, which are representing the friction, with a spring, which is including a preloaded force in the bolt. For the bolt component represented by the preloading force and its slip deformation is the model similar to a conventional bolt model. The shear characteristic is shown in Figure 4. The initial linear shear stiffness is determined from the stiffness of the cylinder under the head of the bolt and it is practically rigid. The shear force limit includes the external tensile load to the bolt in accordance with EN 1993-1-8:2006. The advantage of this simplified model is its computational stability and low demands of the FE model and the consistence of results with cl. 3.6.2.2. The model does not respect the actual distribution of contact pressures between the plates and the history of preloading, but models the bolted connection including the bolt and plates deformation well, see Figure 5.

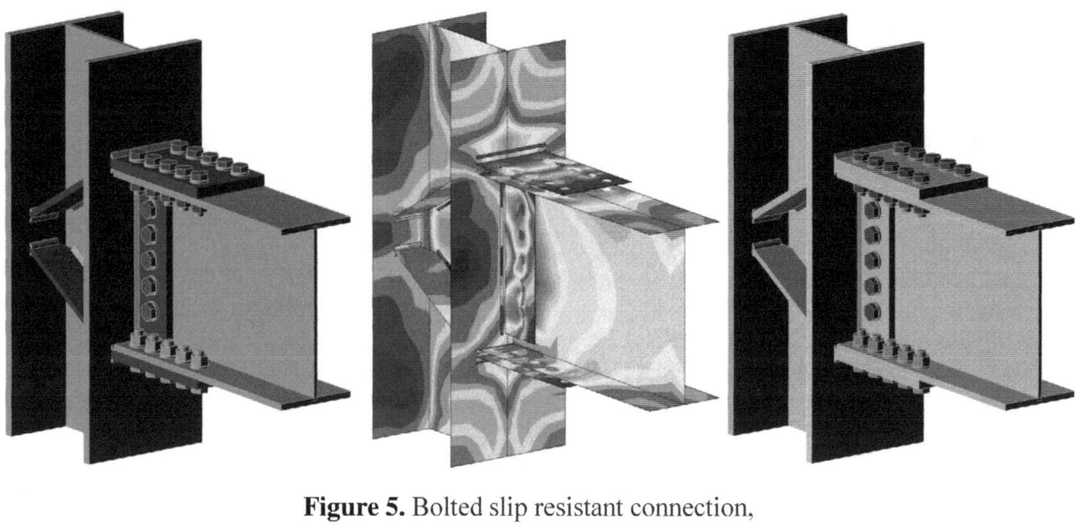

Figure 5. Bolted slip resistant connection,
configuration, von Mises stresses and strains at design stage

3. Application for assessment of old steel bridges

The industrial development in the beginning of the last century would not be successful without the transport infrastructure. At those days, many steel riveted bridges were built in the Czech Republic and also all over the world. Those bridges are usually at the end of their service life, however, the lack of the financial sources forces the infrastructure owners for operating them much longer. To do that, the extensive structural assessment of those bridges have to be organised, in order to evaluate the load capacity according to the existing state, corrosion weakening and possible fatigue damages. As majority of those have used riveting for the connections, the behaviour of the joints is very important for the bridge numerical model.

Traditionally, truss bridges were analysed under certain assumptions, such as the members are loaded only by axial forces. This assumption is only true if the connections allow the rotation between elements, which in reality is not the case. There are always certain moments transmitted depending on the stiffness of the connection. Unfortunately, the riveted joints are mostly complicated and hard to analyse by standard FEM software. However, CBFEM models those joints and obtains the initial stiffness, which is an essential input to the global numerical model.

In last two years, several bridges were analysed with the help of the CBFEM method and software tool [11]. The results of the impact of the connection stiffness on the behaviour of a historical steel railway bridge are shown here together with the main results [12]. The Figure 6 shows the old steel heritage bridge in Prague, Vyšehrad, where a number of models were created, such as the connection between top chord and bearing vertical, show on the Figure 7 from [13].

Figure 6. General view of the steel railway bridge in Prague.

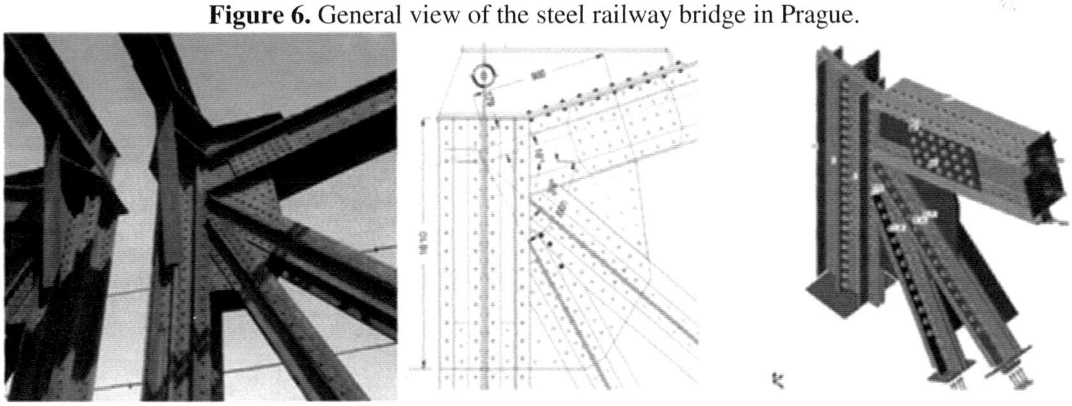

Figure 7. Connection C-02 on railway bridge in Prague, its drawing and its CBFEM model.

The Figure 8 shows the typical detail of the stringer to cross beam joint that is present on many steel bridges. Beams are riveted I profiles, with flanges made from riveted angles. Also connection was done by riveting of angles to the web of both elements. The model allows to analyse the rotational initial stiffness in both directions that can be directly inserted in the global numerical model. Especially this type of joint and also the connection between cross beam and main girder are essential for the load capacity assessment and the stiffness significantly influence the final result.

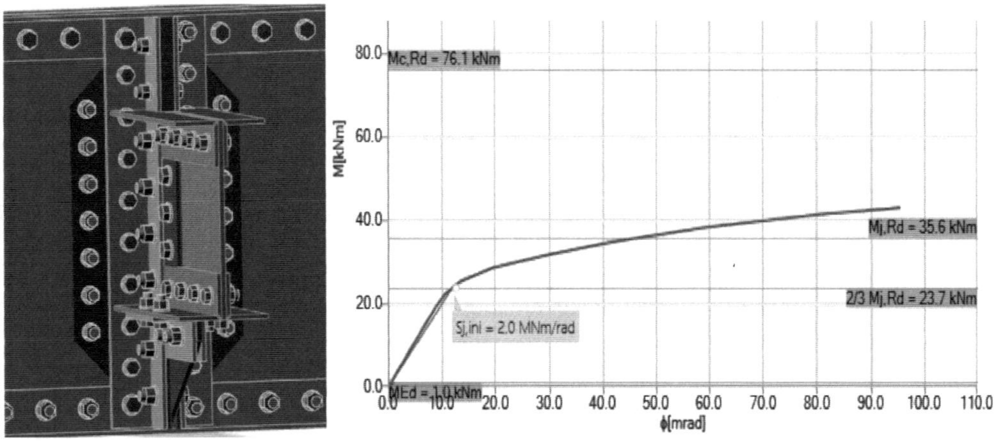

Figure 8. Stringer to cross beam connection on railway bridge in Prague and its moment rotational diagram by CBFEM model.

Seven bridges were analysed with all modelled details on each structure. This amount of data can be plotted in the summary diagram, as shown on the Figure 9. This graph can help to roughly predict the joint stiffness even without the time consuming analysis.

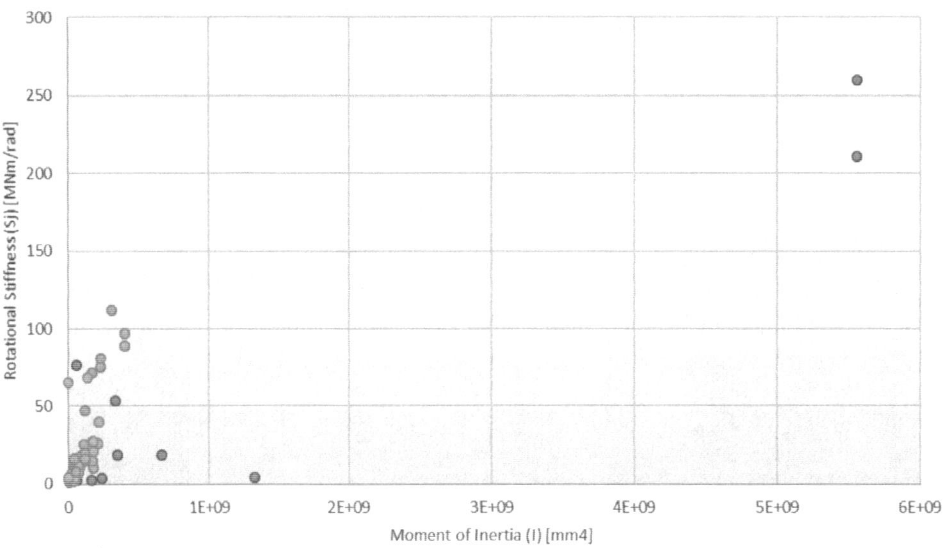

Figure 9. Summary of the results from seven analysed bridges, different symbols represents individual bridges.

4. Conclusion

The global analyses of steel structures is today carried out by FEA and all the traditional procedures are not used any more. In new generation of structural Eurocodes after 2021 is expected to stress the principles of safety of applications of FEA by System response quantity process in EN1993-1-14:2021. The FEA analyses of the structural steel connections is replacing the curve fitting and component design methods. For its proper use is necessary to apply a good Validation and Verification procedures with well-defined hierarchy to allow a safe use.

The presented results show the good accuracy of CBFEM verified to CM and to advanced calculations/experiments in cases where the CBFEM gives higher stiffness, resistance, or deformation capacity, see [8]. The prediction of bolts as component by spring models allow accurate prediction of bolted connection resistance, stiffness and deformation capacity takin into account the initial elastic and post elastic stages.

That CBFEM helped to improve the global numerical models of old steel bridges and keep them in the service for the longer time. This is a beneficial result for the budget of the owners and lovers.

5. Announcement

The work was prepared under the R&D project MERLION II supported by Technology Agency of the Czech Republic, project No TH02020301.

References

[1] Zoetemeijer P., *Summary of the Researches on Bolted Beam-to-Column Connections*. Report 6-85-7, University of Technology, Delft, 1985

[2] Steenhuis M., Gresnigt N., Weynand K., Pre-Design of Semi-Rigid Joints in Steel Frames, *Proceedings of the Second State of the Art Workshop on Semi-Rigid Behaviour of Civil Engineering Structural Connections*, COST C1, Prague, pp131-140 1994

[3] Jaspart J.P., "Design of structural joints in building frames", *Progress in Structural Engineering and Materials* 4 pp18–34, 2002

[4] Wald F., Sokol Z., Steenhuis M. and Jaspart, J.P., Component Method for Steel Column Bases, *Heron* 53, pp 3-2, 2008

[5] EN1993-1-8, *Eurocode 3: Design of steel structures - Part 1-8: Design of joints*, CEN, Brussels, 2006.

[6] EN1994-1-1, *Eurocode 4: Design of composite steel and concrete structures- Part 1-1: General rules and rules for buildings*, CEN, Brussels, 2010

[7] Da Silva Simoes L., Towards a consistent design approach for steel joints under generalized loading, *Journal of Constructional Steel Research* 64 pp 1059-1075 2008

[8] Wald F., Sabatka L., et al. *Benchmark cases for advanced design of structural steel connections*, Prague, Česká technika, 2016

[9] Wald F.; Gödrich L. et al. Generally loaded connections by component based finite element method, *Proceedings of the Eighth International Workshop,* AISC, pp 143-152 2016

[10] ISO 898-1, Mechanical properties of fasteners made of carbon steel and alloy steel, Part 1, *Bolts, screws and studs with specified, property classes*, Coarse thread and fine pitch thread, Geneva, 2009

[11] IDEA RS: *Product documentation*, www.idea-rs.com, 2018

[12] García, O.M., The Impact of the Connection Stiffness on the Behaviour of a Historical Steel Railway Bridge, Master Thesis, Prague, 2017

[13] Flores Pazmiño, M., B., The Prediction of the Joint Stiffness in Riveted Steel Bridges, Master Thesis, Prague, 2017